Ultrashort Laser Pulse Phenomena

Fundamentals, Techniques,
and Applications on a
Femtosecond Time Scale

OPTICS AND PHOTONICS
(formerly Quantum Electronics)

EDITED BY

PAUL F. LIAO

Bell Communications Research, Inc.
Red Bank, New Jersey

PAUL L. KELLEY

Tufts University
Medford, Massachusetts

IVAN KAMINOW

AT&T Bell Laboratories
Holmdel, New Jersey

A complete list of titles in this series appears at the end of this volume.

Ultrashort Laser Pulse Phenomena

Fundamentals, Techniques, and Applications on a Femtosecond Time Scale

Jean-Claude Diels

Department of Physics and Astronomy
The University of New Mexico
Albuquerque, New Mexico

Wolfgang Rudolph

Department of Physics and Astronomy
The University of New Mexico
Albuquerque, New Mexico

ACADEMIC PRESS

San Diego New York Boston London Sydney Tokyo Toronto

- 2.5.2 Space-time distribution of the pulse intensity at the focus of a lens 62
- 2.5.3 Achromatic doublets 67
- 2.5.4 Mirrors 68
- 2.6 Elements with angular dispersion 70
  - 2.6.1 Tilting of pulse fronts 71
  - 2.6.2 GVD through angular dispersion 75
- 2.7 Wave-optical description of angular dispersion 87
- 2.8 Optical matrices for dispersive systems 92
- 2.9 Problems 98

3 Light–Matter Interaction 101
- 3.1 Density matrix equations 102
- 3.2 Pulse shaping with resonant particles 112
  - 3.2.1 General 112
  - 3.2.2 Pulses much longer than the phase relaxation time ($\tau_p \gg T_2$) 114
  - 3.2.3 Pulse duration comparable with or longer than the phase relaxation time ($\tau_p \geq T_2$) 122
- 3.3 Nonlinear, nonresonant optical processes 123
  - 3.3.1 General 123
  - 3.3.2 Second harmonic generation (SHG) 125
  - 3.3.3 Optical parametric interaction 134
  - 3.3.4 Nonlinear phase modulation 138
- 3.4 Continuum generation 144
- 3.5 Nonlinear space-time effects 145
  - 3.5.1 Self-focusing of continuous beams 146
  - 3.5.2 Self-focusing of ultrashort pulses 148
- 3.6 Problems 151

4 Coherent Phenomena 153
- 4.1 From coherent to incoherent interactions 153
- 4.2 Coherent interactions with two-level systems 157
  - 4.2.1 Maxwell-Bloch equations 157
  - 4.2.2 Rate equations 161
  - 4.2.3 Evolution equations 162
  - 4.2.4 Steady-state pulses 172
- 4.3 Multiphoton coherent interaction 176
  - 4.3.1 Introduction 176

This book is printed on acid-free paper.

Copyright © 1996 by ACADEMIC PRESS, INC.

All Rights Reserved.
No part of this publication may be reproduced or transmitted in any form or by any means, electronic or mechanical, including photocopy, recording, or any information storage and retrieval system, without permission in writing from the publisher.

Academic Press, Inc.
A Division of Harcourt Brace & Company
525 B Street, Suite 1900, San Diego, California 92101-4495

United Kingdom Edition published by
Academic Press Limited
24-28 Oval Road, London NW1 7DX

Library of Congress Cataloging-in-Publication Data

Diels, Jean-Claude.
　　Ultrashort laser pulse phenomena : fundamentals, techniques, and applications on a femtosecond time scale / by Jean-Claude Diels.
　　　　p.　　cm. -- (Optics and photonics series)
　　Includes index.
　　ISBN 0-12-215492-4
　　1. Laser pulses, Ultrashort.　I. Rudolph, Wolfgang. date.
II. Title.　III. Series.
TA1677.D54　1995
621.36'6--dc20
　　　　　　　　　　　　　　　　　　　　　　94-48778
　　　　　　　　　　　　　　　　　　　　　　CIP

PRINTED IN THE UNITED STATES OF AMERICA
96　97　98　99　00　01　BC　9　8　7　6　5　4　3　2　1

Contents

PREFACE   xiii

1 Fundamentals   **1**
  1.1 Characteristics of fs light pulses 1
      1.1.1 Complex representation of the electric field 1
      1.1.2 Power, energy, and related quantities 4
      1.1.3 Pulse duration and spectral width 7
  1.2 Pulse propagation 12
      1.2.1 The reduced wave equation 13
      1.2.2 Complex dielectric constant 23
  1.3 Interaction of light pulses with linear optical elements 26
  1.4 Generation of phase modulation 29
  1.5 Beam propagation 30
      1.5.1 General 30
      1.5.2 Analogy between pulse and beam propagation 32
      1.5.3 Analogy between spatial and temporal imaging 35
  1.6 Numerical modeling of pulse propagation 36
  1.7 Space–time effects 39
  1.8 Problems 40

2 Femtosecond Optics   **43**
  2.1 Introduction 43
  2.2 White light and short pulse interferometry 43
  2.3 Mirror dispersion 50
  2.4 Other interferometric structures 53
  2.5 Focusing elements 59
      2.5.1 Singlet lenses 59

	4.3.2	Multiphoton multilevel transitions	179
	4.3.3	Simplifying a n-level system to a two-level transition	191
	4.3.4	Four photon resonant coherent interaction	196
	4.3.5	Miscellaneous applications	202
4.4	Problems		207

5 Ultrashort Sources — 209

5.1	Introduction		209
	5.1.1	Mode-locking	209
	5.1.2	Elements of a fs laser	211
	5.1.3	Pulse formation	212
	5.1.4	Pulse compression mechanisms	213
5.2	Round-trip model of a fs laser		217
	5.2.1	General	217
	5.2.2	Approximate analytical solutions	218
	5.2.3	The continuous model	220
5.3	Evolution of the pulse energy		221
5.4	Specific pulse shaping mechanisms		225
	5.4.1	Passive amplitude modulation	225
	5.4.2	Saturation	227
	5.4.3	Induced grating effects for $\tau_p \ll T_1$	231
	5.4.4	Phase modulation and dispersion	235
	5.4.5	Self-lensing	239
5.5	Initiation of mode-locking		244
5.6	Passively mode-locked lasers		247
	5.6.1	Generalities	247
	5.6.2	Study of the passively mode-locked dye laser with the round-trip model	248
5.7	Synchronous and hybrid mode-locking		254
	5.7.1	Synchronous pumping, stability, noise	254
	5.7.2	Stabilization	256
	5.7.3	Hybrid mode-locking	257
5.8	Solitons in femtosecond lasers		260
5.9	Negative feedback		263
	5.9.1	Introduction	263
	5.9.2	Feedback mechanisms	264
5.10	Additive pulse mode-locking		268
	5.10.1	Generalities	268
	5.10.2	Analysis of APML	269

- 5.11 Femtosecond laser cavities ... 270
  - 5.11.1 Cavity modes ... 271
  - 5.11.2 The geometrical cavity ... 274
- 5.12 Generation of synchronized wavelengths ... 281
  - 5.12.1 Nonlinear mixing ... 281
  - 5.12.2 Other techniques ... 289
- 5.13 Miniature dye lasers ... 291
  - 5.13.1 Distributed feedback dye lasers ... 292
  - 5.13.2 Short cavity lasers ... 293
- 5.14 Semiconductor lasers ... 294
  - 5.14.1 External cavity semiconductor laser ... 294
  - 5.14.2 Integrated circuit fs lasers ... 297
  - 5.14.3 Semi-integrated circuit fs lasers ... 299
- 5.15 Fiber lasers ... 300
  - 5.15.1 Introduction ... 300
  - 5.15.2 Raman soliton fiber lasers ... 300
  - 5.15.3 Doped fiber lasers ... 301
- 5.16 Problems ... 305

6 Femtosecond Pulse Amplification 309
- 6.1 Introduction ... 309
- 6.2 Fundamentals ... 310
  - 6.2.1 Gain factor and saturation ... 310
  - 6.2.2 Shaping in amplifiers ... 314
  - 6.2.3 Amplified spontaneous emission (ASE) ... 318
- 6.3 Nonlinear refractive index effects ... 321
  - 6.3.1 General ... 321
  - 6.3.2 Self-focusing ... 323
  - 6.3.3 Thermal noise ... 324
  - 6.3.4 Combined pulse amplification and chirping ... 325
  - 6.3.5 Chirped pulse amplification ... 326
- 6.4 Amplifier design ... 329
  - 6.4.1 Gain media and pump pulses ... 329
  - 6.4.2 Amplifier configurations ... 331
  - 6.4.3 Single-stage, multi-pass amplifiers ... 333
  - 6.4.4 Regenerative amplifiers ... 335
  - 6.4.5 Traveling wave amplification ... 337
- 6.5 Problems ... 340

7 Pulse Shaping — 343
- 7.1 Pulse compression — 343
  - 7.1.1 General — 343
  - 7.1.2 The fiber compressor — 347
  - 7.1.3 Pulse compression using bulk materials — 358
- 7.2 Shaping through spectral filtering — 360
- 7.3 Problems — 362

8 Diagnostic Techniques — 365
- 8.1 Intensity correlations — 365
  - 8.1.1 General properties — 365
  - 8.1.2 The intensity autocorrelation — 366
  - 8.1.3 Intensity correlations of higher order — 367
- 8.2 Interferometric correlations — 367
  - 8.2.1 General expression — 367
  - 8.2.2 Interferometric autocorrelation — 369
- 8.3 Measurement techniques — 373
  - 8.3.1 Nonlinear optical processes for measuring fs pulse correlations — 373
  - 8.3.2 Recurrent signals — 373
  - 8.3.3 Single shot measurements — 376
- 8.4 Pulse amplitude and phase reconstruction — 380
  - 8.4.1 Introduction — 380
  - 8.4.2 Iterative fitting, analytical expressions — 381
  - 8.4.3 Cross-correlation with downchirped pulses — 382
  - 8.4.4 Cross-correlation for arbitrary pulses — 384
  - 8.4.5 Method based on a higher order nonlinearity — 387
  - 8.4.6 Detection in the frequency domain — 392
- 8.5 Problems — 399

9 Measurement Techniques of Femtosecond Spectroscopy — 401
- 9.1 Introduction — 401
- 9.2 Data deconvolutions — 403
- 9.3 Beam geometry and temporal resolution — 404
- 9.4 Transient absorption spectroscopy — 407
- 9.5 Transient polarization rotation — 411
- 9.6 Transient grating techniques — 413
  - 9.6.1 General technique — 413
  - 9.6.2 Degenerate four-wave mixing (DFWM) — 415

9.7 Femtosecond resolved fluorescence 419
9.8 Photon echoes 421
9.9 Zero-area pulse propagation 425
9.10 Impulsive stimulated Raman scattering 430
    9.10.1 General description 430
    9.10.2 Detection 431
    9.10.3 Theoretical framework 433
    9.10.4 Single pulse shaping versus mode-locked train 435
9.11 Self-action experiments 437
9.12 Problems 438

10 Examples of Ultrafast Processes in Matter   441
10.1 Introduction 441
10.2 Ultrafast transients in atoms 442
    10.2.1 The classical limit of the quantum mechanical atom . 442
    10.2.2 The radial wave packet 442
    10.2.3 The angularly localized wave packet 444
10.3 Ultrafast processes in molecules 446
    10.3.1 Observation of molecular vibrations 446
    10.3.2 Chemical reactions 451
    10.3.3 Molecules in solution 454
10.4 Ultrafast processes in solid state materials 454
    10.4.1 Excitation across the band gap 454
    10.4.2 Excitons 456
    10.4.3 Intraband relaxation 456
    10.4.4 Phonon dynamics 458
    10.4.5 Laser-induced surface disordering 460
10.5 Primary steps in photo-biological reactions 461
    10.5.1 Femtosecond isomerization of rhodopsin 461
    10.5.2 Photosynthesis 463

11 Generation of Extreme Wavelengths   465
11.1 Generation of terahertz radiation 466
11.2 Generation of ultrafast x-ray pulses 472
11.3 Generation of ultrashort acoustic pulses 475
11.4 Generation of ultrafast electric pulses 478

12 Miscellaneous Applications — 485
   12.1 Imaging 485
      12.1.1 Introduction 485
      12.1.2 Range gating with ultrashort pulses 486
      12.1.3 Imaging through scatterers 490
      12.1.4 Prospects for four-dimensional imaging 492
      12.1.5 Microscopy 493
   12.2 Femtosecond laser gyroscopes 496
      12.2.1 Introduction 496
      12.2.2 Rotation sensing 498
      12.2.3 Measurement of changes in index 500
   12.3 Solitons 501
      12.3.1 Temporal solitons 503
      12.3.2 Spatial solitons 505
      12.3.3 Spatial and temporal solitons 506
   12.4 Problem 507

APPENDICES — 509

A Phase Shifts upon Transmission—Reflection — 509
   A.1 The symmetrical interface 509
   A.2 Coated interface between two different dielectrics 510

B Fs Dye Laser Configurations — 513

C Four Photon Coherent Interaction — 519

D Dyes and Solvents — 523

E Reconstruction of the Spectral Phase — 525

BIBLIOGRAPHY — 527

INDEX — 576

Preface

What do we understand under "ultrashort laser pulse phenomena"? It really takes a whole book to define the term. By ultrashort we mean femtosecond (fs), which is a unit of time equal to 10^{-15} s. This time scale becomes accessible due to progress in the generation, amplification and measurement of ultrashort light pulses. Ultrashort phenomena involve more than just the study of ultrashort lived events. Because of the large energy concentration in a fs optical pulse, this topic encompasses the study of the interaction of intense laser light with matter, as well as transient response of atoms and molecules, and basic properties of the fs radiation itself.

This book is intended as an introduction to ultrashort phenomena to researchers, graduate and senior undergraduate students in optics, physics, chemistry, and engineering. A preliminary version of this book has been used at the Universities of New Mexico, Jena and Pavia, as a course for graduate and advanced undergraduate students. The "femtosecond light" gives a different illumination to some classical problems in electromagnetism, optics, quantum mechanics, and electrical engineering. We believe therefore that this book can provide useful illustrations for instructors in these fields.

It is *not* the goal of this book to represent a complete overview of *the latest* progress in the field. We wish to apologize in advance for all the important and pioneering fs work that we failed to cite. For space limitation, we have chosen to present only a few examples of application in the various fields. We are not offering different theoretical aspects of any particular phenomenon, but rather choose to select a description that is consistent throughout the book. Our aim is to cover the basic techniques and applications, rather than enter into details of the most fashionable "topic of the day." We have attempted to use simple notations, and to remain within the MKS system of units.

Consistent with the instructional goal of this book, the first chapter is an extensive review of propagation properties of light, in time and frequency domain. Classical optics is reviewed in the next chapter, in light of the particular propagation properties of fs pulses. Some aspects of white light optics—such as coherence, focusing—can be explained in the simplest manner by picturing incoherent radiation as a random sequence of fs pulses. Femtosecond pulses are generally meant to interact with matter. Therefore, a review of this aspect is given in Chapter 3. The latter serves as introduction to the most startling, unexpected, complex, properties of transient interaction of coherent fs pulses with resonant physical and chemical systems (Chapter 4). This is a sub-field of which the basic foundations are well understood, but it is still open to numerous experimental demonstrations and applications. Chapters 5 through 9 review practical aspects of femtosecond physics, such as sources, amplifiers, pulse shapers, diagnostic techniques, and measurement techniques.

The last three chapters are examples of application of ultrafast techniques. In Chapter 10, the frontier between quantum mechanics and classical mechanics is being probed with fs pulses. New techniques make it possible to "visualize" electrons in Rydberg orbits, or the motion of atoms in molecules. The examples of ultrafast processes in matter are presented in this chapter by order of increasing system complexity (from the orbiting electron to the biological complex).

Femtosecond pulses of very high peak powers lead to the generation of extremely short wavelengths electron and x-ray pulses, as well as to extremely long wavelengths. Some of these techniques are reviewed in Chapter 11. On the long wavelength end of the spectrum, fs pulses are used as Dirac delta function on antennas for submillimeter radiation (frequencies in the THz range). This is a recent application of ultrafast solid state photoconductive switches.

A few applications which exploit either the short duration (range gating imaging), the short direction and low duty cycle (gyroscopes), or the high intensity (filamentation in air) have been selected for the final Chapter 12.

Problems are given at the end of most chapters. Some are typical "textbook" problems with a straightforward solution. Other problems are designed to put the student in a realistic "research situation."

We would like to express our gratitude to all our colleagues and students who have supported us with numerous suggestions and corrections. In particular, we are indebted to S. Diddams, M. Dennis, M. Kempe, A. Knorr, J. Nicholson, X. Zhao, Professors G. Reali, B. Wilhelmi, K. Wodkiewicz, A. Zewail and W. Zinth.

We are grateful also to the contribution of all the students who took a course in the development stage of this book.

Last but not least, we are grateful to our wives who watched the years go by as our life became hostage of this endeavor.

Why ultrashort pulse phenomena?

Yes, you are right! You can be happy without femtosecond pulses and, maybe, consider yourself lucky enough not to be involved with it too deeply. Nevertheless it is a fascinating as well as challenging task to observe and to control processes in nature on a time scale of several femtoseconds. Note, one femtosecond (1 fs) is the $1:10^{15}$th part of a second and corresponds to about half a period of red light. The ratio of one fs to one second is about the ratio of five minutes to the age of the earth. During one fs visible light travels over a distance of several hundred nanometers, which is hardly of any concern to us in our daily routine. However, this pathlength corresponds to several thousand elementary cells in a solid which is quite a remarkable number of atomic distances. This suggests the importance the fs time scale might have in the microcosm. Indeed, various essential processes in atoms and molecules as well as interactions among them, proceed faster than what can be resolved on a picosecond time scale (1 ps = $10^{-12}s$). Their relevance results simply from the fact that these events are the primary steps for most (macroscopic) reactions in physics, chemistry and biology.

To illustrate the latter point, let us have a look at the most simple atom — the hydrogen atom — consisting of a positively charged nucleus and a negatively charged electron. Quantum mechanics tells us that an atomic system exists in discrete energy states described by a quantum number n. In the classical picture this corresponds to an electron (wave packet) circulating around the proton on paths with radius $R \propto n^2$. From simple textbook physics [1, 2] the time T_R necessary for one round trip can be estimated with $T_R = 4n^3 h^3 \epsilon_0^2/(e^4 m_e)$ where h is Planck's constant, ϵ_0 is the permittivity of free space, and m_e, e are the electron mass and charge. For $n = 10$, for instance, we obtain a period of about 150 fs. Consequently, an (hydrogen) atom excited to a high Rydberg state is expected to show some macroscopic properties changing periodically on a fs time scale.

Let us consider next atoms bound in a molecule. Apart from translation, the isolated molecule has various internal degrees of freedom for periodical motion - rotation and vibration as well as for conformation changes. Depending on the binding forces, potentials and masses of the constituents, the corresponding periods may range from the ps to the fs scale. Another example of ultrafast dynamics in the molecular world is the chemical reaction which shall be discussed by the simple dissociation $(AB)^* \rightarrow A + B$. Here the breaking of the bond is accompanied by a geometrical separation of the two components caused by a repulsive potential. Typical recoil velocities are of the order of 1 km/s, which implies that the transition from the bound state to the isolated complexes proceeds within 100 fs. Similar time intervals, of course, can be expected if separated particles undergo a chemical reaction.

Additional processes come into play if the particle we look at is not isolated but under the influence of surrounding atoms or molecules, which happens in a gas (mixture) or solution. Strong effects are expected as a result of collisions. Moreover, even a simple translation or rotation which alters the relative position of the molecule to the neighboring particles may lead to a variation of the molecular properties due to a changed local field. The characteristic time constants depend on the particle density and the translation/rotation velocity which in turn is determined by the temperature and strength of interaction with the neighbors. The characteristic times can be comparatively long in diluted gases (ns to μs) and can be very short in solutions at room temperature (fs).

Finally, let us have a look at a solid where the atomic particles are usually trapped at a relatively well-defined position in the lattice. Their motion is restricted usually to lattice vibrations (phonons) with possible periods in the order of 100 fs which corresponds to phonon energies of several tens of milli-electronvolts (for instance, the longitudinal optical or LO phonon in GaAs has an energy of about 35 meV).

The fundamental problem to be solved is to find tools and techniques which allow us to observe and manipulate on a fs time scale. At present, speaking about tools and techniques for fs physics means dealing with laser physics, in particular with ultrashort light pulses produced in lasers. Shortly after the invention of the laser in 1960, methods were developed to use them for the generation of light pulses. In the sixties the microsecond (μs) and nanosecond (ns) range were extensively studied. In the seventies, progress in laser physics opened up the ps range, and the eighties were characterized by the broad introduction of fs techniques [3, 4, 5] (Extrapolating this dramatic development we may expect the attosecond physics in the late nineties...). Optical methods have been taken precedence over electronics in time resolving fast events, ever since light pulses shorter than a few ps have become available. It should also be mentioned that the shortest electrical and x-ray pulses are now being produced by means of fs light pulses, which in turn enlarges the application field of ultrafast techniques.

Femtosecond technology opens up new fascinating possibilities based on the unique properties of femtosecond light pulses:

- Energy can be concentrated in a temporal interval as short as several 10^{-15} s which corresponds to only a few optical cycle in the visible range.

- The pulse peak power can be extremely large even at modulate pulse energies. For instance a 50 fs pulse with an energy of 1 mJ ($\approx 3 *$

10^{15} "red" photons) exhibits a peak power of 20 Gigawatt. Focusing this pulse to a 100 μm^2 spot yields an intensity of 10 Petawatt/cm^2 (20 10^{15}W/cm^2!), which means an electric field strength of about 3 GV/cm. This value is larger than a typical inner-atomic field of 1 GV/cm.

- The geometrical length of a fs pulse amounts only to several micrometer (10 fs corresponds to 3 μm in vacuum). Such a coherence length is usually associated with incoherent light. The essential difference is that incoherent light is generally spread over a much longer distance.

The attractiveness of fs light pulses not only lies in the possibility to trace processes in their ultrafast dynamics but also in the fact that one simply can do things faster. Of course only few, but essential parts, in modern technology can be accelerated by using ultrashort (fs) light pulses. Of primary importance are data transfer and data processing utilizing the high carrier frequency of light and the subsequent large possible bandwidths. In this respect one of the most spectacular goals is to create an optical computer. Moreover, techniques are being developed which allow distortionless propagation of ultrashort light pulses over very long distances (several thousand kilometer) through optical fibers, a pre-condition for a future Terahertz information transfer.

A variety of nonlinear processes, reversible as well as irreversible ones, become accessible thanks to the large intensities of fs pulses. There are proposals to use such pulses for laser fusion. To reach TW intensities, table top devices are replacing the building size high energy facilities previously required. First attempts to generate short x-ray pulses by using fs pulse induced plasmas have already proven successful.

The short geometrical lengths of fs light pulses suggests interesting applications for optical ranging with micrometer resolution as well as for combinations of micrometer spatial resolution with femtosecond temporal resolution.

The ultrashort phenomena to which this book refers are created by *light pulses*, which are wave packets of electromagnetic waves oscillating at optical frequencies. The emphasis of this book is not on the optical frequency range but on physical phenomena associated with *ultrafast* electromagnetic pulses. The latter will be ephemeral when consisting of only a small number of optical periods, and spatially confined when made up of a small number of wavelengths. Another criterion for "short" is that the length of the pulse be small compared with the distance over which it propagates, particularly when large changes of shape and modulation take place. In the particular area of light-matter interaction, a pulse is generally considered as a "δ" function excitation, when its duration is small compared to that of all atomic or molecular relaxations.

Chapter 1

Fundamentals

1.1 Characteristics of fs light pulses

Electromagnetic waves are fully described by the time and space dependent electric field. In the frame of a semiclassical treatment the propagation of such fields and the interaction with matter are governed by Maxwell's equations with the material response given by a macroscopic polarization. In this first chapter we will summarize the essential notations and definitions used throughout the book. The pulse is characterized by measurable quantities which can be directly related to the electric field. A complex representation of the field amplitude is particularly convenient in dealing with propagation problems of electromagnetic pulses. The next section expands on the choice of field representation.

1.1.1 Complex representation of the electric field

Let us consider first the temporal dependence of the electric field neglecting its spatial dependence, i.e., $\mathbf{E}(x,y,z,t) = E(t)$. A complete description can be given either in the time or the frequency domain. Even though the measured quantities are real, it is generally more convenient to use complex representation. For this reason, starting with the real $E(t)$, one defines the complex spectrum of the field strength $\tilde{E}(\Omega)$, through the complex Fourier transform ($\mathcal{F}$):

$$\tilde{E}(\Omega) = \mathcal{F}\left\{E(t)\right\} = \int_{-\infty}^{\infty} E(t)e^{-i\Omega t} dt = |\tilde{E}(\Omega)|e^{i\Phi(\Omega)} \qquad (1.1)$$

In the definition (1.1), $|\tilde{E}(\Omega)|$ denotes the spectral amplitude and $\Phi(\Omega)$ is the spectral phase. Here and in what follows, complex quantities related to the field are written with a tilde.

Since $E(t)$ is a real function, $\tilde{E}(\Omega) = \tilde{E}^*(-\Omega)$ holds. Given $\tilde{E}(\Omega)$, the time dependent electric field is obtained through the inverse Fourier transform ($\mathcal{F}^{-1}$):

$$E(t) = \mathcal{F}^{-1}\left\{\tilde{E}(\Omega)\right\} = \frac{1}{2\pi}\int_{-\infty}^{\infty} \tilde{E}(\Omega)e^{i\Omega t}d\Omega \qquad (1.2)$$

For practical reasons it may not be convenient to use functions which are non-zero for negative frequencies, as needed in the evaluation of Eq. (1.2). Frequently a complex representation of the electric field, also in the time domain, is desired. Both aspects can be satisfied by introducing an electric field as

$$\tilde{E}^+(t) = \frac{1}{2\pi}\int_{0}^{\infty} \tilde{E}(\Omega)e^{i\Omega t}d\Omega \qquad (1.3)$$

and a corresponding spectral field strength that contains only positive frequencies:

$$\tilde{E}^+(\Omega) = |\tilde{E}(\Omega)|e^{i\Phi(\Omega)} = \begin{cases} \tilde{E}(\Omega) & \text{for } \Omega \geq 0 \\ 0 & \text{for } \Omega < 0 \end{cases} \qquad (1.4)$$

$\tilde{E}^+(t)$ and $\tilde{E}^+(\Omega)$ are related to each other through the complex Fourier transform defined in Eq. (1.1) and Eq. (1.2), i.e.

$$\tilde{E}^+(t) = \frac{1}{2\pi}\int_{-\infty}^{\infty} \tilde{E}^+(\Omega)e^{i\Omega t}d\Omega \qquad (1.5)$$

and

$$\tilde{E}^+(\Omega) = \int_{-\infty}^{\infty} \tilde{E}^+(t)e^{-i\Omega t}dt. \qquad (1.6)$$

The real physical electric field $E(t)$ and its complex Fourier transform can be expressed in terms of the quantities derived in Eq. (1.5) and Eq. (1.6) and the corresponding quantities $\tilde{E}^-(t)$, $\tilde{E}^-(\Omega)$ for the negative frequencies. These quantities relate to the real electric field:

$$E(t) = \tilde{E}^+(t) + \tilde{E}^-(t) \qquad (1.7)$$

and its complex Fourier transform:

$$\tilde{E}(\Omega) = \tilde{E}^+(\Omega) + \tilde{E}^-(\Omega) \qquad (1.8)$$

1.1. CHARACTERISTICS OF FS LIGHT PULSES

It can be shown that $\tilde{E}^+(t)$ can also be calculated through analytic continuation of $E(t)$

$$\tilde{E}^+(t) = E(t) + iE'(t) \tag{1.9}$$

where $E'(t)$ and $E(t)$ are Hilbert transforms of each other. In this sense $\tilde{E}^+(t)$ can be considered as the complex analytical correspondent of the real function $E(t)$. The complex electric field $\tilde{E}^+(t)$ is usually represented by a product of an amplitude function and a phase term:

$$\tilde{E}^+(t) = \frac{1}{2}\mathcal{E}(t)e^{i\Gamma(t)} \tag{1.10}$$

In most practical cases of interest here the spectral amplitude will be centered around a mean frequency ω_ℓ and will have appreciable values only in a frequency interval $\Delta\omega$ small compared to ω_ℓ. In the time domain this suggests the convenience of introducing a carrier frequency ω_ℓ and of writing $\tilde{E}^+(t)$ as:

$$\tilde{E}^+(t) = \frac{1}{2}\mathcal{E}(t)e^{i\varphi(t)}e^{i\omega_\ell t} = \frac{1}{2}\tilde{\mathcal{E}}(t)e^{i\omega_\ell t} \tag{1.11}$$

where $\varphi(t)$ is the time dependent phase, $\tilde{\mathcal{E}}(t)$ is called the complex field envelope and $\mathcal{E}(t)$ the field envelope, respectively. The description of the field given by Eqs. (1.9) through (1.11) is quite general. However, the usefulness of the concept of an envelope and carrier frequency as defined in Eq. (1.11) is limited to the cases where the bandwidth is only a small fraction of the carrier frequency:

$$\frac{\Delta\omega}{\omega_\ell} \ll 1 \tag{1.12}$$

For inequality (1.12) to be satisfied, the temporal variation of $\mathcal{E}(t)$ and $\varphi(t)$ within an optical cycle $T = 2\pi/\omega_\ell$ ($T \approx 2$ fs for visible radiation) has to be small. The corresponding requirement for the complex envelope $\tilde{\mathcal{E}}(t)$ is

$$\left| \frac{d}{dt}\tilde{\mathcal{E}}(t) \right| \ll \omega_\ell |\tilde{\mathcal{E}}(t)| \tag{1.13}$$

Keeping in mind that today the shortest light pulses contain only a few optical cycles, one has to carefully check whether a slowly varying envelope and phase can describe the pulse behavior satisfactorily. Given the spectral description of a signal, $\tilde{E}^+(\Omega)$, the complex envelope $\tilde{\mathcal{E}}(t)$ is simply the inverse transform of the translated spectral field:

$$\tilde{\mathcal{E}}(t) = \frac{1}{2\pi}\int_{-\infty}^{\infty} 2\tilde{E}^+(\Omega + \omega_\ell)e^{i\Omega t}d\Omega \tag{1.14}$$

and the modulus of Eq. (1.14) gives the real envelope. The optimum "translation" ω_ℓ is the one that gives the envelope $\tilde{\mathcal{E}}(t)$ with the least amount of modulation. Spectral translation of Fourier transforms is a standard technique to reconstruct the envelope of interference patterns, and will be used extensively in the chapter on diagnostic techniques.

Let us now discuss more carefully the physical meaning of the phase function $\varphi(t)$. The choice of carrier frequency in Eq. (1.11) should be such as to minimize the variation of phase $\varphi(t)$. The first derivative of the phase factor in $\Gamma(t)$ in Eq. (1.10) establishes a time dependent carrier frequency (instantaneous frequency):

$$\omega(t) = \omega_\ell + \frac{d}{dt}\varphi(t) \qquad (1.15)$$

For $d\varphi/dt = b = $ const., a non-zero value of b just means a correction of the carrier frequency which is now $\omega = \omega_\ell + b$. For $d\varphi/dt = f(t)$, the carrier frequency varies with time and the corresponding pulse is said to be frequency modulated or chirped. For $d^2\varphi/dt^2 < (>)0$, the carrier frequency decreases (increases) along the pulse, which then is called down(up)chirped. From Eq.(1.10) it is obvious that the decomposition of $\Gamma(t)$ into ω_ℓ and $\varphi(t)$ is not unique. The most useful decomposition is one that ensures the smallest $d\varphi/dt$ during the intense portion of the pulse. A common practice is to identify ω_ℓ with the carrier frequency at the pulse peak. A better definition — which is consistent in the time and frequency domains — is to use the intensity weighted *average* frequency:

$$\omega_\ell = \frac{\int_{-\infty}^{\infty} |\tilde{\mathcal{E}}|^2 \omega(t) dt}{\int_{-\infty}^{\infty} |\tilde{\mathcal{E}}|^2 dt} = \frac{\int_{-\infty}^{\infty} |\tilde{E}^+(\Omega)|^2 \Omega d\Omega}{\int_{-\infty}^{\infty} |\tilde{E}^+(\Omega)|^2 d\Omega} \qquad (1.16)$$

The various notations are illustrated in Fig. 1.1 where a linearly upchirped pulse is taken as an example. The temporal dependence of the real electric field is sketched on top of Fig 1.1. A complex representation in the time domain is illustrated with the amplitude and instantaneous frequency of the field. The positive and negative frequency components of the Fourier transform are shown in amplitude and phase in the bottom part of the figure.

1.1.2 Power, energy, and related quantities

Let us imagine the practical situation in which the pulse propagates in a beam with a cross section A, and with $E(t)$ as the relevant component of the

1.1. CHARACTERISTICS OF FS LIGHT PULSES

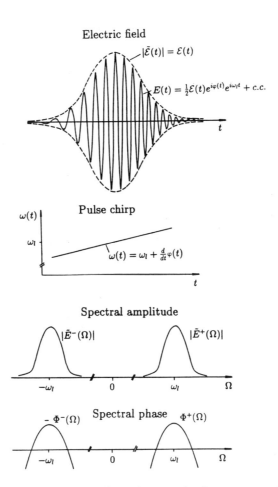

Figure 1.1: Electric field, time dependent carrier frequency, and spectral amplitude of an upchirped pulse.

electric field. The (instantaneous) pulse power (in Watts) in a dispersionless material of refractive index n can be derived from the Poynting theorem of electrodynamics [6] and is given by

$$\mathcal{P}(t) = \epsilon_0 c n \int_A dS \frac{1}{T} \int_{t-T/2}^{t+T/2} E^2(t') dt' \tag{1.17}$$

where c is the velocity of light in vacuum, ϵ_0 is the dielectric permittivity and $\int_A dS$ stands for integration over the beam cross section. The power can be measured by a detector (photodiode, photomultiplier etc.) which integrates over the beam cross section. The temporal response (τ_R) of this device must be short as compared to the speed of variations of the field envelope to be measured. The temporal averaging is performed over one optical period T. Note that the instantaneous power as introduced in Eq. (1.17) is then just a convenient theoretical quantity. In a practical measurement T has to be replaced by the actual response time τ_R of the detector. Therefore, even with the fastest detectors available today ($\tau_R \approx 10^{-13} - 10^{-12}$s), details of the envelope of fs light pulses can not be resolved directly.

A temporal integration of the power yields the energy $\mathcal{W}$ (in Joules):

$$\mathcal{W} = \int_{-\infty}^{\infty} \mathcal{P}(t') dt' \tag{1.18}$$

where the upper and lower integration limits essentially mean "before" and "after" the pulse under investigation.

The corresponding quantities per unit area are the intensity (W/cm^2):

$$\begin{aligned} I(t) &= \epsilon_0 c n \frac{1}{T} \int_{t-T/2}^{t+T/2} E^2(t') dt' \\ &= \frac{1}{2} \epsilon_0 c n \mathcal{E}^2(t) = 2\epsilon_0 c n \tilde{E}^+(t) \tilde{E}^-(t) = \frac{1}{2} \epsilon_0 c n \tilde{\mathcal{E}}(t) \tilde{\mathcal{E}}^*(t) \end{aligned} \tag{1.19}$$

and the energy density per unit area (J/cm^2):

$$W = \int_{-\infty}^{\infty} I(t') dt' \tag{1.20}$$

Sometimes it is convenient to use quantities which are related to photon numbers, such as the photon flux $\mathcal{F}$ (photons/s) or the photon flux density F (photons/s/cm^2):

$$\mathcal{F}(t) = \frac{\mathcal{P}(t)}{\hbar \omega_\ell} \quad \text{and} \quad F(t) = \frac{I(t)}{\hbar \omega_\ell} \tag{1.21}$$

where $\hbar \omega_\ell$ is the energy of one photon at the carrier frequency.

1.1. CHARACTERISTICS OF FS LIGHT PULSES

The spectral properties of the light are typically obtained by measuring the intensity of the field, without any time resolution, at the output of a spectrometer. The quantity, called spectral intensity, that is measured is:

$$S(\Omega) = |\eta(\Omega)\tilde{E}^+(\Omega)|^2 \tag{1.22}$$

where η is a scaling factor which accounts for losses, geometrical influences, and the finite resolution of the spectrometer. Assuming an ideal spectrometer, $|\eta|^2$ can be determined from the requirement of energy conservation:

$$|\eta|^2 \int_{-\infty}^{\infty} |\tilde{E}^+(\Omega)|^2 \, d\Omega = 2\epsilon_0 c n \int_{-\infty}^{\infty} \tilde{E}^+(t)\tilde{E}^-(t) dt \tag{1.23}$$

and Parseval's theorem [7]:

$$\int_{-\infty}^{\infty} |\tilde{E}^+(t)|^2 dt = \frac{1}{2\pi} \int_0^{\infty} |\tilde{E}^+(\Omega)|^2 \, d\Omega \tag{1.24}$$

Thus we have $|\eta|^2 = \epsilon_0 c n/\pi$. Figure 1.2 gives examples of typical pulse shapes and the corresponding spectra.

The complex quantity $\tilde{E}^+$ will be used most often throughout the book to describe the electric field. Therefore, to simplify notations, we will omit the superscript "+" whenever this will not cause confusion.

1.1.3 Pulse duration and spectral width

The shorter the pulse duration, the more difficult it becomes to assert its detailed characteristics. In the femtosecond domain, even the simple concept of pulse duration seems to fade away in a cloud of mushrooming definitions. Part of the problem is that it is difficult to determine the exact pulse shape. For single pulses, the typical representative function that is readily accessible to the experimentalist is the intensity autocorrelation:

$$A_{\text{int}}(\tau) = \int_{-\infty}^{\infty} I(t)I(t-\tau) dt \tag{1.25}$$

The Fourier transform of the correlation of Eq. (1.25) is the real function:

$$A_{\text{int}}(\Omega) = \tilde{\mathcal{I}}(\Omega)\tilde{\mathcal{I}}^*(\Omega) \tag{1.26}$$

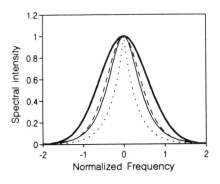 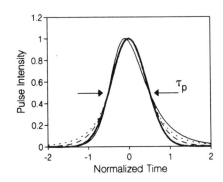

Figure 1.2: Temporal pulse profiles and the corresponding spectra (normalized).

———— Gaussian pulse	$\mathcal{E}(t) \propto$	$\exp[-1.385(t/\tau_p)^2]$
– – – – – sech - pulse	$\mathcal{E}(t) \propto$	$\mathrm{sech}[1.763(t/\tau_p)]$
········· Lorentzian pulse	$\mathcal{E}(t) \propto$	$[1 + 1.656(t/\tau_p)^2]^{-1}$
———— asymm. sech pulse	$\mathcal{E}(t) \propto$	$[\exp(t/\tau_p) + \exp(-3t/\tau_p)]^{-1}$

where the notation $\tilde{\mathcal{I}}(\Omega)$ is the Fourier transform of the function $I(t)$, which should not be confused with the spectral intensity $S(\Omega)$. The fact that the autocorrelation function $A_{\mathrm{int}}(\tau)$ is symmetric, hence its Fourier transform is real [7], implies that little information about the pulse shape can be extracted from such a measurement. Furthermore, the intensity autocorrelation (1.25) contains no information about the pulse phase or its coherence. This point is discussed in detail in Chapter 8.

The quantity that is measured most often, in the case of a train of ultrashort pulses, is the ensemble average:

$$\overline{A_{\mathrm{int}}(\tau)} = \frac{1}{N} \sum_{j=1}^{N} A_{\mathrm{int},j}(\tau). \qquad (1.27)$$

In Eq. (1.27) $A_{\mathrm{int},j}(\tau)$ refers to the autocorrelation of the pulse with number j of a train of N pulses. If the pulse characteristics, such as the pulse duration, are fluctuating during the train, the averaged intensity autocorrelation will reflect the train statistics rather than the properties of the individual pulses.

It is beyond the scope of this introductory chapter to discuss the various methods of pulse measurement. Of more immediate importance at this stage is to clarify the various definitions of pulse "duration" and "spectral width"

1.1. CHARACTERISTICS OF FS LIGHT PULSES

used to characterize short pulses. Unless specified otherwise, we define the pulse duration τ_p as the full width at half maximum (FWHM) of the intensity profile and the spectral width $\Delta\omega_p$ as the FWHM of the spectral intensity. Making that statement is an obvious admission that other definitions exist. Precisely because of the difficulty of asserting the exact pulse shape, standard waveforms have been selected. The most commonly cited are the Gaussian, for which the temporal dependence of the field is:

$$\tilde{\mathcal{E}}(t) = \tilde{\mathcal{E}}_0 \exp\{-(t/\tau_G)^2\} \tag{1.28}$$

and the secant hyperbolic:

$$\tilde{\mathcal{E}}(t) = \tilde{\mathcal{E}}_0 \operatorname{sech}(t/\tau_s). \tag{1.29}$$

The parameters $\tau_G = \tau_p/\sqrt{2\ln 2}$ and $\tau_s = \tau_p/1.76$ are generally more convenient to use in theoretical calculations involving pulses with these assumed shapes.

Since the temporal and spectral characteristics of the field are related to each other through Fourier transforms, the bandwidth $\Delta\omega_p$ and pulse duration τ_p cannot vary independently of each other. There is a minimum duration-bandwidth product:

$$\Delta\omega_p \tau_p = 2\pi \Delta\nu_p \tau_p \geq 2\pi c_B. \tag{1.30}$$

c_B is a numerical constant on the order of 1, depending on the actual pulse shape. Some examples are shown in Table 1.1. The equality holds for pulses without frequency modulation (unchirped) which are called "bandwidth limited" or "Fourier limited". Such pulses exhibit the shortest possible duration at a given spectral width and pulse shape. If there is a frequency variation across a pulse, its spectrum will contain additional spectral components. Consequently, this pulse possesses a spectral width which is larger than what would correspond to the Fourier limit given in Eq. (1.30). The relation (1.30) can also be interpreted as a consequence of the uncertainty principle, which expresses here the limit for simultaneous temporal and spectral resolution.

Sometimes pulse duration and spectral width defined by the FWHM values are not suitable measures. This is, for instance, the case in pulses with substructure or broad wings causing a considerable part of the energy to lie outside the range given by the FWHM. In these cases one may use favorably averaged values derived from the appropriate second–order moments. In the

Field envelope	Intensity profile	τ_p (FWHM)	Spectral profile	$\Delta\omega_p$ (FWHM)	c_B
Gauss	$e^{-2(t/\tau_G)^2}$	$1.177\tau_G$	$e^{-(\Omega\tau_G)^2/2}$	$2.355/\tau_G$	0.441
sech	$\text{sech}^2(t/\tau_s)$	$1.763\tau_s$	$\text{sech}^2(\pi\Omega\tau_s/2)$	$1.122/\tau_s$	0.315
Lorentz	$[1+(t/\tau_L)^2]^{-2}$	$1.287\tau_L$	$e^{-2|\Omega|\tau_L}$	$0.693/\tau_L$	0.142
asym. sech	$\left[e^{t/\tau_a}+e^{-3t/\tau_a}\right]^{-2}$	$1.043\tau_a$	$\text{sech}(\pi\Omega\tau_a/2)$	$1.677/\tau_a$	0.278
rectang.	1 for $|t/\tau_r|\le 1$, 0 else	τ_r	$\text{sinc}^2(\Omega\tau_r)$	$2.78/\tau_r$	0.443

Table 1.1: Examples of standard pulse profiles.

time domain, the second moment of the temporal dependence is defined as:

$$\bar{\tau}_p = \left[\frac{1}{W}\int_{-\infty}^{\infty} t^2 I(t)dt - \frac{1}{W^2}\left(\int_{-\infty}^{\infty} tI(t)dt\right)^2\right]^{\frac{1}{2}} \qquad (1.31)$$

The corresponding definition in the frequency domain is:

$$\overline{\Delta\omega}_p = \left[\frac{1}{W}\int_{-\infty}^{\infty} \Omega^2 S(\Omega)d\Omega - \frac{1}{W^2}\left(\int_{-\infty}^{\infty} \Omega S(\Omega)d\Omega\right)^2\right]^{\frac{1}{2}} \qquad (1.32)$$

The difference between the FWHM τ_p and the second moment $\bar{\tau}_p$ is illustrated in Fig. 1.3.

Gaussian pulses

Having introduced essential pulse characteristics, it seems convenient to discuss an example to which we can refer to in later chapters. We choose a Gaussian pulse with linear chirp. This choice is one of analytical convenience: The Gaussian shape is *not* the most commonly encountered temporal shape. The electric field is given by

$$\tilde{\mathcal{E}}(t) = \mathcal{E}_0 e^{-(1+ia)(t/\tau_G)^2} \qquad (1.33)$$

with the pulse duration

$$\tau_p = \sqrt{2\ln 2}\,\tau_G. \qquad (1.34)$$

1.1. CHARACTERISTICS OF FS LIGHT PULSES

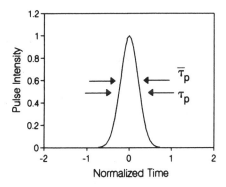

Figure 1.3: Comparison of pulse durations defined in different ways.

Note that with the definition (1.33) the chirp parameter a is positive for a downchirp $(d\varphi/dt = -2at/\tau_G^2)$. The Fourier transform of (1.33) yields

$$\tilde{\mathcal{E}}(\Omega) = \frac{\mathcal{E}_0\sqrt{\pi}\tau_G}{\sqrt[4]{1+a^2}} \exp\left\{i\Phi - \frac{\Omega^2 \tau_G^2}{4(1+a^2)}\right\} \quad (1.35)$$

with the spectral phase given by:

$$\Phi(\Omega) = -\frac{1}{2}\arctan(a) + \frac{a\tau_G^2}{4(1+a^2)}\Omega^2 \quad (1.36)$$

It can be seen from Eq. (1.35) that the spectral intensity is the Gaussian:

$$S(\Omega) = \frac{|\eta|^2 \pi \mathcal{E}_0^2 \tau_G^2}{\sqrt{1+a^2}} \exp\left\{-\frac{\Omega^2 \tau_G^2}{2(1+a^2)}\right\} \quad (1.37)$$

with a FWHM given by:

$$\Delta\omega_p = \frac{1}{\tau_G}\sqrt{8\ln 2(1+a^2)} = 2\pi\Delta\nu_p \quad (1.38)$$

For the pulse duration-bandwidth product we find

$$\Delta\nu_p \tau_p = \frac{2\ln 2}{\pi}\sqrt{1+a^2} \quad (1.39)$$

Obviously, the occurrence of chirp ($a \neq 0$) results in additional spectral components which enlarge the spectral width and lead to a duration bandwidth product exceeding the Fourier limit ($2\ln 2/\pi \approx 0.44$) by a factor $\sqrt{1+a^2}$.

We also want to point out that the spectral phase given by Eq. (1.36) changes quadratically with frequency if the input pulse is linearly chirped. While this is exactly true for Gaussian pulses as can be seen from Eq. (1.36), it holds approximately for other pulse shapes.

1.2 Pulse propagation

So far we have considered only temporal and spectral characteristics of light pulses. In this subsection we shall be interested in the propagation of such pulses through matter. This is the situation one has always to face when working with electromagnetic wave packets (at least until somebody succeeds in building a suitable trap). The electric field, now considered in its temporal and spatial dependence, is again a suitable quantity for the description of the propagating wave packet. In view of the optical materials that will be investigated, we can neglect external charges and currents and confine ourselves to nonmagnetic permeabilities and uniform media. A wave equation can be derived for the electric field vector $\mathbf{E}$ from Maxwell equations (see for instance Ref. [8]) which in Cartesian coordinates reads

$$\left(\frac{\partial^2}{\partial x^2} + \frac{\partial^2}{\partial y^2} + \frac{\partial^2}{\partial z^2} - \frac{1}{c^2}\frac{\partial^2}{\partial t^2}\right)\mathbf{E}(x,y,z,t) = \mu_0 \frac{\partial^2}{\partial t^2}\mathbf{P}(x,y,z,t) \qquad (1.40)$$

where μ_0 is the magnetic permeability of free space. The source term of Eq. (1.40) contains the polarization $\mathbf{P}$ and describes the influence of the medium on the field as well as the response of the medium. Usually the polarization is decomposed into two parts:

$$\mathbf{P} = \mathbf{P}^L + \mathbf{P}^{NL} \qquad (1.41)$$

The decomposition of Eq. (1.41) is intended to distinguish a polarization that varies linearly ($\mathbf{P}^L$) and from one that varies nonlinearly ($\mathbf{P}^{NL}$) with the field. Historically, $\mathbf{P}^L$ represents the medium response in the frame of "ordinary" optics, e.g., classical optics [9], and is responsible for such effects as diffraction, dispersion, refraction, linear losses and linear gain. Frequently, these processes can be attributed to the action of a host material which in turn may contain sources of a nonlinear polarization $\mathbf{P}^{NL}$. The latter is responsible for nonlinear optics [10, 11] which includes, for instance, nonlinear absorption and gain, harmonic generation and Raman processes.

As will be seen in subsequent chapters, both $\mathbf{P}^L$ and in particular $\mathbf{P}^{NL}$ are often related to the electric field by complicated differential equations.

1.2. PULSE PROPAGATION

One reason is that no physical phenomenon can be truly instantaneous. We will initially omit $\mathbf{P}^{NL}$. Depending on the actual problem under consideration, $\mathbf{P}^{NL}$ will have to be specified and added to the wave equation as a source term.

1.2.1 The reduced wave equation

Equation (1.40) is of rather complicated structure and in general can solely be solved by numerical methods. However, by means of suitable approximations and simplifications, one can derive a "reduced wave equation" which will enable us to deal with many practical pulse propagation problems in a rather simple way. For this we assume the electric field to be linearly polarized and propagating in the z-direction as a plane wave, i.e., the field is uniform in the transverse x, y direction. The wave equation has now been reduced to:

$$\left(\frac{\partial^2}{\partial z^2} - \frac{1}{c^2}\frac{\partial^2}{\partial t^2}\right) E(z,t) = \mu_0 \frac{\partial^2}{\partial t^2} P^L(z,t) \qquad (1.42)$$

As known from classical electrodynamics [8] the linear polarization of a medium is related to the field through the dielectric susceptibility χ. In the frequency domain we have

$$\tilde{P}^L(\Omega, z) = \epsilon_0 \, \chi(\Omega) \tilde{E}(z, \Omega) \qquad (1.43)$$

which is equivalent to a convolution integral in the time domain

$$P^L(t, z) = \epsilon_0 \int_{-\infty}^{t} dt' \, \chi(t') E(z, t - t'). \qquad (1.44)$$

The finite upper integration limit, t, expresses the fact that the response of the medium must be causal. For a nondispersive medium (which implies an "infinite bandwidth" of χ) the medium response is instantaneous, i.e., memory free. In general, $\chi(t)$ describes a finite response time of the medium which, in the frequency domain, means nonzero dispersion. This simple fact has important implications for the propagation of short pulses and time varying radiation in general. We will refer to this point several times in later chapters — in particular when dealing with coherent interaction.

The Fourier transform of (1.42) together with (1.43) yields

$$\left[\frac{\partial^2}{\partial z^2} + \frac{\Omega^2}{c^2}\epsilon(\Omega)\right] \tilde{E}(z, \Omega) = 0 \qquad (1.45)$$

where we have introduced the dielectric constant

$$\epsilon(\Omega) = [1 + \chi(\Omega)]. \tag{1.46}$$

Note that we have assumed a real susceptibility and dielectric constant, respectively. Later we will discuss effects associated with complex quantities. The general solution of (1.45) for the propagation in the $+z$ direction is

$$\tilde{E}(\Omega, z) = \tilde{E}(\Omega, 0)e^{-ik(\Omega)z} \tag{1.47}$$

where the propagation constant $k(\Omega)$ is determined by the dispersion relation of linear optics

$$k^2(\Omega) = \frac{\Omega^2}{c^2}\epsilon(\Omega) = \frac{\Omega^2}{c^2}n^2(\Omega). \tag{1.48}$$

For further consideration we expand $k(\Omega)$ about the carrier frequency ω_ℓ

$$k(\Omega) = k_\ell + \left.\frac{dk}{d\Omega}\right|_{\omega_\ell}(\Omega - \omega_\ell) + \frac{1}{2}\left.\frac{d^2k}{d\Omega^2}\right|_{\omega_\ell}(\Omega - \omega_\ell)^2 + \ldots = k_\ell + \delta k \tag{1.49}$$

and write Eq. (1.47) as

$$\tilde{E}(\Omega, z) = \tilde{E}(\Omega, 0)e^{-ik_\ell z}e^{-i\delta k\, z} \tag{1.50}$$

where $k_\ell^2 = \omega_\ell^2 \epsilon(\omega_\ell)/c^2 = \omega_\ell^2 n^2(\omega_\ell)/c^2$. In most practical cases of interest, the Fourier amplitude will be centered around a mean wave vector k_ℓ, and will have appreciable values only in an interval Δk small compared to k_ℓ. In analogy to the introduction of an envelope function slowly varying in time after the separation of a rapidly oscillating term, cf. Eqs. (1.11)–(1.13), we can define now an amplitude which is slowly varying in the spatial coordinate

$$\tilde{\mathcal{E}}(\Omega', z) = \tilde{E}(\Omega, 0)e^{-i\delta k\, z} \tag{1.51}$$

where $\Omega' = \Omega + \omega_l$. Again, for this concept to be useful we must require that

$$\left|\frac{d}{dz}\tilde{\mathcal{E}}(\Omega', z)\right| \ll k_\ell \left|\tilde{\mathcal{E}}(\Omega', z)\right| \tag{1.52}$$

which implies a sufficiently small wave number spectrum

$$\left|\frac{\Delta k}{k_\ell}\right| \ll 1. \tag{1.53}$$

1.2. PULSE PROPAGATION

In other words, the pulse envelope must not change significantly while traveling through a distance comparable with the wavelength $\lambda_\ell = 2\pi/k_\ell$. Fourier transforming of Eq. (1.50)) into the time domain gives

$$\tilde{E}(t,z) = \frac{1}{2\pi}\left\{\int_{-\infty}^{\infty} d\Omega\, \tilde{E}(\Omega,0)e^{-i\delta k\, z}e^{i(\Omega-\omega_\ell)t}\right\}e^{i(\omega_\ell t - k_\ell z)} \quad (1.54)$$

which can be written as

$$\tilde{E}(t,z) = \frac{1}{2}\tilde{\mathcal{E}}(t,z)e^{i(\omega_\ell t - k_\ell z)} \quad (1.55)$$

where $\tilde{\mathcal{E}}(t,z)$ is now the envelope varying slowly in space and time, defined by the term in the curled brackets in Eq. (1.54).

Further simplification of the wave equation requires a corresponding equation for $\tilde{\mathcal{E}}$ utilizing the envelope properties. Only a few terms in the expansion of $k(\Omega)$ and $\epsilon(\Omega)$, respectively, will be considered. To this effect we expand $\epsilon(\Omega)$ as series around ω_ℓ, leading to the following form for the linear polarization (1.43)

$$\tilde{P}^L(\Omega,z) = \epsilon_0\left(\epsilon(\omega_\ell) - 1 + \sum_{n=1}^{\infty}\frac{1}{n!}\frac{d^n\epsilon}{d\Omega^n}\bigg|_{\omega_\ell}(\Omega-\omega_\ell)^n\right)\tilde{E}(\Omega,z). \quad (1.56)$$

In terms of the pulse envelope, the above expression corresponds in the time domain to

$$\begin{aligned}\tilde{P}^L(t,z) &= \frac{1}{2}\bigg\{\epsilon_0[\epsilon(\omega_\ell)-1]\tilde{\mathcal{E}}(t,z) \\ &+ \epsilon_0\sum_{n=1}^{\infty}(-i)^n\frac{\epsilon^{(n)}(\omega_\ell)}{n!}\frac{\partial^n}{\partial t^n}\tilde{\mathcal{E}}(t,z)\bigg\}e^{i(\omega_\ell t - k_\ell z)}\end{aligned} \quad (1.57)$$

where $\epsilon^{(n)}(\omega_\ell) = \frac{\partial^n \epsilon}{\partial\Omega^n}\big|_{\omega_\ell}$. The term in the curled brackets defines the slowly varying envelope of the polarization $\tilde{\mathcal{P}}$. The next step is to replace the electric field and the polarization in the wave equation (1.42) by Eq. (1.54) and Eq. (1.57), respectively. We transfer thereafter to a coordinate system (η,ξ) moving with the group velocity $v_g = \left(\frac{dk}{d\Omega}\big|_{\omega_\ell}\right)^{-1}$, which is the standard transformation to a "retarded" frame of reference:

$$\xi = z \qquad \eta = t - \frac{z}{v_g} \quad (1.58)$$

and
$$\frac{\partial}{\partial z} = \frac{\partial}{\partial \xi} - \frac{1}{v_g}\frac{\partial}{\partial \eta} \qquad \frac{\partial}{\partial t} = \frac{\partial}{\partial \eta}. \tag{1.59}$$

A straightforward calculation leads to the final result:

$$\frac{\partial}{\partial \xi}\tilde{\mathcal{E}} - \frac{i}{2}k_\ell''\frac{\partial^2}{\partial \eta^2}\tilde{\mathcal{E}} + \mathcal{D} = -\frac{i}{2k_\ell}\frac{\partial}{\partial \xi}\left(\frac{\partial}{\partial \xi} - \frac{2}{v_g}\frac{\partial}{\partial \eta}\right)\tilde{\mathcal{E}} \tag{1.60}$$

The quantity

$$\begin{aligned}\mathcal{D} &= \frac{i}{3k_\ell c^2}\sum_{n=3}^{\infty}\frac{(-i)^n}{n!}\left[\omega_\ell^2 \epsilon^{(n)}(\omega_\ell) + 2n\omega_\ell \epsilon^{(n-1)}(\omega_\ell)\right.\\ &\quad + \left. n(n-1)\epsilon^{(n-2)}(\omega_\ell)\right]\frac{\partial^n}{\partial \eta^n}\tilde{\mathcal{E}}\end{aligned} \tag{1.61}$$

contains dispersion terms of higher order and

$$\begin{aligned}k_\ell'' &= \left.\frac{\partial^2 k}{\partial \Omega^2}\right|_{\omega_\ell} = -\frac{1}{v_g^2}\left.\frac{dv_g}{d\Omega}\right|_{\omega_\ell}\\ &= \frac{1}{2k_\ell}\left[\frac{2}{v_g^2} - \frac{2}{c^2}\epsilon(\omega_\ell) - \frac{4\omega_\ell}{c^2}\epsilon^{(1)}(\omega_\ell) - \frac{\omega_\ell^2}{c^2}\epsilon^{(2)}(\omega_\ell)\right]\end{aligned} \tag{1.62}$$

is the group velocity dispersion (GVD) parameter. It should be mentioned that the GVD is usually defined as the derivative of v_g with respect to λ, $dv_g/d\lambda$, related to k'' through

$$\frac{dv_g}{d\lambda} = \frac{\Omega^2 v_g^2}{2\pi c}\frac{d^2 k}{d\Omega^2}. \tag{1.63}$$

So far we have not made any approximations and the structure of Eq. (1.60) is rather complex. However, we can exploit at this point the envelope properties (1.13) and (1.52), which, in this particular situation, imply:

$$\left|\frac{1}{k_\ell}\left(\frac{\partial}{\partial \xi} - \frac{2}{v_g}\frac{\partial}{\partial \eta}\right)\tilde{\mathcal{E}}\right| = \left|\frac{1}{k_\ell}\left(\frac{\partial}{\partial z} - \frac{1}{v_g}\frac{\partial}{\partial t}\right)\tilde{\mathcal{E}}\right| \ll |\tilde{\mathcal{E}}| \tag{1.64}$$

The right-hand side of (1.60) can thus be neglected if the prerequisites for introducing pulse envelopes are fulfilled. This procedure is called slowly varying envelope approximation (SVEA) and reduces the wave equation to first-order derivatives with respect to the spatial coordinate.

1.2. PULSE PROPAGATION

If the propagation of very short pulses is computed over long distances, the cumulative error introduced by neglecting the right hand side of Eq. (1.60) may be significant. In those cases, a direct numerical treatment of the second order wave equation is required. In writing the general solution of the wave equation (1.45) in the form of a *forward* propagating wave (1.47), we have implicitly neglected the amplitude of the wave reflected by the linear or nonlinear polarizability of the medium. There are situations — involving, for instance, highly concentrated absorbing or amplifying dyes — where this basic approximation breaks down.

Further simplifications are possible for a very broad class of problems of practical interest, where the dielectric constant changes slowly over frequencies within the pulse spectrum. In those cases, terms with $n \geq 3$ can be omitted too, leading to a greatly simplified reduced wave equation:

$$\frac{\partial}{\partial \xi}\tilde{\mathcal{E}}(\eta,\xi) - \frac{i}{2}k_\ell'' \frac{\partial^2}{\partial \eta^2}\tilde{\mathcal{E}}(\eta,\xi) = 0 \tag{1.65}$$

which describes the evolution of the complex pulse envelope as it propagates through a loss-free medium with GVD. The reader will recognize the structure of the one–dimensional Schrödinger equation.

In the case of zero GVD ($k_\ell'' = 0$), the pulse envelope does not change at all in the system of local coordinates (η, ξ). Hence the usefulness of introducing a coordinate system moving at the group velocity. In the laboratory frame, the pulse travels at the group velocity without any distortion.

In dealing with short pulses as well as in dealing with white light (see Chapter 2) the appropriate "retarded frame of reference" is moving at the *group* rather than at the *wave* velocity. Indeed, while a monochromatic wave of frequency Ω travels at the phase velocity $v_p(\Omega) = c/n(\Omega)$, it is the superposition of many such waves with differing phase velocities that leads to a wave packet (pulse) propagating with the group velocity. The importance of the frame of reference moving at the group velocity is such that, in the following chapters, the notation z and t will be substituted for ξ and η, unless the laboratory frame is explicitly specified.

For nonzero GVD ($k'' \neq 0$) the propagation problem (1.65) can be solved either directly in the time or in the frequency domain. In the first case, the solution is given by a Poisson-integral [12] which here reads

$$\tilde{\mathcal{E}}(t,z) = \frac{1}{\sqrt{2\pi i k_\ell'' z}} \int_{-\infty}^{t} \tilde{\mathcal{E}}(t', z=0) \exp\left(i\frac{(t-t')^2}{2k_\ell'' z}\right) dt' \tag{1.66}$$

In the second case, we find after Fourier transformation of (1.65) a solution of the form

$$\tilde{\mathcal{E}}(\Omega, z) = \tilde{\mathcal{E}}(\Omega, 0) e^{-\frac{i}{2} k'' \Omega^2 z}. \tag{1.67}$$

Thus we have for the temporal envelope

$$\tilde{\mathcal{E}}(t, z) = \mathcal{F}^{-1} \left\{ \tilde{\mathcal{E}}_0(\Omega) e^{-\frac{i}{2} k''_\ell \Omega^2 z} \right\}. \tag{1.68}$$

When propagating through the sample, an initially bandwidth-limited pulse develops a spectral phase with a quadratic frequency dependence, resulting in chirp. From the example at the end of the previous subsection, we know that this is accompanied by the occurrence of additional frequency components. Since, however, the spectrum of the pulse $|\tilde{\mathcal{E}}(\Omega, z)|^2$ remains constant [see Eq. (1.51)], the spectral components responsible for chirp must appear at the expense of the envelope shape, which has to become broader.

At this point we want to introduce some useful relations for the characterization of the dispersion. The dependence of a dispersive parameter can be given as a function of either the frequency Ω or the vacuum wavelength λ. The first, second and third order derivatives are related to each other by

$$\frac{d}{d\Omega} = -\frac{\lambda^2}{2\pi c} \frac{d}{d\lambda} \tag{1.69}$$

$$\frac{d^2}{d\Omega^2} = \frac{\lambda^2}{(2\pi c)^2} \left(\lambda^2 \frac{d^2}{d\lambda^2} + 2\lambda \frac{d}{d\lambda} \right) \tag{1.70}$$

$$\frac{d^3}{d\Omega^3} = -\frac{\lambda^3}{(2\pi c)^3} \left(\lambda^3 \frac{d^3}{d\lambda^3} + 6\lambda^2 \frac{d^2}{d\lambda^2} + 6\lambda \frac{d}{d\lambda} \right) \tag{1.71}$$

The dispersion of the material is described by either the frequency dependence $n(\Omega)$ or the wavelength dependence $n(\lambda)$ of the index of refraction. The derivatives of the propagation constant used most often in pulse propagation problems, expressed in terms of the index n, are:

$$\frac{dk}{d\Omega} = \frac{n}{c} + \frac{\Omega}{c} \frac{dn}{d\Omega} = \frac{1}{c} \left(n - \lambda \frac{dn}{d\lambda} \right) \tag{1.72}$$

$$\frac{d^2 k}{d\Omega^2} = \frac{2}{c} \frac{dn}{d\Omega} + \frac{\Omega}{c} \frac{d^2 n}{d\Omega^2} = \left(\frac{\lambda}{2\pi c} \right) \frac{1}{c} \left(\lambda^2 \frac{d^2 n}{d\lambda^2} \right) \tag{1.73}$$

$$\frac{d^3 k}{d\Omega^3} = \frac{3}{c} \frac{d^2 n}{d\Omega^2} + \frac{\Omega}{c} \frac{d^3 n}{d\Omega^3} = -\left(\frac{\lambda}{2\pi c} \right)^2 \frac{1}{c} \left(3\lambda^2 \frac{d^2 n}{d\lambda^2} + \lambda^3 \frac{d^3 n}{d\lambda^3} \right) \tag{1.74}$$

1.2. PULSE PROPAGATION

For a more quantitative picture of the influence that GVD has on the pulse propagation we consider the linearly chirped Gaussian pulse of Eq. (1.33) entering the sample. From Eqs. (1.67), (1.35), and (1.36) we find for the spatially (z) dependent field spectrum

$$\tilde{\mathcal{E}}(\Omega, z) = \tilde{A}_0 e^{-x\Omega^2} e^{iy\Omega^2} \tag{1.75}$$

where

$$x = \frac{\tau_G^2}{4(1+a^2)} \tag{1.76}$$

and

$$y(z) = \frac{a\tau_G^2}{4(1+a^2)} - \frac{k_\ell'' z}{2} \tag{1.77}$$

and $\tilde{A}_0$ is a complex amplitude factor which we will not consider in what follows. The time dependent electric field that we obtain by Fourier transforming Eq. (1.75) can be written as

$$\tilde{\mathcal{E}}(t, z) = \tilde{A}_1 \exp\left\{-\left(1 + i\frac{y(z)}{x}\right)\left(\frac{t}{\sqrt{\frac{4}{x}[x^2 + y^2(z)]}}\right)^2\right\}. \tag{1.78}$$

Obviously, this describes again a linearly chirped Gaussian pulse. For the "pulse duration" (note $\tau_p = \sqrt{2\ln 2}\, \tau_G$) and chirp we find

$$\tau_G(z) = \sqrt{\frac{4}{x}[x^2 + y^2(z)]} \tag{1.79}$$

and

$$\varphi(t, z) = -\frac{y(z)}{4[x^2 + y^2(z)]} t^2. \tag{1.80}$$

Let us consider first an initially unchirped input pulse ($a = 0$). The pulse duration and phase develop as:

$$\tau_G(z) = \tau_{G0}\sqrt{1 + \left(\frac{z}{L_d}\right)^2} \tag{1.81}$$

$$\frac{\partial^2}{\partial t^2}\varphi(t, z) = \left(\frac{1}{\tau_{G0}^2}\right) \frac{2z/L_d}{1 + (z/L_d)^2}, \tag{1.82}$$

where $\tau_{G0} = \tau_G(z=0)$, and we have defined a characteristic length:

$$L_d = \frac{\tau_{G0}^2}{2|k_\ell''|}. \tag{1.83}$$

For later reference let us introduce a so-called dispersive length defined as

$$L_D = \frac{\tau_{p0}^2}{|k_\ell''|} \tag{1.84}$$

where for Gaussian pulses $L_D \approx 2.77 L_d$. Bandwidth limited Gaussian pulses double their length after propagation of about $0.6 L_D$.

There is a complete analogy between the propagation (diffraction) effects of a spatially Gaussian beam and the temporal evolution of a Gaussian pulse in a dispersive medium. For instance, the pulse duration and the slope of the chirp follow the same evolution with distance as the waist and curvature of a Gaussian beam, as detailed at the end of this chapter.

A linearly chirped Gaussian pulse in a dispersive medium is completely characterized by the position and (minimum) duration of the unchirped pulse, just as a spatially Gaussian beam is uniquely defined by the position and size of its waist. To illustrate this point, let us consider a linearly chirped pulse of "duration" τ_G at a point z_1. The position z_c of the minimum duration (unchirped pulse) is found by setting $y = 0$ in Eq. (1.77):

$$z_c = z_1 + \frac{\tau_G^2}{2k_\ell''}\frac{a}{1+a^2} = z_1 + a\frac{\tau_{Gmin}^2}{2k_\ell''}. \tag{1.85}$$

The minimum pulse duration (FWHM) at $z = z_c$ is $\tau_{pmin} = \sqrt{2\ln 2}\,\tau_{Gmin}$ with

$$\tau_{Gmin} = \frac{\tau_G}{\sqrt{1+a^2}}. \tag{1.86}$$

The position z_c is after z_1 if a and k_ℓ'' have the same sign[1]; before z_1 if they have opposite sign. All the temporal characteristics of the pulse are most conveniently defined in terms of the distance $L = z - z_c$ to the point of zero chirp, and the minimum duration τ_{Gmin}. The chirp parameter a and the pulse "duration" τ_G at any point are then simply given by

$$a = L/L_d \tag{1.87}$$
$$\tau_G = \tau_{Gmin}\sqrt{1+a^2} \tag{1.88}$$

where the dispersion parameter $L_d = \tau_{Gmin}^2/(2k_\ell'')$.

With an initial chirp a, the minimum pulse duration is reached when the spectral phase in Eq. (1.75) vanishes, i.e., when the input spectral phase is

---

[1] For instance, an initially downchirped ($a > 0$) pulse at $z = z_c$ will be compressed in a medium with positive dispersion ($k_\ell'' > 0$).

1.2. PULSE PROPAGATION

compensated by the GVD. The larger the input chirp ($|a|$) is, the shorter the minimum pulse duration that can be obtained in this manner [see Eq. (1.86)]. The underlying reason is that the excess bandwidth of a chirped pulse is converted into a narrowing of the envelope by chirp compensation until the Fourier limit is reached. The whole procedure including the impression of chirp on a pulse will be treated in Chapter 7 (pulse shaping) in more detail. To summarize, Fig. (1.4) illustrates the behavior of a linearly chirped Gaussian pulse as it propagates through a dispersive sample.

Simple physical consideration can lead directly to a crude approximation for the maximum broadening a bandwidth limited pulse of duration τ_p and spectral width $\Delta\omega_p$ experiences. Each spectral component Ω travels with its own group velocity $v_g(\Omega)$. The difference of group velocities over the pulse spectrum becomes then:

$$\Delta v_g = \left[\frac{dv_g}{d\Omega}\right]_{\omega_\ell} \Delta\omega_p. \tag{1.89}$$

Accordingly, after a travel distance L the pulse spread can be as large as

$$\Delta\tau_p = \Delta\left(\frac{L}{v_g}\right) \tag{1.90}$$

which, by means of Eq. (1.62), yields:

$$\Delta\tau_p = L k_\ell'' \Delta\omega_p. \tag{1.91}$$

A characteristic length after which a pulse has approximately doubled its duration can now be written as:

$$L_D' = \frac{1}{|k_\ell''|\Delta\omega_p^2}. \tag{1.92}$$

Measuring the length in m and the spectral width in nm the GVD of materials is sometimes given in fs/(m nm) which pictorially describes the pulse broadening per unit travel distance and unit spectral width. From Eq. (1.91) we find for the corresponding quantity

$$\frac{\Delta\tau_p}{L\Delta\lambda} = -2\pi\frac{c}{\lambda_\ell^2}k_\ell''. \tag{1.93}$$

For BK7 glass at 620 nm, $k_\ell'' \approx 1.02\times10^{-25}\text{s}^2/\text{m}$, and the GVD as introduced above is about -500 fs per nm spectral width and m propagation length.

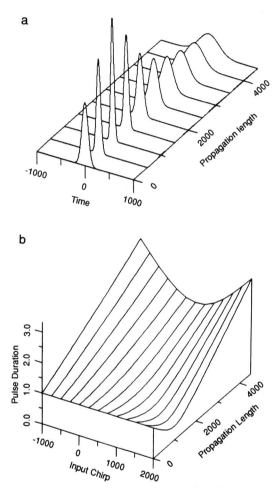

Figure 1.4: Propagation of a linearly chirped Gaussian pulse in a GVD medium [pulse shape (a), pulse duration for different input chirp (b)].

1.2.2 Complex dielectric constant

In general, the dielectric constant, which was introduced in Eq. (1.45) as a real quantity, is complex. Indeed a closer inspection of Eq. (1.44) shows that the finite memory time of matter requires not only ϵ, χ to be frequency dependent but also that they be complex. The real and imaginary part of $\tilde{\epsilon}$, $\tilde{\chi}$ are not independent of each other but related through a Kramers–Kronig relation. The consideration of a real $\epsilon(\Omega)$ is justified as long as we can neglect (linear) losses or gain. This is valid for transparent samples or propagation lengths which are too short for these processes to become essential for the pulse shaping. For completeness we will modify the reduced wave equation (1.65) by taking into account a complex dielectric constant $\tilde{\epsilon}(\Omega)$ represented as

$$\tilde{\epsilon}(\Omega) = \epsilon(\Omega) + i\epsilon_i(\Omega). \tag{1.94}$$

Let us assume $\tilde{\epsilon}(\Omega)$ to be weakly dispersive. The same procedure introduced to derive Eq. (1.65) can be used after inserting the complex dielectric constant $\tilde{\epsilon}$ into the expression of the polarization Eq. (1.56). Now the reduced wave equation becomes

$$\frac{\partial}{\partial z}\tilde{\mathcal{E}}(t,z) - \frac{i}{2}k''\frac{\partial^2}{\partial t^2}\tilde{\mathcal{E}}(t,z) = \kappa_1\tilde{\mathcal{E}}(t,z) + i\kappa_2\frac{\partial}{\partial t}\tilde{\mathcal{E}}(t,z) + \kappa_3\frac{\partial^2}{\partial t^2}\tilde{\mathcal{E}}(t,z) \tag{1.95}$$

where

$$\kappa_1 = \frac{\omega_\ell}{2c}\epsilon_i(\omega_\ell) \tag{1.96}$$

$$\kappa_2 = \frac{1}{2c}\left[2\epsilon_i(\omega_\ell) + \omega_\ell\frac{d}{d\Omega}\epsilon_i(\Omega)\bigg|_{\omega_\ell}\right] \tag{1.97}$$

$$\kappa_3 = \frac{1}{4c\omega_\ell}\left[2\epsilon_i(\omega_\ell) + 4\omega_\ell\frac{d}{d\Omega}\epsilon_i(\Omega)\bigg|_{\omega_\ell} + \omega_\ell^2\frac{d^2}{d\Omega^2}\epsilon_i(\Omega)\bigg|_{\omega_\ell}\right]. \tag{1.98}$$

For zero–GVD, and neglecting the two last terms in the right-hand side of Eq. (1.95), the pulse evolution with propagation distance z is described simply by

$$\frac{\partial}{\partial z}\tilde{\mathcal{E}}(t,z) - \kappa_1\tilde{\mathcal{E}}(t,z) = 0 \tag{1.99}$$

which has the solution

$$\tilde{\mathcal{E}}(t,z) = \tilde{\mathcal{E}}(t,0)e^{\kappa_1 z}. \tag{1.100}$$

Obviously, the pulse experiences losses or gain depending on the sign of κ_1 and does not change its shape. Equation (1.100) states simply the Lambert-Beer law of linear optics.

An interesting situation is that in which there would be neither gain nor loss, i.e., $\epsilon_i(\omega_\ell) = 0$ and $\frac{d}{d\Omega}\epsilon_i(\Omega)\big|_{\omega_\ell} \neq 0$, could occur between an absorption and amplification line. Neglecting the terms with the second temporal derivative of $\tilde{\mathcal{E}}$, the propagation problem is governed by the equation

$$\frac{\partial}{\partial z}\tilde{\mathcal{E}}(t,z) - i\kappa_2 \frac{\partial}{\partial t}\tilde{\mathcal{E}}(t,z) = 0. \tag{1.101}$$

The solution of this equation is simply

$$\tilde{\mathcal{E}}(t,z) = \tilde{\mathcal{E}}(t + i\kappa_2 z, 0). \tag{1.102}$$

To get an intuitive picture on what happens with the pulse according to Eq. (1.102), let us choose an unchirped Gaussian pulse $\tilde{\mathcal{E}}(t,0)$ [see (1.33) for $a = 0$], entering the sample at $z = 0$. From Eq. (1.102) we find:

$$\tilde{\mathcal{E}}(t,z) = \tilde{\mathcal{E}}(t,0)\exp\left[\kappa_2^2(z/\tau_G)^2\right]\exp\left[-i2\kappa_2 tz/\tau_G^2\right]. \tag{1.103}$$

The pulse is amplified, and simultaneously its center frequency is shifted with propagation distance. The latter shift is due to the amplification of one part of the pulse spectrum (the high (low) – frequency part if $\kappa_2 < (>)0$) while the other part is absorbed. The result is a continuous shift of the pulse spectrum in the corresponding direction and a net gain while the pulse shape is preserved.

In the beginning of this section we mentioned that there is always an imaginary contribution of the dielectric constant leading to gain or loss. The question arises whether a wave equation such as Eq. (1.65), where only the real part of $\tilde{\epsilon}$ was considered, is of any practical relevance for describing pulse propagation through matter. The answer is yes, because in (almost) transparent regions the pulse change due to dispersion can be much larger than the change caused by losses. An impressive manifestation of this fact is pulse propagation through optical fibers. High-quality fibers made from fused silica can exhibit damping constants as low as 1 dB/km at wavelengths near 1 μm, where the GVD term is found to be $k_\ell'' \approx 75$ ps^2/km, see for example [13]. Consequently, a 100 fs pulse launched into a 10 m fiber loses just about 2% of its energy while it broadens by about a factor of 150. To illustrate the physics underlying the striking difference between the action of

1.2. PULSE PROPAGATION

damping and dispersion, let us consider a dielectric constant $\tilde{\epsilon}(\Omega)$ originating from a single absorption line. In the frame of the harmonic oscillator model $\tilde{\epsilon}$ is now given by

$$\tilde{\epsilon}(\Omega) = 1 + \frac{e^2 \bar{N}}{\epsilon_0 m} \frac{1}{\omega_0^2 - \Omega^2 + i\Omega\gamma_D} \tag{1.104}$$

where e, m is the electron charge and mass, $\bar{N}$ is the number of oscillators (dipoles) per unit volume, ω_0 is the resonance frequency and γ_D is the damping constant. The complex propagation constant $\tilde{k}$ and refractive index $\tilde{n}$ are related to $\tilde{\epsilon}$ through:

$$\begin{aligned}\tilde{\epsilon}(\Omega) &= \tilde{n}^2(\Omega) = [n_r(\Omega) + i n_i(\Omega)]^2 \\ &= \frac{c^2}{\Omega^2}\tilde{k}^2(\Omega) = \frac{c^2}{\Omega^2}[k(\Omega) + i k_i(\Omega)]^2. \end{aligned} \tag{1.105}$$

From the solution of the reduced wave equation (1.47) it is evident that $k_i(\Omega)$ is responsible for loss or gain depending on its sign. For frequencies Ω being sufficiently far from resonance, i.e. $|(\omega_0 - \Omega)/\gamma_D| \gg 1$, but with $|\omega_\ell - \Omega| \ll \omega_\ell$, the propagation constant is given by:

$$k(\Omega) \simeq \frac{\Omega}{c} + B\bar{N}\frac{1}{[2(\omega_0 - \Omega)/\gamma_D]^2} \tag{1.106}$$

$$k_i(\Omega) \simeq -B\bar{N}\frac{1}{[2(\omega_0 - \Omega)/\gamma_D]^2}. \tag{1.107}$$

For the GVD term we find

$$k''(\Omega) \simeq \frac{8B\bar{N}}{\gamma_D^2} \frac{1}{[2(\omega_0 - \Omega)/\gamma_D]^3} \tag{1.108}$$

where $B = e^2/(2\epsilon_0 m c \gamma_D)$. For small travel distances L the relative change of pulse energy can be estimated from Eq. (1.47) and Eq. (1.105) to be:

$$\Delta \mathcal{W}_{rel} = \frac{\mathcal{W}(L)}{\mathcal{W}(0)} - 1 \approx -2k_i L. \tag{1.109}$$

The relative change of pulse duration due to GVD can be evaluated from Eq. (1.81) and we find:

$$\Delta \tau_{rel} = \frac{\tau_G(L)}{\tau_{G0}} - 1 \approx 2\left(\frac{k''_\ell L}{\tau_{G0}^2}\right)^2. \tag{1.110}$$

To compare both pulse distortions we consider their ratio:

$$\frac{\Delta \tau_{rel}}{\Delta W_{rel}} = \Delta W_{rel} \frac{8}{\gamma_D^2 (\omega_0 - \Omega)^2 \tau_{G0}^4}. \tag{1.111}$$

Obviously, at given material parameters and carrier frequency, shorter pulses always lead to a dominant pulse spreading. For $\gamma_D = 10^{10} s^{-1}$ (typical value for a single electronic resonance), and a detuning $2(\omega_0 - \omega_\ell)/\gamma_D = 10^4$, we find for example:

$$\frac{\Delta \tau_{rel}}{\Delta W_{rel}} \approx \Delta W_{rel} \left(\frac{2800 \, \text{fs}}{\tau_{p0}} \right)^4. \tag{1.112}$$

To summarize, a resonant transition of certain spectral width γ_D influences short pulse propagation outside resonance mainly due to dispersion. Therefore, the consideration of a transparent material ($\epsilon_i \approx 0$) with a frequency dependent, real dielectric constant $\epsilon(\Omega)$, which was necessary to derive Eq. (1.65), is justified in many practical cases.

1.3 Interaction of light pulses with linear optical elements

Even though this topic is treated in detail in the chapter on fs optics, we want to discuss here some general aspects of pulse distortions induced by linear optical elements. These elements comprise typical optical components, such as mirrors, prisms, and gratings, which one usually finds in all optical setups. Here we shall restrict ourselves to the temporal and spectral changes the pulse experiences and shall neglect a possible change of the beam characteristics. A linear optical element of this type can be characterized by a complex optical transfer function

$$\tilde{H}(\Omega) = R(\Omega) e^{-i\Psi(\Omega)} \tag{1.113}$$

that relates the incident field spectrum $\tilde{E}_{in}(\Omega)$ to the field at the sample output $\tilde{E}(\Omega)$

$$\tilde{E}(\Omega) = R(\Omega) e^{-i\Psi(\Omega)} \tilde{E}_{in}(\Omega) \tag{1.114}$$

where $R(\Omega)$ is the (real) amplitude response and $\Psi(\Omega)$ is the (real) phase response. The reflectivity of a mirror (in amplitude and phase) can be measured as function of the frequency by means of a Michelson interferometer, following the procedure detailed in Chapter 2, Section 3. As can be seen

1.3. LINEAR OPTICAL ELEMENTS

from Eq. (1.114), the influence of $R(\Omega)$ is that of a frequency filter. The phase factor $\Psi(\Omega)$ can be interpreted as the phase delay which a spectral component of frequency Ω experiences. To get an insight of how the phase response affects the light pulse, we assume that $R(\Omega)$ does not change over the pulse spectrum whereas $\Psi(\Omega)$ does. Thus, we obtain for the output field from Eq. (1.114):

$$\tilde{E}(t) = \frac{1}{2\pi} R \int_{-\infty}^{+\infty} \tilde{E}_{in}(\Omega) e^{-i\Psi(\Omega)} e^{i\Omega t} \, d\Omega. \tag{1.115}$$

Replacing $\Psi(\Omega)$ by its Taylor expansion around the carrier frequency ω_ℓ

$$\Psi(\Omega) = \sum_{n=0}^{\infty} b_n (\Omega - \omega_\ell)^n \tag{1.116}$$

with the expansion coefficients

$$b_n = \frac{1}{n!} \left. \frac{d^n \Psi}{d\Omega^n} \right|_{\omega_\ell} \tag{1.117}$$

we obtain for the pulse

$$\begin{aligned}
\tilde{E}(t) &= \frac{1}{2} \tilde{\mathcal{E}}(t) e^{i\omega_\ell t} \\
&= \frac{1}{2\pi} R e^{-ib_0} e^{i\omega_\ell t} \int_{-\infty}^{+\infty} \tilde{E}_{in}(\Omega) \\
&\quad \times \exp\left(-i \sum_{n=2}^{\infty} b_n (\Omega - \omega_\ell)^n\right) e^{i(\Omega - \omega_\ell)(t - b_1)} \, d\Omega.
\end{aligned} \tag{1.118}$$

By means of Eq. (1.118) we can easily interpret the effect of the various expansion coefficients b_n. The term e^{-ib_0} is a constant phase shift (phase delay) having no effect on the pulse. A nonvanishing b_1 leads solely to a shift of the pulse on the time axis t; the pulse would obviously keep its position on a time scale $t' = t - b_1$. The term b_1 determines a group delay in a similar manner as the first–order expansion coefficient of the propagation constant k defined a group velocity in Eq. (1.58). The higher–order expansion coefficients produce a nonlinear behavior of the spectral phase which changes the pulse envelope and chirp. The action of the term with $n = 2$, for example, producing a quadratic spectral phase, is analogous to that of GVD in transparent media.

If we decompose the input field spectrum into modulus and phase $\tilde{E}_{in}(\Omega) = |\tilde{E}_{in}(\Omega)| \exp(i\Phi_{in}(\Omega))$, we obtain from Eq. (1.118) for the spectral phase at the output

$$\Phi(\Omega) = \Phi_{in}(\Omega) - \sum_{n=0}^{\infty} b_n (\Omega - \omega_\ell)^n. \qquad (1.119)$$

It is interesting to investigate what happens if the linear optical element is chosen, so that it compensates for the phase of the input field, i.e.,

$$b_n = \frac{1}{n!} \left. \frac{d^n}{d\Omega^n} \Phi_{in}(\Omega) \right|_{\omega_\ell}. \qquad (1.120)$$

The terms $n < 2$ correspond merely to a temporal translation of the pulse. The linear optical element fulfills a physically meaningful function if Eq. (1.120) is satisfied for $n \geq 2$. A closer inspection of Eq. (1.118) shows that when Eq. (1.120) is satisfied, all spectral components are in phase for $t - b_1 = 0$, leading to a pulse with maximum peak intensity. We will come back to this important point when discussing pulse compression. We want to point out the formal analogy between the solution of the linear wave equation (1.47) and Eq.(1.114) for $R(\Omega) = 1$ and $\Psi(\Omega) = k(\Omega)z$. This analogy expresses simply the fact that a dispersive transmission object is just one example of a linear element. In this case we obtain for the spectrum of the complex envelope

$$\tilde{\mathcal{E}}(\Omega, z) = \tilde{\mathcal{E}}_{in}(\Omega, 0) \exp\left[-i \sum_{n=0}^{\infty} \frac{1}{n!} k_\ell^{(n)} (\Omega - \omega_\ell)^n z\right] \qquad (1.121)$$

where $k_\ell^{(n)} = (d^n/d\Omega^n) k(\Omega)|_{\omega_\ell}$.

Next let us consider a sequence of m optical elements. The resulting transfer function is given by the product of the individual contributions $\tilde{H}_j(\Omega)$

$$\tilde{H}(\Omega) = \prod_{j=1}^{m} \tilde{H}_j(\Omega) = \left(\prod_{j=1}^{m} R_j(\Omega)\right) \exp\left[-i \sum_{j=1}^{m} \Psi_j(\Omega)\right] \qquad (1.122)$$

which means an addition of the phase responses in the exponent. Subsequently, by a suitable choice of elements, one can reach a zero-phase response so that the action of the device is through the amplitude response only. In particular, the quadratic phase response of an element (e.g., dispersive glass path) leading to pulse broadening can be compensated with an element having an equal phase response of opposite sign (e.g., grating arrangement, cf. Chapter 2) which automatically would recompress the pulse to its original duration. Such methods are of great importance for the handling of ultrashort light pulses.

1.4 Generation of phase modulation

At this point let us briefly discuss essential physical mechanisms to produce a time dependent phase of the pulse, i.e., a chirped light pulse. Processes resulting in a phase modulation can be divided into those that increase the pulse spectral width and those that leave the spectrum unchanged. The latter can be attributed to the action of linear optical processes. As we have seen in the section on linear optical elements, the phase modulation results from the different phase delays which different spectral components experience upon interaction. The result is a temporally broadened pulse with a certain frequency distribution across the envelope. For an element to act in this manner its phase response $\Psi(\Omega)$ must have non-zero derivatives of at least second order.

A phase modulation that leads to a spectral broadening is most easily discussed in the time domain. Let us assume that the action of a corresponding optical element on an unchirped input pulse can be formally written as:

$$\tilde{E}(t) = T(t)e^{i\Phi(t)}\tilde{E}_{in}(t) \qquad (1.123)$$

where T and Φ define a time dependent amplitude and phase response, respectively. For our simplified discussion here let us further assume that $T = $ const., leaving the pulse envelope unaffected. Since the output pulse has an additional phase modulation $\Phi(t)$ its spectrum must have broadened during the interaction. If the pulse under consideration is responsible for the time dependence of Φ, then we call the process self-phase modulation. If additional pulses cause the temporal change of the optical properties we will refer to it as cross-phase modulation. Often, phase modulation occurs through a temporal variation of the index of refraction n of a medium during the passage of the pulse. For a medium of length d the corresponding phase is:

$$\Phi(t) = -k(t)d = -\frac{2\pi}{\lambda_\ell}n(t)d. \qquad (1.124)$$

In later chapters we will discuss in detail several nonlinear optical interaction schemes with short light pulses that can produce a time depenedence of n.

A time dependence of n can also be achieved by applying a voltage pulse at an electro-optic material for example. However, with the view on phase shaping of femtosecond light pulses the requirements for the timing accuracy of the voltage pulse make this technique almost impracticable.

1.5 Beam propagation

1.5.1 General

So far we have considered light pulses as propagating as plane waves which allowed us to describe the time varying field with only one spatial coordinate. This simplification implies that the intensity across the beam is constant and, moreover, that the beam diameter is infinitely large. Obviously, both features hardly fit what we know from laser beams. Despite this fact such a description has been successfully applied for many practical applications and we will use it in this book whenever possible. This simplified treatment is justified if the processes under consideration either do not influence the transverse beam profile (e.g., sufficiently short sample length) or allow one to discuss the change of beam profile and pulse envelope as if they occur independently from each other. The general case, where both dependencies mix, is often more complicated and, frequently, requires extensive numerical treatment. Here we will discuss solely the situation where the change of such pulse characteristics as duration, chirp, and bandwidth can be separated from the change of the beam profile. Again we restrict ourselves to a linearly polarized field which now has to be considered in its complete spatial dependence. Assuming a propagation in the z-direction, we can write the field in the form:

$$E = E(x,y,z,t) = \frac{1}{2}\tilde{u}(x,y,z)\tilde{\mathcal{E}}(t,z)e^{i(\omega_\ell t - k_\ell z)} + c.c. \qquad (1.125)$$

In the definition (1.125) the scalar $\tilde{u}(x,y,z)$ is to describe the transverse beam profile and $\tilde{\mathcal{E}}(t,z)$ is the slowly varying complex envelope introduced in Eq. (1.54). Note that the rapid z-dependence of E is contained in the exponential function. Subsequently, $\tilde{u}$ is assumed to vary slowly with z. Under these conditions the insertion of Eq. (1.125) into the wave equation (1.40) yields after separation of the time dependent part in paraxial approximation [14]:

$$\left(\frac{\partial^2}{\partial x^2} + \frac{\partial^2}{\partial y^2} - 2ik_\ell \frac{\partial}{\partial z}\right)\tilde{u}(x,y,z) = 0. \qquad (1.126)$$

1.5. BEAM PROPAGATION

Paraxial approximation means that the transverse beam dimensions remain sufficiently small compared with typical travel distances of interest. Relevant solutions of this equation describing beam propagation are so-called Gaussian beams (see, e.g., [15]), which can be written in the form:

$$\tilde{u}(x,y,z) = \frac{u_0}{\sqrt{1+z^2/\rho_0^2}} e^{-i\Theta(z)} e^{-ik_\ell(x^2+y^2)/2R(z)} e^{-(x^2+y^2)/w^2(z)} \qquad (1.127)$$

where

$$R(z) = z + \rho_0^2/z \qquad (1.128)$$
$$w(z) = w_0\sqrt{1+z^2/\rho_0^2} \qquad (1.129)$$
$$\Theta(z) = \arctan(z/\rho_0) \qquad (1.130)$$
$$\rho_0 = n\pi w_0^2/\lambda_\ell. \qquad (1.131)$$

Sometimes it is convenient to write Eq. (1.127) as

$$\tilde{u}(x,y,z) = \frac{u_0}{\sqrt{1+z^2/\rho_0^2}} e^{-i\Theta(z)} e^{-ik_\ell(x^2+y^2)/2\tilde{q}(z)} \qquad (1.132)$$

where $\tilde{q}(z)$ is the complex beam parameter which is defined by:

$$\frac{1}{\tilde{q}(z)} = \frac{1}{R(z)} - \frac{i\lambda_\ell}{\pi w^2(z)} = \frac{1}{\tilde{q}(0)+z}. \qquad (1.133)$$

Obviously, optical beams described by Eq. (1.127) exhibit a Gaussian intensity profile transverse to the propagation direction with $w(z)$ as a measure of the beam diameter, see also Fig. 1.5. The origin of the z-axis ($z = 0$) is chosen to be the position of the beam waist $w_0 = w(z = 0)$. $R(z)$ is the radius of curvature of planes of constant phase[2], it is infinity at the beam waist (plane phase front). The length ρ_0 is called the Rayleigh range, $2\rho_0$ is the confocal parameter. For $-\rho_0 \geq z \leq \rho_0$, the beam size is within the limits $w_0 \leq w \leq \sqrt{2}w_0$. Given the amplitude u_0 at a given beam waist and wavelength λ_ℓ, the field at an arbitrary position (x, y, z) is completely predictable by means of Eq. (1.127).

Instead of using the differential equation (1.126) one can favorably describe the field propagation by an integral approach too. The physics behind

---

[2]The phase term $\Theta(z)$ in Eq. (1.127) takes on a constant value and need not be considered for $z \gg \rho_0$.

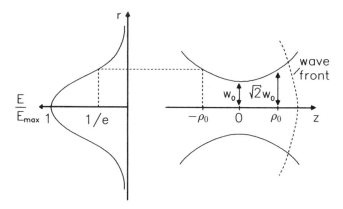

Figure 1.5: Parameters of Gaussian beams

this is to start with Huygens' principle and apply the Fresnel approximation assuming paraxial wave propagation [14]. Suppose we know the field distribution (or beam profile) $\tilde{u}(x', y', z') = \tilde{u}_0(x', y')$ at a plane $z' = \text{const.}$ the field distribution $\tilde{u}(x, y, z)$ at a plane $z = z' + L$ is then given by:

$$\tilde{u}(x,y,z) = \frac{ie^{ik_\ell L}}{\lambda L} \int_{-\infty}^{\infty} \int_{-\infty}^{\infty} \tilde{u}_0(x', y') e^{-ik_\ell[(x'-x)^2 + (y'-y)^2]/2L} dx' dy'. \quad (1.134)$$

Note that both ways of describing the field variation due to diffraction are equivalent. One can easily show that the field (1.134) is a solution of the paraxial wave equation (1.126).

1.5.2 Analogy between pulse and beam propagation

Comparing the paraxial wave equation (1.126) and the reduced wave equation (1.65) describing pulse propagation through a GVD medium we notice an interesting correspondence. Both equations are of similar structure. In terms of the reduced wave equation the transverse space coordinates x, y in Eq. (1.126) seem to play the role of the time variable. This space-time analogy suggests the possibility of translating simply the effects related to dispersion into beam propagation properties. For instance, we may compare the temporal broadening of an unchirped pulse due to dispersion with the change of beam size due to diffraction. In this sense free-space propagation plays a similar role for the beam characteristics as a GVD medium does for the pulse envelope. To illustrate this in more detail let us start with the

1.5. BEAM PROPAGATION

diffraction integral (1.134) and, for simplicity, consider the one-dimensional problem. The field is given in the form

$$\tilde{u}(x, L) \propto \int_{-\infty}^{\infty} \tilde{u}_0(x') e^{-ik_\ell(x'-x)^2/2L} dx' \qquad (1.135)$$

which obviously can be interpreted as the convolution of the input field $\tilde{u}_0(x')$ with a phase function $\exp(-ik_\ell x'^2/2L)$ in x'. Since the Fourier transform of a convolution equals the product of the Fourier transform of the individual functions, the output field can be represented as:

$$\tilde{u}(x, L) \propto \mathcal{F}^{-1}\left\{\tilde{u}_0(\tilde{x})e^{-iL\tilde{x}^2/2k_\ell}\right\} \qquad (1.136)$$

where

$$\tilde{u}_0(\tilde{x}) = \mathcal{F}\left\{\tilde{u}_0(x')\right\} \qquad (1.137)$$

and

$$e^{-iL\tilde{x}^2/2k_\ell} = \mathcal{F}\left\{e^{-ik_\ell x'^2/2L}\right\}. \qquad (1.138)$$

The quantity $\tilde{x}$ is the space–frequency coordinate. Let us next recall Eq. (1.68) which described the temporal pulse envelope after a GVD medium of length L

$$\tilde{\mathcal{E}}(\eta, L) = \mathcal{F}^{-1}\left\{\tilde{\mathcal{E}}_0(\Omega)e^{-\frac{i}{2}k_\ell''\Omega^2 L}\right\}. \qquad (1.139)$$

A comparison with Eq. (1.136) clearly shows the similarity between the diffraction and the dispersion problem. This is to be illustrated in more detail for Gaussian pulse profiles and Gaussian beams. As we have seen in the previous section the quadratic phase factor in Eq. (1.139) broadens an unchirped input pulse and leads to a (linear) frequency sweep across the pulse (chirp) while the pulse spectrum remains unchanged. In an analogous manner we can interpret Eq. (1.136) for the beam profile. A "bandwidth-limited" Gaussian beam means a beam without phase variation across the beam, which, in terms of Eq. (1.127), requires a radius of curvature of the phase front $R = \infty$. Thus, a Gaussian beam is "bandwidth-limited" at its waist where it takes on its minimum possible size (at a given frequency spectrum). Multiplication with a quadratic phase factor to describe the beam propagation, cf. Eq. (1.136), leads to beam broadening and "chirp." The latter simply accounts for a finite phase front curvature. Roughly speaking, the spatial frequency components which are not needed to form the broadened beam profile are responsible for the beam divergence. Table 1.2 summarizes our discussion comparing the characteristics of Gaussian beam and pulse propagation.

Gaussian pulse	Gaussian beam
Bandwidth-limited pulse (unchirped pulse) $$\tilde{\mathcal{E}}_0(t) \propto e^{-(t/\tau_{G0})^2}$$ $$\tilde{\mathcal{E}}_0(\Omega) \propto e^{-(\tau_{G0}\Omega/2)^2}$$	beam waist (plane phase fronts) $$\tilde{u}_0(x) \propto e^{-(x/w_0)^2}$$ $$\tilde{u}_0(\tilde{x}) \propto e^{-(\tilde{x}w_0/2)^2}$$
Propagation through a medium of length L (dispersion) $$\tilde{\mathcal{E}}(\Omega,L) \propto \exp[-(\frac{\tau_{G0}\Omega}{2})^2 i \frac{k''_\ell L \Omega^2}{2}]$$ $$\tilde{\mathcal{E}}(t,L) \propto \exp[-(1+i\bar{a})(\frac{t}{\tau_G})^2]$$ $$\propto \exp[i\omega_\ell \frac{t^2}{2\tilde{p}(L)}]$$ $$\bar{a} = L/L_d$$ $$\tau_G = \tau_{G0}\sqrt{1+\bar{a}^2}$$	Free space propagation over distance L (diffraction) $$\tilde{u}(\tilde{x},L) \propto \exp[-(\frac{w_0\tilde{x}}{2})^2 - i\frac{L\tilde{x}^2}{2k_\ell}]$$ $$\tilde{u}(x,L) \propto \exp[-(1+i\bar{b})(\frac{x}{w_L})^2]$$ $$\propto \exp[-ik_\ell \frac{x^2}{2\tilde{q}(L)}]$$ $$\bar{b} = z/\rho_0$$ $$w_L = w_0\sqrt{1+\bar{b}^2}$$
Chirp coefficient (slope) $$\ddot{\varphi} = \frac{2\bar{a}}{1+\bar{a}^2}\frac{1}{\tau_{G0}^2}$$	Wavefront curvature $$\frac{1}{R} = \frac{\bar{b}}{1+\bar{b}^2}\frac{1}{\rho_0}$$
Characteristic (dispersion) length $$L_d = \frac{\tau_{G0}^2}{2|k''_\ell|}$$	Characteristic (Rayleigh) length $$\rho_0 = \frac{n\pi w_0^2}{\lambda_\ell} = \frac{k_\ell w_0^2}{2}$$
Complex pulse parameter $$\frac{1}{\tilde{p}(L)} = \frac{\ddot{\varphi}(L)}{\omega_\ell} + \frac{2i/\omega_\ell}{\tau_G^2(L)}$$	Complex beam parameter $$\frac{1}{\tilde{q}(L)} = \frac{1}{R(L)} + \frac{i\lambda}{\pi w^2(L)}$$

Table 1.2: Comparison of dispersion and diffraction

1.5. BEAM PROPAGATION

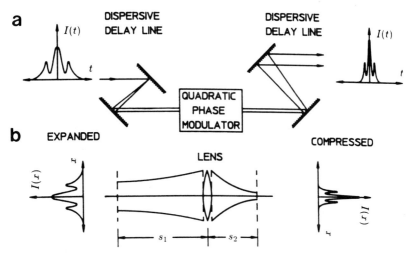

Figure 1.6: Space–time analogy. (a) Temporal imaging configuration. (b) Spatial imaging configuration (from [16]).

1.5.3 Analogy between spatial and temporal imaging

The analogy between pulse and beam propagation was applied to establish a time–domain analog of an optical imaging system by Kolner and Nazarathy [16]. Optical microscopy, for example, serves to magnify tiny structures so that they can be observed by a (relatively) low–resolution system such as our eyes. The idea of the "time lens" is to magnify ultrafast (fs) transients so that they can be resolved, for example, by a relatively slow oscilloscope. Of course, the opposite direction is also possible, which would lead to data compression in space or time. While Table 1.2 illustrates the space–time duality for free–space propagation we now need to look for devices resembling imaging elements such as lenses. From Fourier optics it is known that a lens introduces a quadratic phase factor, thus transforming a (Fourier–limited) input beam (parallel beam) into a spatially chirped (focused) beam. The time equivalent is a quadratic phase modulator. The dispersive (GVD) element in the time domain used to compress the pulse is the free space propagation leadig to a focus. Figure 1.6 shows a comparison of spatial and temporal imaging. It is interesting to note that, with some approximations, one can also derive the time–domain equivalent of the

Gaussian lens formula [16]:

$$\left(L_1 \frac{d^2 k_1}{d\omega^2}\right)^{-1} + \left(L_2 \frac{d^2 k_2}{d\omega^2}\right)^{-1} = (f_T/\omega_0)^{-1} \qquad (1.140)$$

where $L_{1,2}(d^2 k_{1,2}/d\omega^2)$ are the dispersion characteristics of the object and image side, respectively, and $\omega_0/f_T = \partial^2 \phi/\partial t^2$ is the parameter of the quadratic phase modulation impressed by the modulator. As in optical imaging, to achieve large magnification with practical devices, short focal lengths are desired. For time imaging this translates into a short focal time f_T which in turn requires a suitably large phase modulation.

1.6 Numerical modeling of pulse propagation

The generation and application of femtosecond light pulses requires one to study their propagation through linear and nonlinear optical media. Those studies have been undertaken not only to satisfy theorists. They are very necessary to design and optimize experiments, and to save time and money. Because of the complexity of interactions taking place numerical methods have to be mostly used. From the mathematical point of view it is desirable to develop a numerical model optimized with respect to computer time and accuracy for each experimental situation to be described. In this section we will present a procedure that allows one to study pulse propagation through a variety of materials. This model is optimized neither with respect to computer time nor with respect to accuracy. However, it is very universal and is directly associated with the physics of the problem. Moreover, it has been successfully applied to various situations. Among them are, for instance, pulse propagation through nonlinear optical fibers and amplifiers and pulse evolution in fs lasers. Without going into the numerical details, we will briefly describe the main features of this concept. In the course of the book we will then present various examples.

In the frame of approximations discussed in the section of beam propagation the electric field can be represented as

$$\begin{aligned} E(x,y,z,t) &= \frac{1}{2}\tilde{u}(x,y,z)\tilde{\mathcal{E}}(z,t)e^{i(\omega_\ell t - k_\ell z)} + c.c. \qquad (1.141) \\ &= \frac{1}{2}\mathcal{U}(x,y,z,t)e^{i(\omega_\ell t - k_\ell z)} + c.c. \end{aligned}$$

1.6. NUMERICAL MODELING OF PULSE PROPAGATION

where $\tilde{\mathcal{E}}$ is the complex pulse envelope and $\tilde{u}$ describes the transverse beam profile. The medium through which the pulse travels is not to be specified. In general, it will respond linearly as well as nonlinearly to the electric field. For example, the pulse changes shape and chirp due to dispersion while it is amplified or absorbed nonlinearly because of a time dependent gain coefficient. Therefore, the wave equation derived before for the case of linear dispersive media must be supplemented by certain nonlinear interaction terms. In following chapters we will discuss those nonlinear processes in detail. For the moment we will introduce them only formally. Let us first assume that a change in the beam profile can be neglected. Then the behavior of the field is fully described by its complex envelope $\tilde{\mathcal{E}}$. The propagation equation in local coordinates reads

$$\frac{\partial}{\partial z}\tilde{\mathcal{E}} = \frac{1}{2}ik''_\ell \frac{\partial^2}{\partial t^2}\tilde{\mathcal{E}} - \mathcal{D} + \mathcal{B}_1 + \mathcal{B}_2 + \cdots + \mathcal{B}_n \qquad (1.142)$$

where the terms $\mathcal{B}_i$ stand for contributions from nonlinear light matter interaction. A direct numerical evaluation of Eq. (1.142) often requires solving a set of nonlinear, partial differential equations. Note that for the determination of the $\mathcal{B}_i$, additional (differential) equations describing the medium must be considered. As with partial differential equations in general, the numerical procedures are rather complicated. Moreover, they may differ largely from each other even when the problems seem to be similar from the physical point of view.

A more intuitive approach can be chosen, as outlined next. The sample of length L is divided into M slices of length $\Delta z = L/M$; each slice is assumed to induce only a small change in the pulse parameters. Assuming that the complex envelope at propagation distance $z = m\Delta z$ ($m = 0, 1, \cdots, M-1$) is given by $\tilde{\mathcal{E}}(t, z)$, the envelope at the output of the next slice $(z + \Delta z)$ can be obtained from Eq (1.142) as

$$\begin{aligned}\tilde{\mathcal{E}}(t, z + \Delta z) &= \tilde{\mathcal{E}}(t, z) + \left[\frac{1}{2}ik''_\ell \frac{\partial^2}{\partial t^2}\tilde{\mathcal{E}}(t, z) - \mathcal{D} + \mathcal{B}_1(t, z, \tilde{\mathcal{E}})\right. \\ &\left. + \mathcal{B}_2(t, z, \tilde{\mathcal{E}}) + \cdots + \mathcal{B}_n(t, z, \tilde{\mathcal{E}})\right]\Delta z\end{aligned} \qquad (1.143)$$

which we can be written formally as

$$\begin{aligned}\tilde{\mathcal{E}}(t, z + \Delta z) &= \tilde{\mathcal{E}}(t, z) + \delta_{k''}\tilde{\mathcal{E}}(t, z) + \delta_\mathcal{D}\tilde{\mathcal{E}}(t, z) + \delta_1\tilde{\mathcal{E}}(t, z) \\ &\quad + \delta_2\tilde{\mathcal{E}}(t, z) +, \cdots, + \delta_n\tilde{\mathcal{E}}(t, z).\end{aligned} \qquad (1.144)$$

The quantities $\delta_i\tilde{\mathcal{E}}(z,t)$ represent the (small) envelope changes due to the various linear and nonlinear processes. For their calculation the envelope at z only is required. The action of the individual processes is treated as if they occur successively and independently in each slice. The pulse envelope at the end of each slice is then the sum of the input pulse plus the different contributions. The resulting envelope $\tilde{\mathcal{E}}(t, z + \Delta z)$ serves as input for the next slice, and so on until $z + \Delta z = L$.

The methods which can be applied to determine $\delta_i\tilde{\mathcal{E}}$ depend on the specific kind of interaction. For example, it may be necessary to solve a set of differential equations, but only with respect to the time coordinate. As mentioned before, the discussion of nonlinear optical processes will be the subject of following chapters.

This type of numerical calculation is critically dependent on the number of slices. It is the strongest interaction affecting the propagating pulse which determines the length of the slices. As a rule of thumb, the envelope distortion in each slice must not exceed a few percent, and doubling and halving of M must not change the results more than the required accuracy allows.

Many propagation problems have been investigated already with ps and ns light pulses, theoretically as well as experimentally. The severe problem when dealing with fs light pulses is dispersion, which enters Eq (1.144) through $\delta_{k''}\tilde{\mathcal{E}}$ (GVD) and $\delta_\mathcal{D}\tilde{\mathcal{E}}$ (higher order dispersion). From the discussion in the preceding sections we can easily derive expressions for these quantities. If only GVD needs to be considered, we can start from Eq (1.67)

$$\tilde{\mathcal{E}}(\Omega, z + \Delta z) = \tilde{\mathcal{E}}(\Omega, z)e^{-ik''_\ell \Omega^2 \Delta z/2} \tag{1.145}$$

which, for sufficiently small Δz, can be approximated as

$$\tilde{\mathcal{E}}(\Omega, z + \Delta z) \approx \tilde{\mathcal{E}}(\Omega, z) - \frac{1}{2}ik''_\ell \Omega^2 \Delta z \tilde{\mathcal{E}}(\Omega, z). \tag{1.146}$$

Thus we have for $\delta_{k''}\tilde{\mathcal{E}}(t,z)$

$$\delta_{k''}\tilde{\mathcal{E}}(t,z) \approx \mathcal{F}^{-1}\left\{-\frac{1}{2}ik''_\ell \Omega^2 \Delta z \tilde{\mathcal{E}}(\Omega, z)\right\}. \tag{1.147}$$

If additional dispersion terms matter, we can utilize Eq. (1.121) and obtain

$$\delta_\mathcal{D}\tilde{\mathcal{E}}(t,z) = \mathcal{F}^{-1}\left\{-i\sum_{n=3}^{\infty}\frac{1}{n!}k_\ell^{(n)}\Omega^n \Delta z \tilde{\mathcal{E}}(\Omega, z)\right\}. \tag{1.148}$$

1.7. SPACE–TIME EFFECTS

Next, let us consider a change in the beam profile. This must be taken into account if the propagation length through the material is long as compared with the confocal length. In addition, beam propagation effects can play a role if the setup to be modeled consists of various individual elements separated from each other by air or vacuum. This is the situation that is, for instance, encountered in lasers. It is the evolution of $\tilde{\mathcal{U}} = \tilde{u}\tilde{\mathcal{E}}$ rather than only that of $\tilde{\mathcal{E}}$ that has to be modeled now. The change of $\tilde{\mathcal{U}}$ from z to $z + \Delta z$ is

$$\tilde{\mathcal{U}}(x,y,z+\Delta z,t) = \tilde{\mathcal{U}}(x,y,z,t) + \delta\tilde{\mathcal{U}} \qquad (1.149)$$

where

$$\delta\tilde{\mathcal{U}} = \tilde{u}\delta\tilde{\mathcal{E}} + \tilde{\mathcal{E}}\delta\tilde{u}. \qquad (1.150)$$

The change of the pulse envelope $\delta\tilde{\mathcal{E}}$ can be derived as described above. For the determination of $\delta\tilde{u}$ we can evaluate the diffraction integral (1.134) over a propagation length[3] Δz. For Gaussian beams we may simply use Eq. (1.132).

1.7 Space–time effects

For very short pulses a coupling of spatial and temporal effects becomes important even for propagation in a nondispersive medium. The physical reason is that self–diffraction of a beam of finite transverse size (e.g., Gaussian beam) is wavelength dependent. A separation of time and frequency effects according to Eqs. (1.125) and (1.126) is clearly not feasible if such processes matter. One can construct a solution by solving the diffraction integral (1.134) for each spectral component. The superposition of these solutions and an inverse Fourier–transform then yields the temporal field distribution. Starting with a field $\tilde{E}(x',y',\Omega) = \mathcal{F}\left\{\tilde{E}(x',y',t)\right\}$ in a plane $\Sigma'(x',y')$ at $z = 0$ we find for the field in a plane $\Sigma(x,y)$ at $z = L$:

$$\tilde{E}(x,y,L,t) = \mathcal{F}^{-1}\left\{\frac{i\Omega e^{-ikL}}{2\pi cL}\int\int dx'dy'\,\tilde{E}(x',y',\Omega)\right.$$
$$\left.\times \exp\left[-i\frac{\Omega}{2Lc}\left((x-x')^2 + (y-y')^2\right)\right]\right\} \qquad (1.151)$$

where we have assumed a nondispersive medium. Solutions can be found by solving numerically Eq. (1.151) starting with an arbitrary pulse and beam

---
[3]Note that Eq. (1.134) was valid for paraxial beam propagation. Among others this determines a lower limit for Δz.

profile at a plane $z = 0$. Properties of these solutions were discussed by Christov [17]. They revealed that the pulse becomes phase modulated in space and time with a pulse duration that changes across the beam profile. Due to the stronger diffraction of long–wavelength components the spectrum on axis shifts to shorter wavelengths.

For a Gaussian beam and pulse profile at $z = 0$, i.e., $\tilde{E}(x', y', 0, t) \propto \exp(-r'^2/w'^2)\exp(-t^2/\tau_{G0}^2)exp(i\omega_\ell t)$ with $r'^2 = x'^2 + y'^2$, the time–space distribution of the field at $z = L$ is of the form [17]:

$$\tilde{E}(r, z = L, t) \propto \exp\left(-\frac{\eta^2}{\tau_G^2}\right) \exp\left[\left(-\frac{w'\omega_\ell \tau_{G0}}{2Lc\tau_G}r\right)^2\right] \exp\left(i\frac{\omega_\ell \tau_{G0}^2}{\tau_G^2}\eta\right) \quad (1.152)$$

where $\tau_G^2 = \tau_{G0}^2 + (w'r/Lc)^2$ and $\eta = [t - L/c - r^2/(2Lc)]$.

1.8 Problems

1. A polarization — to second order in the electric field — is defined as $P^2(t) \propto \chi^{(2)} E^2(t)$. We have seen that the preferred representation for the field is the complex quantity $E^+(t) = \frac{1}{2}\mathcal{E}(t)\exp[i(\omega_\ell t + \varphi(t)]$. Give a convenient description of the nonlinear polarization in terms of $E^+(t)$, $\mathcal{E}(t)$ and $\varphi(t)$. Consider in particular second harmonic generation and optical rectification. Explain the physics associated with the various terms of $P^{(2)}$ (or $P^{+(2)}$, if you can define one).

2. Starting from the one–dimensional wave equation (1.42), show that the slowly–varying envelope approximation corresponds essentially to neglecting self–induced reflection.

3. Verify the c_B factors of the pulse–duration–bandwidth–product of a Gaussian and sech-pulse as given in Table 1.1.

4. Calculate the pulse duration $\bar{\tau}_p$ defined as the second moment in Eq. (1.31) for a Gaussian pulse and compare with τ_p (FWHM).

5. Consider a medium consisting of particles that can be described by harmonic oscillators so that the dielectric constant in the vicinity of a resonance is given by Eq. (1.104). Investigate the behavior of the phase and group velocity in the absorption region. You will find a region where $v_g > v_p$. Is the theory of relativity violated here?

1.8. PROBLEMS

6. Assume a Gaussian pulse which is linearly chirped in a phase modulator that leaves its envelope unchanged. The chirped pulse is then sent through a spectral amplitude–only filter of spectral width (FWHM) $\Delta\omega_F$. Calculate the duration of the filtered pulse and determine an optimum spectral width of the filter for which the shortest pulses are obtained. (<u>Hint:</u> For simplification you may assume an amplitude only filter of Gaussian profile, i.e., $\tilde{H}(\omega - \omega_\ell) = \exp\left[-\ln 2 \left(\frac{\omega - \omega_\ell}{\Delta\omega_F}\right)^2\right]$.)

7. Derive the general expression for $d^n/d\Omega^n$ in terms of derivatives with respect to λ.

TO CHIRP OR NOT TO CHIRP ...

... THAT IS THE CHALLENGE

Chapter 2

Femtosecond Optics

2.1 Introduction

Whether short pulses or continuous radiation, light should follow the rules of classical optics. There are, however, some properties related to the bending, focalization of light that are specific to fs pulses. Ultrashort pulses are more "unforgiving" of some "defects" of optical systems, as compared to ordinary light of large spectral bandwidth, i.e., white light.[1] Studying optical systems with fs pulses helps in turn to improve the understanding and performances of these systems in white light. We will study properties of basic elements (coatings, lenses, prisms, gratings) and some simple combinations thereof. The dispersion of the index of refraction is the essential parameter for most of the effects to be discussed in this chapter. Some values are listed for selected optical materials in Table 2.1.

We shall start this chapter with an analysis of a simple Michelson interferometer.

2.2 White light and short pulse interferometry

Incoherent radiation has received increasing attention as the poor man's fs source (even the wealthiest experimentalist will now treat bright incoherent sources with a certain amount of deference). The similarities between white

---

[1]Such light can be regarded as superposition of random fluctuations (short light pulses), the mean duration of which determines the spectral width. A measurement of the light intensity, however, averages over these fluctuations.

material	λ_ℓ [nm]	$n(\omega_\ell)$	$n'(\omega_\ell)$ 10^{-2} [fs]	$n'(\lambda_\ell)$ 10^{-2} [μm^{-1}]	$n''(\omega_\ell)$ 10^{-3} [fs^2]	$n''(\lambda_\ell)$ 10^{-1} [μm^{-2}]	$n'''(\omega_\ell)$ 10^{-4} [fs^3]	$n'''(\lambda_\ell)$ [μm^{-3}]
BK7	400	1.5307	1.13	-13.	3.0	11.0	6.9	-12.
	500	1.5213	0.88	-6.6	2.3	3.96	7.7	-3.5
	620	1.5154	0.75	-3.6	1.6	1.50	13	-1.1
	800	1.5106	0.67	-2.0	0.06	0.50	39	-.29
	1000	1.5074	0.73	-1.4	-3.2	0.16	114	-.09
SF6	400	1.8674	5.8	-67.	30	74.0	214	-120
	500	1.8236	3.7	-28.	16	20.0	86	-21.
	620	1.8005	2.7	-13.	12	7.0	50	-5.3
	800	1.7844	2.0	-5.9	8	2.2	56	-1.2
	1000	1.7757	1.71	-3.2	4	0.8	115	-.36
SF10	400	1.7784	4.6	-54.	24	59.	183	-98.
	500	1.7432	3.0	-22.	13	16.	69	-17.
	620	1.7244	2.2	-11.	9	5.6	42	-4.2
	800	1.7112	1.7	-5.0	6	1.7	58	-1.0
	1000	1.7038	1.5	-2.8	2	0.6	132	-0.3
SF14	400	1.8185	5.3	-62.	27	68.	187	-10.9
	500	1.7786	2.8	-25.	15	19.	85	-2.0
	620	1.7576	2.5	-12.	10	6.3	50	-4.8
	800	1.7430	1.8	-5.5	7	2.0	54	-1.1
	1000	1.7349	1.6	-3.0	3.4	0.72	110	-.33
SQ1	248	1.5121	2.36	-72.	11	150.	76	-520.
	308	1.4858	1.35	-27.	4.1	33.	23	-66.0
	400	1.4701	0.93	-11.	2.3	8.6	6	-9.80
	500	1.4623	0.73	-5.5	1.8	3.2	6	-2.80
	620	1.4574	0.62	-3.0	1.2	1.3	13	-0.89
	800	1.4533	0.58	-1.7	-0.4	0.4	41	-0.24
	1000	1.4504	0.67	-1.3	-3.8	0.12	121	-0.08
	1300	1.4469	1.0	-1.1	-14	-.003	446	-0.02
	1500	1.4446	1.4	-1.2	-27	-.031	915	-0.01
ZnSe	620	2.586	14.23	-69.	117.3	4.1		
	800	2.511	8.35	-24.6	63.18	1.1		

Table 2.1: Dispersion parameters for some optical materials. BK7 is the most commonly used optical glass. The SF... are dispersive heavy flint glasses. SQ1 is fused silica. The dispersion parameters for the glasses were calculated with Sellmeir's equations and data from various optical catalogs. The data for the UV wavelengths must be considered as order of magnitude approximations. The ZnSe data are taken from Ref. [18]. Using Eqs. (1.70)–(1.74), the values given in terms of $n(\Omega)$ can easily be transformed into the corresponding values for $k(\Omega)$.

2.2. WHITE LIGHT AND SHORT PULSE INTERFEROMETRY

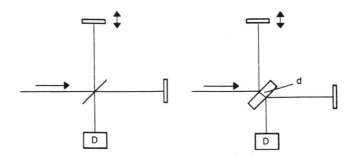

Figure 2.1: Balanced (left) and unbalanced (right) Michelson interferometer

light and femtosecond light pulses are most obvious when studying coherence properties, but definitely transcend the field of coherent interactions.

Let us consider the basic Michelson interferometer sketched in Fig. 2.1. The *real* field on the detector, resulting from the interferences of E_1 and E_2, is $E = E_1(t - \tau) + E_2(t)$ with τ being the delay parameter. The intensity at the output of the interferometer is given by the electric field squared averaged over one light period T [Eq. (1.19)]:

$$\begin{aligned}
I(t,\tau) &= \epsilon_0 cn \frac{1}{T} \int_{t-T/2}^{t+T/2} [E_1(t'-\tau) + E_2(t')]^2 dt' \\
&= 2\epsilon_0 cn [\tilde{E}_1^+(t-\tau) + \tilde{E}_2^+(t)][\tilde{E}_1^-(t-\tau) + \tilde{E}_2^-(t)] \\
&= \frac{1}{2} \epsilon_0 cn \left\{ \mathcal{E}_1^2(t-\tau) + \mathcal{E}_2^2(t) \right. \\
&\quad \left. + \tilde{\mathcal{E}}_1^*(t-\tau)\tilde{\mathcal{E}}_2(t)e^{i\omega_\ell \tau} + \tilde{\mathcal{E}}_1(t-\tau)\tilde{\mathcal{E}}_2^*(t)e^{-i\omega_\ell \tau} \right\}
\end{aligned} \quad (2.1)$$

Here again, we have chosen to decompose the field in an amplitude function $\tilde{\mathcal{E}}$ and a phase function centered around a somewhat arbitrary average frequency of the radiation, ω_ℓ, as in Eqs. (1.10) and (1.11).

The actual signal recorded at the output of the interferometer is the intensity, $\bar{I}$, averaged over the response time τ_{res} of the detector. In the case of ultrashort pulses $\tau_{res} \gg \tau_p$ holds and what is being measured is the time integral $\int_{-\infty}^{+\infty} I(t', \tau) dt'$. In the case of continuous radiation, such as white light, the measurement is an average over a time T_f which is the mean fluctuation time of the radiation or the response time of the detector (whichever is larger) $\frac{1}{T_f} \int_{t-T_f/2}^{t+T_f/2} I(t', \tau) dt'$. We will use the notation $\langle \; \rangle$ for either integration or averaging, which results in a quantity that is time independent. Assuming thus that all fluctuations of the signal are averaged out by the detector's slow response, the measured signal reduces to the following expression:

$$\begin{aligned}
\bar{I}(\tau) &= \frac{\epsilon_0 c n}{4}\left\{\langle \tilde{\mathcal{E}}_1^2\rangle + \langle \tilde{\mathcal{E}}_2^2\rangle + \langle \tilde{\mathcal{E}}_1^*(t-\tau)\tilde{\mathcal{E}}_2(t)e^{i\omega_\ell \tau}\rangle + \langle \tilde{\mathcal{E}}_1(t-\tau)\tilde{\mathcal{E}}_2^*(t)e^{-i\omega_\ell \tau}\rangle\right\} \\
&= \epsilon_0 c n \left\{A_{11}(0) + A_{22}(0) + \tilde{A}_{12}^+(\tau) + \tilde{A}_{12}^-(\tau)\right\}
\end{aligned} \quad (2.2)$$

On the right hand side of the first line in Eq. (2.2) we recognize correlation functions similar to that in Eq. (1.25), except that it involves the electric fields rather than the intensities. In complete analogy with the definitions of the complex electric fields, the two complex functions correspond to positive and negative spectral components[2] of a correlation function $A_{12}(\tau) = \tilde{A}_{12}^+(\tau) + \tilde{A}_{12}^-(\tau)$, where, e.g., the positive frequency component is defined as:

$$\begin{aligned}
\tilde{A}_{12}^+(\tau) &= \frac{1}{4}\langle \tilde{\mathcal{E}}_1^*(t-\tau)\tilde{\mathcal{E}}_2(t)e^{i\omega_\ell \tau}\rangle \\
&= \frac{1}{2}\tilde{\mathcal{A}}_{12}(\tau)e^{i\omega_\ell \tau}
\end{aligned} \quad (2.3)$$

The Fourier transform of the correlation of two functions is the product of the Fourier transforms [7]:

$$\begin{aligned}
\tilde{A}_{12}^+(\Omega) &= \int_{-\infty}^{\infty} \tilde{A}_{12}^+(\tau)e^{-i\Omega \tau}d\tau \\
&= \frac{1}{4}\tilde{\mathcal{E}}_1^*(\Omega-\omega_\ell)\tilde{\mathcal{E}}_2(\Omega-\omega_\ell) \\
&= \tilde{E}_1^*(\Omega)\tilde{E}_2(\Omega)
\end{aligned} \quad (2.4)$$

In the ideal case of infinitely thin beam splitter, nondispersive broadband reflectors and beam splitters, $\tilde{E}_1 = \tilde{E}_2$, and the Fourier transform (2.4) is real. Correspondingly, the correlation defined by Eq. (2.3) is an electric field autocorrelation which is a symmetric function with respect to the delay origin $\tau = 0$. This fundamental property is of little practical importance when manipulating data from a real instrument, because, in the optical time domain, it is difficult to determine exactly the "zero delay" point, which requires measurement of the relative delays of the two arms with an accuracy better than 100 Å. It is therefore more convenient to use an arbitrary origin

---

[2]Spectrum is defined here with respect to the delay parameter τ.

2.2. WHITE LIGHT AND SHORT PULSE INTERFEROMETRY

for the delay τ, and use the generally complex Fourier transformation of Eq. (2.4).

For an ideally balanced interferometer, the output from the two arms is identical, and the right–hand side of Eq. (2.4) is simply the spectral intensity of the light. This instrument is therefore referred to as a Fourier spectrometer.

Let us turn our attention to the slightly "unbalanced" Michelson interferometer. For instance, with a single beam splitter of finite thickness d, the beam 2 will have traversed $L = 2d$ more glass than beam 1 (Fig. 2.1). It is well known that the "white light" interference fringes are particularly elusive, because of the short coherence length of the radiation, which translates into a very restricted range of delays over which a fringe pattern can be observed. How will that fringe pattern be modified and shifted by having one beam traverse a path of length $2d$ in glass rather than in air? Let $\tilde{E}_1(t)$ refer to the field amplitude at the detector, corresponding to the beam that has passed through the unmodified arm with the least amount of glass. Using Eq. (1.114) with $R = 1$, $\Psi(\Omega) = k(\Omega)L$ and considering only terms with $n \leq 2$ in the expansion of Ψ, cf. Eq. (1.116), we find the second beam through the simple transformation:

$$\begin{aligned}\tilde{E}_2(\Omega) &= \tilde{E}_1(\Omega)\exp\left\{-iL\left[k(\Omega) - \Omega/c\right]\right\} \\ &\approx \tilde{E}_1(\Omega)\exp\left\{-i\left[\left(k_\ell - \frac{\Omega}{c}\right)L + k'_\ell L(\Omega - \omega_\ell) + \frac{k''_\ell L}{2}(\Omega - \omega_\ell)^2\right]\right\}\end{aligned} \quad (2.5)$$

where, as outlined earlier, $(k'_\ell)^{-1} = \left(\left[\frac{dk}{d\Omega}\right]_{\omega_\ell}\right)^{-1}$ determines the group velocity of a wave packet centered around ω_ℓ and $k''_\ell = \left[\frac{d^2k}{d\Omega^2}\right]_{\omega_\ell}$ is responsible for group velocity dispersion (GVD). The time dependent electric field is given by the Fourier transform of Eq. (2.5). Neglecting GVD we find for the complex field envelope:

$$\tilde{\mathcal{E}}_2(t) = e^{-ik_\ell L}\tilde{\mathcal{E}}_1\left[t - (k'_\ell - 1/c)L\right] \quad (2.6)$$

Apart from an unimportant phase factor the obvious change introduced by the glass path in one arm of the interferometer is a shift of time origin, i.e., a shift of the maximum of the correlation. This is a mere consequence of the longer time needed for light to traverse glass instead of air. The shift in "time origin" measured with the "unbalanced" versus "balanced" Michelson is

$$\begin{aligned}\Delta\tau &= \left(k'_\ell - \frac{1}{c}\right)L \\ &= \frac{L}{c}\left\{(n-1) + \omega_\ell\left[\frac{dn}{d\Omega}\right]_{\omega_\ell}\right\} \\ &= \frac{L}{c}\left\{(n-1) - \lambda_\ell\left[\frac{dn}{d\lambda}\right]_{\lambda_\ell}\right\}\end{aligned} \quad (2.7)$$

where we replaced k'_ℓ by Eq. (1.73). The first term in the right-hand side represents the temporal delay resulting from the difference of the optical pathlength in air ($n \approx 1$) and glass. The second term contains the contribution from the group velocity in glass. In the above derivation, we have not specifically assumed that the radiation consists of short pulses. It is also the case for white light cw radiation that the group velocity contributes to the shift of zero delay introduced by an unbalance of dispersive media between the two arms of the interferometer. The third (and following) terms of the expansion of $k(\Omega)$ account for the deformation of the fringe pattern observed in the recording of Fig. 2.2. The propagation can be more easily visualized in the time domain for fs pulses. The group velocity delay is due to the pulse envelope "slipping" with respect to the waves. The group velocity *dispersion* causes different parts of the pulse to travel at different velocity, resulting in pulse deformation. The result of the Michelson interferogram is a cross-correlation between the field amplitude of the "original" pulse and the signal propagated through glass.

The same considerations can be applied to white light, which can be viewed as a temporal random distribution of ultrashort pulses. The concept of incoherent radiation being constructed out of a statistical time sequence of ultrashort pulses is also useful for studying coherence in light–matter interactions, as will be studied in detail in the chapter on coherent interactions. The correlation product (2.3) is maximum for exactly overlapping statistical phase and intensity fluctuations from both arms of the interferometer. These fluctuations have a duration of the order of the inverse bandwidth of the radiation, hence can be in the fs range for broad bandwidth light. Each of these individual fs spikes will travel at the group velocity. Dispersion in group velocity causes individual frequency components of these spikes to travel at different speeds, resulting most often in pulse stretching. If the source for the interferogram of Fig. 2.2 had been a fs pulse, the recording on the right of the figure would represent the cross–correlation between the

2.2. WHITE LIGHT AND SHORT PULSE INTERFEROMETRY 49

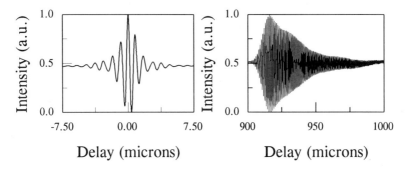

Figure 2.2: "White light" Michelson interferogram. The fringes of the balanced interferometer are shown on the left. The fringe pattern shifts to the right and is broadened by the insertion of a thin quartz plate in one arm of the interferometer.

field of the stretched–out pulse $\tilde{\mathcal{E}}_2(t)$ with the original (shorter) pulse $\tilde{\mathcal{E}}_1(t)$ (of which the autocorrelation is shown on the left of the figure). Such a measurement can be used to determine the shape of the field $\tilde{\mathcal{E}}_2(t)$. The limiting case of a cross–correlation between a δ function and an unknown function yields the function directly. Indeed, the unbalanced Michelson is a powerful tool leading to a complete determination of the shape of fs signals, in amplitude and phase, as will be seen in Chapter 8.

In the case of the incoherent radiation used for the recording of Fig. 2.2, the broadened signal on the right merely reflects the "stretching" of the statistical fluctuations of the white light. This measurement however provides important information on material properties essential in fs optics. We will see in the next section how the displacement of the "zero delay" point in the interferogram of Fig. 2.2 can be used to determine the first terms of an expansion of the transfer function of linear optical elements.

According to Eq. (2.4), the Fourier transform of the correlation function $\tilde{A}_{11}^+(\tau)$ measured with the balanced interferometer[3] is simply the spectral field intensity of the source. It is difficult, and not essential, to determine exactly the zero point, and therefore the measurement generally provides $\tilde{A}_{11}^+(\tau + \tau_e)\exp(i\varphi_e)$, which is the function $\tilde{A}_{11}^+(\tau)$ with an unknown phase (φ_e) and delay (τ_e) error. Similarly, the cross-correlation measured after addition of a dielectric sample in one arm of the interferometer (right hand

---

[3] $\tilde{A}_{11}^+(\tau)$ corresponds to the third term $\tilde{A}_{12}^+(\tau)$ in Eq. (2.2) taken for identical beams (subscript 1 = subscript 2), not to be confused with the first term $A_{11}^+(\tau)$ in that same equation.

side of Fig. 2.2) is $\tilde{A}_{12}^+(\tau+\tau_f)\exp(i\varphi_f)$, which is the function $\tilde{A}_{12}^+(\tau)$ with an unknown phase (φ_f) and delay (τ_f) error. The ratio of the Fourier transforms of both measurements is:

$$\frac{\tilde{A}_{12}^+(\Omega)}{\tilde{A}_{11}^+(\Omega)} \frac{e^{-i(\Omega\tau_f-\varphi_f)}}{e^{-i(\Omega\tau_e-\varphi_e)}} = \frac{\tilde{E}_2(\Omega)}{\tilde{E}_1(\Omega)} e^{-i[\Omega(\tau_f-\tau_e)-(\varphi_f-\varphi_e)]}$$
$$= e^{-i[k(\Omega)L+\Omega(\tau_f-\tau_e)-(\varphi_f-\varphi_e)]}, \qquad (2.8)$$

where we have made use of Eqs. (2.4) and (2.5). Unless special instrumental provisions have been made to make $(\varphi_f = \varphi_e)$, and $(\tau_f = \tau_e)$, this measurement will not provide the first two terms of a Taylor expansion of the dispersion function $k(\Omega)$. This is generally not a serious limitation, since, physically, the undetermined terms are only associated with a phase shift and delay of the fs pulses.

Equation (2.8) is not limited to dielectric samples. Instead, any optical transfer function $\tilde{H}$ which can be described by an equation similar to Eq. (1.113), can be determined from such a procedure. For instance, the preceding discussion remains valid for absorbing materials, in which case the wave vector is complex, and Eq. (2.8) leads to a complete determination of the real and imaginary part of the index of refraction of the sample versus frequency. Another example is the response of an optical mirror, as we will see in the following subsection.

2.3 Mirror dispersion

In optical experiments mirrors are used for different purposes and are usually characterized only in terms of their reflectivity at a certain wavelength. The latter gives a measure about the percentage of incident light intensity which is reflected. In dealing with femtosecond light pulses, one has however to consider the dispersive properties of the mirror [19, 20]. This can be done by analyzing the optical transfer function which, for a mirror, is given by

$$\tilde{H}(\Omega) = R(\Omega)e^{-i\Psi(\Omega)}. \qquad (2.9)$$

It relates the spectral amplitude of the reflected field $\tilde{E}_r(\Omega)$ to the incident field $\tilde{E}_0(\Omega)$

$$\tilde{E}_r(\Omega) = R(\Omega)e^{-i\Psi(\Omega)}\tilde{E}_0(\Omega). \qquad (2.10)$$

Here $|R(\Omega)|^2$ is the reflection coefficient and $\Psi(\Omega)$ is the phase response of the mirror. As mentioned earlier a nonzero $\Psi(\Omega)$ in a certain spectral range is unavoidable if $R(\Omega)$ is frequency dependent. Depending on the functional

2.3. MIRROR DISPERSION

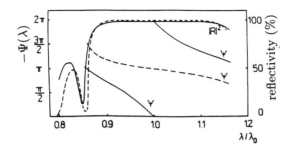

Figure 2.3: Amplitude and phase response for a high reflection multilayer mirror (dashed line) and a weak output coupler (solid line) as a function of the wavelength (from [20]).

behavior of $\Psi(\Omega)$ (cf. Section 1.3 in Chapter 1), reflection at a mirror not only introduces a certain intensity loss but may also lead to a change in the pulse shape and to chirp generation or compensation. These effects are usually more critical if the corresponding mirror is to be used in a laser. This is because its action is multiplied by the number of effective cavity round trips of the pulse. Such mirrors are mostly fabricated as dielectric multilayers on a substrate. By changing the number of layers and layer thickness a desired transfer function, i.e., reflectivity and phase response, in a certain spectral range can be realized. As an example, Fig. 2.3 shows the amplitude and phase response of a broadband high–reflection mirror and a weak output coupler. Note, although they have very similar reflection coefficients around a center wavelength λ_0, the phase response differs greatly. The physics behind it is that $R(\Omega)$ [or $R(\lambda)$] far from $\omega_0 = 2\pi c/\lambda_0$ (not shown) influences the behavior of $\Psi(\Omega)$ [or $\Psi(\lambda)$] near ω_0, which is expressed by the Kramers–Kronig relation.

Before dealing with the influence of other optical components on fs pulses, let us discuss some methods to determine experimentally the mirror characteristics. In this respect the Michelson interferometer is not only a powerful instrument to analyze a sample in transmission, but it can also be used to determine the dispersion and reflection spectrum of a mirror. The reference spectrum is as shown on the left of Fig. 2.2. Such a symmetric interference pattern can only be achieved in a well compensated Michelson interferometer (left part of Fig. 2.1) with identical (for symmetry) mirrors in both arms, which are also broadband (to obtain a narrow correlation pattern). The white light interferogram obtained after substitution of one of the mirrors by a sample coating can be used to measure its phase dispersion. As in the example of the transmissive sample, substitution of the reflective sample can in general not be done without losing the relative phase and delay references. The cross-correlation measured after substitution of the sample mirror in one arm of the interferometer (right hand side of Fig. 2.2) is $\tilde{A}_{12}^+(\tau+\tau_f)\exp(i\varphi_f)$, which is the function $\tilde{A}_{12}^+(\tau)$ with an unknown phase (φ_f) and delay (τ_f) error. The ratio of the Fourier transforms of both measurements is in anology with Eq (2.8):

$$\frac{\tilde{A}_{12}^+(\Omega)}{\tilde{A}_{11}^+(\Omega)} \frac{e^{-i(\Omega\tau_f-\varphi_f)}}{e^{-i(\Omega\tau_e-\varphi_e)}} = \frac{\tilde{E}_2(\Omega)}{\tilde{E}_1(\Omega)} e^{-i[\Omega(\tau_f-\tau_e)-(\varphi_f-\varphi_e)]}$$
$$= R(\Omega)e^{-i[\Psi(\Omega)+\Omega(\tau_f-\tau_e)-(\varphi_f-\varphi_e)]}. \quad (2.11)$$

Both the amplitude and phase of the transfer function $\tilde{H}(\Omega)$ can be extracted from the measurement, with the limitation that, in general, this measurement will not provide the first two terms of a Taylor expansion of the phase function $\Psi(\Omega)$. Again, this is not a serious limitation, since, physically, the undetermined terms are only associated with a phase shift and delay of the fs pulses. If – as will be generally the case – the bandwidth of the reflection spectrum of the coating exceeds that of the radiation, $R(\Omega) \approx 1$, and the phase of the ratio of Fourier transforms in Eq. (2.11) is simply the phase shift upon reflection $\Psi(\Omega)$.

The Michelson interferometer in white light is one of the simplest and most powerful tools to measure the dispersion of transmissive and reflective optics. Its potential has only been partly utilized in previous works. Knox et al. [21] used it to measure directly the group velocity by measuring the delay induced by a sample, at selected wavelengths (the wavelength selection was accomplished by filtering white light). Naganuma et al. [22] used essentially the same method to measure group delays, and applied the technique to the measurement of "alpha parameters" (current dependence of the index of refraction in semiconductors) [23]. In fs lasers, the frequency dependence of the complex reflection coefficient of the mirrors contributes to an overall cavity dispersion. Such a dispersion can be exploited for optimal pulse compression, provided there is a mechanism for matched frequency modulation in the cavity. It will simply contribute to pulse broadening of initially unmodulated pulses, if no intensity or time dependent index is affecting the pulse phase, as will be discussed later. It is therefore important to diagnose the fs response of dielectric mirrors used in a laser cavity.

A direct method is to measure the change in shape of a fs pulse, after reflection on a dielectric mirror, as proposed and demonstrated by Weiner

2.4. OTHER INTERFEROMETRIC STRUCTURES

et al. [24]. It is clear in the *frequency domain* that the phases of the various frequency components of the pulse are being scrambled, and therefore the pulse shape should be affected. What is physically happening in the time domain is that the various dielectric layers of the coating accumulate more or less energy at different frequencies, resulting in a delay of some parts of the pulse. Therefore, significant pulse reshaping with broadband coatings occurs only when the pulse length is comparable to the coating thickness. Pulses of less than 30 fs duration were used in Refs. [25, 24]. Clearly, the use of a white light source and a simple Michelson interferometer is more economical to measure the dispersive properties [$\Psi(\Omega)$] of a coating.

An alternate method, advantageous for its sensitivity, but limited to the determination of the group velocity dispersion, is to compare glass and coating dispersion inside a fs laser cavity. As will be seen Chapter 5, an adjustable thickness of glass is generally incorporated in the cavity of mode-locked dye lasers, to tune the amount of group velocity dispersion for minimum pulse duration. The dispersion of mirrors can be measured by substituting mirrors with different coatings in one cavity position, and noting the change in the amount of glass required to compensate for the additional dispersion [26, 20]. The method is very sensitive, because the effect of the sample mirror is multiplied by the mean number of cycles of the pulse in the laser cavity. It is most useful for selecting mirrors for a particular fs laser cavity.

2.4 Other interferometric structures

So far we have introduced (Michelson) interferometers only as a tool to split a pulse and to generate a certain delay between the two partial pulses. In general, however, the action of an interferometer is more complex. This is particularly true for multiple-beam devices such as a Fabry–Perot interferometer. The complex transmission function can be found for example in [9] (within the approximation of infinitely thin mirrors) and reads

$$\tilde{H}(\Omega) = -\frac{t_1 t_2 e^{i\delta/2}}{1 - r_1 r_2 e^{i\delta}} \qquad (2.12)$$

with

$$\delta = -2\frac{\Omega d n(\Omega)}{c} \cos\theta \qquad (2.13)$$

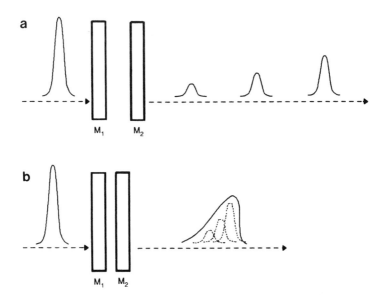

Figure 2.4: Effect of a Fabry-Perot interferometer on a light pulse if the mirror spacing is larger (a) and shorter (b) than the geometrical length of the incident pulse.

where d is the spacing between the two mirrors having (real) amplitude reflection and transmission coefficients r_1, r_2 and t_1, t_2, respectively and θ is the angle of incidence. Intuitively, we expect the transmission and/or reflection of a Fabry-Perot to consist of a train of pulses of decreasing intensity if the spacing d between the two mirrors is larger than the geometrical pulse length, [Fig. 2.4(a)]. The latter condition prevents interference between field components of successive pulses. The free spectral range of the Fabry–Perot interferometer is much smaller than the spectral width of the pulse. On the other hand if d is smaller than the pulse length the output field is determined by interference, as illustrated in Fig. 2.4(b). An example of a corresponding device was the dielectric multilayer mirror discussed before which can be considered as a sequence of many Fabry–Perot interferometers having equal resonance frequencies. Here the free spectral range is much broader than the pulse spectrum and it is the behavior around a resonance which determines the shape of pulse envelope and phase. The actual pulse characteristics can easily be determined by multiplying the field spectrum of the incident pulse with the corresponding transfer function (2.12). For a multilayer mir-

2.4. OTHER INTERFEROMETRIC STRUCTURES

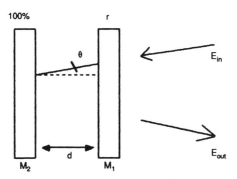

Figure 2.5: Schematic diagram of a Gires–Tournois interferometer.

ror this function can be obtained from a straightforward multiplication of matrices for the individual layers [15].

Among the various types of interferometers that can be used for pulse shaping, we choose to detail here the Gires–Tournois interferometer [27]. Its striking feature is a very high and almost constant amplitude transmission while the spectral phase can be tuned continuously. This device can be used to control the GVD in a fs laser in a similar manner as gratings and prisms. A sketch of the principle of the Gires–Tournois interferometer is shown in Fig. 2.5. It is a special type of a Fabry-Perot interferometer with one mirror (mirror M_2) having a reflection coefficient of (almost) 1. Consequently the device is used in reflection. In this case the transfer function is given by

$$R(\Omega)e^{-i\Psi(\Omega)} = \frac{-r + e^{i\delta}}{1 - re^{i\delta}} \quad (2.14)$$

where δ is the phase delay[4] between two successive partial waves that leave the interferometer and r is the (real) amplitude reflection of M_1 (assumed to be nondispersive). It can easily be shown that the reflectivity of the device is $|R| = 1$, i.e., there is practically no change in the pulse energy. The phase response determined by Eq. (2.14) can be written as

$$\Psi(\Omega) = -\arctan\left[\frac{(r^2 - 1)\sin\delta}{2r - (r^2 + 1)\cos\delta}\right] \quad (2.15)$$

---

[4]In the definition of the phase delay (2.13) applied to the Gires-Tournois, θ is the *internal* angle.

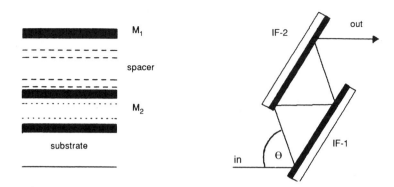

Figure 2.6: Gires-Tournois interferometer for fs light pulses using dielectric multilayers. By rotating two parallel interferometers the overall dispersion can be adjusted through a change of the angle of incidence Θ and the number of reflections. Note, the beam direction is not changed. The lateral displacement can be compensated by a second pair of interferometers (from [28]).

which gives for the quantity responsible for GVD

$$\left.\frac{d^2}{d\Omega^2}\Psi(\Omega)\right|_{\omega_\ell} = \frac{2r(r^2-1)\sin\delta}{(1+r-2r\cos\delta)^2}\left(\left.\frac{d\delta}{d\Omega}\right|_{\omega_\ell}\right)^2. \qquad (2.16)$$

Obviously the GVD can be tuned continuously by adjusting δ which can be either through a change of the mirror separation d or through a change of the angle of incidence Θ. Gires and Tournois [27] conceived this interferometer to adapt to optical frequencies the pulse compression technique used in radar (sending a frequency modulated pulse through a dispersive delay line). Duguay and Hansen were the first to apply this device for the compression of pulses from a He-Ne laser [29]. Since typical pulse durations are in the order of several hundred ps the mirror spacing need to be in the order of few mm. To use the interferometer for the shaping of fs pulses the corresponding mirror spacing has to be in the order of few microns. Heppner and Kuhl [30] overcame this obvious practical difficulty by designing a Gires–Tournois interferometer on the basis of dielectric multilayer systems, as illustrated in Fig. 2.6(a). The 100% mirror M_2 is a sequence of dielectric coatings with alternating refractive index deposited on a substrate. A certain spacer of optical thickness d consisting of a series of $\lambda/2$ layers of one and the same material is placed on top of M_2. The partially reflective surface M_1 is realized by one $\lambda/4$ layer of high refractive index. The dispersion of this compact device can be tuned by changing the angle of incidence and/or

2.4. OTHER INTERFEROMETRIC STRUCTURES

the number of passes through the interferometer. A possible arrangement which was successfully applied for GVD adjustments in fs lasers [28] is shown in Fig. 2.6(b).

Let us inspect in more detail the actual transfer function of the multilayer Gires-Tournois taking into account the mirror dispersion. The first reflecting face of the Gires Tournois is a multi- or single quarter wave layer dielectric coating, with complex reflection coefficient $\tilde{r}_i = r\exp(i\varphi_{r,i})$ and transmission coefficient $\tilde{t}_i = t\exp(i\varphi_{t,i})$. The subscript $i = 1, 2$ refers to reflection (transmission) outside (1) or inside (2) the interferometer. "Inside" refers to a spacer between the first reflecting layer and the 100% reflector filled with a dielectric of index of refraction $n(\Omega)$. It can be shown from simple energy conservation considerations (cf. Appendix A) that, for any angle of incidence, complex reflection and transmission coefficients satisfy the relation $\tilde{t}_1\tilde{t}_2 - \tilde{r}_1\tilde{r}_2 = 1$. Let us designate by δ the phase shift accumulated by the wave having propagated from the first reflecting layer to the total reflector and back to the first layer:

$$\delta = -2k_\ell d\cos\theta + \varphi_{r2} = -\frac{2\Omega n(\Omega) d\cos\theta}{c} + \varphi_{r2}, \quad (2.17)$$

where φ_{r2} is the phase shift upon reflection at the totally reflecting layer(s) (mirror M_2), and θ is the internal angle of incidence on the reflecting interfaces. The complex (field) reflection coefficient of the structure is:

$$\begin{aligned} R(\Omega)e^{-i\Psi(\Omega)} &= \frac{\tilde{r}_1 + (\tilde{t}_1\tilde{t}_2 - \tilde{r}_1\tilde{r}_2)e^{i\delta}}{1 - \tilde{r}_2 e^{i\delta}} \\ &= \frac{\tilde{r}_1 + e^{i\delta}}{1 - \tilde{r}_2 e^{i\delta}}, \end{aligned} \quad (2.18)$$

where the last equality result from the relation between phase shift upon reflection and transmission mentioned above. The reflectivity of the device is:

$$|R(\Omega)|^2 = \frac{|\tilde{r}_1|^2 + 1 + \tilde{r}_1 e^{-i\delta} + \tilde{r}_1^* e^{i\delta}}{|\tilde{r}_2|^2 + 1 - \tilde{r}_2^* e^{-i\delta} - \tilde{r}_2 e^{i\delta}}, \quad (2.19)$$

which is only equal to unity if the phase on reflection obey the relation $\tilde{r}_2 = -\tilde{r}_1^*$. This relation is consistent with the phase shift upon reflection on a dielectric interface. The expression for the complex reflection can be re-written in terms of the reflection coefficient inside the interferometer $\tilde{r}_2$:

$$R(\Omega)e^{-i\Psi(\Omega)} = \frac{-\tilde{r}_2^* + e^{i\delta}}{1 - \tilde{r}_2 e^{i\delta}}. \quad (2.20)$$

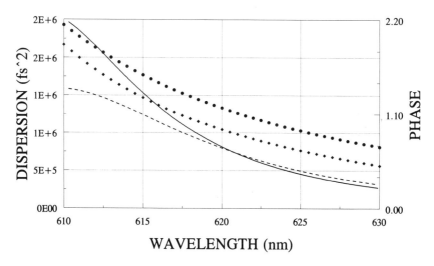

Figure 2.7: Comparison of the exact phase function $\psi(\lambda_\ell)$ (diamonds) and its second derivative $d^2\psi/d\Omega^2$ (solid line) of a Gires-Tournois with the approximation of assuming constant φ_r and φ_{r2} in Eq. (2.21) (circles for ψ, dashed line for the second derivative). The Gires-Tournois has been designed for a central wavelength $\lambda_0 = 620$ nm. The curves are calculated for an external angle of incidence of $20°$. For this particular example, the high reflector is made of 11 layers of TiO_2 (thickness $\lambda_0/4n = 67.4$ nm) alternating with layers of SiO_2 (thickness $\lambda_0/4n = 106.9$ nm) on a glass substrate ($n = 1.5$). The spacer consists of 5 half wave spacing of SiO_2, for a total thickness of 1068.9 nm. The top reflector ($r = 0.324$) consists of a quarter wave layer of TiO_2 (thickness $\lambda_0/4n = 67.4$ nm). In applying Eq. (2.21), the values of phase shifts upon reflection were $\varphi_r = -0.0192$ (top layer) and $\varphi_{r2} = -0.0952$ (high reflector).

Let us express the reflectivity $\tilde{r}_2$ in terms of its amplitude and phase: $\tilde{r}_2 = r\exp(i\varphi_r)$. The phase response can now be calculated:

$$\Psi(\Omega) = -\arctan\frac{(r^2 - 1)\sin\delta(\Omega) - r\sin\varphi_r(\Omega)}{2r\cos\varphi_r(\Omega) - (1 + r^2)\cos\delta(\Omega)}, \qquad (2.21)$$

which is a generalization of Eq. (2.15) to the more complex multilayer Gires-Tournois structure. Only when $\varphi_r = 0$ and φ_{r2} (in δ) is frequency independent in the range of interest are the dispersions described by Eqs. (2.15) and (2.21) equal. The error is, however, small in real situation, as can be seen from the comparison of the approximation (2.15) with the exact phase $\Psi(\Omega)$ shown in Fig. 2.7. The latter functional dependence can be calculated

2.5 Focusing elements

2.5.1 Singlet lenses

One main function of fs pulses is to concentrate energy in time and space. The ability to achieve extremely high peak power densities depends partly on the ability to keep pulses short in time, and concentrate them in a short volume by focusing. The difference between group and phase velocity in the lens material can reduce the peak intensity on focus by delaying the time of arrival of the rays on axis, with respect to the time of arrival of peripheral rays. The group velocity dispersion leads to reduction of peak intensity by stretching the pulse in time. As pointed out by Bor [31, 32], when simple focusing singlet lenses are used, the former effect can lead to several picosecond lengthening of the time required to deposit the energy of a fs pulse on focus.

Let us assume a plane pulse and phase front at the input of a spherical lens as sketched in Fig. 2.8. According to Fermat's principle the optical path along rays from the input phase front to the focus is independent of the radius coordinate r. The lens transforms the plane phase front into a spherical one which converges in the (paraxial) focus. Assimilating air as vacuum, it is only while propagating through the lens that the pulses experience a group

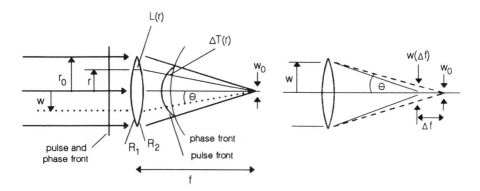

Figure 2.8: Left: delay of the temporal front with respect to the phase front, in the case of a singlet lens. Right: spread of the pulse energy due to chromatic aberration.

velocity v_g different from the phase velocity $v_p = c/n$. The result is a pulse front that is delayed with respect to the (spherical) phase front, depending on the amount of glass traversed. As we have seen in Chapter 1, the group velocity is:

$$v_g = \left(\frac{dk}{d\Omega}\right)^{-1} = \frac{c}{n - \lambda_\ell \frac{dn}{d\lambda}}, \tag{2.22}$$

where λ_ℓ is the wavelength in vacuum. The difference in propagation time between the phase front and pulse front is:

$$\Delta T(r) = \left(\frac{1}{v_p} - \frac{1}{v_g}\right) L(r) \tag{2.23}$$

which is also the difference of the time of arrival at the focus of pulses traversing the lens at distance r from the axis and peripheral rays touching the lens rim. For a spherical thin lens, the thickness L is given by

$$L(r) = \frac{r_0^2 - r^2}{2} \left(\frac{1}{R_1} - \frac{1}{R_2}\right) \tag{2.24}$$

where $R_{1,2}$ are the radii of curvature of the lens and r_0 is the radius of the lens aperture.[5] Substituting the expressions for the group velocity (2.22) and for the lens thickness (2.24) into Eq. (2.23) yields for the difference in time of arrival between a pulse passing through the lens at the rim and at r:

$$\begin{aligned}\Delta T(r) &= \frac{r_0^2 - r^2}{2c}\left(\frac{1}{R_1} - \frac{1}{R_2}\right)\left(\lambda \frac{dn}{d\lambda}\right) \\ &= \frac{r_0^2 - r^2}{2c} \lambda \frac{d}{d\lambda}\left(\frac{1}{f}\right)\end{aligned} \tag{2.25}$$

where the focal length f has been introduced by $1/f = (n-1)(R_1^{-1} - R_2^{-1})$. Equation (2.26) illustrates the connection between the radius dependent pulse delay and the chromaticity $d/d\lambda(1/f)$ of the lens. For an input beam of radius r_b the pulse broadening in the focus can be estimated with the difference in arrival time $\Delta T'$ of a pulse on an axial ray and a pulse passing through the lens at r_b:

$$\Delta T'(r_b) = \frac{r_b^2}{2c} \lambda \frac{d}{d\lambda}\left(\frac{1}{f}\right). \tag{2.26}$$

---

[5]Regarding sign considerations we will use positive (negative) $R_{1,2}$ for refracting surfaces which are concave (convex) towards the incident side.

2.5. FOCUSING ELEMENTS

To illustrate the effects of group velocity delay and dispersion, let us assume that we would like to focus a 50 fs pulse at the excimer laser wavelength of 248 nm (KrF) down to a spot size of 0.6 μm, using a fused silica lens (singlet) of focal distance $f = 30$ mm. Let us further assume that the input beam profile is Gaussian. Since the half divergence angle in the focused beam is $\theta = \lambda/(\pi w_0)$, the radius w of the Gaussian beam [radial dependence of the electric field: $\tilde{\mathcal{E}}(r) = \tilde{\mathcal{E}}(0) \exp\{-r^2/w^2\}$] incident on the lens should be approximately $\theta f = (\lambda/\pi w_0)f \approx 4$ mm. To estimate the pulse delay we evaluate $\Delta T'$ at $r_b = w$:

$$\begin{aligned}\Delta T'(r_b = w) &= \frac{w^2}{2c}\lambda\frac{d}{d\lambda}\left(\frac{1}{f}\right) \\ &= -\frac{w^2}{2cf(n-1)}\left(\lambda\frac{dn}{d\lambda}\right) \\ &= -\frac{\theta^2 f}{2c(n-1)}\left(\lambda\frac{dn}{d\lambda}\right).\end{aligned} \quad (2.27)$$

For the particular example chosen, $n \approx 1.51$, $\lambda dn/d\lambda \approx -0.17$, and the difference in time of arrival (at the focus) of the rays at $r = 0$ and $r_b = w$ is ≈ 300 fs, which can be used as a rough measure of the pulse broadening for Gaussian illumination.

The direct effect of the chromaticity of the lens is a spatial spread of the optical energy near the focus, because different spectral components of the pulse are focused at different points on axis. For a bandwidth limited Gaussian pulse of duration $\tau_p = \sqrt{2\ln 2}\,\tau_{G0}$ with spectral width $\Delta\lambda = 0.441\lambda^2/c\tau_p$, the focus spreads by the amount $\Delta f = -f^2 \frac{d(1/f)}{d\lambda}\Delta\lambda$:

$$\Delta f = -\frac{f\lambda^2}{c(n-1)}\frac{0.441}{\tau_p}\frac{dn}{d\lambda}. \quad (2.28)$$

Applying Eq. (2.28) to our example of a 30 mm fused silica lens to focus a 50 fs pulse, we find a spread of $\Delta f = 60$ μm, which is large compared to the Rayleigh range of the beam $\rho_0 = w_0/\theta \approx 5$ μm. We can therefore write the following approximation for the broadening of the beam: $w(\Delta f)/w_0 = \sqrt{1+(\Delta f/2\rho_0)^2} \approx (\Delta f/2\rho_0)$. Substituting the value for Δf from Eq. (2.28):

$$w(\Delta f)/w_0 = -\frac{0.441\pi}{2\tau_p}\frac{f\lambda\theta^2}{c(n-1)}\frac{dn}{d\lambda}. \quad (2.29)$$

We note that the spatial broadening of the beam due to the spectral extension of the pulse, as given by Eq. (2.29), is (within a numerical factor) the same expression as the group velocity delay [Eq. (2.27)] relative to the pulse duration. In fact, neither expression is correct, in the sense that they do not give a complete description of the spatial and temporal evolution of the pulse near the focus. An exact calculation of the focalization of a fs pulse by a singlet is presented in the subsection that follows.

In addition to the group delay effect, there is a direct temporal broadening of the pulse *in the lens itself* due to GVD in the lens material, as discussed in Section 1.5. Let us take again as an example the fused silica singlet of 30 mm focal length and of 16 mm diameter used to focus a 248 nm laser beam to a 0.6 μm spot size. The broadening will be largest for the beam on axis, for which the propagation distance through glass is $L(r = 0) = d_0 = r_0^2/\{2f(n-1)\} = 2.1$ mm. Using for the second-order dispersion an index $\lambda d^2n/d\lambda^2 \approx 2.1$ μm^{-1} [31], we find from Eq. (1.81) that a 50 fs (FWHM) unchirped Gaussian pulse on axis will broaden to about 60 fs. At a wavelength of 800 nm, where the dispersion is much smaller than in the UV (see table 2.1), the pulse would only broaden to 50.4 fs.

The example above illustrates the differences between peak intensity reduction at the focal point of a lens because of group velocity effects, and effects of group velocity dispersion. The latter is strongly chirp dependent, while the former is not. The delay in pulse energy due to the difference between group and phase velocity is independent of the pulse duration and is directly related to the spot size that will be achieved (the effect is larger for optical arrangements with a large F-number). The group velocity dispersion effect is pulse width dependent, and, in typical materials, becomes significant only for pulse durations well below 100 fs in the VIS and NIR spectral range.

2.5.2 Space-time distribution of the pulse intensity at the focus of a lens

The geometrical optical discussion of the focusing of ultrashort light pulses presented above gives a satisfactory order of magnitude estimate for the temporal broadening effects in the focal plane of a lens. We showed this type of broadening to be associated with chromatic aberration. Frequently the experimental situation requires an optimization not only with respect to the temporal characteristics of the focused pulse, but also with respect to the achievable spot size. To this aim we need to analyze the space-time distribution of the pulse intensity in the focal region of a lens in more detail. The general procedure is to solve either the wave equation (1.40), or better

2.5. FOCUSING ELEMENTS

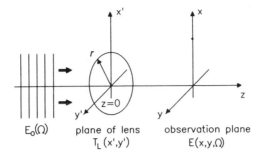

Figure 2.9: Diffraction geometry for focusing.

the corresponding diffraction integral, which in Fresnel approximation was given in Eq. (1.134). However, we cannot simply separate space and time dependence of the field with a product ansatz (1.125) since we expect the chromaticity of the lens to induce an interplay of both. Instead we will solve the diffraction integral for each "monochromatic" Fourier component of the input field $\tilde{E}_0(\Omega)$ which will result in the field distribution in a plane (x, y, z) behind the lens, $\tilde{E}(x, y, z, \Omega)$. The time-dependent field $\tilde{E}(x, y, z, t)$ then is obtained through the inverse Fourier transform of $\tilde{E}(x, y, z, \Omega)$ so that we have for the intensity distribution:

$$I(x, y, z, t) \propto |\mathcal{F}^{-1}\{\tilde{E}(x, y, z, \Omega)\}|^2. \tag{2.30}$$

The geometry of this diffraction problem is sketched in Fig. 2.9. Assuming plane waves of amplitude $E_0(\Omega) = E_0(x', y', z' = 0, \Omega)$ at the lens input, the diffraction integral to be solved reads, apart from normalization constants:

$$E(x, y, z, \Omega) \propto \frac{\Omega}{c} \int\int E_0(\Omega) T_L(x', y') T_A(x', y') e^{-i\frac{k}{2z}[(x'-x)^2 + (y'-y)^2]} dx' dy' \tag{2.31}$$

where T_L and T_A are the transmission function of the lens and the aperture stop, respectively. The latter can be understood as the lens rim in the absence of other beam limiting elements. The lens transmission function describes a radially dependent phase delay which in case of a thin, spherical lens can be written:

$$T_L(x', y') = \exp\left\{-i\frac{\Omega}{c}[nL(r') + d_0 - L(r')]\right\} \tag{2.32}$$

with $r'^2 = x'^2 + y'^2$ and

$$L(r') = d_0 - \frac{r'^2}{2}\left(\frac{1}{R_1} - \frac{1}{R_2}\right) = d_0 - \frac{r'^2}{2(n-1)f} \tag{2.33}$$

where d_0 is the thickness in the lens center. Note that because of the dispersion of the refractive index n, the focal length f becomes frequency dependent. For a spherical opening of radius r'_0 the aperture function T_A is simply:

$$T_A(r') = \begin{cases} 1 & \text{for} \quad x'^2 + y'^2 = r'^2 \leq r'^2_0 \\ 0 & \text{otherwise} \end{cases} \quad (2.34)$$

If we insert Eq. (2.33) into Eq. (2.32) we can rewrite the lens transmission function as:

$$T_L(x', y') = \exp\left\{-i\left[k_g(\Omega)d_0 - \left(k_g(\Omega) - \frac{\Omega}{c}\right)\frac{r'^2}{2}\left(\frac{1}{R_1} - \frac{1}{R_2}\right)\right]\right\} \quad (2.35)$$

where

$$k_g(\Omega) = \frac{\Omega}{c}n(\Omega) \quad (2.36)$$

is the wave vector in the glass material. Substituting this transmission function in the diffraction integral Eq.(2.31) we find for the field distribution in the focal plane:

$$E(\Omega) \propto \frac{\Omega}{c} e^{-ik_g(\Omega)d_0} \iint T_A E_0(\Omega) \exp\left[i\left(k_g(\Omega) - \frac{\Omega}{c}\right)\frac{r'^2}{2}\left(\frac{1}{R_1} - \frac{1}{R_2}\right)\right]$$
$$\times e^{-i\frac{k}{2z}[(x'-x)^2+(y'-y)^2]}dx'dy' \quad (2.37)$$

The exponent of the second exponential function is radially dependent and is responsible for the focusing, while the first one describes propagation through a dispersive material of length d_0. For a closer inspection let us assume that the glass material is only weakly dispersive so that we may expand $k_g(\Omega)$ and $[k_g(\Omega) - \Omega/c]$ up to second order. In both exponential functions this will result in a sum of terms proportional to $(\Omega - \omega_\ell)^m$ ($m = 0, 1, 2$). According to our discussion in the section about linear elements, optical transfer functions which have the structure $\exp[-ib_1(\Omega - \omega_\ell)]$ give rise to a certain pulse delay. Since b_1 is a function of r' this delay becomes radius dependent, a result which has already been expected from our previous ray-optical discussion. The next term of the expansion ($m = 2$) is responsible for pulse broadening in the lens material. A numerical evaluation of Eq. (2.37) and subsequent inverse Fourier transform [Eq. (2.30)] allows one to study the complex space-time distribution of the pulse intensity behind a lens. An example is shown in Fig. 2.10. In the aberration-free case we recognize a spatial distribution

2.5. FOCUSING ELEMENTS 65

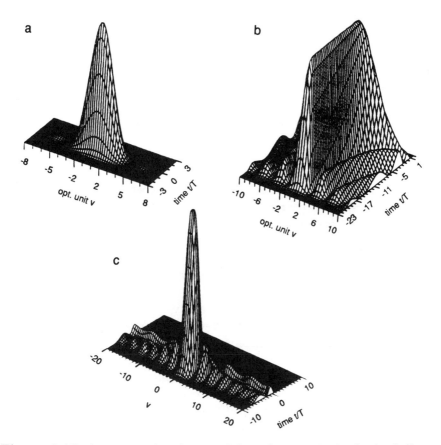

Figure 2.10: Space-time distribution of the pulse intensity in the focal plane of a lens: (a) Focusing without chromatic or spherical aberration, (b) Focusing with chromatic aberration $\tau/T = 20$. The input pulse was chosen to vary as $e^{-(t/T)^2}$. v is the optical coordinate defined as $v = r'_0 k_\ell \sqrt{x^2 + y^2}/f_\ell$ and $\tau = T'(r_0) = \left| \frac{r'^2_0 \lambda_\ell}{2fc(n-1)} n'(\lambda_\ell) \right|$ is a measure for the dispersion (from [33]). (c) Focusing with chromatic and spherical aberration. The intensity distribution in the plane of the marginal focus is shown (from [34]).

corresponding to the Airy disc and no temporal distortion. The situation becomes more complex if chromaticity plays a part. We see spatial as well as temporal changes in the intensity distribution. At earlier times the spatial distribution is narrower. This can easily be understood if we remember that pulses from the lens rim (or aperture edge) arrive first in the focal plane and are responsible for the field distribution. At later times pulses from inner parts of the lens arrive. The produced spot becomes larger since the effective aperture size is smaller. If we use achromatic doublets (cf. next subsection) the exponential proportional to $(\Omega - \omega_\ell)$ in the corresponding diffraction integral does not appear. The only broadening then is due to GVD in the glass material.

Interesting effects occur if spherical aberration is additionally taken into account [34] which is essential to correctly model strong focusing with singlet lenses. As known from classical optics, spherical aberration results in different focal planes for beams passing through. Since ultrashort pulses passing through different lens annuli experience the same delay, almost no temporal broadening occurs for the light which is in focus, as illustrated in Fig. 2.10(c). The space-time distribution in the focal area can differ substantially from that obtained with a purely chromatic lens.

To measure the interplay of chromatic and spherical aberration in focusing ultrashort light pulses, one can use an experimental setup as shown in Fig. 2.11(a). The beam is expanded and sent into a Michelson interferometer. One arm contains the lens to be characterized, which can be translated so as to focus light passing through certain lens annuli onto mirror M_1. Provided the second arm has the proper length, an annular interference pattern can be observed at the output of the interferometer. The radius of this annulus is determined by the setting of Δf. If no spherical aberration is present, an interference pattern is observable only for $\Delta f \approx 0$ and a change of the time delay by translating M_2 would change the radius of the interference pattern. With spherical aberration present, at a certain Δf, an interference pattern occurs only over a delay corresponding to the pulse duration while the radius of the annulus remains constant. This can be proved by measuring the cross-correlation, i.e., by measuring the second harmonic signal as function of the time delay. The width of the cross-correlation does not differ from the width of the autocorrelation which is measured without the lens in the interferometer arm. Figure 2.11(b) shows the position of the peak of the cross-correlation as function of Δf and the corresponding r, respectively. For comparison, the delay associated with chromatic aberration alone is also shown.

2.5. FOCUSING ELEMENTS

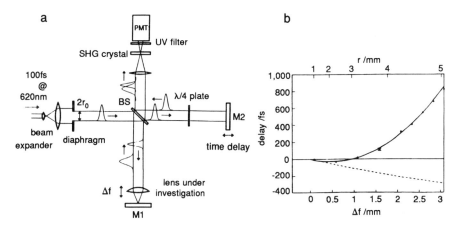

Figure 2.11: (a) Correlator for measuring the effect of chromatic and spherical aberration on the focusing of fs pulses. (b) Measured pulse delay (data points) as a function of the lens position (Δf — deviation from the paraxial focus) and the corresponding radial coordinate r of light in focus. The solid line is obtained with ray–pulse tracing; the dashed curve shows the effect of chromatic aberration only. Lens parameters: $f_0 = 12.7$ mm, BK7 glass (from [35]).

2.5.3 Achromatic doublets

The chromaticity of a lens was found to be the cause for a radius dependent pulse arrival time in the focal plane, cf., Eq.(2.26). Therefore one should expect achromatic optics to be free of this undesired pulse lengthening. To verify that this is indeed the case, let us consider the doublet shown in Fig. 2.12. The thicknesses of glass traversed by the rays in the media of index n_1 and n_2 are L_1 and L_2 and are given by:

$$L_1 = d_1 - \frac{r^2}{2}\left(\frac{1}{R_1} - \frac{1}{R_2}\right) \tag{2.38}$$

and

$$L_2 = d_2 - \frac{r^2}{2}\left(\frac{1}{R_2} - \frac{1}{R_3}\right) \tag{2.39}$$

where $d_{1,2}$ is the center thickness of lens $1, 2$. The inverse of the focal length of the doublet lens is:

$$\frac{1}{f} = (n_1 - 1)\left(\frac{1}{R_1} - \frac{1}{R_2}\right) + (n_2 - 1)\left(\frac{1}{R_2} - \frac{1}{R_3}\right). \tag{2.40}$$

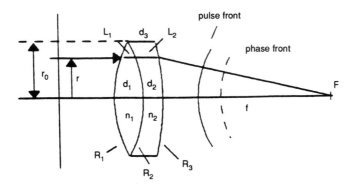

Figure 2.12: Ray tracing in an achromat (from [31]).

The condition of achromaticity $\frac{d}{d\lambda}(1/f) = 0$ gives an additional relation between the radii of curvature R_i and the indices n_i. The expression for the transit time in glass [31] in which we have inserted the chromaticity of the doublet is:

$$T(r) = \frac{d_1}{c}\left\{n_1 - \lambda\frac{dn_1}{d\lambda}\right\} + \frac{d_2}{c}\left\{n_2 - \lambda\frac{dn_2}{d\lambda}\right\} + \frac{\lambda r^2}{2c}\frac{d}{d\lambda}\left(\frac{1}{f}\right). \quad (2.41)$$

Equation (2.41) indicates that, for an achromatic doublet for which the third term on the right hand side vanishes, the transit time has no more radial dependence. The phase front and wave front are thus parallel, as sketched in Fig. 2.12. In this case, the only mechanism broadening the energy at the focus is group velocity dispersion. The latter can be larger than with singlet lenses since achromatic doublets usually contain more glass.

2.5.4 Mirrors

Another way to avoid the chromatic aberration and thus pulse broadening is to use mirrors for focusing. With spherical mirrors and on-axis focusing the first aberration to be considered is the spherical one. The analysis of spherical aberration of mirrors serves also as a basis to the study of spherical aberration applied to lenses.

Let us consider the situation of Fig. 2.13, where a plane pulse- and wavefront impinge upon a spherical mirror of radius of curvature R. The reflected rays are the tangents of a caustic — the curve commonly seen as light reflect off a coffee cup. Rays that are a distance r off-axis intersect the optical axis at point T which differs from the paraxial focus F. The difference in arrival

2.5. FOCUSING ELEMENTS

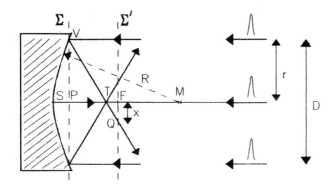

Figure 2.13: Focusing of light pulses by a spherical mirror.

time between pulses traveling along off-axis rays and on-axis pulses in the paraxial focal plane Σ is:

$$\Delta T = \frac{1}{c}\left[\overline{VQ} - \left(\overline{PS} + \frac{R}{2}\right)\right]. \qquad (2.42)$$

Through simple geometrical considerations on can find an expression for ΔT in the form of an expansion in powers of (r/R). The first non-zero term of that expansion is:

$$\Delta T = \frac{3}{4}\frac{R}{c}\left(\frac{r}{R}\right)^4. \qquad (2.43)$$

Likewise, one obtains for the geometrical deviation from the paraxial focus in Σ':

$$x = \frac{R}{2}\left(\frac{r}{R}\right)^3. \qquad (2.44)$$

For a beam diameter $D = 3$ mm and a focal length $f = 25$ mm the arrival time difference amounts to about 1.6 fs and is thus negligibly small. However, ΔT increases rapidly with increasing beam diameter. For $D = 1$ cm for instance, we obtain $\Delta T = 200$ fs. The deviations from the paraxial focus corresponding to $D = 3$ mm and $D = 1$ cm amount to 0.5 μm and 160 μm, respectively. In experimental situations where even a small aberration should be avoided, parabolic mirrors can advantageously be used. An example requiring such optics is upconversion experiments where fluorescence with fs rise time from a large solid angle has to be focused tightly, without modifying its temporal behavior.

2.6 Elements with angular dispersion

Besides focusing elements there are various other optical components which modify the temporal characteristics of ultrashort light pulses through a change of their spatial propagation characteristics.

Even a simple prism can provide food for thought in fs experiments. Let us consider an expanded parallel beam of short light pulses incident on a prism, and diffracted by the angle $\beta = \beta(\Omega)$, as sketched in Fig. 2.14. The beam is assumed to be broad, which allows us to neglect the distribution of diffraction angles $\Delta\beta = \frac{d\beta}{d\Omega}\Big|_{\omega_\ell} \Delta\omega$ caused by the finite pulse spectrum. This approximation is justified as long as we are interested in the beam properties at a distance L_p behind the prism much shorter than $D'/\Delta\beta$. As discussed by Bor et al. [32], the prism introduces a tilt of the pulse front with respect to the phase front. As in lenses, the physical origin of this tilt is the difference between group and phase velocity. According to Fermat's principle the prism transforms a phase front $\overline{AB}$ into a phase front $\overline{A'B'}$. The transit times for the phase and pulse fronts along the marginal ray $\overline{BOB'}$ are equal ($v_p \sim v_g$ in air). In contrast the pulse is delayed with respect to the phase in any part of the ray that travels through a certain amount of glass. This leads to an increasing delay across the beam characterized by a certain tilt angle α. The maximum arrival time difference in a plane perpendicular to the propagation direction is $(D'/c)\tan\alpha$.

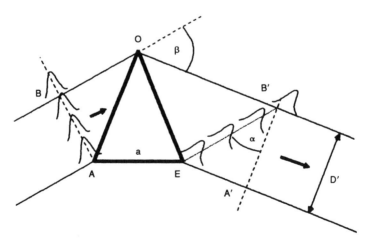

Figure 2.14: Pulse front tilt introduced by a prism. The position of the (plane) wavefronts is indicated by the dashed lines $\overline{AB}$ and $\overline{A'B'}$.

2.6. ELEMENTS WITH ANGULAR DISPERSION

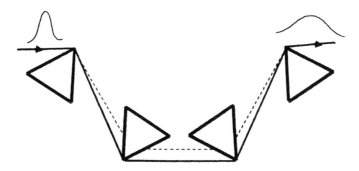

Figure 2.15: Pulse broadening in a four prism sequence.

Before discussing pulse front tilt more thoroughly, let us briefly mention another possible prism arrangement where the above condition for L_p is not necessary. Let us consider for example the symmetrical arrangement of four prisms sketched in Fig. 2.15. During their path through the prism sequence, different spectral components travel through different optical distances. At the output of the fourth prism all these components are again equally distributed in one beam. The net effect of the four prisms is to introduce a certain amount of GVD leading to broadening of an unchirped input pulse. We will see later in this chapter that this particular GVD can be interpreted as a result of angular dispersion and has a sign opposite to that of the GVD introduced by the glass material constituting the prisms.

2.6.1 Tilting of pulse fronts

In an isotropic material the direction of energy flow — usually identified as ray direction — is always orthogonal to the surfaces of constant phase (*wave fronts*) of the corresponding propagating wave. In the case of a beam consisting of ultrashort light pulses, one has to consider in addition planes of constant intensity (*pulse fronts*). For most applications it is desirable that these pulse fronts be parallel to the phase fronts and thus orthogonal to the propagation direction. In the section on focusing elements we have already seen how lenses cause a radially dependent difference between pulse and phase fronts. This leads to a temporal broadening of the intensity distribution in the focal plane. There are a number of other optical components which introduce a tilt of the pulse front with respect to the phase front and to the normal of the propagation direction, respectively. Another example was the prism discussed in the introduction of this section. As a general

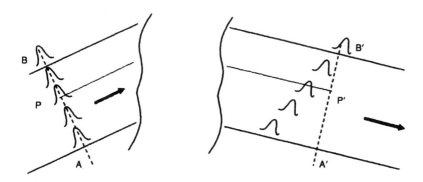

Figure 2.16: Delay of the pulse front with respect to the phase front.

rule, the pulse front tilting should be avoided whenever an optimum focalization of the pulse energy is sought. There are situations where the pulse front tilt is desirable to transfer a temporal delay to a transverse coordinate. Applications exploiting this property of the pulsefront tilt are pulse diagnostics [36, 37] and traveling wave amplification [38].

The general approach for tilting pulse fronts is to introduce an optical element in the beam path which retards the pulse fronts as a function of a coordinate transverse to the beam direction. This is schematically shown in Fig. 2.16 for an element that changes only the propagation direction of a (plane) wave. Let us assume that a wavefront $\overline{AB}$ is transformed into a wavefront $\overline{A'B'}$. From Fermat's principle it follows that the optical pathlength P_{OL} between corresponding points at the wavefronts $\overline{AB}$ and $\overline{A'B'}$ must be equal:

$$P_{OL}(BB') = P_{OL}(PP') = P_{OL}(AA'). \tag{2.45}$$

Since the optical pathlength corresponds to a phase change $\Delta\Phi = 2\pi P_{OL}/\lambda$, the propagation time of the wavefronts can be written as

$$T_{phase} = \frac{\Delta\Phi}{\omega_\ell} \tag{2.46}$$

where we referred to the center frequency of the pulse. This phase change is given by

$$\Delta\Phi = \int_P^{P'} k(s)ds = \frac{\omega_\ell}{c}\int_P^{P'} n(s)ds = \omega_\ell \int_P^{P'} \frac{ds}{v_p(s)} \tag{2.47}$$

2.6. ELEMENTS WITH ANGULAR DISPERSION

where s is the coordinate along the beam direction. In terms of the phase velocity the propagation time is

$$T_{phase} = \int_P^{P'} \frac{ds}{v_p(s)}. \tag{2.48}$$

The propagation time of the pulse fronts however, T_{pulse}, is determined by the group velocity

$$T_{pulse} = \int_P^{P'} \frac{ds}{v_g(s)} = \int_P^{P'} \left.\frac{dk}{d\Omega}\right|_{\omega_\ell} ds. \tag{2.49}$$

From Eqs. (2.48) and (2.49) the difference in propagation time between phase front and pulse front becomes

$$\Delta T(d) = T_{phase} - T_{pulse} = \int_P^{P'} \left(\frac{1}{v_p} - \frac{1}{v_g}\right) ds = \int_P^{P'} \left[\frac{k}{\omega_\ell} - \left.\frac{dk}{d\Omega}\right|_{\omega_\ell}\right] ds, \tag{2.50}$$

which can be regarded as a generalization of Eq. (2.23).

A simple optical arrangement to produce pulse front tilting is an interface separating two different optical materials – for instance air (vacuum) and glass (Fig. 2.17). At the interface F the initial beam direction is changed by an angle $\beta = \gamma - \gamma'$ where γ and γ' obey Snell's law $\sin\gamma = n(\omega_\ell)\sin\gamma'$. The point A of an incident wavefront $\overline{AB}$ is refracted at time $t = t_0$. It takes the time interval T_{phase} to recreate the wavefront $\overline{A'B'}$ in medium 2, which propagates without distortion with a phase velocity $v_p = c/n(\omega_\ell)$. The time interval T_{phase} is given by

$$T_{phase} = \frac{n\overline{AA'}}{c} = \frac{\overline{BB'}}{c} = \frac{\overline{PF} + n\overline{FP'}}{c} = \frac{D\tan\gamma}{c}. \tag{2.51}$$

The beam path from B to B' is through a nondispersive material and thus pulse front and wavefront coincide at B'. In contrast the phase front and pulse front propagate different distances during the time interval T_{phase} in medium 2 and thus become separated. Since in optical materials the group velocity is smaller than the phase velocity the pulse front is delayed with respect to the phase front. In our case this delay increases linearly over the beam cross section. The characteristic tilt angle α is given by

$$\tan\alpha = \frac{\overline{EA'}}{D'}. \tag{2.52}$$

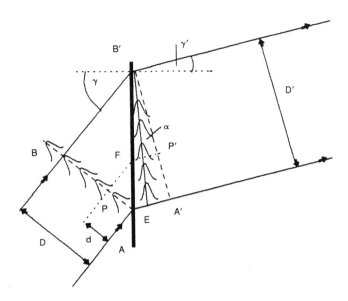

Figure 2.17: Pulse front tilt through refraction at an interface.

From simple geometrical considerations we find for the two distances

$$\overline{EA'} = \left(\frac{c}{n} - v_g\right) T_{phase} = \left(\frac{c}{n} - v_g\right) \frac{D}{c} \tan\gamma \qquad (2.53)$$

and

$$D' = D\frac{\cos\gamma'}{\cos\gamma} = D\frac{\sqrt{n^2 - \sin^2\gamma}}{n\cos\gamma}. \qquad (2.54)$$

Inserting Eqs. (2.53) and (2.54) in Eq. (2.52) and using the expression for the group velocity, we obtain for α

$$\tan\alpha = \frac{\omega_\ell n'(\omega_\ell)}{\omega_\ell n'(\omega_\ell) + n(\omega_\ell)} \frac{\sin\gamma}{\sqrt{n^2 - \sin^2\gamma}}. \qquad (2.55)$$

Following this procedure we can also analyze the pulse front at the output of a prism, cf. Fig. 2.14. The distance $\overline{EA'}$ is the additional pathlength over which the phase has traveled as compared to the pulse path. Thus, we have

$$\overline{EA'} = v_p\left[\frac{a}{v_g} - \frac{a}{v_p}\right] = a\omega_\ell n'(\omega_\ell) \qquad (2.56)$$

2.6. ELEMENTS WITH ANGULAR DISPERSION

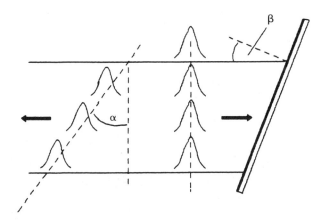

Figure 2.18: A beam is diffracted at a grating into diffraction order -1 which produces pulse front tilt.

which results in a tilt angle

$$\tan\alpha = \frac{a}{b}\omega_\ell n'(\omega_\ell) = -\frac{a}{b}\lambda_\ell \left.\frac{dn}{d\lambda}\right|_{\lambda_\ell} \quad (2.57)$$

where $b = D'$ is the beam width. As pointed out by Bor [32], there is a general relation between pulse front tilt and the angular dispersion $d\beta/d\lambda$ of a dispersive device which reads

$$|\tan\alpha| = \lambda\left|\frac{d\beta}{d\lambda}\right|. \quad (2.58)$$

The latter equation can be proven easily for a prism, by using the deviation equation $d\beta/d\lambda = (a/b)(dn/d\lambda)$ in Eq. (2.57). Similarly to prisms, gratings produce a pulse front tilt, as can be verified easily from the sketch of Fig. 2.18. To determine the tilt angle we just need to specify Eq. (2.58) using the grating equation.

2.6.2 GVD through angular dispersion

General

Angular dispersion has been advantageously used for a long time to resolve spectra or for spectral filtering, utilizing the spatial distribution of the frequency components behind the dispersive element (e.g., prism, grating). In connection with fs optics, angular dispersion has the interesting property of introducing GVD. At first glance this seems to be an undesired effect. However, optical devices based on angular dispersion can be designed which allow a continuous tuning of GVD. This idea was first implemented in [39] for the compression of chirped pulses with diffraction gratings. The concept was later generalized to prisms and prism sequences [40]. Simple expressions for two and four prism sequences are given in [41]. From a general point of view, the diffraction problem can be treated by solving the corresponding Fresnel integrals [39, 42, 43]. We will sketch this procedure at the end of this chapter. Another successful approach is to analyze the sequence of optical elements by ray-optical techniques and calculate the optical beam path P as a function of Ω. From our earlier discussion we expect the response of any linear element to be of the form:

$$R(\Omega)e^{-i\Psi(\Omega)} \tag{2.59}$$

where the phase delay Ψ is related to the optical pathlength P_{OL} through

$$\Psi(\Omega) = \frac{\Omega}{c} P_{OL}(\Omega). \tag{2.60}$$

$R(\Omega)$ is assumed to be constant over the spectral range of interest and thus will be neglected. We know that non-zero terms $[(d^n/d\Omega^n)\Psi \neq 0]$ of order $n \geq 2$ are responsible for changes in the complex pulse envelope. In particular

$$\frac{d^2}{d\Omega^2}\Psi(\Omega) = \frac{1}{c}\left(2\frac{dP_{OL}}{d\Omega} + \Omega\frac{d^2 P_{OL}}{d\Omega^2}\right) = \frac{\lambda^3}{2\pi c^2}\frac{d^2 P}{d\lambda^2} \tag{2.61}$$

is related to the GVD parameter.

The relation between angular dispersion and GVD can be derived from the following intuitive approach. Let us consider a light ray which is incident onto an optical element at point Q, as in Fig. 2.19. At this point we do not specify the element, but just assume that it causes angular dispersion. Thus, different spectral components originate at Q under different angles. Two rays corresponding to the center frequency ω_ℓ of the spectrum and to an arbitrary frequency Ω, respectively, are shown in Fig. 2.19. The planes S, S_0 and S', S_0' are perpendicular to the ray direction and represent (plane) wave fronts of the incident light and diffracted light, respectively. Let P_0 be our point of reference and be located on $\vec{r}(\omega_\ell)$ where $\overline{QP_0} = \ell$. A wavefront S' of $\vec{r}(\Omega)$ at

2.6. ELEMENTS WITH ANGULAR DISPERSION

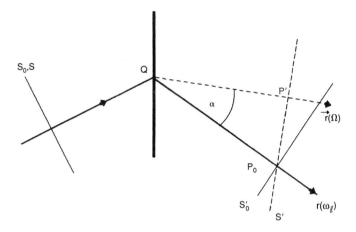

Figure 2.19: Angular dispersion causes GVD.

P' is assumed to intersect $\vec{r}(\omega_\ell)$ at P_0. The optical pathlength $\overline{QP'}$ is thus

$$\overline{QP'} = P_{OL}(\Omega) = P_{OL}(\omega_\ell)\cos\alpha = \ell\cos\alpha \qquad (2.62)$$

which gives for the phase delay

$$\Psi(\Omega) = \frac{\Omega}{c}\ell\cos\alpha \qquad (2.63)$$

The dispersion constant responsible for GVD is obtained by twofold derivation with respect to Ω:

$$\begin{aligned}\left.\frac{d^2\Psi}{d\Omega^2}\right|_{\omega_\ell} &= \left.-\frac{\ell}{c}\left\{\sin\alpha\left[2\frac{d\alpha}{d\Omega}+\Omega\frac{d^2\alpha}{d\Omega^2}\right]+\Omega\cos\alpha\left(\frac{d\alpha}{d\Omega}\right)^2\right\}\right|_{\omega_\ell} \\ &\approx -\frac{\ell\omega_\ell}{c}\left(\left.\frac{d\alpha}{d\Omega}\right|_{\omega_\ell}\right)^2 \end{aligned} \qquad (2.64)$$

where ω_ℓ can be viewed as the center frequency of the pulse ω_ℓ for which $\sin\alpha \ll 1$. The quantity $(d\alpha/d\Omega)|_{\omega_\ell}$, responsible for angular dispersion, is a characteristic of the actual optical device to be considered. It is interesting to note that the dispersion parameter is always negative independently of the sign of $d\alpha/d\Omega$ and that the dispersion increases with increasing distance from the diffraction point. Therefore angular dispersion always results in negative

GVD. Differentiation of Eq. (2.64) results in the next higher dispersion order

$$\left.\frac{d^3\Psi}{d\Omega^3}\right|_{\omega_\ell} = -\frac{\ell}{c}\left\{\cos\alpha\left[3\left(\frac{d\alpha}{d\Omega}\right)^2 + 3\Omega\frac{d\alpha}{d\Omega}\frac{d^2\alpha}{d\Omega^2}\right]\right.$$
$$\left.+ \sin\alpha\left[3\frac{d^2\alpha}{d\Omega^2} + \Omega\frac{d^3\alpha}{d\Omega^3} - \Omega\left(\frac{d\alpha}{d\Omega}\right)^3\right]\right\}\bigg|_{\omega_\ell}$$
$$\approx -\frac{3\ell}{c}\left[\left(\frac{d\alpha}{d\Omega}\right)^2 + \Omega\frac{d\alpha}{d\Omega}\frac{d^2\alpha}{d\Omega^2}\right]\bigg|_{\omega_\ell} \quad (2.65)$$

The most widely used optical devices for angular dispersion are prisms and gratings. To determine the dispersion introduced by them we just need to specify the quantity $\alpha(\Omega)$ in the expressions derived above.

Group velocity dispersion introduced by prisms

In most cases, one wishes to control GVD, and at the same time avoid the beam divergence introduced by angular dispersion. The simplest practicable solution to this problem consists of two identical prisms adjusted for the angular dispersion to be compensated, as in Fig. 2.20. The first prism serves as a disperser producing GVD as discussed above, while the second prism produces a parallel output ray. The apex of the second prism is assumed to be located at the reference point P_0, cf. Fig. 2.19, so that Eq. (2.64) can be applied for the group dispersion. In order to specify $d\alpha/d\Omega$, we use the identity

$$\frac{d\alpha}{d\Omega} = \frac{d\alpha}{dn}\frac{dn}{d\Omega} \quad (2.66)$$

where n is the refractive index of the prism material and

$$\frac{dn}{d\Omega} = -\frac{\lambda^2}{2\pi c}\frac{dn}{d\lambda} \quad (2.67)$$

describes the material dispersion. The quantity $d\alpha/dn$ must be determined from the actual geometry, i.e., apex angle β and angle of incidence φ_0. At symmetrical beam path through the prism for the component ω_ℓ (minimum deviation) and with the apex angle being chosen so that the Brewster condition is satisfied (minimum reflection losses),

$$\frac{d\alpha}{dn} = -2 \quad (2.68)$$

2.6. ELEMENTS WITH ANGULAR DISPERSION

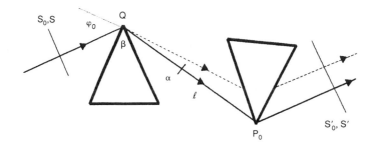

Figure 2.20: Two prism sequence to introduce GVD without net angular dispersion.

and

$$\frac{d^2\alpha}{dn^2} = \frac{2}{n^3} - 4n \tag{2.69}$$

holds (see for example, [9]). Inserting Eqs. (2.66) and (2.68) into Eq. (2.64) we obtain

$$\left.\frac{d^2\Psi}{d\Omega^2}\right|_{\omega_\ell} \approx -4\frac{\omega_\ell}{c}\ell\left(\left.\frac{dn}{d\Omega}\right|_{\omega_\ell}\right)^2 \tag{2.70}$$

or in terms of wavelengths

$$\Psi''(\omega_\ell) = \left.\frac{d^2\Psi}{d\Omega^2}\right|_{\omega_\ell} \approx -4\ell\frac{\lambda_\ell^3}{2\pi c^2}\left(\left.\frac{dn}{d\lambda}\right|_{\lambda_\ell}\right)^2. \tag{2.71}$$

Likewise, for the next higher-order dispersion term we find for this particular beam geometry, from Eq. (2.65):

$$\begin{aligned}\Psi'''(\omega_\ell) &= \left.\frac{d^3\Psi}{d\Omega^3}\right|_{\omega_\ell} \\ &\approx \left.\frac{12\ell\lambda_\ell^4}{(2\pi)^2 c^3}\left\{\left(\frac{dn}{d\lambda}\right)^2\left[1 - \lambda_\ell\frac{dn}{d\lambda}(n^{-3} - 2n)\right] + \lambda_\ell\left(\frac{dn}{d\lambda}\frac{d^2n}{d\lambda^2}\right)\right\}\right|_{\lambda_\ell}.\end{aligned} \tag{2.72}$$

In deriving Eq. (2.72) we neglected terms which are usually of much smaller magnitude than those considered above. To decide whether this prism sequence can be considered to simply introduce GVD, we calculate the ratio of the third- and second-order phase term in the Taylor expansion [Eq. (1.116)] of the phase response:

$$R_P = \left|\frac{b_3(\Omega-\omega_\ell)^3}{b_2(\Omega-\omega_\ell)^2}\right| = \left|\frac{\Psi'''(\omega_\ell)}{3\Psi''(\omega_\ell)}\right||\Omega-\omega_\ell|$$

$$\approx \left[1 - \lambda_\ell\frac{dn}{d\lambda}\bigg|_{\lambda_\ell}\left(\frac{1}{n^3}-2n\right) + \lambda_\ell\frac{\frac{d^2n}{d\lambda^2}|_{\lambda_\ell}}{\frac{dn}{d\lambda}|_{\lambda_\ell}}\right]\frac{\Delta\lambda_\ell}{\lambda_\ell}. \qquad (2.73)$$

Here, $|\Omega-\omega_\ell|$ has been approximated by the spectral width of the pulse $\Delta\omega_\ell$, and we have used the identity $|\Delta\omega_\ell| = (2\pi c)|\Delta\lambda_\ell|/(\lambda_\ell^2)$. Obviously, both the pulse characteristics (relative spectral width $\Delta\lambda_\ell/\lambda_\ell$) and the material parameters determine to what extent the prism sequence can be regarded as a GVD element. As an example, one finds $R_p \approx 0.1$ for 20 fs pulses at 620 nm, and fused silica. In general, the influence of higher-order dispersion is more critical in dispersive devices in the laser resonator, as compared to extracavity application.

So far we have regarded a prism (sequence) as introducing GVD through angular dispersion only. In practice, however, the beam also passes through a certain amount of glass in each prism which contributes to the dispersion. Let the cumulative mean glass path be L; then the additional second-order dispersion parameter is

$$\Psi_L''(\omega_\ell) = \frac{d^2}{d\Omega^2}\left(\frac{\Omega}{c}nL\right)\bigg|_{\omega_\ell}, \qquad (2.74)$$

or, expressed as a function of wavelength:

$$\Psi_L''(\omega_\ell) = \frac{\lambda_\ell^3}{2\pi c^2}L\frac{d^2n}{d\lambda^2}\bigg|_{\lambda_\ell}. \qquad (2.75)$$

The third-order dispersion term for glass is:

$$\Psi_L'''(\omega_\ell) = \frac{d^3}{d\Omega^3}\left(\frac{\Omega}{c}nL\right) = \left(3\frac{d^2n}{d\Omega^2} + \Omega\frac{d^3n}{d\Omega^3}\right)\frac{L}{c}, \qquad (2.76)$$

which is most often expressed as a function of wavelength:

$$\Psi_L'''(\omega_\ell) = -\frac{\lambda_\ell^2}{(2\pi)^2c^3}\left[3\lambda_\ell^2\frac{d^2n}{d\lambda^2} + \lambda_\ell^3\frac{d^3n}{d\lambda^3}\right]L. \qquad (2.77)$$

The total dispersion of the prism sequence is:

$$\Psi_{\text{tot}}''(\omega_\ell) \approx \frac{\lambda_\ell^3}{2\pi c^2}[Ln'' - 4\ell n'^2], \qquad (2.78)$$

2.6. ELEMENTS WITH ANGULAR DISPERSION

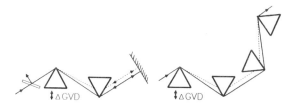

Figure 2.21: Set-ups for adjustable GVD without transverse displacement of spectral components.

and

$$\Psi'''_{\text{tot}}(\omega_\ell) \approx \frac{\lambda_\ell^4}{(2\pi c)^2 c} \left[12\ell \left(n'^2 \left[1 - \lambda_\ell n'(n^{-3} - 2n) \right] + \lambda_\ell n' n'' \right) - L(3n'' + \lambda_\ell n''') \right]. \tag{2.79}$$

To simplify the notation, we have introduced n', n'' and n''' for the derivatives of n with respect to λ taken at λ_ℓ. While the angular dispersion [second term in Eq. (2.78)] always results in negative group velocity dispersion, the finite glass path (first term) gives rise to positive GVD in the visible and near infrared spectral range where $d^2n/d\lambda^2 > 0$. This sign difference in the types of group velocity dispersion offers the possibility of tuning the GVD by changing L. A convenient method is to simply translate one of the prisms perpendicularly to its base, which alters the glass path while keeping the beam deflection constant. It will generally be desirable to avoid a transverse displacement of spectral components at the output (group velocity) of the dispersive device. Two popular prism arrangements which do not separate the spectral components of the pulse are sketched in Fig. 2.21. The beam is either sent through two prism, and retro-reflected by a plane mirror, or sent directly through a sequence of four prisms. In these cases the dispersion doubles.

In this section we have derived approximate expressions for dispersion terms of increasing order. It is also possible by methods of pulse tracing through the prisms to determine the exact phase factor at any frequency and angle of incidence [41, 44, 45, 46, 47]. The more complex studies revealed that the GVD and the ratio R depend on the angle of incidence and apex angle of the prism. In addition, any deviation from the Brewster condition increases the reflection losses. An example is shown in Fig. 2.22. The

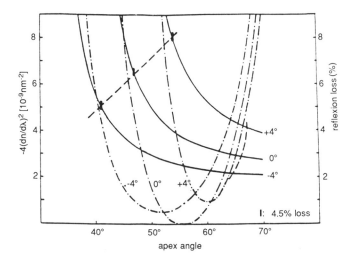

Figure 2.22: GVD parameter (−) and reflection losses (− · −) for a two prism sequence (SQ1-glas/s) and $\lambda = 620$ nm as a function of the apex angle and for three values of the angle of incidence (the deviation from the apex angle corresponding to the Brewster condition at minimum deviation is indicated at the curves: $-4°$, $0°$, $4°$). In the region between the dashed lines the reflection losses are below 4.5% (from [46]).

values of Ψ'', Ψ''', etc. that are best suited to a particular experimental situation can be predetermined through a selection of the optimum prism separation ℓ, the glass pathlength L, and the material (cf. Table 2.1). Such optimization methods are particularly important for the generation of sub-20 fs pulses in lasers [48, 49].

GVD introduced by gratings

Gratings can produce larger angular dispersion than prisms. The resulting negative GVD was first utilized by Treacy to compress pulses of a Nd:glass laser [39]. In complete analogy with prisms, the simplest practical device consists of two identical elements arranged as in Fig. 2.23 for zero net angular dispersion. The dispersion introduced by a pair of parallel gratings can be determined by tracing the frequency dependent ray path. The optical pathlength P_{OL} between A and an output wavefront $\overline{PP_o}$ is frequency dependent and can be determined with help of Fig. (2.23) to be:

$$P_{OL}(\Omega) = \overline{ACP} = \frac{b}{\cos(\beta' + \alpha)} \left[1 + \cos(\beta' + \beta + \alpha)\right] \quad (2.80)$$

2.6. ELEMENTS WITH ANGULAR DISPERSION

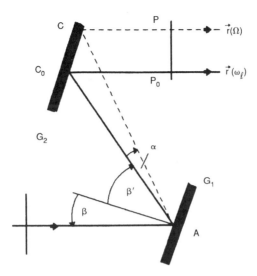

Figure 2.23: Two parallel gratings produce GVD without net angular dispersion. For convenience a reference wavefront is assumed so that the extension of $\overline{PP_0}$ intersects G_1 at A.

where β is the angle of incidence, $\beta' + \alpha$ is the diffraction angle for the frequency component Ω and b is the normal separation between G_1 and G_2. If we restrict our consideration to first-order diffraction, the angle of incidence and the diffraction angle are related through the grating equations

$$\sin\beta - \sin\beta' = -\frac{2\pi c}{\omega_\ell d} \tag{2.81}$$

and

$$\sin\beta - \sin(\beta' + \alpha) = -\frac{2\pi c}{\Omega d} \tag{2.82}$$

where d is the grating constant. The particularity of a grating is to introduce an additional phase shift of 2π at each ruling of G_2 [39]. Because only the relative phase shift across $\overline{PP_0}$ matters, we may simply count the rulings from the (virtual) intersection of the normal in A with G_2. Thus, we find for $\Psi(\Omega)$:

$$\Psi(\Omega) = \frac{\Omega}{c}P_{OL}(\Omega) + 2\pi\frac{b}{d}\tan(\alpha + \beta'). \tag{2.83}$$

Keeping in mind that $\alpha = \alpha(\Omega)$ with $\alpha(\omega_\ell) = 0$ the second derivative of Ψ can be determined from Eqs. (2.80)–(2.83) and reads

$$\left.\frac{d^2\Psi}{d\Omega^2}\right|_{\omega_\ell} = -\frac{\lambda_\ell}{2\pi c^2}\left(\frac{\lambda_\ell}{d}\right)^2 \frac{b}{\sqrt{r}}\frac{1}{r}, \qquad (2.84)$$

where we have introduced the parameter $r = 1 - [2\pi c/(\omega_\ell d) - \sin\beta]^2 = \cos^2\beta'$. In Eq. (2.84), $b/\sqrt{r}$ is the distance between the two gratings along the ray at $\Omega = \omega_\ell$. The third derivative can be written as

$$\left.\frac{d^3\Psi}{d\Omega^3}\right|_{\omega_\ell} = -\frac{3\lambda_\ell}{2\pi cr}\left[r + \frac{\lambda_\ell}{d}\left(\frac{\lambda_\ell}{d} - \sin\beta\right)\right]\left.\frac{d^2\Psi}{d\Omega^2}\right|_{\omega_\ell}. \qquad (2.85)$$

To decide when the third term in the expansion of the phase response of the grating needs to be considered we evaluate the ratio

$$R_G = \left|\frac{b_3(\Omega-\omega_\ell)^3}{b_2(\Omega-\omega_\ell)^2}\right| = \left|\frac{\Psi'''(\omega_\ell)}{3\Psi''(\omega_\ell)}\right||\Omega-\omega_\ell| \approx \frac{\Delta\omega_\ell}{\omega_\ell}\left[1 + \frac{\lambda_\ell/d(\lambda_\ell/d - \sin\beta)}{1 - (\lambda_\ell/d - \sin\beta)^2}\right]. \qquad (2.86)$$

Obviously it is possible to minimize (or tune) the ratio of second- and third-order dispersion by changing the grating constant and the angle of incidence. For instance, with $\Delta\omega_\ell/\omega_\ell = 0.05$, $\lambda_\ell/d = 0.5$ and $\beta = 0°$ we obtain $R_G \approx 0.07$.

Let us next compare Eq. (2.84) with Eq. (2.64), which related GVD to angular dispersion in a general form. From Eq. (2.82) we obtain for the angular dispersion of a grating

$$\left.\frac{d\alpha}{d\Omega}\right|_{\omega_\ell} = -\frac{2\pi c}{\omega_\ell^2 \cos\beta' d} \qquad (2.87)$$

If we insert Eq. (2.87) in Eq. (2.64) and remember that $\ell = b/\cos\beta'$, we also obtain Eq. (2.84).

In practice, gratings rather than prisms are used for the shaping of high-power pulses. This is to avoid undesired nonlinear optical effects in the glass material. However, the total transmission through a grating pair is considerably smaller as compared to a pair of prisms. Usually it does not exceed 80%. Therefore, and because of the easy tunability, prism sequences are used intracavity.

The transverse displacement of the spectral components at the output of the second grating can be compensated by using two pairs of gratings in

2.6. ELEMENTS WITH ANGULAR DISPERSION

Device	λ_ℓ [nm]	ω_ℓ [fs^{-1}]	Ψ'' [fs^{-2}]	Ψ''' [fs^{-3}]
SQ1 ($L = 1$ cm)	620	3.04	550	240
	800	2.36	362	280
Brewster prism pair, SQ1	620	3.04	-760	-1300
$\ell = 50$ cm	800	2.36	-523	-612
grating pair $b = 20$ cm; $\beta = 0°$	620	3.04	-8.2 10^4	1.1 10^5
$d = 1.2$ μm	800	2.36	-3 10^6	6.8 10^6

Table 2.2: Values of second and third order dispersion for typical devices.

sequence or by sending the beam once more through the first grating pair. As with prisms, the overall dispersion then doubles. Tunability is achieved by changing the grating separation b. Unlike with prisms, however, the GVD is always negative. The order of magnitude of the dispersion parameters of some typical devices is compiled in Table 2.2.

Combination of focusing and angular dispersive elements

A disadvantage of prism and grating sequences is that for achieving large GVD the length ℓ between two diffraction elements becomes rather large, cf. Eq.(2.64). As proposed by Martinez et al. [50] the GVD of such devices can be considerably increased (or decreased) by using them in connection with focusing elements such as telescopes. Let us consider the optical arrangment of Fig. 2.24, where a telescope is placed between two gratings, so that its object focal point is at the grating G_1. According to [50] the effect of a telescope can then be understood as follows. Neglecting lens aberrations, the optical beam path between focal points F and F' is independent of α, α'. Therefore, the length which has to be considered for the dispersion reduces from ℓ to

$$z' = \ell - \overline{FF'} = \ell - 2(f + f'). \tag{2.88}$$

In addition, the telescope introduces an angular magnification of $M = f/f'$ which means that the angular dispersion of G_1 is magnified by M to $M(d\alpha/d\lambda)$. For the second grating to produce a parallel output beam its dispersion must be M times larger than that of G_1. With Eq. (2.64) the

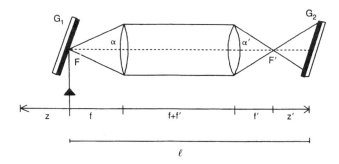

Figure 2.24: Combination of a grating pair and a telescope. In the particular arrangement shown, the object focal point F is at the grating. In general grating 1 can be a distance z away from F. Note: z and z' can be positive as well as negative depending on the relative positions of the gratings and focal points.

overall dispersion is now given by

$$\left.\frac{d^2\Psi}{d\Omega^2}\right|_{\omega_\ell} = -\frac{\omega_\ell}{c}\left(\left.\frac{d\alpha}{d\Omega}\right|_{\omega_\ell}\right)^2 z' M^2 = -\frac{\omega_\ell}{c}\left(\left.\frac{d\alpha}{d\Omega}\right|_{\omega_\ell}\right)^2 [\ell - 2(f + f')]\left(\frac{f}{f'}\right)^2. \tag{2.89}$$

Since it is not very practicable to use a second grating of higher dispersion, one can fold the arrangement by means of a roof prism as is done in standard grating compressors. Also, it is not necessary that the object focal point of the telescope coincide with the diffraction point Q at the grating. A possible separation z between Q and F has then to be added to $z'M^2$ to obtain the overall dispersion from Eq. (2.89):

$$\left.\frac{d^2\Psi}{d\Omega^2}\right|_{\omega_\ell} = -\frac{\omega_\ell}{c}\left(\left.\frac{d\alpha}{d\Omega}\right|_{\omega_\ell}\right)^2 \left(z'M^2 + z\right). \tag{2.90}$$

Equation (2.90) suggests another interesting application of telescopic systems. For $z'M^2 + z < 0$ the dispersion changes sign. The largest amount of positive $d^2\Psi/d\Omega^2$ is achieved for $z = -f$ and $z' = -f'$. Since the sign of the angular dispersion is changed by the telescope we have to tilt the second grating to recollimate the beam. For the folded geometry we can use a mirror instead of a roof prism.

In summary, the use of telescopes in connection with grating or prism pairs allows us to increase or decrease the amount of GVD as well as to change the sign of the GVD. As will be discussed later, interesting applications of such devices include the recompression of pulses after very long

2.7. WAVE-OPTICAL DESCRIPTION

optical fibers [50] and extreme pulse broadening (> 1000) before amplification [51]. A more detailed discussion of this type of dispersers, including the effects of finite beam size, can be found in [43].

2.7 Wave-optical description of angular dispersive elements

Because our previous discussion of pulse propagation through prisms, gratings, and other elements was based on ray-optical considerations, it failed to give details about the influence of a finite beam size. These effects can be included by a wave-optical description which is also expected to provide new insights into the spectral, temporal, and spatial field distribution behind the optical elements. We will follow the procedure developed by Martinez [43], and use the characteristics of Gaussian beam propagation, i.e., remain in the frame of paraxial optics. First, let us analyze the effect of a single element with angular dispersion as sketched in Fig. 2.25. The electric field at the disperser can be described by a complex amplitude $\tilde{U}(x,y,z,t)$ varying slowly with respect to the spatial and temporal coordinate. It is related to the electric field through the relation

$$E(x,y,z,t) = \frac{1}{2}\tilde{U}(x,y,z,t)e^{i(\omega_\ell t - k_\ell z)} + c.c. \qquad (2.91)$$

Using Eq. (1.132) the amplitude at the disperser can be written as

$$\tilde{U}(x,y,t) = \tilde{\mathcal{E}}_0(t)\exp\left[-\frac{ik_\ell}{2\tilde{q}(d)}(x^2+y^2)\right] = \tilde{U}(x,t)\exp\left[\frac{-ik_\ell y^2}{2\tilde{q}(d)}\right] \qquad (2.92)$$

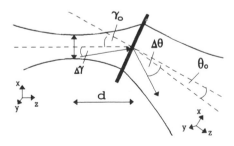

Figure 2.25: Interaction of a Gaussian beam with a disperser.

where $\tilde{q}$ is the complex beam parameter, d is the distance between beam waist and disperser, and $\tilde{\mathcal{E}}_0$ is the amplitude at the disperser. Our convention shall be that x and y refer to coordinates transverse to the respective propagation direction z. Further, we assume the disperser to act only on the field distribution in the x direction, so that the field variation with respect to y is the same as for free space propagation. Hence, propagation along a distance z changes the last term in Eq. (2.92) simply through a change of the complex beam parameter $\tilde{q}$. According to Eq. (1.133) this change is given by

$$\tilde{q}(d+z) = \tilde{q}(d) + z. \qquad (2.93)$$

To discuss the variation of $\tilde{U}(x,t)$ it is convenient to transfer to frequencies $\bar{\Omega}$ and spatial frequencies ρ applying the corresponding Fourier transforms

$$\tilde{U}(x,\bar{\Omega}) = \int_{-\infty}^{\infty} \tilde{U}(x,t) e^{-i\bar{\Omega}t} dt \qquad (2.94)$$

and

$$\tilde{U}(\rho,\bar{\Omega}) = \int_{-\infty}^{\infty} \tilde{U}(x,\bar{\Omega}) e^{-i\rho x} dx. \qquad (2.95)$$

A certain spatial frequency spectrum of the incident beam means that it contains components having different angles of incidence. In terms of Fig. 2.25 this is equivalent to a certain distribution $\Delta\gamma$. The spatial frequency ρ is related to $\Delta\gamma$ through

$$\Delta\gamma = \frac{\rho}{k_\ell}. \qquad (2.96)$$

For a plane wave, $\tilde{U}(\rho,\bar{\Omega})$ exhibits only one non-zero spatial frequency component which is at $\rho = 0$. The disperser not only changes the propagation direction ($\gamma_0 \to \theta_0$) but also introduces a new angular distribution $\Delta\theta$ of beam components which is a function of the angle of incidence γ and the frequency $\bar{\Omega}$

$$\begin{aligned}
\Delta\theta &= \Delta\theta(\gamma,\Omega) \\
&= \left.\frac{\partial\theta}{\partial\gamma}\right|_{\gamma_0} \Delta\gamma + \left.\frac{\partial\theta}{\partial\Omega}\right|_{\omega_\ell} \bar{\Omega} \\
&= \alpha\Delta\gamma + \beta\bar{\Omega}.
\end{aligned} \qquad (2.97)$$

Note that here $\bar{\Omega}$ is the variable describing the spectrum of the envelope while $\Omega = \bar{\Omega} + \omega_\ell$ is the actual frequency of the field. The quantities α and β are characteristics of the disperser and can easily be determined, for

2.7. WAVE-OPTICAL DESCRIPTION

example, from the prism and grating equations.[6] By means of Eq. (2.96) the change of the angular distribution $\Delta\gamma \to \Delta\theta$ can also be interpreted as a transformation of spatial frequencies ρ into spatial frequencies $\rho' = \Delta\theta k_\ell$ where

$$\rho' = \alpha k_\ell \Delta\gamma + k_\ell \beta\bar{\Omega} = \alpha\rho + k_\ell \beta\bar{\Omega}. \tag{2.98}$$

Just behind the disperser we have an amplitude spectrum $\tilde{U}_T(\rho', \bar{\Omega})$ given by

$$\tilde{U}_T(\rho', \bar{\Omega}) = C_1 \tilde{U}\left(\frac{1}{\alpha}\rho' - \frac{k_\ell}{\alpha}\beta\bar{\Omega}, \bar{\Omega}\right) \tag{2.99}$$

where C_1 and further constants C_i to be introduced are factors necessary for energy conservation which shall not be specified explicitly. In spatial coordinates the field distribution reads

$$\begin{aligned}
\tilde{U}_T(x,\bar{\Omega}) &= \int_{-\infty}^{\infty} \tilde{U}_T(\rho',\bar{\Omega}) e^{i\rho' x} d\rho' \\
&= C_1 \int_{-\infty}^{\infty} \tilde{U}\left(\frac{1}{\alpha}\rho' - \frac{k_\ell}{\alpha}\beta\bar{\Omega}, \bar{\Omega}\right) e^{i\rho' x} d\rho' \\
&= C_1 \int_{-\infty}^{\infty} \tilde{U}(\rho,\bar{\Omega}) e^{i\alpha\rho x} e^{ik_\ell \beta\bar{\Omega}x} d(\alpha\rho) \\
&= C_2 e^{ik_\ell \beta\bar{\Omega}x} \tilde{U}(\alpha x, \bar{\Omega}).
\end{aligned} \tag{2.100}$$

The disperser introduces a phase factor $\exp(ik_\ell \beta\bar{\Omega}x)$ and a magnification factor α. For the overall field distribution we obtain with Eqs. (2.92), (2.93), and (2.100)

$$\tilde{U}_T(x,y,\bar{\Omega}) = C_3 \tilde{\mathcal{E}}_0(\bar{\Omega}) e^{ik_\ell \beta\bar{\Omega}x} \exp\left[-\frac{ik_\ell}{2\tilde{q}(d)}(\alpha^2 x^2 + y^2)\right]. \tag{2.101}$$

The field a certain distance ℓ away from the disperser is connected to the field distribution Eq. (2.101) through a Fresnel transformation which describes the free space propagation. Thus, it can be written as

$$\begin{aligned}
\tilde{U}_T(x,y,\ell,\bar{\Omega}) &= C_4 \exp\left[\frac{-ik_\ell y^2}{2\tilde{q}(d+\ell)}\right] \int \tilde{\mathcal{E}}_0(\bar{\Omega}) e^{ik_\ell \beta\bar{\Omega}x'} \\
&\quad \times \exp\left[-i\frac{k_\ell}{2\tilde{q}(d)}\alpha^2 x'^2\right] \exp\left[-\frac{i\pi}{\ell\lambda}(x-x')^2\right] dx'.
\end{aligned} \tag{2.102}$$

---

[6] For a Brewster prism adjusted for minimum deviation we find $\alpha = 1$ and $\beta = -(\lambda^2/\pi c)(dn/d\lambda)$. The corresponding relations for a grating used in diffraction order m are $\alpha = \cos\gamma_0/\cos\theta_0$ and $\beta = -m\lambda^2/(2\pi c d \cos\theta_0)$.

Solving this integral yields an analytical expression for the spectral amplitude

$$\tilde{U}_T(x,y,\ell,\bar{\Omega}) = C_5\tilde{\mathcal{E}}_0(\bar{\Omega})\exp\left[-ik_\ell\frac{x^2}{2\ell}\right]$$
$$\times \exp\left[-ik_\ell\frac{y^2}{2\tilde{q}(d+\ell)}\right]\exp\left\{\frac{ik_\ell\ell}{2}\frac{\tilde{q}(d)}{\tilde{q}(d+\alpha^2\ell)}\left[\frac{x^2}{\ell^2} + 2\frac{\beta x}{\ell}\bar{\Omega} + \beta^2\bar{\Omega}^2\right]\right\}. \quad (2.103)$$

The phase term proportional to $\bar{\Omega}^2$ is responsible for GVD according to our discussion in the section on linear elements. As expected from our ray-optical treatment, this term increases with increasing distance ℓ and originates from angular dispersion β. The term linear in $\bar{\Omega}$ varies with the transverse coordinate x. It describes a frequency variation across the beam and accounts for different propagation directions of different spectral components. We know that exponentials proportional to $b_1\bar{\Omega}$ result in a pulse delay, as discussed previously following Eq. (1.118). Since $b_1 \propto x$ the pulse delay changes across the beam — a feature which we have called tilt of pulse fronts. This proves the general connection between angular dispersion and pulse front tilting introduced earlier in a more intuitive way.

For a collimated input beam and $\alpha = 1$ we can estimate

$$\frac{\tilde{q}(d)}{\tilde{q}(d+\alpha^2\ell)} \approx 1 \quad (2.104)$$

and the temporal delay becomes $b_1 = k_\ell\beta x$. Looking at the beam at a certain instant the corresponding spatial delay is $k_\ell\beta xc$. Thus, we find for the tilt angle α

$$|\tan\alpha| = \left|\frac{d}{dx}(k_\ell\beta xc)\right| = \omega_\ell\left|\frac{d\theta}{d\bar{\Omega}}\right|_{\Omega=\omega_\ell} = \lambda_\ell\left|\frac{d\theta}{d\lambda}\right|_{\lambda_\ell} \quad (2.105)$$

which confirms our previous results. With the same approximation we obtain for the GVD term:

$$\frac{d^2\Psi}{d\bar{\Omega}^2} = 2b_2 = -k_\ell\ell\beta^2 = -\frac{\ell\omega_\ell}{c}\left(\frac{d\theta}{d\Omega}\bigg|_{\omega_\ell}\right)^2 \quad (2.106)$$

in agreement with Eq. (2.64).

For compensating the remaining angular dispersion we can use a properly aligned second disperser which has the parameters $\alpha' = 1/\alpha$ and $\beta' = \beta/\alpha$.

2.7. WAVE-OPTICAL DESCRIPTION

According to our general relation for the action of a disperser (2.100) the new field distribution is given by

$$\begin{aligned}\tilde{U}_F &= C_2 e^{ik_\ell \frac{\beta}{\alpha}\bar{\Omega}x} \tilde{U}_T(\frac{x}{\alpha}, y, \ell, \bar{\Omega}) \\ &= C_6 \tilde{\mathcal{E}}_0(\bar{\Omega}) e^{\frac{i}{2}k_\ell \bar{\Omega}^2 \beta^2 \ell} \exp\left\{\frac{-ik_\ell}{2}\left[\frac{(x+\alpha\beta\bar{\Omega}\ell)^2}{\tilde{q}(d+\alpha^2\ell)} + \frac{y^2}{\tilde{q}(d+\ell)}\right]\right\}\end{aligned} \quad (2.107)$$

which again exhibits the characteristics of a Gaussian beam. Hence, to account for an additional propagation over a distance ℓ', we just have to add ℓ' in the arguments of $\tilde{q}$. As discussed by Martinez [43] $\alpha \neq 1$ gives rise to astigmatism (the position of the beam waist is different for the x and y directions) and only for a well-collimated input beam does the GVD not depend on the travel distance ℓ' after the second disperser. For $\alpha = 1$ and $q(d + \ell + \ell') \approx q(0)$ (collimated input beam) the field distribution becomes

$$\tilde{U}_F(x,y,\ell,\bar{\Omega}) = C_6 \tilde{\mathcal{E}}_0(\bar{\Omega}) e^{\frac{i}{2}k_\ell \beta^2 \ell \bar{\Omega}^2} \exp\left\{-\frac{(x+\beta\bar{\Omega}\ell)^2 + y^2}{w_0^2}\right\}. \quad (2.108)$$

The first phase function is the expected GVD term. The $\bar{\Omega}$ dependence of the second exponential indicates the action of a frequency filter. Its influence increases with increasing $(\beta\bar{\Omega}\ell/w_0)^2$, i.e., with the ratio of the lateral displacement of different frequency components and the original beam waist. The physics behind is that after the second disperser, not all frequency components can interfere over the entire beam cross-section, leading to an effective bandwidth reduction and thus to pulse broadening. If the experimental situation requires even this to be compensated, the beam can be sent through an identical second pair of prisms. Within the approximations introduced above we just have to replace ℓ by 2ℓ in Eqs. (2.107),(2.108). For a well collimated beam $(\beta\bar{\Omega}\ell/w_0 \ll 1)$ this results in

$$\tilde{U}_{F2}(x,y,\ell,\bar{\Omega}) = C_7 \tilde{\mathcal{E}}_0(\bar{\Omega}) e^{ik_\ell \beta^2 \ell \bar{\Omega}^2} e^{-(x^2+y^2)/w_0^2}. \quad (2.109)$$

In this (ideal) case the only modification introduced by the dispersive element is the phase factor leaving the beam characteristics unchanged. It is quite instructive to perform the preceding calculation with temporally chirped input pulse as in Eq. (1.33) having a Gaussian spatial as well as temporal profile [43]:

$$\tilde{\mathcal{E}}_0(t) = \mathcal{E}_0 e^{-(1+ia)(t/\tau_G)^2} e^{-(x^2+y^2)/w_0^2}. \quad (2.110)$$

The (temporal) Fourier transform yields

$$\tilde{\mathcal{E}}_0(\bar{\Omega}) = C_8 \exp\left(i\frac{\bar{\Omega}^2 \tilde{\tau} a}{4}\right) \exp\left(-\frac{\bar{\Omega}^2 \tilde{\tau}^2}{4}\right) \exp\left(-\frac{x^2 + y^2}{w_0^2}\right) \quad (2.111)$$

with $\tilde{\tau}^2 = \tau_G^2/(1+a^2)$, where according to our discussion following Eq. (1.39) $\tau_G/\tilde{\tau}$ is the maximum possible shortening factor after chirp compensation. This pulse is to travel through an ideal two-prism sequence described by Eq. (2.108) where $\beta\ell$ has to be chosen so as to compensate exactly the quadratic phase term of the input pulse Eq. (2.111). Under this condition the insertion of Eq. (2.111) into Eq. (2.108) yields

$$\tilde{U}_F(x, y, \bar{\Omega}, l) = C_9 e^{-\bar{\Omega}^2 \tilde{\tau}^2/4} \exp\left[-\frac{(x + \beta\bar{\Omega}\ell)^2 + y^2}{w_0^2}\right]. \quad (2.112)$$

The time dependent amplitude obtained from Eq. (2.112) after inverse Fourier transform is

$$\tilde{\mathcal{E}}(t) = C_{10} \exp\left(\frac{-t^2}{\tilde{\tau}^2(1+u^2)}\right) \exp\left[\left(\frac{-x^2}{(1+u^2)w_0^2} - \frac{y^2}{w_0^2}\right)\right] \exp\left[\frac{-iu^2 xt}{(1+u^2)\beta\ell}\right], \quad (2.113)$$

where $u = 2\beta\ell/(\tilde{\tau}w_0)$. The last exponential function accounts for a frequency sweep across the beam which prevents the different frequency components from interfering completely. As a result, the actual shortening factor is $\sqrt{1+u^2}$ times smaller than the theoretical one, as can be seen from the first exponential function. The influence of such a filter can be decreased by using a large beam size. A measure of this frequency filter, i.e., the magnitude of the quantity $(1+u^2)$, can be derived from the second exponent. Obviously the quantity $(1+u^2)$ is responsible for a certain ellipticity of the output beam which can be measured.

2.8 Optical matrices for dispersive systems

In Chapter 1 we pointed out the similarities between Gaussian beam propagation and pulse propagation. Even though this fact has been known for many years [52, 39], it was only recently that optical matrices have been introduced to describe pulse propagation through dispersive systems [53, 54, 55, 56, 57] in analogy to optical ray matrices. The advantage of such an approach is that the propagation through a sequence of optical elements can

2.8. OPTICAL MATRICES FOR DISPERSIVE SYSTEMS

be described using matrix algebra. Dijaili et al. [56] defined a 2 × 2 matrix for dispersive elements which relates the complex pulse parameters (cf. Table 1.2) of input and output pulse, $\tilde{p}$ and $\tilde{p}'$, to each other. Döpel [53] and Martinez [55] used 3 × 3 matrices to describe the interplay between spatial (diffraction) and temporal (dispersion) mechanisms in a variety of optical elements, such as prisms, gratings and lenses, and in combinations of them. The advantage of this method is the possibility to analyze complicated optical systems such as femtosecond laser cavities with respect to their dispersion — a task of increasing importance, as attempts are being made to propagate ultrashort pulses near the bandwidth limit through complex optical systems. The actual analysis is difficult since the matrix elements contain information about the optical system as well as of the pulse. To date the most comprehensive approach to describe ray and pulse characteristics in optical elements by means of matrices is that of Kostenbauder [57]. He defined 4 × 4 matrices which connect the input and output ray and pulse coordinates to each other. As in ray optics, all information about the optical system is carried in the matrix while the spatial and temporal characteristics of the pulse are represented in a ray-pulse vector (x, Θ, t, ν). Its components are defined by position x, slope Θ, time t and frequency ν. These coordinates have to be understood as difference quantities with respect to the coordinates of a reference pulse. The spatial coordinates are similar to those known from the ordinary $\begin{pmatrix} A & B \\ C & D \end{pmatrix}$ ray matrices. However, the origin of the coordinate system is defined now by the path of a diffraction limited reference beam at the average pulse frequency. This reference pulse has a well-defined arrival time at any reference plane; the coordinate t, for example, is then the difference in arrival time of the pulse under investigation. In terms of such coordinates and using a 4 × 4 matrix, the action of an optical element can be written as

$$\begin{pmatrix} x \\ \Theta \\ t \\ \nu \end{pmatrix}_{out} = \begin{pmatrix} A & B & 0 & E \\ C & D & 0 & F \\ G & H & 1 & I \\ 0 & 0 & 0 & 1 \end{pmatrix} \begin{pmatrix} x \\ \Theta \\ t \\ \nu \end{pmatrix}_{in} \quad (2.114)$$

where A, B, C, D are the components of the ray matrix and the additional elements are

$$E = \frac{\partial x_{out}}{\partial \nu_{in}}, \quad F = \frac{\partial \Theta_{out}}{\partial \nu_{in}}, \quad G = \frac{\partial t_{out}}{\partial x_{in}}, \quad H = \frac{\partial t_{out}}{\partial \Theta_{in}}, \quad I = \frac{\partial t_{out}}{\partial \nu_{in}}. \quad (2.115)$$

The occurrence of the zero-elements can easily be explained using simple physical arguments, namely (i) the center frequency must not change in a linear (time invariant) element and (ii) the ray properties must not depend on t_{in}. It can be shown that only six elements are independent of each other [57] and therefore three additional relations between the nine nonzero matrix elements exists. They can be written as

$$AD - BC = 1$$
$$BF - ED = \lambda_\ell H \qquad (2.116)$$
$$AF - EC = \lambda_\ell G.$$

Using the known ray matrices [14] and Eq. (2.115), the ray-pulse matrices for a variety of optical systems can be calculated. Some important examples are given in Table 2.3. More examples can be found in [57, 58]. An arbitrary sequence of these elements can be described by the ordered product of the corresponding matrices defining the system matrix. As we have seen in our previous discussion an important feature of a system of dispersive elements is the frequency dependent optical beam path P, and the corresponding phase delay Ψ, respectively. In particular, this information was sufficient for geometries which did not introduce a change in the beam parameters. Examples which have been discussed in this respect are four-prism and four-grating sequences illuminated by a well-collimated beam.

As shown in Ref. [57], Ψ can be expressed in terms of the coordinates of the system matrix as

$$\Psi = \frac{\pi \nu^2}{B}(EH - BI) - \frac{\pi}{B\lambda_\ell}L(\nu) \qquad (2.117)$$

where

$$L(\nu) = \begin{pmatrix} x_{in} & x_{out} \end{pmatrix} \begin{pmatrix} A & -1 \\ -1 & D \end{pmatrix} \begin{pmatrix} x_{in} \\ x_{out} \end{pmatrix} + 2\begin{pmatrix} E & \lambda_0 H \end{pmatrix} \begin{pmatrix} x_{in} \\ x_{out} \end{pmatrix} \qquad (2.118)$$

and x_{in}, x_{out} are the position coordinates of the input and output vectors, respectively. The argument ν of L and Ψ is the cyclic frequency coordinate relative to the pulse central frequency $\nu = (\Omega - \omega_\ell)/2\pi$. The calculations according to Eq. (2.117) have to be repeated for a set of frequencies to obtain $\Psi(\nu)$. From $\Psi(\nu)$ we can then determine chirp and temporal behavior of the output pulses using relation (1.114) for linear elements with $R = 1$. For

2.8. OPTICAL MATRICES FOR DISPERSIVE SYSTEMS

Lens or Mirror ($\mathcal{M}_L$)	**Brewster Prism** ($\mathcal{M}_{BP}$)
$\begin{pmatrix} 1 & 0 & 0 & 0 \\ -1/f & 1 & 0 & 0 \\ 0 & 0 & 1 & 0 \\ 0 & 0 & 0 & 1 \end{pmatrix}$	$\begin{pmatrix} 1 & L/n^3 & 0 & -\frac{RL}{n^3} \\ 0 & 1 & 0 & -2R \\ -\frac{2R}{\lambda_\ell} & -\frac{LR}{n^3\lambda_\ell} & 1 & \frac{LR^2}{n^3\lambda_\ell} + 2\pi L k_\ell'' \\ 0 & 0 & 0 & 1 \end{pmatrix}$
f — focal length	$R = 2\pi \left.\frac{\partial n}{\partial \Omega}\right|_{\omega_\ell}$, L — mean glass path
Dispersive Slab ($\mathcal{M}_{DS}$)	**Grating** ($\mathcal{M}_G$)
$\begin{pmatrix} 1 & L/n & 0 & 0 \\ 0 & 1 & 0 & 0 \\ 0 & 0 & 1 & 2\pi L k_\ell'' \\ 0 & 0 & 0 & 1 \end{pmatrix}$	$\begin{pmatrix} -\frac{\cos\beta'}{\cos\beta} & 0 & 0 & 0 \\ 0 & -\frac{\cos\beta}{\cos\beta'} & 0 & \frac{c(\sin\beta' - \sin\beta)}{\lambda_\ell \sin\beta'} \\ \frac{\sin\beta - \sin\beta'}{c\sin\beta} & 0 & 1 & 0 \\ 0 & 0 & 0 & 1 \end{pmatrix}$
$k_\ell'' = \left.\frac{d^2 k}{d\Omega^2}\right|_{\omega_\ell}$, L — thickness of slab	β — angle of incidence, β' — diffraction angle

Table 2.3: Examples of Ray-Pulse Matrices

pulses incident on-axis ($x_{in} = 0$) Eq. (2.118) yields for the phase response

$$\Psi_M(\nu) = \frac{1}{4\pi B}\left[\left(EH - BI - \frac{1}{\lambda_\ell}DE^2\right)\nu^2 - 4\pi EH\nu\right], \qquad (2.119)$$

where the index M is to express the derivation of the phase response from the ray-pulse matrix. Information about the temporal broadening can also be gained directly from the matrix element I since $t_{out} = t + I\, t_{in}$ where $I = I(\nu)$. Different frequency components need different times to travel from the input to the exit plane which gives an approximate broadening of $\frac{dI}{d\nu}\Delta\nu$ for a bandwidth limited input pulse with a spectral width $\Delta\omega = 2\pi\Delta\nu$. For on-axis propagation ($x_{in} = x_{out} = 0$) we find $L(\nu) = 0$ and the dispersion is given by the first term in Eq. (2.117). For a dispersive slab, for example, we find from Table 2.3:

$$\Psi_M = \frac{1}{2} L k_\ell''(\Omega - \omega_\ell)^2 \tag{2.120}$$

which agrees with Eq. (1.121) and the corresponding discussion.

As another example let us discuss the action of a Brewster prism at minimum deviation and analyze the ray-pulse at a distance ℓ behind it. The system matrix is the product of $(\mathcal{M}_{BP})$ and $(\mathcal{M}_{DS})$ for free space, which is given by

$$\begin{pmatrix} 1 & B+\ell & 0 & E+F\ell \\ 0 & 1 & 0 & F \\ G & H & 1 & I \\ 0 & 0 & 0 & 1 \end{pmatrix}. \tag{2.121}$$

For the sake of simplicity the elements of the prism matrix have been noted $A, B, \ldots, H$. For the new position and time coordinate we obtain

$$x_{out} = x_{in} + (B+\ell)\Theta_{in} + (E+F\ell)\nu \tag{2.122}$$

and

$$t_{out} = G x_{in} + H \Theta_{in} + t_{in} + I \nu. \tag{2.123}$$

Let us next verify the tilt of the pulse fronts derived earlier. The pulse front tilt can be understood as an arrival time difference t_{out} which depends on the transverse beam coordinate x_{out}. The corresponding tilt angle α' is then

$$\tan \alpha' = \frac{\partial (c t_{out})}{\partial x_{out}} = c \frac{\partial t_{out}}{\partial x_{in}} \frac{\partial x_{in}}{\partial x_{out}} = cG. \tag{2.124}$$

After we insert G for the Brewster prism, the tilt angle becomes (cf. Table 2.2):

$$\tan \alpha' = -2\omega_\ell \left. \frac{\partial n}{\partial \Omega} \right|_{\omega_\ell} = \frac{2}{\lambda_\ell} \left. \frac{\partial n}{\partial \lambda} \right|_{\lambda_\ell}. \tag{2.125}$$

This result is equivalent to Eq. (2.57) if we use $a/b = 2$, which is valid for Brewster prisms. The different signs result from the direction of the x-axis chosen here.

As a final example we want to apply the matrix formalism to discuss the field distribution behind a two-prism sequence used for pulse compression, such as the one sketched in Fig. 2.20. We assume that one prism is traversed at the apex while the second is responsible for a mean glass path L. The

2.8. OPTICAL MATRICES FOR DISPERSIVE SYSTEMS

corresponding system matrix is obtained by multiplying matrix (2.121) from the left with the transposed[7] matrix of a Brewster prism. The result is

$$\begin{pmatrix} 1 & \frac{L}{n^3}+\ell & 0 & -R\left[\frac{L}{n^3}+2\ell\right] \\ 0 & 1 & 0 & 0 \\ 0 & \frac{R}{\lambda_\ell}\left[\frac{L}{n^3}+2\ell\right] & 1 & -\frac{R^2}{\lambda_\ell}\left[\frac{L}{n^3}+4L\right]-2\pi k''_\ell L \\ 0 & 0 & 0 & 1 \end{pmatrix}. \quad (2.126)$$

To get an expression for the dispersion consistent with our ealier derivation, we make the assumption that $L \ll \ell$. This allows us to neglect terms linear in L in favor of those linear in ℓ whenever they appear in a summation. For the second derivative of the phase response (2.119) we then find

$$\Psi''(\omega_\ell) = L k''_\ell - \frac{8\pi}{\lambda_\ell}\ell\left(\left.\frac{dn}{d\Omega}\right|_{\omega_\ell}\right)^2 \quad (2.127)$$

in agreement with Eq. (2.78).

It is well known that ray matrices can be used to describe Gaussian beam propagation, e.g., [14], where the beam parameter of the output beam is connected to the input parameter by

$$\tilde{q}_{out} = \frac{A\tilde{q}_{in}+B}{C\tilde{q}_{in}+D}. \quad (2.128)$$

Kostenbauder [57] showed that, in a similar manner, the ray-pulse matrices contain all information which is necessary to trace a generalized Gaussian beam through the optical system. Using a 2 × 2 complex "beam" matrix $(\tilde{Q}_{in})$, the amplitude of a generalized Gaussian beam is of the form

$$\exp\left[-\frac{i\pi}{\lambda_\ell}\begin{pmatrix} x_{in} & x_{out} \end{pmatrix}\left(\tilde{Q}_{in}\right)^{-1}\begin{pmatrix} x_{in} \\ t_{in} \end{pmatrix}\right] \quad (2.129)$$

which explicitly varies as

$$\exp\left[-\frac{i\pi}{\lambda_\ell}\left(\tilde{Q}^r_{xx}x_{in}^2 + 2\tilde{Q}^r_{xt}x_{in}t_{in} - \tilde{Q}^r_{tt}t_{in}^2\right)\right]$$
$$\times \exp\left[\frac{\pi}{\lambda_\ell}\left(\tilde{Q}^i_{xx}x_{in}^2 + 2\tilde{Q}^i_{xt}x_{in}t_{in} - \tilde{Q}^i_{tt}t_{in}^2\right)\right], \quad (2.130)$$

---

[7]Note that the second prism has an orientation opposite to the first one.

where $\tilde{Q}_{ij}^r$, $\tilde{Q}_{ij}^i$ are the real and imaginary coordinates of the matrix $(\tilde{Q}_{in})^{-1}$ and $\tilde{Q}_{xt} = -\tilde{Q}_{tx}$. The first factor in Eq.(2.130) expresses the phase behavior and accounts for the wave front curvature and chirp. The second term describes the spatial and temporal beam (pulse) profile. Note that unless $\tilde{Q}_{xt}^{r,i} = 0$ the diagonal elements of $(\tilde{Q}_{in})$ do not give directly such quantities as pulse duration, beam width, chirp parameter, and wave front curvature. One can show that the field at the output of an optical system is again a generalized Gaussian beam where in analogy to (2.128) the generalized beam parameter $(\tilde{Q}_{out})$ can be written as

$$(\tilde{Q}_{out}) = \frac{\begin{pmatrix} A & 0 \\ G & 1 \end{pmatrix}(\tilde{Q}_{in}) + \begin{pmatrix} B & E/\lambda_\ell \\ H & I/\lambda_\ell \end{pmatrix}}{\begin{pmatrix} C & 0 \\ 0 & 0 \end{pmatrix}(\tilde{Q}_{in}) + \begin{pmatrix} D & F/\lambda_\ell \\ 0 & 1 \end{pmatrix}}. \qquad (2.131)$$

The evaluation of such matrix equations is quite complex since it generally gives rather large expressions. However, the use of advanced algebraic formula-manipulation computer codes makes this approach practicable.

2.9 Problems

1. Consider the Gires–Tournois interferometer. (a) As explained in the text the reflectivity is $R = \text{constant} = 1$, while the phase shows a strong variation with frequency. Does this violate the Kramers-Kronig relation? Explain your answer. (b) Derive the transfer function (2.18).

2. Derive an expression for the space-time intensity distribution of a pulse in the focal plane of a chromatic lens of focal length $f(\lambda)$. To obtain an analytical formula make the following assumptions. The input pulse is bandwidth-limited and exhibits a Gaussian temporal and transverse spatial profile. The lens has an infinitely large aperture and the GVD can be neglected. [Hint: You can apply Gaussian beam analysis for each spectral component to obtain the corresponding field in the focal plane. Summation over spectral contributions (Fourier back-transform) gives then the space-time field distribution.]

3. Suppose you have to implement a pair of prisms into the cavity of a femtosecond pulse laser at 620 nm. This is to provide GVD tunability of 20% around -800 fs^2 where the next higher-order dispersion

2.9. PROBLEMS

should be as small as possible. With the help of Table 2.1 choose a suitable prism material, calculate the apex angle of the prisms for the Brewster condition at symmetric beam path, and determine the prism separation. If needed, assume a beam diameter of 2 mm to estimate a minimum possible glass path through the prisms.

4. Derive the ray-pulse matrix (2.126) for a pair of Brewster prisms and verify the second-order dispersion given in relation (2.127).

5. Derive the delay and aberration parameter of a spherical mirror as given in Eqs. (2.42) and (2.43). Explain physically what happens if a parallel input beam impinges on the mirror with a certain angle α.

6. A parallel beam with plane pulse fronts impinges on a circular aperture with radius R centered on the optic axis. The pulse is unchirped and Gaussian. Estimate the frequency shift that the diffracted pulse experiences if measured with a detector placed on the optic axis. Give a physical explanation of this shift. Make a numerical estimate for a 100 fs and a 10 fs pulse. Can this effect be used to obtain ultrashort pulses in new spectral regions by placing diffracting apertures in series? [Hint: you can start with Eq. (2.37) and take out the lens terms. For mathematical ease you can let $R \to \infty$).] Note that a frequency shift (of the same origin) occurs when the on-axis pulse spectrum of a Gaussian beam is monitored along its propagation path.

Chapter 3

Light–Matter Interaction

The generation and application of fs light pulses revolve around light-matter interaction. In the preceding chapter we considered situations in which the propagating field does not change the material response. The result was a linear dependence of the output field on the input field, a feature attributed to linear optical elements. The medium could be described by a complex dielectric constant $\tilde{\epsilon}(\Omega)$ and the element by a transfer function $\tilde{H}(\Omega)$. The real part of $\tilde{\epsilon}(\Omega)$ is responsible for dispersion, determining phase and group velocity, for example. The imaginary part describes (frequency dependent) loss or gain. In many cases these linear optical media can be considered as host materials for sources of nonlinear polarization. It is the latter which shall be discussed in this section.

Let us consider a pulse propagating through a (linear) optical material, for instance glass. In addition to the processes described above, we will have to consider nonlinear optical interaction if the electric field strength is high. This can result from high pulse power or/and tight focusing. The decomposition of the polarization according to Eq. (1.41), $P = P^L + P^{NL}$, accounts then for different optical properties of one and the same material. Another possible situation is that the host material affects the pulse through the linear optical response only, but contains additional substances interacting nonlinearly with the pulse. This, for example, can occur if the light pulse is at resonance with a dopand material, which considerably increases the interaction strength. Examples of such composite materials are glass doped with ions, dye molecules dissolved in a (transparent) solvent, and resonant gas molecules surrounded by a (nonresonant) buffer gas. In these cases the contributions, P^L and P^{NL}, belong to different materials.[1] Finally, there can be situations where the nonresonant (host) material as well as a resonant (second) material produce a nonlinear polarization. Here P^{NL} may be decomposed into a nonresonant $P^{N,NL}$ and resonant part $P^{R,NL}$:

$$P^{NL} = P^{N,NL} + P^{R,NL}. \tag{3.1}$$

3.1 Density matrix equations

The semiclassical treatment has been most widely applied to discuss the interaction of short light pulses with matter. The electromagnetic field is used as a classical quantity, whereas the matter is described in the frame of quantum mechanics. This both allows a simple interpretation of the results in terms of measurable quantities and accounts for the quantum properties of matter. We will restrict ourselves to atomic particles (atoms, molecules, etc.) with a transition at resonance with the incident light pulse and we will idealize the resonant medium as an ensemble of two-level systems. By "resonant" we mean that there is sufficient overlap of the pulse spectrum and the transition profile for the pulse to experience absorption or gain. If not specified differently, we will refer to a volume element at a certain space coordinate z. This allows us to consider only the time dependence of the sample response.

Of course, what we have to take into account is the interaction of the resonant particles with each other and/or with surrounding nonresonant particles. The physical nature of such interactions can be manyfold; for instance, collisions or changing local fields due to particle motion may contribute. Generalizing, one says that the resonant system is coupled to a dissipative system which in turn is regarded as a reservoir with a large number of degrees of freedom. The result is a time dependence of certain physical properties of the resonant particle, which implies a time dependence of the interaction strength with an incident electric field.

We should keep in mind that, typically, the light pulse interacts with a large number of particles ($\approx 10^5$–10^{23}cm^{-3}). We can therefore exclude the possibility of calculating the dynamic behavior of each individual ensemble member and summing up for the macroscopic sample response. But do we really need to consider individual particles? The answer to this question depends upon the time scale on which we look at the medium and,

---

[1] Of course, resonant matter also may interact purely linearly with the light pulse if the field strength is sufficiently small (linear loss and gain).

3.1. DENSITY MATRIX EQUATIONS

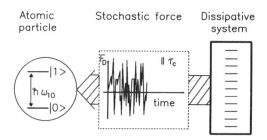

Figure 3.1: Resonant atomic particle under the influence of a dissipative system exerting a stochastic force with correlation time τ_c.

fortunately, in many cases is "no." The physical reason is the random character of the forces acting on the resonant atoms, as illustrated in Fig. 3.1, which is quite understandable if we think of the Brownian molecular motion. Mathematically, this force can be represented by a stochastic function $F_D(t)$ characterized by a certain correlation time τ_c. The latter means that the correlation of first order

$$A(\tau) = \int_{-\infty}^{+\infty} F_D(t') F_D(t' - \tau) \, dt' \qquad (3.2)$$

has appreciable values only for $|\tau| < \tau_c$. In other words, there is no net force on the particle under investigation on a time scale $t > \tau_c$; a situation we will refer to as stationary state. If we assume homogeneity of the medium, all resonant particles will experience similar distortions by the dissipative system. In order to decide whether the particles have identical properties, we have to perform a measurement, for example, to determine the absorptivity or reflectivity. Since each measurement requires a certain time T_M, the result is a sample property, $\bar{X}$, averaged over T_M:

$$\bar{X}(t) = \frac{1}{T_M} \int_{t-T_M/2}^{t+T_M/2} X(t') \, dt'. \qquad (3.3)$$

This averaging makes evident the difficulties of defining identical properties. For measuring times (e.g., given by the interaction time with a light pulse) $T_M > \tau_c$, we expect to see identical behavior for all ensemble members. For $T_M < \tau_c$ the stochastic influence of the dissipative system does not average to zero and the atomic particles show "individuality." The magnitude of τ_c depends on the actual medium; in condensed matter, at room temperature, it is mostly below 10^{-14} s.

In general, it is impossible to calculate the behavior of the resonant particle under the influence of the dissipative system exactly, no matter what time scale is of interest. The reasons are the complexity of the dissipative system and its large number of degrees of freedom. In terms of quantum mechanics we have to deal with a weakly prepared system [59] which can be described favorably by its density operator $\hat{\rho}$ and the corresponding density matrix equations, respectively. If the action of the dissipative system can be regarded as a perturbation for the two-level system, the elements of the density matrix obey the following equations of motion, see e.g., [11]:

$$\frac{d}{dt}\rho_{10}(t) = -i\omega_{10}\rho_{10}(t) - \frac{1}{T_2}\rho_{10}(t) \tag{3.4}$$

$$\frac{d}{dt}\rho_{11}(t) = -\frac{1}{T_1}[\rho_{11}(t) - \rho_{11}^{(e)}] \tag{3.5}$$

$$\rho_{01}(t) = \rho_{10}^*(t) \tag{3.6}$$

$$\rho_{00}(t) = 1 - \rho_{11}(t) \tag{3.7}$$

where T_1, T_2 are the energy (longitudinal) and phase (transverse) relaxation time, respectively. This system of equations describes the evolution of the two-level system towards a stationary state (e.g., thermal equilibrium) denoted by the superscript (e). The elements of the density matrix, ρ_{kl} ($k, l = 0, 1$), can be calculated by means of the eigenstates of the (unperturbed) two-level system $1\rangle$ and $|2\rangle$ and read

$$\rho_{kl} = \langle k|\hat{\rho}|l\rangle \qquad k, l = 0, 1. \tag{3.8}$$

With the density operator $\hat{\rho}$ given any measurable quantity X can be determined as an expectation value of the corresponding observable $\hat{X}$

$$X = Tr(\hat{\rho}\hat{X}). \tag{3.9}$$

Assuming a homogeneous distribution of two-level systems of number density $\bar{N}$, the occupation number density of level k is

$$\gamma_k = \bar{N}Tr(\hat{\rho}\hat{\gamma}_k) = \bar{N}\rho_{kk} \qquad k = 0, 1 \tag{3.10}$$

and the polarization can be written as

$$P = \bar{N}Tr(\hat{\rho}\hat{p}) = \bar{N}(p_{10}\rho_{01} + p_{01}\rho_{10}). \tag{3.11}$$

Here $\hat{\gamma}_{0,1}$ is the occupation number operator and $\hat{p}$ is the atomic dipole operator with the matrix elements p_{10} and p_{01}. Obviously, the diagonal

3.1. DENSITY MATRIX EQUATIONS

elements of the density matrix ρ_{11} and ρ_{00} determine the probability of finding the system in state $|0\rangle$ and $|1\rangle$, respectively.

Let us assume that at $t = 0$ the system can be described by the initial conditions $\gamma_{0,1}(0)$ and $P(0)$. For $t > 0$ the evolution of the occupation numbers and polarization according to Eqs. (3.4)–(3.7) is then given by

$$P(t) \propto P(0)e^{-2t/T_2} \cos(\omega' t) \tag{3.12}$$

and

$$\gamma_{0,1}(t) = \gamma_{0,1}^{(e)} + \left[\gamma_{0,1}(0) - \gamma_{0,1}^{(e)}\right] e^{-t/T_1}. \tag{3.13}$$

As can be seen the polarization behaves like a damped harmonic oscillator where T_2^{-1} plays the role of a damping constant and $\omega' = \sqrt{\omega_{10}^2 + T_2^{-2}} \approx \omega_{10}$ is the oscillation frequency. Thus in the stationary state $(t \to \infty)$ the polarization is zero.

The relaxation of the polarization is not necessarily associated with a relaxation of the energy of the two-level system. Instead, T_2 can be regarded as the phase memory time of the two-level system, i.e., as the time interval in which the two-level system remembers the phase of the oscillation at $t = 0$ (excitation event). It is T_1 which describes the recovery of the occupation numbers towards their equilibrium values $\gamma_{0,1}^{(e)}$ and thus determines an energy relaxation. The convenience of the density matrix equations (3.4)–(3.7) is obvious; the dissipative system enters through two characteristic relaxation constants only. However, to derive Eqs. (3.4)–(3.7), it is necessary to perform an averaging over time intervals τ_c which limits the range of validity. Therefore, with the view on ultrashort pulse interaction, we have to be aware that the temporal resolution with which we can trace the dynamics of the two-level system with Eqs. (3.4) – (3.7) is always worse than τ_c.

What determines these relaxation rates and what are typical values? For the reasons mentioned above, an exact calculation of T_1, T_2 is rather difficult if not impossible for many systems of practical relevance. These parameters have a complex dependence on particle density, type of interaction, particle velocity etc.. As will be discussed later, measurements with ultrashort light pulses can yield the desired information directly. A limiting case occurs for isolated two-level systems which are not influenced by surrounding particles. The only dissipative system acting on the atomic particle is the vacuum field, causing natural line broadening which implies $T_2 = 2T_1$. Generally speaking, whenever the interaction with the dissipative system is such that each event leads to an occupation change, $T_2 = 2T_1$ holds. Here the dephasing time is determined by the energy relaxation time. However, there are

Medium	T_1 [s]	T_2 [s]	$\sigma_{01}^{(0)}$ [cm^2]
solids doped with resonant atomic systems	$10^{-3} - 10^{-6}$	$10^{-11} - 10^{-14}$	$10^{-19} - 10^{-21}$
dye molecules solved in an organic solvent	$10^{-8} - 10^{-12}$	$10^{-13} - 10^{-14}$	10^{-16}
semiconductors	$10^{-4} - 10^{-12}$	$10^{-12} - 10^{-14}$	

Table 3.1: Typical energy and phase relaxation times and interaction cross sections of some materials

many other interaction processes which do not change the occupation numbers, but which may affect the phase of the oscillations. These additional relaxation channels for the phase are responsible for the fact that most often (particularly in condensed matter) $T_2 < (<)T_1$. Table 3.1 offers some examples. It should also be noted that the transition profile resulting from the steady–state solution of Eqs. (3.4)–(3.7) implies a Lorentzian line shape with a FWHM given by $2T_2^{-1}$.

So far we have considered an ensemble of identical atoms, which enabled us to calculate the macroscopic quantities by multiplying the contribution of one particle by the particle number (density), as shown in Eqs. (3.10) and (3.11). The result is a transition profile which has exactly the same shape and width as those of a single atom. Such a medium is referred to as *homogeneously broadened*. However, in many real situations, different particles from an ensemble of identical atoms have (slightly) different resonance frequencies ω'_{10}. Among the various causes for such a distribution of frequencies, the most common are Doppler shift in gases, and different local surroundings in solids (e.g. defects, impurities). The total response of the medium is now given by the sum of the responses of all sub-ensembles with polarization $P'(\omega'_{10})$ and occupation numbers $\gamma'_{0,1}(\omega'_{10})$. A sub-ensemble characterized by a certain transition frequency ω'_{10} is to contain particles with resonance frequencies which fall within a homogeneous linewidth. Thus, for the total polarization we obtain

$$P(t) = \sum_{\omega'_{10}} P'(t, \omega'_{10}), \qquad (3.14)$$

3.1. DENSITY MATRIX EQUATIONS

and likewise the total number of particles in state $|0\rangle$ or $|1\rangle$ is

$$\gamma_{0,1}(t) = \sum_{\omega'_{10}} \gamma'_{0,1}(t, \omega'_{01}). \tag{3.15}$$

Frequently the distribution of atomic species with a certain transition frequency can be described by a distribution function $g_{inh}(\omega'_{10} - \omega_{ih})$ centered at ω_{ih} and having a FWHM of $\Delta\omega_{inh}$. This line shape is referred to as *inhomogeneous* (as opposed to the homogeneous broadening cited earlier). The polarization and occupation number densities which result from particles having resonance frequencies in the interval $(\omega'_{10}, \omega'_{10} + d\omega'_{10})$ can be written as

$$P'(t,\omega'_{10}) = \bar{N} g_{inh}(\omega'_{10} - \omega_{ih})[p_{01}\rho_{10}(t,\omega'_{10}) + p_{10}\rho_{01}(t,\omega'_{10})]d\omega'_{10} \tag{3.16}$$

and

$$\gamma'_{0,1}(t,\omega'_{10}) = \bar{N} g_{inh}(\omega'_{10} - \omega_{ih})\rho_{00,11}(t,\omega'_{10})d\omega'_{10} \tag{3.17}$$

where g_{inh} must satisfy the normalization condition

$$\int_0^\infty g_{inh}(\omega'_{10} - \omega_{ih})d\omega'_{10} = 1. \tag{3.18}$$

The density matrix elements can be determined from the set of density matrix equations (3.4)–(3.7) for each frequency ω'_{10}. In terms of g_{inh} the total polarization and occupation numbers are

$$\begin{aligned} P(t) &= \int_0^\infty P'(t,\omega'_{10})d\omega'_{10} \\ &= \int_0^\infty \bar{N} g_{inh}(\omega'_{10} - \omega_{ih})[p_{01}\rho_{10}(t,\omega'_{10}) + p_{10}\rho_{01}(t,\omega'_{10})]d\omega'_{10} \end{aligned} \tag{3.19}$$

and

$$\begin{aligned} \gamma_{0,1}(t) &= \int_0^\infty \gamma'(t,\omega'_{10})d\omega'_{10} \\ &= \int_0^\infty \bar{N} g_{inh}(\omega'_{10} - \omega_{ih})\rho_{00,11}(t,\omega'_{10})d\omega'_{10}. \end{aligned} \tag{3.20}$$

The use of Eqs. (3.20) and (3.19) in connection with Eqs. (3.4)–(3.7), and (3.9) for the determination of $P(t)$ and $\gamma_{0,1}(t)$ requires that the composition of the sub-ensembles remains unchanged during the time range of interest. In other words, each atom possesses a certain (constant) transition frequency ω'_{10}. Surely this concept of "static" inhomogeneous broadening is justified where the inhomogeneity is caused by a time independent crystal disorder etc.; it becomes questionable when the resonance frequency ω'_{10} is a function of time. The latter, for example, is true if ω'_{10} is influenced by particle motion as is the case with Doppler broadening associated with velocity changing collisions. A similar situation arises in liquid dye solutions, where the absorbing molecules move through changing local fields, which (through Stark shifts) also causes random drifts of the resonance frequency. Such statistical changes of phase due to drifts in resonance frequency are generally represented by a "cross relaxation time" T_3. The equation of motion for the occupation numbers of an ensemble of particles with resonance frequency $\bar{\omega}'_{10}$ becomes then:

$$\frac{d}{dt}\gamma'_{0,1}(t,\bar{\omega}'_{10}) = -\frac{1}{T_1}[\gamma'_{0,1}(t,\bar{\omega}'_{10}) - \gamma^{(e)}_{0,1}(\bar{\omega}'_{10})] - \frac{1}{T_3}[\gamma'_{0,1}(t,\bar{\omega}'_{10}) - \bar{\gamma}_{0,1}] \quad (3.21)$$

where

$$\bar{\gamma}_{0,1} = g_{inh}(\bar{\omega}'_{10} - \omega_{ih})d\omega'_{10}\int_0^\infty \tilde{N} g_{inh}(\omega'_{10} - \omega_{ih})\rho_{00,11}(t,\omega'_{10})d\omega'_{10} \quad (3.22)$$

Note that the integral in the last equation represents the total number of particles in state $|0\rangle$ and $|1\rangle$, respectively, at a certain instant t. Therefore the second term in the right hand side of Eq. (3.21) is responsible for the relaxation of a disturbed particle distribution towards an equilibrium distribution given by the inhomogeneous line shape. For illustration, let us assume that at $t = t_0$ an initial Gaussian distribution [Fig. 3.2(a)] of particles in state $|0\rangle$ having a certain resonance frequency is distorted into (b) by excitation ($|0\rangle \to |1\rangle$) of a sub-ensemble with resonance frequency $\bar{\omega}'_{10}$. If $\Delta\omega_{inh}$ is much larger than the homogeneous linewidth $\Delta\omega_h = 2/T_2$ a hole

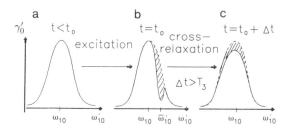

Figure 3.2: Cross relaxation of an inhomogeneously broadened medium (notation: $\omega_{ih} = \omega_{10}$).

3.1. DENSITY MATRIX EQUATIONS

of width $\Delta\omega_h$ will be burnt in the distribution $\gamma_0'(\omega_{10}')$ around the excitation frequency $\bar{\omega}_{10}'$ [Fig. 3.2(b)]. On a time scale $\Delta t > T_3$, these excited particles change their resonance frequency under the influence of the linewidth determining processes, and the Gaussian distribution is re-established (c). For $T_3 \ll T_1$ the hatched areas in Fig. 3.2, representing the number of excited particles, are equal. The initial distribution is then reached after relaxation of the two-level systems into the ground state. The question arises as to what happens if the measuring time T_M is much larger than the cross relaxation time T_3? We expect then to measure a distribution similar to curve (c) in Fig. 3.2. The sample behaves as if it were homogeneously broadened. This shows that the classification of homogeneous and inhomogeneous line broadening depends on the time scale of interest.

Let us next turn to the interaction of an electric field (light pulse) with the two-level system. For the model described by Eqs.(3.4)–(3.7) to be valid, the pulse duration τ_p should be (much) larger than the correlation time of the dissipative system τ_c. Consistently with this requirement, we can simply "add" the field interaction terms to Eqs. (3.4)–(3.7). In the case of dipole interaction, the Hamiltonian is given by $\hat{p}E$, and the density matrix equations read [11]

$$\frac{d}{dt}\rho_{10}(t) + i\omega_{10}\rho_{10}(t) + \frac{1}{T_2}\rho_{10}(t) = i\frac{p}{\hbar}[\rho_{00}(t) - \rho_{11}(t)]E(t) \quad (3.23)$$

$$\rho_{01}(t) = \rho_{10}^*(t) \quad (3.24)$$

$$\frac{d}{dt}\rho_{11}(t) + \frac{1}{T_1}[\rho_{11}(t) - \rho_{11}^{(e)}] = i\frac{p}{\hbar}[\rho_{10}(t) - \rho_{01}(t)]E(t) \quad (3.25)$$

$$\frac{d}{dt}\rho_{00}(t) + \frac{1}{T_1}[\rho_{00}(t) - \rho_{00}^e(t)] = i\frac{p}{\hbar}[\rho_{01}(t) - \rho_{10}(t)]E(t). \quad (3.26)$$

For simplicity, we have introduced $p_{01} = p_{10} = p$. The total population is conserved: $\rho_{00}(t) + \rho_{11}(t) = 1$. For calculating the change of the pulse (electric field) as it propagates through the medium we have to evaluate the wave equation (1.40). Assuming that the resonant atomic particles are embedded in a (transparent) host material with dielectric constant ϵ and applying the SVEA [Eq. (1.64)], the pulse envelope obeys Eq. (1.60) complemented by the resonant polarization. As we have treated the electric field in Eq. (1.55), the polarization of the two-level system is also decomposed into a slowly varying envelope $\tilde{\mathcal{P}}$ and a rapidly oscillating contribution

$$P(t,z) = \frac{1}{2}\tilde{\mathcal{P}}(t,z)e^{i(\omega_\ell t - k_\ell z)} + c.c.. \quad (3.27)$$

The propagation equation becomes

$$\frac{\partial}{\partial z}\tilde{\mathcal{E}} - \frac{i}{2}k_\ell''\frac{\partial^2}{\partial t^2}\tilde{\mathcal{E}} + \mathcal{D} = i\frac{\mu_0}{2k_\ell}\left(\frac{\partial^2}{\partial t^2}\tilde{\mathcal{P}} + 2i\omega_\ell\frac{\partial}{\partial t}\tilde{\mathcal{P}} - \omega_\ell^2\tilde{\mathcal{P}}\right) \quad (3.28)$$

where we have transferred to local coordinates. In the frame of the SVEA, the right-hand side of this equation can be approximated by the last term. Since in this section we shall concentrate on the effects of nonlinear light–matter interaction, we assume the host material to be weakly dispersive and neglect group velocity dispersion and higher order dispersion ($k_\ell'' = \mathcal{D} = 0$). (In later chapters we will study the interplay of dispersion with various nonlinearities.) Hence, we obtain the following propagation equation

$$\frac{\partial}{\partial z}\tilde{\mathcal{E}}(t,z) = -i\frac{\mu_0\omega_\ell^2}{2k_\ell}\tilde{\mathcal{P}}(t,z). \quad (3.29)$$

The polarization needed to solve this equation must be derived from the density matrix equations according to Eq. (3.11) or Eq. (3.19), depending on the kind of line broadening.

Propagation problems which are governed by Eq. (3.29) are associated with pulse shaping and play a decisive role in the pulse generation in lasers and in pulse amplification and shaping. We therefore give this equation a close scrutiny.

First, however, let us discuss briefly some of the important features of the resonant light–matter interaction, under the assumption that the pulse duration is much longer than all relaxation constants, which allows us to find a stationary solution for Eq. (3.23). In addition we will consider propagation only over a small distance Δz. A formal solution of Eq. (3.29) is then

$$\Delta\tilde{\mathcal{E}}(t,z) = -i\frac{\mu_0\omega_\ell^2}{2k_\ell}\tilde{\mathcal{P}}(t,z)\Delta z. \quad (3.30)$$

As will be discussed in the next subsection in detail, using Eq. (3.11) and the stationary solution of Eqs. (3.23) and (3.26), we obtain for the polarization

$$\tilde{\mathcal{P}}(t,z) = i\frac{p^2 T_2/\hbar}{iT_2(\omega_\ell - \omega_{10}) + 1}\,[\rho_{00}(t) - \rho_{11}(t)]\bar{N}\tilde{\mathcal{E}}(t,z). \quad (3.31)$$

This relation inserted into Eq. (3.30) gives

$$\Delta\tilde{\mathcal{E}}(t,z) = \frac{\mu_0\omega_\ell^2 p^2}{2\hbar k_\ell}\frac{T_2}{1 + iT_2(\omega_\ell - \omega_{10})}\,[\rho_{00}(t) - \rho_{11}(t)]\,\bar{N}\tilde{\mathcal{E}}(t,z)\Delta z, \quad (3.32)$$

3.1. DENSITY MATRIX EQUATIONS

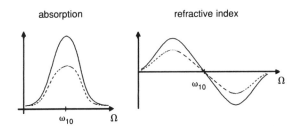

Figure 3.3: Absorption coefficient and refractive index in the vicinity of an optical resonance for two values of the population inversion $\Delta\rho = \rho_{11} - \rho_{00}$.

which has the formal solution

$$\tilde{\mathcal{E}}(t, z + \Delta z) = \tilde{\mathcal{E}}(t, z)e^{A(t)+iB(t)}, \qquad (3.33)$$

where

$$A = \frac{\mu_0 \omega_\ell^2 p^2 T_2/(2\hbar k_\ell)}{T_2^2(\omega_{10} - \omega_\ell)^2 + 1} [\rho_{00}(t) - \rho_{11}(t)]\bar{N}\Delta z \qquad (3.34)$$

and

$$B = A(\omega_{10} - \omega_\ell)T_2. \qquad (3.35)$$

The real part A of the exponential factor in Eq. (3.33) represents absorption or amplification. The imaginary part B has the structure of a phase modulation term. According to our discussion in Chapter 1 [Eq. (1.124)] we can interpret B as resulting from a time dependent optical path length (difference) $(2\pi/\lambda_\ell)\Delta z[n(t) - 1]$, where the time dependence of B and n, respectively, is determined by the inversion density $\Delta\rho = \rho_{11}(t) - \rho_{00}(t)$. Equating both relations yields an expression for the refractive index

$$n(t) = 1 + \frac{Ac}{\omega_\ell \Delta z}(\omega_{10} - \omega_\ell)T_2 \qquad (3.36)$$

in the vicinity of the resonance frequency ω_{10}. In addition to the explicit frequency dependence (ω_ℓ) in Eq. (3.36), the index depends on the laser frequency (ω_ℓ) through A in Eq. (3.34). These dependencies are shown in Fig. 3.3 for an absorbing sample for two different (negative) values of $\Delta\rho$. A smaller $|\Delta\rho|$ results in a smaller absorption coefficient regardless of ω_ℓ while for n we find an increase for $\omega_\ell < \omega_{10}$ and a decrease for $\omega_\ell > \omega_{10}$. At exact resonance $(\omega_\ell = \omega_{10})$ the refractive index does not change at all if $\Delta\rho$ is varied.

3.2 Pulse shaping with resonant particles

3.2.1 General

Let us consider a pulse traveling through a medium of certain length consisting of particles with number density N at resonance with the light. How has the pulse been modified by the medium? In answering this question we assume that only the resonant particles influence the pulse, i.e. we neglect group velocity dispersion of the host material and nonresonant nonlinear contributions. Depending on how the medium is prepared we expect light to be absorbed or amplified. Absorption can occur when $(\rho_{11} - \rho_{00}) < 0$, i.e. if the majority of resonant particles interacting with the pulse are in the ground state. Amplification is expected in the opposite case where $(\rho_{11} - \rho_{00}) > 0$. To reach this situation the particles have to be excited by an appropriate pump mechanism.[2]

At first glance absorption (amplification) seems only to decrease (increase) the pulse energy by a certain factor while leaving the other pulse characteristics unchanged. We will see this to be true only under very specific conditions. In the general case the pulse at the output exhibits a different envelope (shape) as well as a different time-dependent phase, as compared to the input characteristics. The leading part of the pulse changes the properties of the medium, which will then act in a different manner upon the trailing part. In terms of amplification and absorption one can say that these quantities become time dependent because of the change of the inversion density $\rho_{11}(t) - \rho_{00}(t)$. We expect the pulse to be more heavily absorbed (amplified) at its leading part which, of course, modifies the envelope shape. According to our discussion above, a time dependent occupation number means that different pulse parts "see" different optical path lengths, resulting in a time dependent phase change (chirp). To discuss the actual pulse distortion we need to determine the temporal behavior of the occupation numbers by means of Eqs. (3.23)–(3.26). In general, numerical methods are required to analyze this problem. An analytical description can only be made for some limiting cases. The physics of the pulse–matter interaction depends strongly on the ratio of the pulse duration and the characteristic response time of the medium, as well as on the pulse intensity and energy. We will start with the approximation that the phase relaxation time is much shorter than the pulse duration, generally referred to as rate equation approximation. Femtosecond pulses do challenge this often used approximation. Because the

---

[2]Other energy levels have to be taken into account to describe the pumping.

3.2. PULSE SHAPING WITH RESONANT PARTICLES

duration of fs pulses can be comparable or shorter than the phase memory of the medium, the polarization oscillations' excited by the leading part of the pulse can interfere coherently with subsequent pulse parts. We will discuss this situation approximately as a perturbation of the rate equation approximation. Phenomena related to the dominant influence of coherent light–matter interaction will be dealt with in the following chapter.

To illustrate the action of the phase memory, let us inspect the equation of motion for the polarization. Within the SVEA, and assuming $|\omega_\ell - \omega_{10}| \ll \omega_{10}, \omega_\ell$ as well as $\omega_{10}T_2, \omega_\ell T_2 \gg 1$, Eqs. (3.11), (3.23), (3.24) and (3.27) yield a first-order differential equation for the slowly varying envelope component $\tilde{\mathcal{P}}$:

$$\frac{d}{dt}\tilde{\mathcal{P}} + \left[\frac{1}{T_2} + i(\omega_\ell - \omega_{10})\right]\tilde{\mathcal{P}} = -i\frac{\bar{N}p^2}{\hbar}(\rho_{00} - \rho_{11})\tilde{\mathcal{E}}. \tag{3.37}$$

The solution of Eq. (3.37) can be written formally in the form

$$\tilde{\mathcal{P}}(t) = i\frac{\bar{N}p^2}{\hbar}\int_{-\infty}^{t}\tilde{\mathcal{E}}(t')\Delta\rho(t')e^{[i(\omega_\ell - \omega_{10}) + 1/T_2](t'-t)}dt' \tag{3.38}$$

where $\Delta\rho = \rho_{11} - \rho_{00}$ is the population inversion. Obviously the polarization at time t depends on values of the electric field (modulus and phase) and population numbers for all $t' \leq t$, a dependence weighted by the function $e^{-(t-t')/T_2}$. The latter implies that the memory-time is T_2. For later reference we will derive two other representations of the solution of Eq. (3.37). Continuous application of partial integration of Eq. (3.38) leads to

$$\tilde{\mathcal{P}}(t) = i\frac{\bar{N}p^2 T_2}{\hbar}\tilde{L}(\omega_\ell - \omega_{10})\Big\{\tilde{\mathcal{E}}(t)\Delta\rho(t) \\ + \sum_{n=1}^{\infty}\tilde{L}^n(\omega_\ell - \omega_{10})(T_2 d/dt)^n[\tilde{\mathcal{E}}(t)\Delta\rho(t)]\Big\} \tag{3.39}$$

where

$$\tilde{L}(\omega_\ell - \omega_{10}) = \frac{1}{i(\omega_\ell - \omega_{10})T_2 + 1} \tag{3.40}$$

which is called the complex line shape factor. Fourier transforming Eq. (3.39) yields

$$\tilde{\mathcal{P}}(\Omega) = i\frac{\bar{N}p^2 T_2}{\hbar}\tilde{L}(\omega_\ell - \omega_{10})\Big[\mathcal{F}\{\tilde{\mathcal{E}}(t)\Delta\rho(t)\} \\ + \sum_{n=1}^{\infty}\tilde{L}^n(\omega_\ell - \omega_{10})(i\Omega T_2)^n\mathcal{F}\{\tilde{\mathcal{E}}(t)\Delta\rho(t)\}\Big]. \tag{3.41}$$

Note that in Eq. (3.39) and in Eq. (3.41) the first summand corresponds to the zero-order term.

3.2.2 Pulses much longer than the phase relaxation time ($\tau_p \gg T_2$)

If the pulse duration is much longer than the phase relaxation time of the medium there is no coherent superposition of polarization and electric field oscillations. The memory of the medium is only through the change of the occupation numbers. Assuming that the fastest component of the dynamics of the occupation inversion is determined by the pulse under discussion, $\tau_p \gg T_2$ implies that $|T_2(d/dt)| \ll 1$ in Eq. (3.39). In this case we may neglect all terms with $n \geq 1$, a procedure called rate equation approximation, and the polarization reduces to

$$\tilde{\mathcal{P}}(t) = i\frac{p^2 T_2}{\hbar}\tilde{L}(\omega_\ell - \omega_{10})\tilde{\mathcal{E}}(t)\Delta\gamma(t) \tag{3.42}$$

where we have introduced the population inversion density

$$\Delta\gamma = \bar{N}(\rho_{11} - \rho_{00}). \tag{3.43}$$

Note that $\tilde{\mathcal{P}}$ given by Eq. (3.42) corresponds to the stationary solution of Eqs. (3.37) and (3.23), respectively. In terms of Eq. (3.41) the rate equation approximation requires $|(i\Omega T_2)^n \mathcal{F}\{\tilde{\mathcal{E}}(t)\Delta\rho(t)\}| \ll 1$, which means that the spectral width of the pulse is much smaller than $1/T_2$, hence much smaller than the spectral width of the transition. Strictly speaking, this approximation requires a monochromatic wave interacting with an extended transition profile. It does, however, yield satisfactory results in numerous practical cases, even when the above condition is only marginally satisfied.

To study the behavior of modulus and phase of a pulse in propagating through the resonant medium we have to solve the propagation equation (3.29) with the polarization given by Eq. (3.42). Thus, we have

$$\frac{\partial}{\partial z}\tilde{\mathcal{E}} = \frac{p^2 T_2 \mu_0 \omega_\ell^2}{2k_\ell \hbar}\tilde{L}(\omega_\ell - \omega_{10})\Delta\gamma\tilde{\mathcal{E}}. \tag{3.44}$$

The equation of motion for the population inversion density at location z, $\Delta\gamma$, can be obtained from Eqs. (3.23)–(3.26) and reads

$$\frac{\partial}{\partial t}\Delta\gamma + \frac{1}{T_1}(\Delta\gamma - \Delta\gamma^{(e)}) = -\frac{p^2 T_2}{\hbar^2}|\tilde{L}(\omega_\ell - \omega_{10})|^2 \Delta\gamma|\tilde{\mathcal{E}}|^2, \tag{3.45}$$

3.2. PULSE SHAPING WITH RESONANT PARTICLES

where $\Delta\gamma^{(e)}$ is the equilibrium value of $\Delta\gamma$ before the pulse arrives. This equation is often referred to as the rate equation for the population inversion. Representing the complex pulse envelope as $\tilde{\mathcal{E}} = \mathcal{E}\exp(i\varphi)$, Eq. (3.44) yields for the modulus

$$\frac{\partial}{\partial z}\mathcal{E} = \frac{1}{2}\sigma_{01}^{(0)}|\tilde{L}(\omega_\ell - \omega_{10})|^2 \Delta\gamma \mathcal{E} \tag{3.46}$$

and for the phase

$$\frac{\partial}{\partial z}\varphi = -\frac{1}{2}\sigma_{01}^{(0)}|\tilde{L}(\omega_\ell - \omega_{10})|^2 \Delta_\ell \Delta\gamma \tag{3.47}$$

where $\Delta_\ell = (\omega_\ell - \omega_{10})T_2$. The quantity

$$\sigma_{01}^{(0)} = \frac{p^2 T_2 \omega_{10}}{\epsilon_0 c n \hbar} \tag{3.48}$$

is the interaction cross-section at the center of the transition ($\omega_\ell = \omega_{10}$). The interaction cross-section at frequency ω_ℓ is given by

$$\sigma_{01} = \sigma_{01}(\omega_\ell - \omega_{10}) = \sigma_{01}^{(0)}|\tilde{L}(\omega_\ell - \omega_{10})|^2 = \frac{\sigma_{01}^{(0)}}{1 + T_2^2(\omega_\ell - \omega_{10})^2} \tag{3.49}$$

and can be considered as a measure of the interaction strength or of the probability that an absorption (emission) process takes place. Typical values for the interaction cross-sections are listed in Table 3.1. The temporal change of the population inversion, cf. Eq. (3.45), does not depend on the phase of the pulse but on $|\tilde{\mathcal{E}}|^2$. This suggests the convenience of rewriting this equation in terms of the photon flux density F defined in Eq. (1.21). We find

$$\frac{\partial}{\partial t}\Delta\gamma + \frac{1}{T_1}(\Delta\gamma - \Delta\gamma^{(e)}) = -2\sigma_{01}\Delta\gamma F \tag{3.50}$$

and by means of Eq. (3.46) the photon flux density is found to obey

$$\frac{\partial}{\partial z}F = \sigma_{01}\Delta\gamma F. \tag{3.51}$$

From Eq. (3.47) we see that a phase change can only occur if $\Delta\gamma\Delta_\ell \neq 0$, which states that the laser frequency does not coincides with the center frequency of the transition. A phase modulation ($d\varphi/dt$) requires in addition that $(d/dt)\Delta\gamma \neq 0$, i.e. the population difference must change during the interaction with the pulse. For a quantitative discussion we may proceed as follows. We solve Eqs. (3.50) and (3.51) to obtain the pulse envelope change and $\Delta\gamma(t,z)$, which then allows us to determine the phase modulation with Eq. (3.47). In general this requires numerical means. Here we will deal with the two limiting cases where the pulse duration is much shorter and much longer than the energy relaxation time T_1. Combining Eqs. (3.50) and (3.51) to eliminate $\Delta\gamma$ leads to:

$$\frac{\partial^2}{\partial t \partial z}\ln F + 2\sigma_{01}\frac{\partial}{\partial z}F + \frac{1}{T_1}\left(\frac{\partial}{\partial z}\ln F - \sigma_{01}\Delta\gamma^{(e)}\right) = 0 \qquad (3.52)$$

with the boundary condition $F(z=0,t) = F_0(t)$ and the initial condition $(\partial/\partial z)\ln F(t \to -\infty, z) = \sigma_{01}\Delta\gamma^{(e)}(z)$.

For $\tau_p \ll T_1$ the third term in Eq. (3.52) can be neglected and the remaining differential equation can be integrated with respect to z, which yields

$$\frac{\partial}{\partial t}\ln\frac{F(z,t)}{F_0(t)} + 2\sigma_{01}[F(z,t) - F_0(t)] = 0. \qquad (3.53)$$

This equation has a solution of the form [60]:

$$F(z,t) = F_0(t)\frac{e^{2\sigma_{01}\bar{W}_0(t)}}{e^{-a} - 1 + e^{2\sigma_{01}\bar{W}_0(t)}} \qquad (3.54)$$

where $\bar{W}_0(t) = \int_{-\infty}^{t} F_0(t')dt' = 1/(\hbar\omega_\ell)\int_{-\infty}^{t} I_0(t')dt'$ (I_0 intensity of the incident pulse), cf. Eqs. (1.19), (1.20), and

$$a = \sigma_{01}\Delta\gamma^{(e)}z \qquad (3.55)$$

is the absorption ($\Delta\gamma^{(e)} < 0$) or amplification ($\Delta\gamma^{(e)} > 0$) coefficient corresponding to a sample of length z. $\bar{W}_0(t)$ is a measure of the incident pulse energy (area) density until time t in units of (photons)/cm^2. The total incident energy density is $\hbar\omega_\ell\bar{W}_0(t=\infty) = \hbar\omega_\ell\bar{W}_{0,\infty} = W_0$. The transmitted energy density $W(z,t) = \hbar\omega_\ell\bar{W}(z,t)$ is obtained by integrating Eq. (3.54) with respect to time and can be written as

$$W(z,t) = \hbar\omega_\ell\int_{-\infty}^{t} F(z,t')dt' = W_s\ln\left[1 - e^a\left(1 - e^{W_0(t)/W_s}\right)\right], \qquad (3.56)$$

where $W_s = \hbar\omega_\ell/(2\sigma_{01})$ is the saturation energy density of the medium. With Eq. (3.50), in the limit $\tau_p \ll T_1$, we can express the population inversion as

$$\Delta\gamma(z,t) = \Delta\gamma^{(e)}e^{-2\sigma_{01}\bar{W}(z,t)} = \frac{\Delta\gamma^{(e)}}{1 - e^a[1 - e^{W_0(t)/W_s}]}. \qquad (3.57)$$

3.2. PULSE SHAPING WITH RESONANT PARTICLES

It is obvious from Eqs. (3.54) and (3.57) that a modification of the pulse shape is related to a change in the population inversion. The latter is controlled by the magnitude of $2\sigma_{01}\bar{W}_0$. A characteristic quantity is the ratio s of pulse energy density to saturation energy density of the transition:

$$s = 2\sigma_{01}\bar{W}_{0\infty} = \frac{W_0}{W_s}. \tag{3.58}$$

Changes of the occupation numbers and the pulse shape become significant with increasing saturation parameter s. For large s, the population inversion approaches zero (saturation), cf. Eq. (3.57), and the pulse distortion becomes small. The result is that a saturable absorber attenuates the leading edge more than the trailing edge, leading to pulse steepening which is associated with a pulse shortening. In an amplifier the leading part of the pulse experiences higher gain than the trailing part. This can result in pulse shortening as well as broadening, depending on the steepness of the incident pulse. Our qualitative discussion can easily be proven quantitatively by evaluating Eq. (3.54). Figure 3.4 shows some examples. Possible applications of this kind of pulse shaping in lasers and in pulse amplifiers will be discussed later. It should be mentioned that the steepening which can be achieved by a saturable absorber is limited, although this is not described by our approach chosen here. The reason is that for Eq. (3.54) to remain valid (as the pulse shortens) the rise time of the shaped pulse must still be (much) longer than T_2, which, in this manner, sets a lower limit.

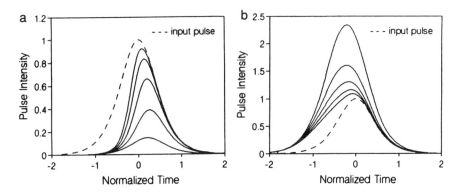

Figure 3.4: Pulse shaping through saturable absorption and amplification. (a) absorber: $a = -10$, s varies from 2 to 8 (increment 1) in the order of increasing intensities. (b) amplifier: $a = 2$, s varies from 0.5 to 8 (increment 0.5) in the order of decreasing intensities, For the input pulse we assumed $F_0(t) = \cosh^{-2}(1.76t/\tau_p)$.

Let us turn next to the second case where $\tau_p \gg T_1$. It should be noted that dealing with fs light pulses this situation is less probable than the previous one. We can now neglect the term with the temporal derivation with respect to the term containing $1/T_1$ in Eq. (3.52). The resulting differential equation is:

$$2\sigma_{01}T_1\frac{\partial}{\partial z}F + \frac{\partial}{\partial z}\ln F = \sigma_{01}\Delta\gamma^{(e)} = \alpha. \tag{3.59}$$

It can be rewritten as

$$\frac{d}{dz}F = \frac{\alpha F}{1 + F/F_s} \tag{3.60}$$

or in terms of intensities as

$$\frac{d}{dz}I = \frac{\alpha I}{1 + I/I_s} \tag{3.61}$$

where we have defined a saturation flux density $F_s = (2\sigma_{01}T_1)^{-1}$ and a saturation intensity $I_s = \hbar\omega_\ell/(2\sigma_{01}T_1)^{-1}$. Equations (3.60) and (3.61) represent Beer's law for a saturable medium. Their integration yields

$$\ln\frac{F(z,t)}{F_0(t)} + \frac{F(z,t) - F_0(t)}{F_s} = a \tag{3.62}$$

and

$$\ln\frac{I(z,t)}{I_0(t)} + \frac{I(z,t) - I_0(t)}{I_s} = a \tag{3.63}$$

where $a = \alpha z$. Equation (3.62) contains the unknown $F(z,t)$ and $I(z,t)$ implicitly. To get an insight into the pulse distortion we will assume $|a| \ll 1$, from which we expect $F = F_0 + \Delta F$ with $|\Delta F/F_0| \ll 1$. Inserting this into Eq. (3.62) and expanding the logarithmic function gives the following relation for the flux at the output of the medium of length z:

$$F(z,t) = F_0(t)\left[1 + \frac{a}{1 + F_0(t)/F_s}\right]. \tag{3.64}$$

The absorption (or amplification) coefficient becomes time dependent, where its value is now controlled by the instantaneous photon flux density rather than by the energy density. A characteristic quantity of the medium is now F_s or I_s. The result of this kind of saturation is that the pulse peak where the intensity takes on a maximum is less absorbed (amplified) than the wings, as illustrated in Fig. 3.5.

3.2. PULSE SHAPING WITH RESONANT PARTICLES

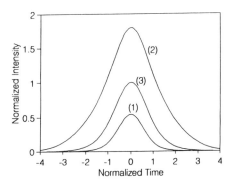

Figure 3.5: Pulse shaping in a saturable absorber (1) and depletable amplifier (2) for $\tau_p \gg T_1$, $a = \mp 3$ and $F_0/F_s = 1.5$ according to Eq. (3.62). For the input pulse (3) we assumed $F_0(t) = \cosh^{-2}(1.76 t/\tau_p)$.

According to our discussion above, the change of the occupation numbers results in a change of the refractive index and we expect a phase modulation to occur. Using Eqs. (3.47) and (3.51) the time dependent frequency change $\delta\omega(t)$ can be written as

$$\delta\omega(t) = \frac{\partial \varphi}{\partial t} = -\frac{1}{2}\sigma_{01}^{(0)}|\tilde{L}|^2 \Delta_\ell \int_0^z \frac{\partial}{\partial t}\Delta\gamma dz = -\frac{1}{2}\Delta_\ell \frac{\partial}{\partial t} \ln \frac{F(z,t)}{F_0(t)}. \qquad (3.65)$$

The sign of $\delta\omega$ depends upon the sign of Δ_ℓ, i.e. on whether the interaction takes place above or below resonance, and on the sign of $\ln(F/F_0)$. The latter is positive (negative) for $F > (<) F_0$ which is true for an amplifier (absorber). For $\tau_p \ll T_1$ the pulse energy controls the dynamics of $\Delta\gamma$, and by means of Eq. (3.54), $\delta\omega(t)$ becomes

$$\delta\omega(t) = -\Delta_\ell \sigma_{01} \frac{e^{-a} - 1}{e^{-a} - 1 + e^{2\sigma_{01}\bar{W}_0(t)}} F_0(t), \qquad (3.66)$$

or equivalently, in terms of the intensity and saturation energy density W_s:

$$\delta\omega(t) = -\frac{\Delta_\ell}{2} \frac{e^{-a} - 1}{e^{-a} - 1 + e^{W(t)/W_s}} \frac{I(t)}{W_s}. \qquad (3.67)$$

The optically thin medium approximation ($|a| \ll 1$) of Eq. (3.66) is:

$$\delta\omega(t) \simeq \Delta_\ell a e^{-2\sigma_{01}\bar{W}_0(t)} \sigma_{01} F_0(t). \qquad (3.68)$$

The other limiting case ($\tau_p \gg T_1$) is associated with a $\delta\omega(t)$ given by

$$\delta\omega(t) = \Delta_\ell \sigma_{01} T_1 \frac{1}{F_s} \frac{\partial}{\partial t} [F(t,z) - F_0(t)], \qquad (3.69)$$

which for small $|a|$ results in

$$\delta\omega(t) = \frac{\Delta_\ell a}{2} \frac{1}{(1 + F_0(t)/F_s)^2} \frac{\partial}{\partial t} \left(\frac{F_0(t)}{F_s}\right), \qquad (3.70)$$

as can easily be verified by inserting Eq. (3.64) into Eq. (3.69). Figure 3.6 shows the time dependent frequency change described by Eq. (3.66) for an amplifier. Equation (3.70) indicates a frequency change towards the leading part of the pulse with increasing saturation. A similar behavior occurs if the pulse passes through an absorber. As a result the frequency change across the FWHM of the pulses becomes maximum at a certain level of saturation.

To get an idea about the order of magnitude of the frequency change let us estimate $\delta\bar{\omega} = \delta\omega(t = -\tau_p/2) - \delta\omega(t = \tau_p/2)$ for a sech pulse using Eq. (3.68) for an absorber. For $\bar{W}_{0,\infty}/\bar{W}_s = W_0/W_s = 1$ we find

$$\tau_p \delta\bar{\omega} \simeq \Delta_\ell a \qquad (3.71)$$

which for $a = -0.1$ and $\Delta_\ell = 1$ yields $\tau_p \delta\tilde{\omega} \approx -0.1$. If we compare this with the pulse–duration–bandwidth product, cf. Table 1.1, $\tau_p \Delta\omega = 2.8$, we

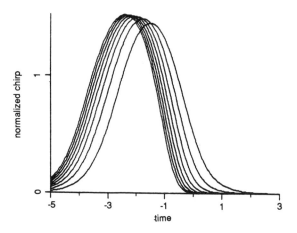

Figure 3.6: Normalized chirp $\tau_p \delta\omega$ versus time after passage through an amplifier according to Eq.(3.66) for a sech^2 input pulse, $e^a = 100$, $s = 0.5,\ldots,4$ ($\Delta s = 0.5$, from right to left), $\Delta_\ell = -1$.

3.2. PULSE SHAPING WITH RESONANT PARTICLES

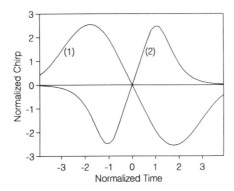

Figure 3.7: Normalized chirp at the output of an absorber (1) and amplifier (2) according to Eq. (3.70) with the parameters of Fig. 3.5 and $\Delta_\ell = 1$.

hardly expect that the induced frequency change is of importance. This is usually true for a single passage; however, it can play a significant role in lasers where the pulse passes through the media many times before it is coupled out [61]. Chirp introduced by a fast ($T_1 \ll \tau_p$) absorber and amplifier is shown in Fig. 3.7 for comparison.

It should be noted that our discussion can easily be expanded to include effects of multi-level systems [62]. This, for example, is necessary to model dye molecules more accurately. For completeness let us briefly discuss an inhomogeneously broadened sample, i.e., a distribution of particles having different resonance frequencies ω'_{10}. We assume that the preconditions for applying the REA are fulfilled. The source terms in the propagation equations for the field components $\mathcal{E}, \varphi$, cf. Eqs. (3.46), (3.47), have to be integrated over the inhomogeneous distribution function g_{inh}. This yields

$$\frac{\partial}{\partial z}\mathcal{E} = \frac{1}{2}\mathcal{E}\int_0^\infty \sigma_{01}^{(0)}|\tilde{L}(\omega_\ell - \omega_0)|^2 \Delta\gamma'(t, \omega'_{10}) d\omega'_{10}$$

$$= \frac{1}{2}\mathcal{E}\int_0^\infty \sigma_{01}(\omega_\ell - \omega'_{10}) g_{inh}(\omega'_{10}) \bar{N}[\rho_{11}(\omega'_{10}) - \rho_{00}(\omega'_{10})] d\omega'_{10} \quad (3.72)$$

and

$$\frac{\partial}{\partial z}\varphi = -\frac{1}{2}\int_0^\infty \Delta'_\ell \sigma_{01}(\omega_\ell - \omega'_{10}) g_{inh}(\omega'_{10}) \bar{N}[\rho_{11}(t, \omega'_{10}) - \rho_{00}(t, \omega'_{10})] d\omega'_{10}. \quad (3.73)$$

The matrix elements ρ_{kk} can be obtained from the density matrix equations. A detailed discussion of chirp generation in such samples can be found in [63].

3.2.3 Pulse duration comparable with or longer than the phase relaxation time ($\tau_p \geq T_2$)

If the incident intensity varies on a time scale which is comparable with the phase relaxation time of the medium, we cannot apply the rate equation approximation as was done before but have to solve the complete set of density matrix equations (3.4)–(3.7) and wave equation (3.29). The physics behind this case is that the pulse "feels" the spectral properties of the resonant medium, that is, different spectral components of the pulse experience different modification during the interaction. This becomes very obvious and simple if the interaction is weak and we may neglect saturation and a change in the occupation numbers, respectively. For $\Delta\rho = $ constant, we find for the polarization after Fourier-transforming Eq. (3.37)

$$\tilde{\mathcal{P}}(\Omega) = i\frac{\bar{N}p^2 T_2 \Delta\rho}{\hbar} \frac{T_2}{1 + iT_2(\Omega + \omega_\ell - \omega_{10})} \tilde{\mathcal{E}}(\Omega). \tag{3.74}$$

Substituting the polarization in the propagation equation (3.29), the transmitted amplitude spectrum can be found after integration with respect to z

$$\tilde{\mathcal{E}}(\Omega, z) = \tilde{\mathcal{E}}(\Omega, 0) e^{\frac{1}{2}a(\Omega)} e^{-\frac{i}{2}a(\Omega)(\Omega + \omega_\ell - \omega_{10})T_2} \tag{3.75}$$

where $a(\Omega)$ is the frequency dependent coefficient of the small signal gain (absorption) which is given by

$$a(\Omega) = \sigma_{01}^{(0)} \bar{N} \Delta\rho z \frac{1}{1 + T_2^2(\Omega + \omega_\ell - \omega_{10})^2}. \tag{3.76}$$

The corresponding relation for the power spectrum reads

$$S(\Omega, z) = S(\Omega, 0) e^{a(\Omega)}. \tag{3.77}$$

As can be seen from Eq. (3.77), different spectral components experience different absorption/gain. Thus the sample acts as filter. For off-resonant interaction ($\omega_\ell \neq \omega_{10}$), the filter also introduces a spectral phase represented by the last term in Eq. (3.75). Therefore, in addition to a change in pulse shape the output pulse can be chirped.

Generally, numerical integration of the density matrix and wave equations is required to deal with pulse–matter interaction in the presence of varying population number changes. For the limiting case of small saturation [$s < (\ll) 1$], a small absorption/gain coefficient [$|a| < (\ll) 1$], and

3.3. NONLINEAR, NONRESONANT OPTICAL PROCESSES

pulse durations being still longer than T_2, we may utilize a perturbation approach [64]. This gives for the pulse amplitude at the output of such an absorber/amplifier

$$\tilde{\mathcal{E}}(t,z) = \left\{1 + \frac{1}{2}a^{(0)}\tilde{L}\left[1 - 2\sigma_{01}\bar{W}_0(t) + \frac{1}{2}(2\sigma_{01}\bar{W}_0(t))^2 + T_2\tilde{L}(2\sigma_{01})F_0(t)\right]\right.$$
$$\left. - \frac{1}{2}a^{(0)}\tilde{L}^2\left[1 - 2\sigma_{01}\bar{W}_0(t)\right]T_2\frac{d}{dt} + \frac{1}{2}a^{(0)}\tilde{L}^3 T_2^2 \frac{d^2}{dt^2}\right\}\tilde{\mathcal{E}}_0(t) \quad (3.78)$$

where $a^{(0)} = \sigma_{01}^{(0)}\Delta\gamma^{(e)}z$ is the absorption/gain coefficient at the resonance frequency of the transition. For $T_2 \to 0$, we obtain a relation which corresponds to Eq. (3.54) if we expand it up to terms linear in a and quadratic in $(2\sigma_{01}\bar{W}(t))$. The additional terms in Eq. (3.78) come into play if $T_2(d/dt)\tilde{\mathcal{E}}_0(t)$ is not vanishingly small, that is if the pulse duration is of the same order of magnitude as T_2. Then the medium not only remembers the number of absorbed/amplified photons but also the phase of the electric field during τ_p.

3.3 Nonlinear, nonresonant optical processes

3.3.1 General

In contrast to the previous section where the interaction was dominated by a resonance, we will be dealing with situations where the light frequency is far away from optical resonances. For cw light of low intensity the medium thus appears completely transparent and merely introduces a phase shift. For pulses, as discussed in Chapter 1, dispersion has to be taken into account, which can lead to pulse broadening and shortening depending upon the input chirp, and to phase modulation effects. The light–matter interaction is linear, i.e., there is a linear relationship between input and output field, which results in a constant spectral intensity. A typical example is the pulse propagation through a piece of glass. The situation becomes much more complex if the pulse intensity is large, which can be achieved by focusing or/and using amplified pulses. The high electric field associated with the propagating pulse is no longer negligibly small as compared to typical local fields inside the material such as inner atomic (inner molecular) fields and crystal fields. The result is that the material properties are changed by the incident field and thus depend on the pulse. The induced polarization which is needed as source term in the wave equation is formally described by the relationship

$$P = \epsilon_0 \chi(E) E = \epsilon_0 \chi^{(1)} E + \epsilon_0 \chi^{(2)} E^2 + \epsilon_0 \chi^{(3)} E^3 + \ldots + \epsilon_0 \chi^{(n)} E^n + \ldots$$
$$= P^{(1)} + P^{(2)} + \ldots + P^{(n)} + \ldots. \tag{3.79}$$

The quantities $\chi^{(n)}$ are known as the nonlinear optical susceptibilities of nth order where $\chi^{(1)}$ is the linear susceptibility introduced in Eq. (1.43). The ratio of two successive terms is roughly given by

$$\left| \frac{P^{(n+1)}}{P^{(n)}} \right| = \left| \frac{\chi^{(n+1)} E}{\chi^{(n)}} \right| \approx \left| \frac{E}{E_{mat}} \right| \tag{3.80}$$

where E_{mat} is a typical value for the inherent electrical field in the material. For simplicity we have taken both E and P as scalar quantities. Generally, $\chi^{(n)}$ is a tensor of order $(n+1)$ which relates an n-fold product of vector components E_j to a certain component of the polarization of nth order,[3] $P^{(n)}$; see, for example, [65, 11, 10]. Moreover, for Eq. (3.79) to be valid in the time domain, we must assume that the sample responds instantaneously to the electric field and that it does not exhibit a memory. In other words, the polarization at an instant $t = t_0$ must depend solely on field values at $t = t_0$. As discussed in the previous section for resonant interaction, a non-instantaneous response and memory effects, respectively, are a result of phase and energy relaxation processes. They become noticeable if they proceed on a time scale of the pulse duration or longer. Fortunately, in non-resonant light–matter interaction many processes are well described by an instantaneous response even when excited by pulses with durations of the order of 10^{-14} s. In particular, this is true for nonlinear effects of electronic origin. If the motion of the much heavier atomic nuclei and molecules contribute to the material response, for example, memory effects are likely to occur on a fs time scale. In this case the nth-order polarization is of the form

$$P^{(n)}(t) = \epsilon_0 \int \int \ldots \int \chi^{(n)}(t_1, t_2, \ldots, t_n) E(t - t_1) E(t - t_1 - t_2) \ldots$$
$$\times E(t - t_1 - \ldots - t_n) dt_1 dt_2 \ldots dt_n \tag{3.81}$$

---

[3] Note that this product can couple up to n different input fields depending upon the conditions of illumination.

3.3. NONLINEAR, NONRESONANT OPTICAL PROCESSES 125

which illustrates the influence of the electric field components at earlier times.

To study pulse propagation in a nonlinear optical medium we can proceed as in the previous section. To the linear wave equation for the electric field, which contains the $\chi^{(1)}$ contribution, we add the nonresonant nonlinear polarization. As result we obtain Eq. (3.28) again, but with the nonlinear polarization as source term:

$$\left(\frac{\partial}{\partial z}\tilde{\mathcal{E}} - \frac{i}{2}k''_\ell \frac{\partial^2}{\partial t^2}\tilde{\mathcal{E}} + \mathcal{D}\right) e^{i(\omega_\ell t - k_\ell z)} + c.c. = i\frac{\mu_0}{k_\ell}\frac{\partial^2}{\partial t^2}P. \qquad (3.82)$$

If we represent the polarization as a product of a slowly varying envelope $\tilde{\mathcal{P}}$ and a term oscillating with an optical frequency ω_p, $e^{i\omega_p t}$, the right-hand side of Eq. (3.82) can be written as

$$\frac{\partial^2}{\partial t^2}\left(\tilde{\mathcal{P}}e^{i\omega_p t} + c.c.\right) = \left(\frac{\partial^2}{\partial t^2}\tilde{\mathcal{P}} + 2i\omega_p \frac{\partial}{\partial t}\tilde{\mathcal{P}} - \omega_p^2 \tilde{\mathcal{P}}\right) e^{i\omega_p t} + c.c.. \qquad (3.83)$$

In order to compare the magnitude of the individual terms we approximate $(\partial/\partial t)\tilde{\mathcal{P}}$ with $\tilde{\mathcal{P}}/\tau_p$ which yields for the ratio of two successive members of the sum in the brackets $\omega_p \tau_p$. Therefore, if the pulse duration is (much) longer than an optical period, that is $\omega_p \tau_p = 2\pi \tau_p / T_p \gg 1$, we may neglect the first two terms in favor of $\omega_p^2 \tilde{\mathcal{P}}$. This will simplify the further evaluation of Eq. (3.82) significantly.

It is beyond the scope of the book to give a detailed description of the various possible nonlinear effects and excitation schemes. Here we shall restrict ourselves to a nonlinearity of second order which is responsible for second harmonic generation (SHG), optical parametric amplification (OPA), and to a nonlinearity of third order describing (self) phase modulation [(S)PM]. The tensor character of the nonlinear susceptibility describes the symmetry properties of the material. For all substances with inversion symmetry, $\chi^{(2n)} = 0$ ($n = 1, 2...$) holds, and therefore no second harmonic processes can be observed in isotropic materials and centrosymmetric crystals for example. In contrast, third-order effects are always symmetry allowed.

3.3.2 Second harmonic generation (SHG)

Second harmonic generation has gained particular importance in ultrashort pulse physics as a means for frequency conversion and nonlinear optical correlation. Owing to the characteristics of ultrashort pulses, a number of new features unknown in the conversion of cw light have to be considered [66, 67, 68, 69, 70]. We will examine first the relatively simple case of type I SHG, in which the fundamental wave propagates as an ordinary (o) or extraordinary (e) wave, producing an extraordinary or ordinary second harmonic (SH) wave, respectively. We will briefly discuss at the end of this section the more complex case of type II SHG, in which the nonlinear polarization, responsible for the generation of a second harmonic propagating as an e wave, is proportional to the product of the e and o components of the fundamental. We will see that, in the case of type I, group velocity mismatch between the fundamental and the SH leads generally to a reduced conversion efficiency and pulse broadening. In the case of type II however, it is possible to have simultaneously high conversion efficiency and efficient compression of the second harmonic in presence of group velocity mismatch.

Type I second harmonic generation

Let us assume a light pulse incident upon a second harmonic generating crystal. The electric field propagating inside the material consists of the original field (subscript 1) and the second harmonic field (subscript 2). The total field obeys a wave equation similar to Eq. (3.82):

$$\left[\left(\frac{\partial}{\partial z} + \frac{1}{v_1}\frac{\partial}{\partial t} - \frac{ik_1''}{2}\frac{\partial^2}{\partial t^2}\right)\tilde{\mathcal{E}}_1 + \mathcal{D}_1\right]e^{i(\omega_1 t - k_1 z)}$$
$$+ \frac{k_2}{k_1}\left[\left(\frac{\partial}{\partial z} + \frac{1}{v_2}\frac{\partial}{\partial t} - \frac{ik_2''}{2}\frac{\partial^2}{\partial t^2}\right)\tilde{\mathcal{E}}_2 + \mathcal{D}_2\right]e^{i(\omega_2 t - k_2 z)} + c.c. = i\frac{\mu_0}{2k_1}\frac{\partial^2}{\partial t^2}P^{(2)} \quad (3.84)$$

where the second-order polarization can be written as

$$P^{(2)} = \epsilon_0 \chi^{(2)} \frac{1}{4}\left[\tilde{\mathcal{E}}_1 e^{i(\omega_1 t - k_1 z)} + \tilde{\mathcal{E}}_2 e^{i(\omega_2 t - k_2 z)}\right]^2 + c.c. \quad (3.85)$$

Since the group velocities v_1 and v_2 are not necessarily equal there is no coordinate frame in which both the fundamental and SH pulses are at rest[4]. Therefore z and t are the (normal) coordinates in the laboratory frame. With the simplifications introduced above for the polarization, and neglecting group velocity dispersion and higher-order dispersion, we obtain two

---

[4]For ease of notation, we have elected to drop the subscript g to specify "group velocity". Unless otherwise specified, v_i will denote group velocities for the remainder of this chapter.

3.3. NONLINEAR, NONRESONANT OPTICAL PROCESSES

coupled differential equations for the amplitude of the fundamental wave

$$\left(\frac{\partial}{\partial z} + \frac{1}{v_1}\frac{\partial}{\partial t}\right)\tilde{\mathcal{E}}_1 = -i\chi^{(2)}\frac{\omega_1^2}{2c^2 k_1}\tilde{\mathcal{E}}_1^* \tilde{\mathcal{E}}_2 e^{i\Delta k z} \qquad (3.86)$$

and for the second harmonic (SH) wave

$$\left(\frac{\partial}{\partial z} + \frac{1}{v_2}\frac{\partial}{\partial t}\right)\tilde{\mathcal{E}}_2 = -i\chi^{(2)}\frac{\omega_2^2}{4c^2 k_2}\tilde{\mathcal{E}}_1^2 e^{-i\Delta k z}. \qquad (3.87)$$

where $\Delta k = 2k_1 - k_2$ is the wave vector mismatch. Since k_1, k_2 are functions of the orientation of the wave vector with respect to the crystallographic axis, it is often possible to find crystals, beam geometry and beam polarizations, for which $\Delta k = 0$ (phase matching) is achieved [65, 10, 11]. In the case of ultrashort pulses, however, k_1 and k_2 vary over the bandwidth of the pulse. The variation of the wave vectors k_i with frequency has only been taken into account to first order in Eqs. (3.86), (3.87) by including the group velocity v_1 or v_2 in the left hand side. The phase matching condition established for the central frequency ω_1 and its second harmonic $(2\omega_1)$ does not hold for terms of second and higher order in the expansion of k_i. To simplify the discussion on effects typical for the conversion of very short light pulses, we will neglect any change in intensity due to focusing effects; approximation which holds for nonlinear materials shorter than the Rayleigh range.

Type I — small conversion efficiencies

Assuming exact phase matching at the pulse carrier frequency, small conversion efficiencies occur at low input intensities and/or small length of the nonlinear medium and nonlinear susceptibility. Under these circumstances we may assume that the fundamental pulse does not suffer losses, $\tilde{\mathcal{E}}_1(z) = \tilde{\mathcal{E}}_1(0)$, and Eq. (3.87) can be integrated directly to yield for the second harmonic field at $z = L$:

$$\tilde{\mathcal{E}}_2\left(t - \frac{L}{v_2}, L\right) = -i\chi^{(2)}\frac{\omega_2^2}{4c^2 k_2}\int_0^L \tilde{\mathcal{E}}_1^2\left[t - \frac{L}{v_2} + \left(\frac{1}{v_2} - \frac{1}{v_1}\right)z\right] dz. \qquad (3.88)$$

The term $(v_2^{-1} - v_1^{-1})z$ in the argument of $\tilde{\mathcal{E}}_1$ describes the walk-off between the second harmonic pulse and the pulse at the fundamental wavelength owing to the different group velocities. The result is a broadening of the second harmonic pulse, as can be seen from Fig. 3.8. Only for crystal lengths

$$L \ll L_D^{SHG} = \frac{\tau_{p1}}{|v_2^{-1} - v_1^{-1}|} \qquad (3.89)$$

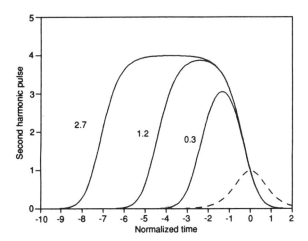

Figure 3.8: Second harmonic pulse at different, normalized crystal lengths, L/L_D^{SHG} according to Eq. (3.88). ($---$ input sech - pulse; the intensity is not to scale)

can the influence of the group velocity mismatch be neglected. In this case the SHG intensity varies with the square of the product of crystal length and intensity of the fundamental, cf. Eq. (3.88). Because of this quadratic dependence, the second harmonic pulse is shorter than the fundamental pulse (by a factor $\sqrt{2}$ for Gaussian pulses). For $L \gg L_D^{SHG}$ the pulse duration is determined by the walk-off and approaches a value of $L/|v_2^{-1} - v_1^{-1}|$, the peak power remains constant, and the energy increases linearly with L. Of course, one needs to avoid this regime if short second harmonic pulses are required. The group velocity mismatch between the fundamental and SH pulse is listed in Table 3.2 for some typical crystals used for SHG.

It is interesting to note what happens when the phase matching condition is not satisfied ($\Delta k \neq 0$). The introduction of $\exp(-i\Delta k z)$ in the integrand of Eq. (3.88) produces a periodically varying second harmonic output. The periodicity length is given by

$$L_P^{SHG} = \frac{2\pi}{\Delta k} \qquad (3.90)$$

if group velocity mismatch can be neglected. In such cases it is recommended to work with crystal lengths $L < L_P^{SHG}$. If, in addition, the group velocity mismatch is significant, evaluation of relation (3.88) yields for the spectral

3.3. NONLINEAR, NONRESONANT OPTICAL PROCESSES

crystal	λ [nm]	θ [°]	$(v_2^{-1} - v_1^{-1})$ [fs/mm]
KDP	550	71	266
	620	58	187
	800	45	77
	1000	41	9
LiIO$_3$	620	61	920
	800	42	513
	1000	32	312
BBO	500	52	680
	620	40	365
	800	30	187
	1000	24	100
	1500	20	5

Table 3.2: Phase matching angle θ and group velocity mismatch $(v_2^{-1} - v_1^{-1})$ for type-I phase matching (oo-e) in some negative uni-axial crystals. The data were obtained from Sellmeier equations, see [71, 72, 73].

intensity of the second harmonic

$$\begin{aligned} S_2(\Omega) &= \frac{\epsilon_0 c n}{4\pi} |\tilde{\mathcal{E}}_2(\Omega, L)|^2 \\ &= \frac{\epsilon_0 c n}{4\pi} \left(\frac{\chi^{(2)} \omega_2^2 L}{4c^2 k_2} \right)^2 \text{sinc}^2 \left\{ \left[(v_2^{-1} - v_1^{-1})\Omega - \Delta k \right] \frac{L}{2} \right\} \\ &\quad \times \left| \int_{-\infty}^{\infty} \tilde{\mathcal{E}}_1(\Omega - \Omega') \tilde{\mathcal{E}}_1(\Omega') d\Omega' \right|^2 . \end{aligned} \qquad (3.91)$$

The last equation indicates that group velocity mismatch causes the SHG process to act as a frequency filter. The bandwidth becomes narrower with increasing crystal length. The sinc^2 term in Eq. (3.91) introduces a modulation of the spectrum of the second harmonic. The period of that modulation can serve to estimate the group velocity mismatch $(v_2^{-1} - v_1^{-1})$ in the particular crystal used.

Type I — large conversion efficiencies

The simple approach of the previous section does no longer apply to conversion efficiencies larger than a few tens of percent. We have to consider

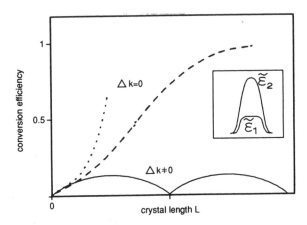

Figure 3.9: Conversion efficiencies neglecting (···) and taking into account (– – –) depletion of the fundamental wave. The inset illustrates the shaping of the SH and fundamental pulse in the crystal.

the depletion of the fundamental pulse as the second harmonic pulse grows according to the complete system of differential equations (3.86), (3.87). In the phase and group velocity matching regime, the second harmonic energy approaches its maximum value asymptotically. Because of their lower intensities, the pulse wings reach this "saturation" regime later and the second harmonic pulse duration τ_{p2} broadens until it reaches a value of about τ_{p1}. Therefore, even a moderate energy conversion requires very high conversion efficiencies for the peak intensities. Figure 3.9 shows schematically the conversion efficiencies in various regimes for zero group velocity mismatch (long pulses).

With the inclusion of group velocity and phase mismatch, the processes involved in SHG become very complex. Numerical studies of Eqs. (3.86) and (3.87) in [74, 75, 76] reveal pulse splitting and a periodical behavior of the conversion efficiency with propagation length under certain circumstances. The complexity results partly from the fact that the phase of the fundamental wave becomes dependent on the conversion process. For cw light, the fundamental phase can be obtained from Eqs. (3.86) and (3.87) and reads [75]

$$\varphi_1(z) = \frac{1}{2} \arccos \left[\frac{c^2 k_1 \tilde{\mathcal{E}}_2(z)}{\chi^{(2)} \omega_1^2 \tilde{\mathcal{E}}_1^2(z)} \Delta k \right] - \frac{\pi - \Delta k z}{4}. \qquad (3.92)$$

3.3. NONLINEAR, NONRESONANT OPTICAL PROCESSES

This phase is responsible for a new phase mismatch $\Delta k_{\text{eff}}(z) = \varphi_2(z) - 2\varphi_1(z)$ which, as opposed to the Δk introduced earlier, is a function of the field amplitudes. The result is that the conversion efficiency drops more rapidly for spectral components for which $\Delta k \neq 0$. Thus, the SH process acts like an intensity dependent spectral filter for short pulses, reducing the conversion efficiency and leading to distortions of the temporal profile. As shown experimentally by Kuehlke and Herpers [77], an optimum input intensity can exist for maximum energy conversion of fs pulses. Usually these efficiencies do not exceed a few tens of percent.

Type I — compensation of the group velocity mismatch

A nonzero group velocity mismatch limits the frequency-doubling efficiency of femtosecond light pulses to a few ten percent. It is interesting to note that the group velocity mismatch is equivalent to the fact that the phase matching condition does not hold over the entire pulse spectrum. We want to leave the actual workout of this fact as one of the problems at the end of this chapter. Generally, it is not possible to match the group velocities by choosing suitable materials while keeping the phase matching condition for the center frequencies, $\Delta k = 0$, as indicated in Table 3.2. However, since phase matching is achieved most often by angular tuning, simultaneous phase matching of an extended spectrum is feasible by realizing different angle of incidence for different spectral components. Corresponding practical arrangements for fs light pulses were suggested in Refs. [78] and [79], and implemented for sum frequency generation to 193 nm [80] (Fig. 3.10). Two gratings are used to disperse and recollimate the beam, respectively. Two achromatic lenses (or telescopes [79]) image A onto the crystal and onto B to ensure zero group velocity dispersion. The combination of L_1 and

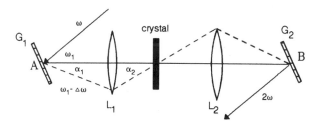

Figure 3.10: Frequency doubler for ultrashort (broadband) light pulses (adapted from [78]).

G_1 enables different angles of incidence for different spectral components. The magnification of lens L_1 is determined by the angular dispersion of the grating $a_1 = d\beta/d\Omega$ and the derivative of the phase matching angle $a_2 = d\theta/d\Omega$

$$M_1 = \frac{d\theta/d\Omega}{d\beta/d\Omega}. \tag{3.93}$$

The magnification of the second lens has to be chosen likewise, taking into account the doubled frequency at the output of the crystal.

Second harmonic generation type II

Group velocity mismatch does not always lead to pulse broadening. Let us consider a phase matched situation where a short second harmonic pulse, of sufficient intensity to deplete the fundamental, is seeded at the trailing edge of a longer fundamental pulse. If the group velocity of the second harmonic v_2 is larger than the velocity of the fundamental v_1, the leading edge of the second harmonic always sees an undepleted fundamental, and is therefore more amplified than the trailing edge. This situation is exploited in second harmonic type II to achieve a compressed SH pulse (as compared to the fundamental pulse duration), with high conversion efficiencies.

Let us extend the system of equations (3.86), (3.87), to second harmonic type II. We choose a retarded time frame of reference traveling with the second harmonic signal at a velocity v_2. The fundamental pulse has a component $\tilde{\mathcal{E}}_o(t) \exp[i(\omega_1 t - k_o z)]$ propagating as an ordinary wave (subscript o) at the group velocity v_o, and a component $\tilde{\mathcal{E}}_e(t) \exp[i(\omega_1 t - k_e z)]$ propagating as an extraordinary wave (subscript e) at the group velocity v_e. The system of equations describing the decay of the fundamental pulses $\tilde{\mathcal{E}}_o$ and $\tilde{\mathcal{E}}_e$, and the growth of the second harmonic wave $\tilde{\mathcal{E}}_2$ is:

$$\left[\frac{\partial}{\partial z} + \left(\frac{1}{v_o} - \frac{1}{v_2}\right)\frac{\partial}{\partial t}\right]\tilde{\mathcal{E}}_o = -i\chi^{(2)}\frac{\omega_1^2}{2c^2 k_o}\tilde{\mathcal{E}}_e^* \tilde{\mathcal{E}}_2 e^{i\Delta k\, z} \tag{3.94}$$

$$\left[\frac{\partial}{\partial z} + \left(\frac{1}{v_e} - \frac{1}{v_2}\right)\frac{\partial}{\partial t}\right]\tilde{\mathcal{E}}_e = -i\chi^{(2)}\frac{\omega_1^2}{2c^2 k_e}\tilde{\mathcal{E}}_o^* \tilde{\mathcal{E}}_2 e^{i\Delta k\, z} \tag{3.95}$$

$$\frac{\partial}{\partial z}\tilde{\mathcal{E}}_2 = -i\chi^{(2)}\frac{\omega_1^2}{c^2 k_2}\tilde{\mathcal{E}}_e \tilde{\mathcal{E}}_o e^{-i\Delta k\, z} \tag{3.96}$$

where $\Delta k = k_o + k_e - k_2$ is the wave vector mismatch. We have so far made no assumption as to the relative magnitude, phase and position of the

3.3. NONLINEAR, NONRESONANT OPTICAL PROCESSES

crystal	θ	L_o (cm)	L_e (cm)	$|m|$
KDP	59.2°	20.9	-15.8	1.32
DKDP	53.5°	28.1	-20.0	1.40

Table 3.3: Relevant constants for second harmonic generation type II in KDP and DKDP at 1.06 μm. m is the ratio of the group lengths.

two fundamental pulses $\tilde{\mathcal{E}}_e(t)$ and $\tilde{\mathcal{E}}_o(t)$. Important parameters in selecting a crystal are the walk-off lengths $L_e = \tau_p/[v_e^{-1} - v_2^{-1}]$, and $L_o = \tau_p/[v_o^{-1} - v_2^{-1}]$, where τ_p is the fundamental pulse duration.

The particularly interesting situation analyzed in detail by Wang and Dragila [70] and Stabinis et al. [81] is that where the walk-off lengths are equal and opposite in sign, and only three or four times longer than the crystal length. The crystal angle θ; the walk-off lengths for a 12 ps pulse at 1.06 μm, and their ratio m are listed in Table 3.3 for second harmonic generation type II in KDP and DKDP.

In order to generate compressed second-harmonic pulses [82], the faster e wave is sent delayed with respect to the o wave in the crystal. A second harmonic "seed" originates from the short overlap region between the two e and o fundamental pulses. As this second harmonic propagates though the crystal it is amplified, while the overlap between all three pulses increases. Because of its faster group velocity, the second harmonic always sees an undepleted o wave at its leading edge. Compression of the second harmonic results from the differential amplification of the leading edge with respect to the trailing edge. This mechanism can only account for a modest compression factor (approximately 5). Much larger compression factors have been observed (in excess of 30 [82]), which involve regeneration of the fundamental.

To model the mechanism leading to larger compression factors, it is necessary to take into account the frequency dependence of the phase mismatch factors Δk in Eqs. (3.94), (3.95) and (3.96). It is left to a problem at the end of this chapter to derive the following set of coupled propagation equations in the frequency domain, without restrictions on the frequency dependence of the k vector:

$$\frac{\partial}{\partial z}\tilde{\mathcal{E}}_o(\Omega) = -i\chi^{(2)}\frac{\omega_1^2}{2c^2 k_o}\int_{-\infty}^{\infty}\tilde{\mathcal{E}}_e^*(\Omega - \Omega')\tilde{\mathcal{E}}_2(\Omega')e^{i\Delta k(\Omega,\Omega')z}d\Omega' \qquad (3.97)$$

$$\frac{\partial}{\partial z}\tilde{\mathcal{E}}_e(\Omega) = -i\chi^{(2)}\frac{\omega_1^2}{2c^2 k_e}e^{i\Delta k(\Omega)z}\int_{-\infty}^{\infty}\tilde{\mathcal{E}}_o^*(\Omega-\Omega')\tilde{\mathcal{E}}_2(\Omega')e^{i\Delta k(\Omega,\Omega')z}d\Omega' \quad (3.98)$$

$$\frac{\partial}{\partial z}\tilde{\mathcal{E}}_2(\Omega) = -i\chi^{(2)}\frac{\omega_1^2}{c^2 k_2}e^{-i\Delta k(\Omega)z}\int_{-\infty}^{\infty}\tilde{\mathcal{E}}_e(\Omega-\Omega')\tilde{\mathcal{E}}_o(\Omega')e^{-i\Delta k(\Omega,\Omega')z}d\Omega' \quad (3.99)$$

where $\Delta k(\Omega,\Omega') = k_o(\omega_1 + \Omega') + k_e(\omega_1 + \Omega - \Omega') - k_2[2(\omega_1 + \Omega)]$ is the wave vector mismatch with the complete frequency dependence. Since exact phase matching cannot be achieved over the full bandwidth of the pulse, regeneration of the fundamental occurs for some spectral components. This particularly complex regime where a high intensity, short pulse duration and long crystal length combine to result in partial regeneration of the pump radiation is sometimes labelled "giant pulse regime". Computer simulations with 10 ps pulses at 1.06 μm predict a compression factor larger than 60 for long KDP crystals ($\geq$ 5 cm) and high high pulse powers ($>$ 30 mJ), leading to conversion efficiencies of the order of 30%. The implementation of this technique and experimental details will be discussed in Chapter 5.

3.3.3 Optical parametric interaction

Coupled field equations

Similar considerations as in the previous section can be made for a number of other nonlinear processes of second order used for generating pulses at new frequencies. Figure 3.11 shows schematically three possible situations. In parametric up-conversion two pulses of frequencies ω_1 and ω_2, respectively, are sent through a nonlinear medium (crystal) and produce a pulse of frequency $\omega_3 = \omega_1 + \omega_2$. In parametric down-conversion a pulse with the difference frequency is generated. In parametric oscillation, a single pulse of frequency ω_3 generates two pulses of frequency ω_1 and ω_2 such that $\omega_1 + \omega_2 = \omega_3$. Which process occurs depends on the realization of the phase matching condition. In principle, to obtain a fs output pulse through

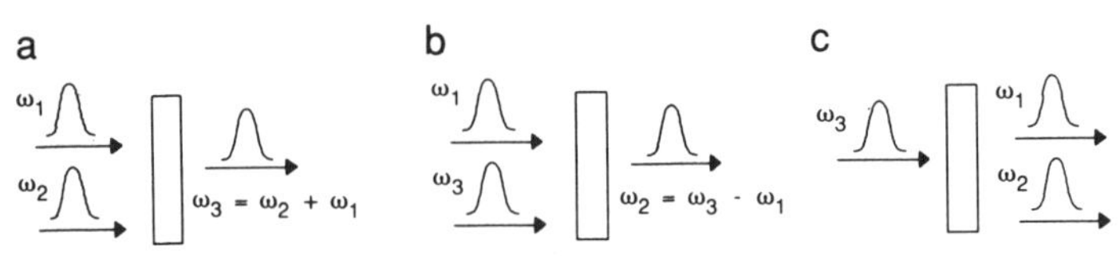

Figure 3.11: Nonlinear optical processes of second order for generating pulses of new frequencies. (a) parametric up-conversion, (b) parametric down-conversion, (c) parametric oscillation.

3.3. NONLINEAR, NONRESONANT OPTICAL PROCESSES 135

up- or down-conversion, it is sufficient to have only one fs input pulse. The second input can be a longer pulse or even cw light. Mokhtari et al. [83], for example, mixed 60 fs pulses at 620 nm from a dye laser with 85 ps pulses from a Nd:YAG laser (1064 nm) to obtain up converted fs pulses at 390 nm. Parametric frequency mixing of two fs input pulses were reported, for example, in [84] and [85]. If the input pulse (pump pulse) is sufficiently strong, two pulses of frequencies ω_1 and ω_2, for which the phase matching condition is satisfied, can arise. In this case, noise photons which are always present in a broad spectral range can serve as seed light. This process is known as optical parametric oscillation; the generated pulses are called idler and signal pulse — the usual convention being that the signal is the generated radiation with the shorter wavelength.

With similar assumptions that allowed us to derive the equations for SHG, we obtain three coupled differential equations for the interaction of the three optical fields as shown in Fig. 3.11:

$$\left(\frac{\partial}{\partial z} + \frac{1}{v_1}\frac{\partial}{\partial t}\right)\tilde{\mathcal{E}}_1 = -i\chi^{(2)}\frac{\omega_1^2}{2c^2 k_1}\tilde{\mathcal{E}}_2^*\tilde{\mathcal{E}}_3 e^{i\Delta kz} \qquad (3.100)$$

$$\left(\frac{\partial}{\partial z} + \frac{1}{v_2}\frac{\partial}{\partial t}\right)\tilde{\mathcal{E}}_2 = -i\chi^{(2)}\frac{\omega_2^2}{2c^2 k_2}\tilde{\mathcal{E}}_1^*\tilde{\mathcal{E}}_3 e^{i\Delta kz} \qquad (3.101)$$

$$\left(\frac{\partial}{\partial z} + \frac{1}{v_3}\frac{\partial}{\partial t}\right)\tilde{\mathcal{E}}_3 = -i\chi^{(2)}\frac{\omega_3^2}{2c^2 k_3}\tilde{\mathcal{E}}_1\tilde{\mathcal{E}}_2 e^{-i\Delta kz} \qquad (3.102)$$

where $\Delta k = k_1 + k_2 - k_3$. The coupling is most effective and the conversion efficiencies are maximum if the phase matching condition, $\Delta k = 0$, is satisfied. The description of the various processes in Fig. 3.11 by Eqs. (3.100)–(3.102) differs only in the initial conditions, that is the field amplitudes at the crystal input. This system of equations is analogous to the ones encountered for second harmonic generation. For relatively weak pulses, conversion efficiencies are low, and the group velocity dispersion contributes to a broadening of the generated radiation.

As in the case of SHG, there is a particularly interesting regime which combines the complexities of short pulse, high intensity and long interaction lengths. The pulses have to be sufficiently short that simultaneous phase matching cannot be achieved over the pulse bandwidth. The crystal length and pulse intensities are sufficiently high for regeneration of the pump to occur. These conditions are also referred as "giant pulse regime", or sometimes "nonlinear parametric generation". These experimental conditions leading to pulse (idler or signal) compression will be discussed in Chapter 5.

Synchronous pumping

Higher efficiencies can be obtained by placing the nonlinear crystal in an optical resonator for the signal pulse. The crystal is then pumped by a sequence of pulses whose temporal separation exactly matches the resonator round trip time. In this manner, the signal pulse passes through the amplifying crystal many times before it is coupled out. Laenen et al. [86] pumped the crystal with a train of 800 fs pulses from a frequency doubled Nd:glass laser and produced 65–260 fs signal pulses which were tunable over a range from 700 to 1800 nm. Edelstein et al. [87] placed the crystal in the resonator of a fs dye laser to utilize the high intracavity pulse power for the pumping. A second resonator was then built around the crystal for the signal pulse. At repetition frequencies of about 80 MHz, the mean output power of the parametric oscillator was on the order of several milliwatts.

Chirp amplification

So far we have been concerned with amplitude modulation effects in nonlinear mixing. Phase modulation introduces an element of complexity that one generally tries to avoid, in particular in the conditions of "giant pulse compression" cited earlier. It has been recognized however that second order interactions can be used to generate or amplify a phase modulation. We will consider here as an example chirp amplification that can take place in parametric processes.

The frequencies of the three interacting pulses obey the relation

$$\omega_3(t) = \omega_1(t) + \omega_2(t). \tag{3.103}$$

If we substitute for the time dependent frequencies $\omega_i(t) = \omega_i + \dot{\varphi}_i(t)$ ($i = 1, 2, 3$), we obtain for the phases

$$\dot{\varphi}_3(t) = \dot{\varphi}_2(t) + \dot{\varphi}_1(t). \tag{3.104}$$

In addition, for efficient parametric oscillation, the phase matching condition must be satisfied, which now implies

$$k_3[\omega_3(t)] = k_2[\omega_2(t)] + k_1[\omega_1(t)]. \tag{3.105}$$

For $|\dot{\varphi}_i| \ll \omega_i$ and linearly chirped pump pulses a Taylor expansion of Eq. (3.105) yields

$$\left.\frac{dk_1}{d\Omega}\right|_{\omega_3} \dot{\varphi}_3(t) = \left.\frac{dk_2}{d\Omega}\right|_{\omega_2} \dot{\varphi}_2(t) + \left.\frac{dk_3}{d\Omega}\right|_{\omega_1} \dot{\varphi}_1(t). \tag{3.106}$$

3.3. NONLINEAR, NONRESONANT OPTICAL PROCESSES

The chirps at the three frequencies are thus related by the *group* velocities v_i. From Eqs. (3.104) and (3.106), a relation between the chirp of idler and signal pulse and pump pulse can be found:

$$\dot{\varphi}_1 = p\dot{\varphi}_3 \tag{3.107}$$
$$\dot{\varphi}_2 = (1-p)\dot{\varphi}_3, \tag{3.108}$$

where p is the chirp enhancement coefficient:

$$p = \left(\frac{v_3^{-1} - v_2^{-1}}{v_1^{-1} - v_2^{-1}} \right), \tag{3.109}$$

and v_i are the group velocities at the respective frequencies ω_i.

Equation (3.109) indicates that chirp amplification is most pronounced in a condition of degenerescence where $\omega_1 \approx \omega_2$. The mechanism of chirp amplification can also be understood graphically on the tuning curves. For instance, Fig. 3.12 shows the tuning curves for phase matching (type I) in KDP. A small change in pump wavelength λ_3 results in a strong change of signal wavelength. As a result, a slightly chirped pump pulse generates signal and idler with enhanced (and opposite) chirp [88] which can also be of opposite sign (chirp reversal). The chirp enhancement can be two orders of magnitude. The signal pulses with enhanced chirp can be compressed in a grating pair compressor. Using the natural chirp of frequency doubled pulses from a Nd:glass laser, pulses of 50 fs at 920 nm were obtained by this technique [88].

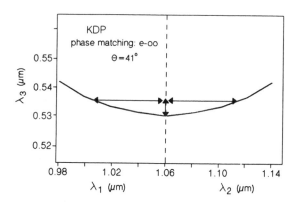

Figure 3.12: Tuning curves of a KDP optical parametric oscillator, pumped at a wavelength of $\lambda_3 = 0.53$ μm (from [88]).

3.3.4 Nonlinear phase modulation

Fundamentals

In the transparent region of many materials, the refractive index depends nonlinearly on the propagating field. Usually the lowest order of this dependence is expressed by one of the following equivalent relations

$$\begin{aligned} n &= n_0 + n_2|\tilde{\mathcal{E}}(t)|^2 \\ &= n_0 + 2n_2\langle E^2(t)\rangle \\ &= n_0 + \bar{n}_2 I(t) \end{aligned} \quad (3.110)$$

where $\bar{n}_2 = 2n_2/(\epsilon_0 c n_0)$. The quantity n_2 is called nonlinear index coefficient and describes the strength of the coupling between the electric field and the refractive index. Many different physical processes can account for an intensity dependent change in index of refraction. Table 3.4 gives some examples. As a rule of thumb, the larger the nonlinearity, the longer the corresponding response time. For fs pulse excitation, it is only a (nonresonant) nonlinearity of electronic origin that can be considered to be without inertia. The corresponding nonlinear refractive index can be described by relations (3.110), and is a result of an optical nonlinearity of third order. If only one pulse is incident on the sample the corresponding polarization reads

$$P^{(3)} = \epsilon_0 \chi^{(3)} E^3 = \epsilon_0 \chi^{(3)} \left(\frac{3}{8}|\tilde{\mathcal{E}}|^2 \tilde{\mathcal{E}} e^{i\omega_\ell t} + \frac{1}{8}\tilde{\mathcal{E}}^3 e^{3i\omega_\ell t} \right) + c.c. \quad (3.111)$$

assuming an instantaneous response. The terms with $3\omega_\ell$ in the argument of the exponential function describe third harmonic generation. In many optical materials which we will consider here this process is not effective

Origin	Example	$\bar{n}_2$ [cm^2/W]	Response time [s]
electronic			
nonresonant	glass	10^{-16}—10^{-15}	10^{-15}—10^{-14}
resonant	semic. doped glass	10^{-10}	10^{-11}
molecular motion	CS_2	10^{-12}	10^{-12}

Table 3.4: Examples of nonlinear refractive index parameters

3.3. NONLINEAR, NONRESONANT OPTICAL PROCESSES 139

and, therefore, will be neglected. From Eq. (3.111) it is then easy to show that

$$n_2 = \frac{3\chi^{(3)}}{8n_0}. \tag{3.112}$$

The intensity dependence of n implies refractive index varying in time and space. The temporal variation, as discussed in Chapter 1 [Eq. (1.124)] leads to a pulse chirp. The (transverse) spatial refractive index dependence leads to lensing effects. These processes are called self-phase modulation (SPM) and self-focusing, respectively. In most cases self-focusing is undesirable and is avoided by minimizing the sample length and/or working with uniform beam profiles whenever possible.

To describe SPM we can substitute Eq. (3.111) into the wave equation (3.82). Let us first recall the approximations needed to derive the simple expression for the second derivative of the polarization used in the previous subsection. These approximations lead to an estimate of the first correction terms. If the nonlinearity is not perfectly inertialess (i.e., does not respond instantaneously to the electric field), we have to compute the polarization using the integral expression (3.81). For response times smaller than the pulse duration, the Fourier transform of Eq. (3.81) can be expanded into a Taylor series about ω_ℓ. Termination of the series after the second term and back-transformation into the time domain yields for the polarization, in terms of the field envelope:

$$P^{(3)}(t) = \frac{3}{8}\epsilon_0 \left[\chi^{(3)} |\tilde{\mathcal{E}}|^2 \tilde{\mathcal{E}} + i \left. \frac{\partial \chi^{(3)}}{\partial} \right|_{\omega_\ell} \frac{\partial}{\partial t}\left(|\tilde{\mathcal{E}}|^2 \tilde{\mathcal{E}} \right) \right] e^{i(\omega_\ell t - k_\ell z)} + c.c. \tag{3.113}$$

The above expression (3.113) restates once more that a non-zero response time τ_r leads to a frequency dependence of the susceptibility. The critical parameter is the spectral variation of this susceptibility over the frequency range covered by the pulse spectrum. To study nonlinear propagation problems, the slowly varying envelope approximation is generally applied to the polarization, as expressed in Eq. (3.83). Inserting the expansion (3.113) into the second derivative of the polarization (3.83) leads to:

$$i\frac{\mu_0}{k_\ell}\frac{d^2}{dt^2}P^{(3)} = \left[-i\frac{n_2 k_\ell}{n_0}|\tilde{\mathcal{E}}|^2\tilde{\mathcal{E}} - \beta\frac{\partial}{\partial t}\left(|\tilde{\mathcal{E}}|^2\tilde{\mathcal{E}}\right) + ... \right] e^{i(\omega_\ell t - k_\ell z)} + c.c. \tag{3.114}$$

where

$$\beta = \frac{n_2}{c}\left(2 - \frac{\omega_\ell}{\chi^{(3)}} \left.\frac{\partial \chi^{(3)}}{\partial \omega}\right|_{\omega_\ell} \right). \tag{3.115}$$

It should be remembered that the temporal derivative in Eq. (3.114) becomes important if the light period is not negligibly short compared to the pulse duration. Equations (3.114) and (3.115) indicate that the first-order correction to the SVEA and the finite response time of the nonlinear susceptibility (the two summands forming β) have the same action on pulse propagation. Corrections to the SVEA may also be important in the spatial (transverse) propagation of the beam.

Pulse propagation through transparent media which is affected by SPM and dispersion has played a key role in fiber optics. For a review we refer to the monograph by Agrawal [13]. In this chapter we will discuss the physics behind SPM and neglect group velocity dispersion. In Chapter 7 we shall describe effects associated with the interplay of SPM and group velocity dispersion.

Short samples with instantaneous response

For interaction lengths much shorter than the dispersion length L_D, and for an instantaneous nonlinearity, the wave equation (3.82) with the source term (3.114) simplifies to

$$\frac{\partial}{\partial z}\tilde{\mathcal{E}}(z,t) = -i\frac{3\omega_\ell^2 \chi^{(3)}}{8c^2 k_\ell}|\tilde{\mathcal{E}}|^2\tilde{\mathcal{E}} = -i\frac{n_2 k_\ell}{n_0}|\tilde{\mathcal{E}}|^2\tilde{\mathcal{E}} \qquad (3.116)$$

For real $\chi^{(3)}$, substituting $\tilde{\mathcal{E}} = \mathcal{E}\exp(i\varphi)$ into Eq. (3.116) and separating the real and imaginary parts results in an equation for the pulse envelope

$$\frac{\partial}{\partial z}\mathcal{E} = 0 \qquad (3.117)$$

and for the pulse phase

$$\frac{\partial \varphi}{\partial z} = -\frac{n_2 k_\ell}{n_0}|\mathcal{E}|^2. \qquad (3.118)$$

Obviously the pulse amplitude $\mathcal{E}$ is constant in the coordinate system traveling with the group velocity, that is, the pulse envelope remains unchanged, $\mathcal{E}(t,z) = \mathcal{E}(t,0) = \mathcal{E}_0(t)$. Taking this into account, we can integrate Eq. (3.118) to obtain for the phase

$$\varphi(t,z) = \varphi_0(t) - \frac{k_\ell n_2}{n_0}z\mathcal{E}_0^2(t) \qquad (3.119)$$

which results in a phase modulation given by

$$\frac{\partial \varphi}{\partial t} = \frac{d\varphi_0}{dt} - \frac{n_2 k_\ell}{n_0}z\frac{d}{dt}\mathcal{E}_0^2(t). \qquad (3.120)$$

3.3. NONLINEAR, NONRESONANT OPTICAL PROCESSES 141

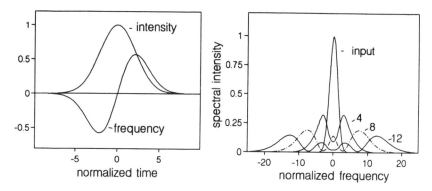

Figure 3.13: Frequency modulation and spectrum of self-phase modulated Gaussian pulses for different propagation lengths z/L_{NL}.

This result can be interpreted as follows. The refractive index change follows the pulse intensity instantaneously. Thus, different parts of the pulse "feel" different refractive indices, leading to a phase change across the pulse. Unlike the phase modulation associated with group velocity dispersion, this self-phase modulation (SPM) produces new frequency components and broadens the pulse spectrum. To characterize the SPM it is convenient to introduce a nonlinear interaction length

$$L_{NL} = \frac{n_0}{n_2 k_\ell \mathcal{E}_{0m}^2} \qquad (3.121)$$

where $\mathcal{E}_{0m}$ is the peak amplitude of the pulse. The quantity z/L_{NL} represents the maximum phase shift which occurs at the pulse peak, as can be seen from Eq. (3.120). Figure 3.13 shows some examples of the chirp and spectrum of self-phase modulated pulses. Because n_2 is mostly positive far from resonances (see problem 5 at the end of this chapter), upchirp occurs in the pulse center. We also see that SPM can introduce a considerable spectral broadening. This process as well as some nonlinear processes of higher order can be used to generate a white light continuum, as discussed in the next section.

For an order of magnitude estimate let us determine the frequency change $\delta\omega$ over the FWHM of a Gaussian pulse. Substituting $\mathcal{E}_{0m} e^{-2\ln 2(t/\tau_p)^2}$ for $\mathcal{E}_0$ in Eq. (3.120) yields

$$\delta\omega\tau_p = \frac{8\ln 2}{\sqrt{2}} \frac{z}{L_{NL}}. \qquad (3.122)$$

For $z = L_{NL}$ the normalized frequency sweep is $\delta\omega\tau_p \approx 4$. Note that the original pulse duration bandwidth product of the unchirped Gaussian input pulse was $\Delta\omega\tau_p \approx 3$.

A pulse can also be phase modulated in the field of a second pulse if both pulses interact in the medium. The phase of pulse 1 is then determined by

$$\frac{\partial \varphi_1}{\partial z} = -\frac{n_2 k_\ell}{n_0} \left(|\tilde{\mathcal{E}}_1|^2 + |\tilde{\mathcal{E}}_2|^2 \right) \tilde{\mathcal{E}}_1. \tag{3.123}$$

This offers the possibility of phase modulating weak pulses by means of a strong pulse. Both pulses can differ in their wavelength, duration, polarization state, etc. This process is known as cross phase modulation and has found several interesting applications [89]. For example, it is possible to transfer information from one pulse train to another by induced spectral changes [90].

Short samples and non-instantaneous response

For very short pulses and/or a non-instantaneous sample response, the source term given by Eq. (3.114) with $\beta \neq 0$ must be incorporated in the wave equation. The pulse propagation is now governed by

$$\frac{\partial}{\partial z}\tilde{\mathcal{E}} + i\frac{n_2 k_\ell}{n_0}|\tilde{\mathcal{E}}|^2\tilde{\mathcal{E}} + \beta\frac{\partial}{\partial t}\left(|\tilde{\mathcal{E}}|^2\tilde{\mathcal{E}}\right) = 0. \tag{3.124}$$

Comparison with Eqs. (1.59) and (1.60) suggests that the term with a time derivative can be interpreted as an intensity dependent group velocity. For $\beta > (<) 0$ the pulse center is expected to travel slower (faster) than the trailing edge. This causes a steepening of the trailing (leading) edge of the pulse, known as "self-steepening." It is similar to the formation of shock waves in acoustics.

To solve Eq. (3.124) we again substitute the complex pulse amplitude by a product of an envelope and a phase function to obtain

$$\frac{\partial}{\partial z}\mathcal{E} + 3\beta\mathcal{E}^2\frac{\partial}{\partial t}\mathcal{E} = 0 \tag{3.125}$$

and

$$\frac{\partial}{\partial z}\varphi + \beta\mathcal{E}^2\frac{\partial}{\partial t}\varphi = -\frac{n_2 k_\ell}{n_0}\mathcal{E}^2. \tag{3.126}$$

The equation for the envelope can be solved independently of the phase equation. Its solution can formally be written as [91]

$$\mathcal{E}(z,t) = \mathcal{E}\left(z=0, t - 3\beta z \mathcal{E}^2(t,z)\right). \tag{3.127}$$

3.3. NONLINEAR, NONRESONANT OPTICAL PROCESSES 143

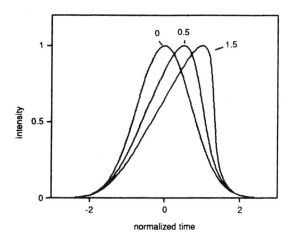

Figure 3.14: Pulse envelope according to Eq. (3.128) for different values of $3\beta z/\tau_G$.

For a Gaussian input pulse, $\mathcal{E}_{0m} e^{-(t/\tau_G)^2}$, we find

$$\mathcal{E}(z,t) = \mathcal{E}_{0m} \exp\left\{-\left[\frac{t - 3\beta z \mathcal{E}^2(z,t)}{\tau_G}\right]^2\right\} \quad (3.128)$$

which contains the envelope implicitly. Figure 3.14 shows the pulse envelope for different values of $3\beta z/\tau_G$. It should be noted that the evolution of a shock pulse with $d\mathcal{E}/dt = \infty$ is not observed in practice because of the unavoidable action of dispersion, which we neglected here for the sake of simplicity. The envelope function can be inserted in Eq. (3.126) to obtain an implicit (analytical) solution for the phase [91], which, however, is rather complex. A numerical evaluation of the pulse spectrum $|\mathcal{F}\{\tilde{\mathcal{E}}(t,z)\}|^2$ reveals an asymmetric behavior, which is expected, since we are now dealing with SPM of asymmetric pulses. Such spectra were reported by Knox et al. [92], for example, as result of SPM of 40 fs pulses in glass. Rothenberg and Grischkowsky directly measured the steepening of pulses in propagating through optical glass fibers [93].

It should also be noted that SPM in connection with saturation, which was discussed earlier, is a particular example of a time dependent sample response. In the case of saturation, there is a memory effect associated with the change of occupation numbers, with a characteristic time determined by the corresponding (energy) relaxation time.

3.4 Continuum generation

One of the most impressive (and simplest) experiments with ultrashort light pulses is the generation of a white light continuum. For a review on this subject see [94]. Provided the pulse is powerful enough, focusing into a transparent material results in a substantial spectral broadening. The output pulse appears on a sheet of paper as a white light flash, even if the exciting pulse is in the near IR or near UV spectral range. However, the continuum does not have a "flat" uniform spectrum. A broad palette of fs laser sources is still desirable to create a continuum with a maximum energy concentration in any particular wavelength range. Continuum generation was first discovered with ps pulses by Alfano and Shapiro [95] and has since been applied to numerous experiments. One of the most attractive fields of application is time-resolved spectroscopy, where the continuum pulse is used as an ultrafast spectral probe.

Spectral super-broadening was observed in many different (preferably transparent) materials including liquids, solids, and gases. Essential processes contributing to the continuum generation are common to all. Figure 3.15 shows as an example a white light continuum generated in a liquid with visible fs pulses [96] and in gas with UV fs pulses [97]. Continuum generation is a rather complex issue which involves changes in the temporal and spatial beam characteristics. With fs pulses, the dominant process and the starting mechanism leading to spectral super-broadening is SPM due to

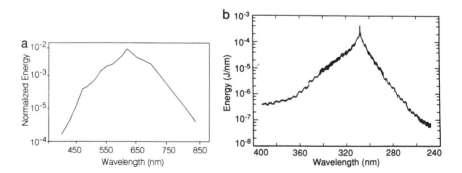

Figure 3.15: Femtosecond spectral super-broadening. (a) In 5 mm CCl_4 with 20 μJ pulses of 100 fs duration focused to a spot size of about 0.01 mm^2 (from [96]). (b) In a 60 cm long Ar cell (40 atm). The pump pulses (4 mJ, 308 nm, 160 fs) were focused with a 50 cm lens (from [97]).

3.5. NONLINEAR SPACE-TIME EFFECTS

an intensity dependent refractive index. However, a number of other nonlinear effects may play a role as well. Indeed, an inspection of Fig. 3.15 shows that not all spectral features can be explained by the action of SPM alone. For example, while the asymmetry can be considered as a result of SPM and self-steepening, the monotonous decrease of the spectral intensity in the wings must have different origins.

In condensed matter SPM is associated with self-focusing which modifies the on-axis intensity. Other nonlinear effects contributing to the continuum can be parametric four-photon mixing and Raman scattering. These processes are particularly efficient if the phase matching condition for the k-vectors of the participating waves is satisfied. Beam distortion occurs in the continuum, since all these k-vectors or the phase matched waves are generally not aligned with the propagation direction of the incident pulse. In general, these nonlinear processes have to be considered in combination with group velocity dispersion. Thin samples $[L \ll L_D]$ reduce the effect of group velocity dispersion and usually also reduce the influence of nonlinearities other than SPM. Owing to the self-phase modulation processes, the different spectral components in the continuum pulse are not uniformly distributed, even if group velocity dispersion has no significant influence. The continuum pulse is strongly chirped and not bandwidth-limited which, for example, was measured by fs frequency-domain interferometry [98]. Sending the continuum pulse through a proper element with group velocity dispersion can therefore result in pulse compression (see Chapter 7). The "ideal" fs continuum pulse is thus a nearly bandwidth-limited fs pulse which is considerably shorter than the original pump pulse. Such continuum pulses enable optical spectroscopy with time resolution better than 10 fs [99, 100].

3.5 Nonlinear space-time effects

The nonlinearities of a medium affect both the temporal and spatial dependences of the electric field of the light. In the previous sections we have avoided this difficulty by assuming a uniform beam profile or neglecting nonlinear space-time coupling effects. However, any nonlinear interaction strong enough to affect the pulse temporal profile will also affect its transverse profile. One example is second harmonic generation for large conversion efficiencies. As sketched in the inset of Fig. 3.9, an initially Gaussian temporal profile will be depleted predominantly in the center, resulting in a flattened shape. The same interaction will also transform an initially Gaussian beam into a square profile.

In this section we will consider as an important example the problem of self-focusing. The intensity dependent index of refraction causes an initially collimated beam to become focused in a medium with $n_2 > 0$. It is the same intensity dependence of the refractive index that causes self-phase modulation of a fs pulse, as we have seen in the previous section. We will first briefly discuss self-focusing of a cw beam, before addressing the effects for dispersion that arise for ultrashort light pulses.

3.5.1 Self-focusing of continuous beams

The action of a nonuniform intensity distribution across the beam profile on a nonlinear refractive index results in a transverse variation of the index of refraction, leading either to focusing or defocusing. Let us assume a cw beam with a Gaussian profile $I = I_0 \exp(-2r^2/w^2)$, and a positive $\bar{n}_2$, as is typical for a nonresonant electronic nonlinearity. The refractive index decreases monotically from the beam center with increasing radial coordinate. One can define a "self-trapping" power $P_{cr,1}$ as the power for which the wavefront curvature (on axis) due to diffraction is exactly compensated by the wavefront curvature due to the self-lensing. We assume that the waist of the Gaussian beam is at the input boundary of the nonlinear medium ($z = 0$). Within the paraxial approximation, diffraction results in a spherical curvature of the wavefront given by the solution of the wave equation [Eq. (1.127)]:

$$\varphi_{\text{diff}} = -\frac{k_\ell}{2R(z)} r^2 \tag{3.129}$$

with $1/R(z) \approx z/\rho_0^2$ near the beam waist [Eq. (1.128) for small z)] and ρ_0 being the Rayleigh range. The action of the nonlinear refractive index results in a radial dependence of the phase:

$$\varphi_{\text{sf}}(r) = -\bar{n}_2 \frac{2\pi}{\lambda_\ell} z I_0 e^{-(2r^2/w_0^2)} \approx -\bar{n}_2 \frac{2\pi}{\lambda_\ell} z I_0 \left(1 - 2\frac{r^2}{w_0^2}\right). \tag{3.130}$$

The critical power $P_{cr,1}$ is reached when the radial parts of Eqs. (3.129) and (3.130) compensate each other:

$$P_{cr,1} = I_0 \frac{\pi w_0^2}{2} = \frac{\lambda_\ell^2}{8\pi n_0 \bar{n}_2} \tag{3.131}$$

where we have made use of $\rho_0 = \pi w_0^2 n_0/\lambda_\ell$. One says that the beam is "self-trapped" because the diffraction characteristic has been modified on axis.

3.5. NONLINEAR SPACE-TIME EFFECTS

This value of critical power is also derived by Marburger [101] by noting that the propagation equation is equivalent to that describing a particle moving in a one-dimensional potential. The condition for which the potential is "attractive" (leads to focusing solutions) is $P \geq P_{cr,1}$.

Another common approach to defining a "self trapping" power is to approximate the radial beam profile by a square distribution of diameter d [65]. If the intensity I_0 inside the tube of light is sufficiently large, the critical angle $\theta_0 = \arccos[n_0/(n_0 + \bar{n}_2 I_0)]$ for total reflection becomes equal to the diffraction angle $\theta_d = 1.22 \lambda_\ell/(2 n_0 d)$. The condition $\theta_0 = \theta_d$ leads to the tube diameter d, hence to the critical power $P_{cr,2} = I_0 \pi d^2/4$:

$$P_{cr,2} = \frac{(1.22)^2 \pi \lambda_\ell^2}{32 n_0 \bar{n}_2} \tag{3.132}$$

The point of agreement between these different definition is the existence of a critical *power* rather than a critical intensity. This result is not surprising, since, for a given power, both the diffraction to be compensated and the lensing effect (nonlinear index) are inversely proportional to the beam diameter. Only numerical calculation can determine the fate of the beam over long propagation distances. These calculations have demonstrated the existence of a critical power P_{cr} [101]:

$$P_{cr} \approx 3.77 P_{cr,1} = 1.03 P_{cr,2} \tag{3.133}$$

For a power that exceeds the critical power the beam reaches a focus after a finite propagation distance z_{SF} given approximately by [65]:

$$z_{SF} = \frac{0.5 \rho_0}{\sqrt{P/P_{cr} - 1}}, \tag{3.134}$$

where it is assumed that the beam enters the nonlinear medium at $z = 0$ at its beam waist. Through the scaling parameter ρ_0 in Eq. (3.134), the self-focusing length is no longer determined by the beam power but depends on the intensity as well. As for the definition of the critical power, numerical calculations led to an improved approximation for z_{SF} [101]:

$$z_{SF} = \frac{0.183 \rho_0}{\sqrt{\left(\sqrt{P/P_{cr}} - 0.852\right)^2 - 0.0219}}. \tag{3.135}$$

Once the beam has reached its focus, higher order linear and nonlinear effects — for instance $\bar{n}_4 I^2$ — come into play. The result can be either total diffraction of the beam, or confinement into a stable waveguide, depending on the sign and magnitude of the higher order terms.

3.5.2 Self-focusing of ultrashort pulses

The self-focusing problem with short pulses is more complex. Self-phase modulation causing spectral broadening and dispersion have to be taken into account, in addition to diffraction and self-lensing. The problem of space-time propagation transcends that of self-focusing: it applies to laser cavities as well, as we will see in Chapter 5.

Assuming a self-focusing nonlinearity, and neglecting nonlinear terms of order higher than $\bar{n}_2 I$, one can derive the following space-time propagation equation by combining Eqs (1.65), (1.126), and (3.116):

$$\left[\frac{\partial}{\partial z} + \frac{i}{2k_\ell} \left(\frac{\partial^2}{\partial x^2} + \frac{\partial^2}{\partial y^2} \right) - \frac{i}{2} k_\ell'' \frac{\partial^2}{\partial t^2} + i \frac{n_2 k_\ell}{n_0} |\tilde{\mathcal{E}}|^2 - \alpha \right] \tilde{\mathcal{E}} = 0. \quad (3.136)$$

The latter equation can represent the propagation of a pulse in a continuous medium with gain/loss coefficient α. As will be explained in Chapter 5, Eq. (3.136) can also be used to approximate the circulation of a pulse through a laser cavity. In this case, each element is considered to produce an infinetisemal change on the pulse, resulting in a differential equation of the form of Eq. (3.136) for the global change in pulse field per round-trip.

In this chapter let us look in more detail to the the complication that the femtosecond time scale brings into the seemingly purely geometric beam propagation problem of self focusing. We have seen that the mere definition of a threshold power for self-focusing of continuous radiation is not a trivial problem. It should not come as a surprise that an ultrashort pulse duration adds an element of complexity. Because GVD leads to pulse broadening (assuming an unchirped input pulse) and to a reduction of peak power, one expects a higher critical power for fs pulses, as opposed to the cw case. The calculation of the complete space-time behavior is very complex and a subject of current research and some controversy. For very short pulses and high intensities some of the approximations that allowed us to derive Eq. (3.136) are questionable. Higher-order linear dispersion, higher-order nonlinearities and the finite response time of the nonlinearities need to be considered, e.g. Refs. [102, 103]. However the major physical mechanism — the interplay of dispersion, diffraction and self-phase modulation — is contained in Eq. (3.136), where we will assume $\alpha = 0$ (medium without gain or absorption) to discuss self-focusing.

Equation (3.136) can be solved numerically, starting for instance with an unchirped Gaussian pulse as initial condition. In a normally dispersive medium ($k_\ell'' > 0$), the numerical calculations carried out near the self-

3.5. NONLINEAR SPACE-TIME EFFECTS

focusing threshold reveal a pulse splitting [104, 105, 106]. First, the pulse broadens due to group velocity dispersion. As the pulse starts to focus, the region near the pulse center undergoes the most self-phase-modulation. Dispersion causes the most modulated portion of the signal to flow away from the pulse center, resulting in the splitting of the initially Gaussian pulse in two pulses.

Given as initial condition a Gaussian pulse and beam profile $\tilde{\mathcal{E}}(x, y, 0, t) = \mathcal{E}_0 \exp(-t^2/\tau_G^2) \exp[-(x^2 + y^2)/w_0^2]$, Luther et al. [106] showed that the solutions of Eq. (3.136) (with $\alpha = 0$) depend only on two parameters:

$$\gamma = \frac{\rho_0}{2L_d} \tag{3.137}$$

$$p = \frac{P_0}{P_{cr}}. \tag{3.138}$$

In Eq. (3.137), L_d is the dispersion length defined in Chapter 1 (see for instance Table 1.2). In Eq. (3.138), P_0 is the peak power of the pulse, and P_{cr} is the critical power for cw radiation defined earlier in Eq. (3.133). These parameters have a simple physical meaning: γ is a measure of the strength of dispersion relative to diffraction, while p describes the strength of the input pulse relative to the nonlinearity. The self-focusing threshold for ultrashort pulses, according to Ref. [106], occurs at a power $P_{TH} = p_{TH} P_{cr}$ at which a characteristic dispersion length scale z_{NLGVD} exceeds the self focusing length z_{SF} introduced in Eq. (3.135). This new dispersion length z_{NLGVD} is defined as the length over which the combined effects of self-phase modulation and normal GVD would reduce the peak power from its input value to the critical power ($p = 1$) in the absence of diffraction. The threshold condition $z_{\text{NLGVD}} = z_{\text{SF}}$ leads to the relations [106]:

$$z_{\text{NLGVD}} \approx 0.5\rho_0 \left[\frac{\sqrt{3.38 + 5.2(p^2 - 1)} - 1.84}{15\gamma p} \right]^{\frac{1}{2}} \tag{3.139}$$

and

$$\gamma \approx \frac{\left[\sqrt{3.38 + 5.2(p_{TH}^2 - 1)} - 1.84\right] \left[(p_{TH}^2 - 0.852)^2 - 0.0219\right]}{2p_{TH}} \tag{3.140}$$

The results of numerical evaluations and comparison with Eqs. (3.139) and (3.140) are summarized in Fig. 3.16.

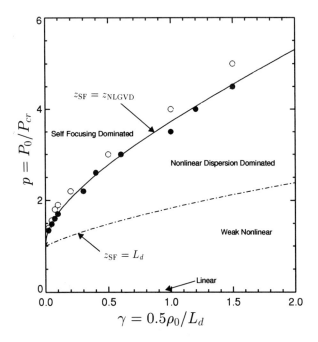

Figure 3.16: Occurence of self-focusing according to Eq. (3.136). The filled (open) circles correspond to those simulations below (above) the threshold P_{TH} for self-focusing. The solid line displays the estimate of Eq. (3.140). The dash-dotted curve gives the estimate taking into account pulse spreading due to linear dispersion only (adapted from [106]).

Similarly as self-focused cw radiation may become trapped in a stable filament, self-focusing of fs pulses may lead to stable "light bullets" able to propagate without distortion in a dispersive, nonlinear medium [107, 108]. As the intensity increases, higher order nonlinear effects take over, as in the cw case. However, a negative $\bar{n}_4 I^2$ which may lead to the formation of a stable filament with cw radiation, will not necessarily stabilize a pulse. In time, the pulse will tend to disperse after a characteristic distance L_d, resulting in a decrease of its peak intensity, hence reducing the self-lensing. The pulse has to be stabilized in time as well as in space to remain a steady state entity. The phase modulation introduced by the $\bar{n}_4 I^2$ term should have a sign such as to cause pulse compression with the normal medium dispersion. We will come back to trapping in space and time in Chapter 12.

3.6 Problems

1. Verify the temporal behavior of the polarization and occupation number as given in Eqs. (3.12) and (3.13).

2. Estimate the refractive index change at a frequency off resonance by $\Delta\omega_F/2$ which is caused by saturation of a homogeneously broadened, absorbing transition with Lorentzian profile. The change in the absorption at resonance was measured to be 0.5. The pulse duration τ_p is much larger than T_1.

3. Show that in the equations governing second harmonic generation [Eqs. (3.86) and (3.87)] the frequency dependence of the wave vector mismatch, $\Delta k(\Omega)$, has been taken into account to first order.

4. Derive Eq. (3.91) for the spectral intensity of a frequency doubled ultrashort light pulse. Using the data given in Table 3.2, estimate the wavelength separation of the two minima in the spectrum next to the maximum if KDP is used. Assume $\Delta k = 0$, a crystal length $L = 1$ mm and 800 nm for the wavelength of the fundamental pulse. Comment on the feasibility to convert fs pulses in this configuration.

5. Show that the nonlinear refractive index coefficient is related to the third-order susceptibility through $n_2 = 3\chi^{(3)}/(8n_0)$ [cf. Eq. (3.112)].

6. Starting from the density matrix equations of a two-level system, find an approximate expression for n_2 which is valid for ω_ℓ far from a single resonance. In particular, comment on the statement following Eq. (3.121) that n_2 is mostly positive.

7. Chirp enhancement through parametric interaction provides an interesting possibility to compress pulses. To illustrate this, let us consider the following simplified model. An initially unchirped Gaussian pulse is sent through a group velocity dispersive element, e.g., piece of glass of length L, leading to linearly chirped output pulses (chirp parameter a). This pulse now serves as a pump in a parametric process producing a (Gaussian) output pulse of the same duration but with an increased chirp parameter (of opposite sign), $a' = -Ra$. A second piece of glass of suitable length can be used to compress this pulse. Calculate the total compression factor in terms of L and R. Note that this configuration would allow us to control the achievable compression factor

simply by changing the group velocity dispersion (e.g., L) of the first linear element which controls the initial chirp.

8. Derive the set of Eqs. (3.97)–(3.99) starting from the propagation equation (1.40) for each of the three fields, within the plane wave approximation. <u>Hint</u>: After taking the Fourier transform of the propagation equation, make the substitution $\tilde{E}_i(z, \omega_i + \Omega) = \tilde{\mathcal{E}}_i(z, \Omega) e^{-ik(\omega_i + \Omega)z}$.

Chapter 4
Coherent Phenomena

This chapter reviews some aspects of coherent interactions between light and matter. By "coherent interaction," it is meant that the exciting pulse is shorter than the phase memory time of the excited medium. Femtosecond pulses have made quite an impact in this field (some selected examples of application are presented in Chapters 9 and 10), because the phase relaxation time of absorbing transitions in condensed matter is generally in the fs range.

Experiments in this field have somewhat contradictory requirements: a source of high coherence, but ultrashort duration. The experimentalist has to walk a tightrope in order to meet the coherence requirements for pulses sometimes shorter than 10 fs.

To be classified in the field of "coherent interactions," the result of a particular experiment should depend on the coherence of the source. This brings us to the question of what constitutes an *incoherent* source? To make a radiation source incoherent, should random phase fluctuations be applied in the time domain or in the frequency domain? Is the temporal coherence always linked to the spatial coherence? These questions are discussed in the first few sections of this chapter.

Coherent interactions with near resonant two-level systems constitute the nucleus of this chapter (Section 3). Section 4 extends the theory of coherent interactions to multilevel molecular or atomic systems.

4.1 From coherent to incoherent interactions

The highest level of coherence is easy to define: monochromatic radiation with a δ-function spectrum, emitted from a point source, has an infinite coherence time and coherence length. Most interesting phenomena of "coherent interaction," however, occur with broad bandwidth radiation, involving temporal variations of either the field amplitude or its phase that are faster than the dephasing time of the system of atoms or molecules being excited. Therefore, a short pulse will be defined as "coherent" if its bandwidth–duration product satisfies the minimum uncertainty relation for that particular pulse shape, as defined in Table 1.1. The lack of coherence can be deterministic. Such is the case for chirped pulses, as discussed in Chapter 1. It can also be of aleatory origin. In the latter case, "incoherence" can be introduced through either

- statistical fluctuations in the time domain; or

- statistical fluctuations in the frequency domain.

In the first approach, given a fixed amount of time during which the radiation is applied, one can introduce "incoherence" through statistical (temporal) fluctuations of the field *amplitude or phase*. The bandwidth of the radiation is increased as a result of the decrease in coherence. In the second approach, the statistical fluctuations are introduced in the amplitude and phase of the field in the frequency domain. The bandwidth of the radiation is fixed.

The distinction between incoherence in the time and frequency domains can best be understood through the experimental implementation of these two concepts. Coherence is a parameter in the interaction of radiation with matter. The influence of coherence on a particular process will be different if incoherence is defined as statistical fluctuations in time or frequency. To illustrate this concept, let us look at the influence of coherence on multiphoton photoionization of atoms. In order to study the influence of coherence on two- and three-photon resonant ionization of Cs and other alkali, Mainfray *et al.* [109] have adjusted the bandwidth of the cavity of a Q-switched Nd:glass laser used to excite a resonant transition. By changing the number of modes let to oscillate simultaneously, the temporal structure of the pulse is also modified. Since the relative phase and amplitude of the oscillating modes are random (the laser is Q-switched, not modelocked), one has introduced statistical fluctuations in the Q-switched pulse as illustrated in Fig. 4.1(a). For the particular interaction being investigated, it was shown in Ref. [110] that, for all practical purposes, the radiation can be considered to be incoherent when 20 or more modes are allowed to oscillate simultaneously. Clearly, the bandwidth of the radiation is modified.

4.1. FROM COHERENT TO INCOHERENT INTERACTIONS 155

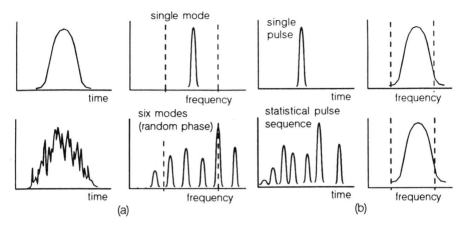

Figure 4.1: Illustration of two different ways to act on the coherence of an optical signal. On the left sides of (a) and (b), the time domain representation is shown. On the right sides, the frequency representation is depicted. Two material resonances (single– or multiphoton) are represented by dashed lines. (a) Modifying coherence by changing the number of oscillating modes of a Q-switched laser. On top, a single oscillating mode, shown off resonance. As the number of oscillating modes, of random amplitude and phase, is increased, resonances can occur (bottom). The corresponding situation in the time domain (left) shows increasing amplitude and phase fluctuations within the pulse duration. The radiation is never completely random, even in the time domain, since the regularity of mode spacing is seen as a repeating pattern of fluctuations. (b) Modifying coherence in the frequency domain by creating a random pulse sequence in the time domain. On the left side, the time domain shows an increasing number of ultrashort pulses of random phase. On the right side, the Fourier amplitude remains within the single pulse bandwidth. The two Fourier transforms have different amplitude and/or phase fluctuations (not shown) under the same envelope. In case (b), with increasing "incoherence," the detuning from resonance with the material transitions (dashed lines) is not modified.

Multiphoton interactions are typically intensity dependent processes. It is not surprising therefore that an increase in multiphoton interaction is observed, because the fluctuations result in larger peak intensities, and hence an increase in nonlinear transition rates. Such an approach is appropriate for the study of non-resonant interactions. However, if the multiphoton process has sharp resonances, such as indicated by the dashed lines in Fig. 4.1(a), then the problem of finding the influence of coherence on the resonance curves is ill defined because the resonance and coherence conditions are not independent. Indeed, in the presence of a sharp resonance, it is not possible to modify independently the parameter "detuning" and the parameter "coherence" [Fig. 4.1(a)]. If the bandwidth of the laser is increased, to allow more modes to oscillate simultaneously, the resonance condition may be modified (if one of the modes coincides with one resonance).

For the study of the influence of source incoherence on near resonant light–matter interaction, it would be desirable to have the radiation confined in a fixed bandwidth. The experimental procedure consists of exciting the sample with a random sequence of pulses [Fig. 4.1(b)]. Each single pulse has the same envelope in the frequency domain, but a different phase corresponding to the random time of arrival in the time domain. Thus, in the frequency domain, within the spectral envelope, the field amplitude and phase become random variables. A comparison of Figs. 4.1(a) and (b) shows that these two approaches are Fourier transforms of each other. The condition of validity of the use of multimode Q-switched pulses to study the influence of coherence on (multiphoton) photo-ionization is that there be no resonance within the broadest bandwidth of excitation. Otherwise, a change in bandwidth will mean a change in resonance condition. The corresponding criterion for the study of coherence in resonant processes with ultrashort statistical pulse sequences is that there has to be a memory time associated with the resonance(s), long enough to establish a correlation between the first and last pulse of the sequence. Otherwise, each pulse of the sequence acts independently, and the result of a two pulse measurement is simply the sum of the results for each individual pulse. This "memory time" is the phase relaxation time of the resonance process, such as was introduced for single photon processes in the previous chapter. As an example, we refer to Ref. [111] for a study of the influence of coherence on multiphoton ionization.

The sequence of pulses represented in Fig. 4.1(b) cannot be defined as "random" out of context. The set of numbers representing the relative phases and delays between pulses are chosen out of a particular statistical distribution, of which the parameters define the "coherence" of the source. A large number of experiments must be averaged to establish the response of a process to the light statistics chosen. Sets of relative phases and delays can be predetermined to match any selected statistical distribution. There are two possible techniques for generating such random pulse sequences. The first one involves simply splitting and delaying pulses with polarizers and mirrors. The second one consists of filtering the Fourier transform of a single pulse.

Should white light be seen as an ensemble of monochromatic sources statistically distributed in amplitude and phase in the frequency domain, or as a random sequence of ultrashort pulses? The question is not purely academic. One or the other aspect dominates, depending on the particular experimental situation. It is the second definition that generally applies to the conditions of coherent interactions discussed in this chapter.

Light is defined as coherent in time if there is a well defined phase relation between the radiation at times t and $t + \delta t$. The typical instrument to measure this type of coherence is the Michelson interferometer. Young's double slit experiment will determine the degree of spatial coherence of a source [6]. A definition of temporal coherence for electromagnetic radiation is the normalized first-order correlation $\langle \tilde{E}(t)\tilde{E}^*(t+\tau)\rangle / \sqrt{\langle I(t)I(t+\tau)\rangle}$. A bandwidth limited ultrashort pulse has thus a coherence of unity. A similar definition can be applied to spatial coherence. A single source of diameter $d \ll \lambda$ has perfect spatial coherence. If the source is an atom, spherical waves are emitted with identical fluctuations at all points at equal distance from the source. Macroscopically, we are used to looking at a volume average of randomly distributed dipoles. Some new sources have the emitted dipoles arranged in a regular fashion. For instance, multiple quantum well lasers are made of "sheets" of emitting atoms which are thinner than the wavelength of the light. This particular arrangement of emitters has important implications for the macroscopic properties of the source [112, 113].

4.2 Coherent interactions with two-level systems

4.2.1 Maxwell-Bloch equations

Whether we are dealing with molecular or atomic transitions, the situation can arise where the ultrashort duration of the optical pulse becomes comparable with – or even less than – the phase relaxation time of the excitation. In the frequency domain, the pulse spectrum is broader than the homogeneous linewidth defined in the first section of Chapter 3. If the pulse is so short that its spectrum becomes much larger than the inhomogeneous linewidth, the medium response becomes similar to that of a single atom. It may seem like a simplified situation when the excitation occurs in a time shorter than all inter-atomic interaction. It is in fact quite to the contrary: in dealing with longer pulses, the faster phase relaxation time of the induced excitation simplifies the light matter response. One is used to dealing with a steady state rather than the "transient" response of light-matter interaction.

We will start from the semi-classical equations for the interaction of near resonant radiation with an ensemble of two-level systems inhomogeneously broadened around a frequency ω_{ih}. The extension to multilevel systems will be discussed in the next section. We refer to the book by Allen and Eberly for more detailed developments [114].

We summarize briefly the results of Chapter 3 for a two-level system, of ground state $|0\rangle$ and upper state $|1\rangle$, excited by the field $E(t)$. The density matrix equation for this two-level system is:

$$\dot{\rho} = \frac{1}{i\hbar}[H_0 - pE, \rho] \tag{4.1}$$

where H_0 is the unperturbed Hamiltonian, and p the dipole moment which is parallel to the polarization direction of the field. Introducing the complex field through $E = \tilde{E}^+ + \tilde{E}^-$ in Eq. (4.1) leads to the following differential equations for the diagonal and off-diagonal matrix elements:

$$\dot{\rho}_{11} - \dot{\rho}_{00} = \frac{2p}{\hbar}\left[i\rho_{01}\tilde{E}^- - i\rho_{10}\tilde{E}^+\right] \tag{4.2}$$

$$\dot{\rho}_{01} = i\omega_0\rho_{01} + \frac{ip\tilde{E}^+}{\hbar}[\rho_{11} - \rho_{00}] \tag{4.3}$$

where ω_0 is the resonance frequency of the two-level system. It is generally convenient to define a complex "pseudo polarization" amplitude $\tilde{Q}$ by

$$i\rho_{01}p\bar{N} = \frac{1}{2}\tilde{Q}\exp(i\omega_\ell t) \tag{4.4}$$

where $\bar{N} = \bar{N}_0 g_{inh}(\omega_0 - \omega_{ih})$ and $\bar{N}_0$ is the total number density of the two-level systems. The real part of $\tilde{Q}$ will describe the attenuation (or amplification for an initially inverted system) of the electric field. Note that $\tilde{Q} = i\tilde{\mathcal{P}}$ where $\tilde{\mathcal{P}}$ is the slowly varying polarization envelope defined in Eq. (3.27). Further we introduce a normalized population inversion:

$$w = p\bar{N}(\rho_{11} - \rho_{00}). \tag{4.5}$$

The complete system of interaction and propagation equations can now be written as:

$$\dot{\tilde{Q}} = i(\omega_0 - \omega_\ell)\tilde{Q} - \kappa\tilde{\mathcal{E}}w - \frac{\tilde{Q}}{T_2} \tag{4.6}$$

$$\dot{w} = \frac{\kappa}{2}[\tilde{Q}^*\tilde{\mathcal{E}} + \tilde{Q}\tilde{\mathcal{E}}^*] - \frac{w - w_0}{T_1} \tag{4.7}$$

$$\frac{\partial\tilde{\mathcal{E}}}{\partial z} = -\frac{\mu_0\omega_\ell c}{2n}\int_0^\infty \tilde{Q}(\omega_0')g_{inh}(\omega_0' - \omega_{ih})d\omega_0'. \tag{4.8}$$

4.2. COHERENT INTERACTIONS WITH TWO-LEVEL SYSTEMS

The quantity $\kappa\mathcal{E}$ with $\kappa = p/\hbar$ is the Rabi frequency. T_1 and T_2 are respectively the energy and phase relaxation times. Most of the energy conserving relaxations are generally lumped in the phase relaxation time T_2. Equation (4.8) has been obtained from Eq. (3.29) by integrating over the polarization of subensembles with resonance frequency ω_0'. The set of Eqs. (4.6)–(4.8) is generally designated as Maxwell–Bloch equations.

Another common set of notations to describe the light-matter interaction uses only real quantities, such as the in-phase (v) and out-of phase (u) components of the pseudo-polarization $\tilde{Q}$, and, for the electric field $\tilde{\mathcal{E}}$, its (real) amplitude $\mathcal{E}$ and its phase φ. Defining

$$\tilde{Q} = (iu + v)e^{i\varphi} \tag{4.9}$$

and substituting in the above system of equations leads to the usual form of Bloch equations[1] for the subensemble of two-level systems having a resonance frequency ω_0.

$$\dot{u} = (\omega_0 - \omega_\ell - \dot{\varphi})v - \frac{u}{T_2} \tag{4.10}$$

$$\dot{v} = -(\omega_0 - \omega_\ell - \dot{\varphi})u - \kappa\mathcal{E}w - \frac{v}{T_2} \tag{4.11}$$

$$\dot{w} = \kappa\mathcal{E}v - \frac{w - w_0}{T_1} \tag{4.12}$$

where the initial value for w at $t = -\infty$ is

$$w_0 = p\bar{N}(\rho_{11}^{(e)} - \rho_{00}^{(e)}). \tag{4.13}$$

The propagation equation Eq. (4.8), in terms of $\tilde{\mathcal{E}}$ and φ, becomes

$$\frac{\partial \mathcal{E}}{\partial z} = -\frac{\mu_0 \omega_\ell c}{2n} \int_0^\infty v(\omega_0')g_{inh}(\omega_0' - \omega_{ih})d\omega_0' \tag{4.14}$$

$$\frac{\partial \varphi}{\partial z} = -\frac{\mu_0 \omega_\ell c}{2n} \int_0^\infty \frac{u(\omega_0')}{\mathcal{E}} g_{inh}(\omega_0' - \omega_{ih})d\omega_0'. \tag{4.15}$$

The vector representation of Feynman et al. [116], for the interaction equations is particularly useful in the description of coherent phenomena.

---

[1]These equations are the electric-dipole analogues of equations derived by F. Bloch [115] to describe spin precession in magnetic resonance.

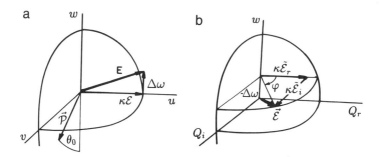

Figure 4.2: Vector model for Bloch's equations. (a) The motion of the pseudopolarization vector $\vec{\mathcal{P}}$ (initially pointing downwards along the w axis) is a rotation around the pseudo-electric field vector $\vec{\mathcal{E}}$ with an angular velocity proportional to the amplitude of that vector. (b) In the complex amplitude representation, the phase of the electric field determines the particular vertical plane containing the pseudo-electric field vector $\vec{\tilde{\mathcal{E}}}$.

The representation is a cinematic representation of the set of equations (4.10), (4.11), and (4.12). For simplicity, we consider first an undamped isolated two-level system ($T_1 = T_2 = T_3 = 0$), and construct a fictitious vector $\vec{\mathcal{P}}$ of components (u, v, w), and a pseudo-electric field vector $\vec{\mathcal{E}}$ of components $(\kappa\mathcal{E}, 0, -\Delta\omega)$. The detuning is defined as $\Delta\omega = \omega_0 - \omega_\ell - \dot{\varphi}$. The system of Eqs. (4.10)–(4.12) are then the cinematic equations describing the rotation of a pseudo-polarization vector $\vec{\mathcal{P}}$ rotating around the pseudo-electric vector $\vec{\mathcal{E}}$ with an angular velocity given by the amplitude of the vector $\vec{\mathcal{E}}$ [Fig. 4.2(a)]. The vectorial form of Eqs. (4.10)- (4.12) is thus:

$$\partial \vec{\mathcal{P}}/\partial t = \vec{\mathcal{E}} \times \vec{\mathcal{P}} \qquad (4.16)$$

Depending on whether the two-level system is initially in the ground state or inverted, the pseudo-polarization vector is initially pointing down or up. Since we have assumed no relaxation, the length of the pseudo-polarization vector is a constant of the motion, and the tip of the vector moves on a sphere. The conservation of length of the pseudo-polarization vector can be verified directly from the set of Bloch's equations. Indeed, the sum of each equation (4.10), (4.11) and (4.12) multiplied by u, v, and w, respectively, yields after integration:

$$u^2 + v^2 + w^2 = w_0^2 \qquad (4.17)$$

which is satisfied for each subensemble of two-level systems. As shown in Fig. 4.2(a), a resonant excitation ($\Delta\omega = 0$)) will tip the pseudo-polarization

4.2. COHERENT INTERACTIONS WITH TWO-LEVEL SYSTEMS

vector by an angle $\theta_0 = \int_{-\infty}^{\infty} \kappa \mathcal{E} dt$ in the (v, w) plane. For a sufficiently intense pulsed excitation, it is possible to achieve complete population inversion when $\theta_0 = \pi$. The effect of phase relaxation (homogeneous broadening) is to shrink the pseudo-polarization vector as it moves around. To take into account inhomogeneous broadening, we have to consider an ensemble of pseudo-polarization vectors, each corresponding to a different detuning $\Delta\omega$.

A similar representation can be made for the system of Eqs. (4.6)–(4.7). The pseudo-polarization vector is then the vector $\vec{\mathcal{Q}}(Q_i, Q_r, w)$ rotating around a pseudo-electric field vector $\vec{\mathcal{E}}(\kappa\tilde{\mathcal{E}}_r, \kappa\tilde{\mathcal{E}}_i, -\Delta\omega)$ [Fig. 4.2(b)]. Physically, the first two components of the pseudo-polarization vector $\vec{\mathcal{Q}}$ represent the dipolar resonant field that opposes the applied external field (and is thus responsible for absorption).

4.2.2 Rate equations

If the light field envelope is slowly varying with respect to T_2, Bloch's equations reduce to the standard rate equations. For pulses longer than the dephasing time T_2, the two first Bloch equations (4.10), (4.11) are stationary on the time scale of the pulse. Solving these equations for u, v, and substituting v into the third equation (4.12) for the population difference, leads to the rate equation:

$$\dot{w} = -\frac{\mathcal{E}^2(\kappa^2 T_1 T_2)}{1 + \Delta\omega^2 T_2^2} \frac{w}{T_1} - \frac{w - w_0}{T_1} \quad (4.18)$$

We note that this equation is identical to the rate equation (3.45) introduced in Chapter 3 in terms of $\Delta\gamma$ ($\Delta\gamma = w/p$). Equation (4.18) defines a saturation field at resonance $\tilde{\mathcal{E}}_{s0} = 1/(\kappa\sqrt{T_1 T_2})$. Off resonance, a larger field $\tilde{\mathcal{E}}_s = \tilde{\mathcal{E}}_{s0}\sqrt{1 + \Delta\omega^2 T_2^2}$ is required to saturate the same transition.

In Chapter 3 we have discussed various cases of pulse propagation through resonant media resulting from the rate equations. For example, for pulses much shorter than the energy relaxation time $\tau_p \ll T_1$ and purely homogeneoulsy broadened media the rate equation (4.18) can be integrated together with the propagation equation (4.8) which yields for the transmitted intensity

$$I(z, t) = I_0(t) \frac{e^{W(t)/W_s}}{e^{-a} - 1 + e^{W(t)/W_s}} \quad (4.19)$$

In this last equation $W(t) = \int_{-\infty}^{t} I_0(t)dt$, and $a = \sigma_{01}^{(0)} w_0 z/p$ is the linear gain/absorption coefficient. Equation (4.19) corresponds to Eq. (3.54) which was written for the photon flux F.

Femtosecond pulse propagation through a homogeneously broadened saturable medium in the limit of $T_2 \ll \tau_p \ll T_1$ is completely determined by two parameters: the saturation energy density W_s and the linear absorption (gain) coefficient a. Equation (4.19) is particularly useful in calculating pulse propagation in amplifiers, as shown in Chapter 3, and further detailed in Chapter 6.

4.2.3 Evolution equations

Energy conservation

The total energy in the resonant light–matter system should be conserved if the pulses are shorter than the energy relaxation time T_1, since no energy is dissipated into the bath. The pulse energy density was defined in Eq. (1.20):

$$W = \frac{1}{2}\epsilon_0 c n \int_{-\infty}^{\infty} \mathcal{E}^2 dt = \frac{1}{2}\sqrt{\epsilon\epsilon_0/\mu_0} \int_{-\infty}^{\infty} \mathcal{E}^2 dt. \qquad (4.20)$$

A simple energy conservation law can be derived by integrating Eq. (4.14) over time, after multiplying both sides by $\mathcal{E}$ and using the third Bloch equation (4.12):

$$\begin{aligned} \frac{dW}{dz} &= \sqrt{\frac{\epsilon\epsilon_0}{\mu_0}} \int_{-\infty}^{\infty} \mathcal{E} \frac{\partial \mathcal{E}}{\partial z} dt \\ &= -\frac{\mu_0 \omega_\ell c}{2} \sqrt{\frac{\epsilon\epsilon_0}{\mu_0}} \int_{-\infty}^{\infty} \int_0^{\infty} v(\omega_0') \mathcal{E} g_{inh}(\omega_0' - \omega_{ih}) dt\, d\omega_0' \\ &= -\frac{\hbar \omega_\ell}{2p} \int_0^{\infty} [w_\infty(\omega_0') - w_0(\omega_0')]\, g_{inh}(\omega_0' - \omega_{ih}) d\omega_0' \end{aligned} \qquad (4.21)$$

The population difference (per unit volume) $(w_\infty - w_0)/p$ integrated over the inhomogeneous transition is a measure of the energy stored in the medium, as a consequence of the energy lost by the pulse, dW/dz.

Area theorem

There are other conservation and evolution laws that can be derived for certain parameters associated with pulses of arbitrary shape. An essential physical parameter for single photon coherent interactions is the *pulse area*

4.2. COHERENT INTERACTIONS WITH TWO-LEVEL SYSTEMS

θ_0, defined as the tipping angle of the pseudopolarization vector (at resonance) as illustrated in Fig. 4.2(a):

$$\theta_0 = \int_{-\infty}^{\infty} \kappa \mathcal{E}(t) dt. \tag{4.22}$$

For a pulse at resonance, the area θ_0 fully describes in which state the medium is left. It can be seen directly from the vector model of Fig. 4.2(a), or by direct integration of Bloch's equations (4.10), (4.11) and (4.12) for exact resonance [$\omega_\ell = \omega_0$ and $\varphi(t) = 0$] and negligible relaxation ($T_1 \approx \infty$ and $T_2 \approx \infty$), that:

$$w_\infty = w(t = \infty) = w_0 \cos \theta_0. \tag{4.23}$$

Similarly, one finds that the polarization (absorptive component) is proportional to $\sin \theta_0$. A "π pulse" is thus a pulse that will completely invert a two-level system, leaving it in a pure state with no macroscopic polarization. A "2π pulse" will leave the system in the ground state, having completed a cycle of population inversion and return to ground state.

Because the area involves the integral of the electric field amplitude rather than the pulse intensity, higher energy densities will be required to achieve the same area with shorter pulses. Therefore, experiments of single photon coherent resonant interactions with fs pulses require intensities at which higher order nonlinear optical effects may have to be taken into account. Another consequence of the electric field amplitude dependence of the area is that, in an absorbing medium, it is possible to have the area conserved, or even growing with distance, while the energy is decreasing. Such a situation arises when the pulse duration increases with distance.

For inhomogeneously broadened media an "area theorem" can be derived which tells us exactly how the pulse area evolves with propagation distance. With the assumptions that the pulses are at resonance ($\omega_\ell = \omega_{ih}$), and shorter than both the energy relaxation time T_1 and the phase relaxation time T_2, a time integration of Eq. (4.14), taking into account Bloch's equations (4.10) through (4.12), yields the *area theorem* [117]:

$$\frac{d\theta_0}{dz} = \frac{\alpha_0}{2} \sin \theta_0. \tag{4.24}$$

where

$$\alpha_0 = \frac{\pi \mu_0 \omega_\ell c p}{\hbar n} w_0(\omega_{ih}) = \frac{\pi \kappa^2 \hbar \omega_\ell}{\epsilon_0 c n p} w_0(\omega_{ih}) \tag{4.25}$$

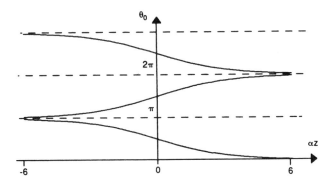

Figure 4.3: Graphic solution of the area theorem equation. The area is plotted as a function of distance (in units of linear absorption length (α_0^{-1}) for an absorbing medium. There is only one point on the set of curves that corresponds to any given area. The origin of distance is the abscissa corresponding to the initial area. From this initial point, a pulse will propagate to the right in an absorbing medium, to the left in an amplifier. Any initial area will evolve asymptotically with distance towards the limits 0, 2π, ..., $2N\pi$ (N integer) in an absorbing medium. The asymptotic limits are π, 3π, ..., $(2N+1)\pi$ in an amplifier.

is the linear absorption coefficient (at resonance) for the inhomogeneously broadened transition. $w_0(\omega_{ih})$ is the initial inversion density at the transition center ($\omega_0' = \omega_{ih}$).

The area theorem applies to amplifying ($w_0 > 0$) as well as to absorbing media ($w_0 < 0$). The derivation is straightforward if we assume a square pulse at resonance, and neglect the reshaping of the pulse. However, Eq. (4.24) can be proven to be quite general and applies to any pulse shape. A graphical representation of the solution of Eq. (4.24) is shown in Fig. 4.3.

The low intensity limit of Eq. (4.24) ($\sin\theta_0 \approx \theta_0$) is the exponential attenuation law (Beer's law). A remarkable feature of Eq. (4.24) is that it points to *area conserving pulses*. Indeed, for any pulse of area equal to a multiple of π, the area is conserved. This is true, for instance, for a "zero-area" pulse, which is not necessarily a zero energy pulse. A signal consisting of two pulses π out of phase has an area equal to zero, but an energy equal to twice the single pulse energy. Such a pulse sequence propagates without loss of "area" through a resonant medium.

The area theorem tells us in addition which of the steady state areas are stable solutions. For a pulse with an initial area smaller than π, the right side

4.2. COHERENT INTERACTIONS WITH TWO-LEVEL SYSTEMS

of Eq. (4.24) is negative, and the area as well as the pulse energy will decay with distance as the pulse is absorbed. On the other hand, if the pulse area is initially between π and 2π, θ *increases* with distance. With such a pulse, a population inversion has been achieved, and the pulse tail is amplified. The stimulated emission at the pulse tail results in pulse stretching, and hence an increase in pulse area with distance. The pulse energy, however, still decreases with distance in the reshaping process. Pulse reshaping proceeds until the area reaches the value of 2π. Once the pulse reshaping is completed, the electric field envelope has acquired a well defined and stable hyperbolic secant (sech) shape and propagates without further distortion or attenuation through the resonant absorbing medium. This phenomenon is called *self-induced transparency*. Pulses of initial area $2n\pi$ break up into n "2π" pulses. It can easily be verified, by substitution into Eqs. (4.10) through (4.12) that the envelope given by:

$$\mathcal{E}(t) = \frac{2}{\kappa \tau_s} \text{sech}\left(\frac{t}{\tau_s} - \frac{z}{\tau_s v_e}\right). \tag{4.26}$$

is a solution of Bloch's equations. The pulse given by Eq. (4.26) has an area of 2π, and a duration (FWHM) of $1.763\tau_s$. This solution is valid on- and off- resonance. In Eq. (4.26), $v_e \ll c$ is the envelope velocity of the 2π pulse. For a pulse duration short compared to the inverse (inhomogeneous) linewidth of the absorber, the envelope velocity is given by $v_e = 2/(\alpha_0 \tau_s)$. This slow velocity essentially expresses that the first half ($\tau_s/2$) of the pulse is absorbed in a distance α_0^{-1}, to be restored to the second half by stimulated emission.

The subscript in θ_0 indicates that the amplitude of the electric field of a pulse at frequency ω_ℓ is used in the above definition (4.22). In Eq. (4.22) the pulse envelope θ_0 was introduced by a time integral. Another definition for the area is related to the amplitude of the Fourier transform of the pulse defined in Eq. (1.6):

$$\theta = \kappa |\tilde{\mathcal{E}}(\Omega - \omega_0 = 0)|. \tag{4.27}$$

Since $\kappa \mathcal{E}(t)$ has the dimension of a frequency, the Fourier transform of that quantity is dimensionless. It is left as a problem at the end of this chapter to show that the definitions (4.22) and (4.27) are equivalent for unchirped pulses at resonance. Chirped or non-resonant pulses of the same energy or of the same area θ_0 will have a smaller area θ, because of the broadening of the Fourier spectrum due to the phase modulation. The distinction is important in determining the threshold for nonlinear propagation phenomena such as *self-induced transparency* [118]. Chirped off- or on-resonance pulses will evolve towards a pure unmodulated "2π" pulse at the original pulse frequency ω_ℓ, provided the initial area θ is larger than π, as has been demonstrated by numerous computer simulations [118].

The coherent absorber is therefore an ideal filter for phase and amplitude fluctuations for pulses of initial area larger than π. Because of the slow propagation velocity of the "2π" pulse, all fast phase and amplitude noise propagates ahead of the pulse as a "precursor" that is ultimately absorbed [118, 119, 120]. An example of the evolution of a small Gaussian disturbance (90° out of phase) initially superimposed on top of a 2π hyperbolic secant pulse is illustrated in Fig. 4.4. The amplitude and phase of the electric field are represented as a function of time and distance. The propagation distance is expressed in units of the linear absorption length. The initial "noise pulse" on top of the 2π pulse is indicated by the dashed circle around the top of the main pulse. The phase of the perturbation is indicated above the amplitude disturbance. The phase modulation is seen to be continuously amplified with propagation distance. This is consistent with the discussion on the average frequency of a pulse off-resonance being shifted further away from resonance with distance. In the spectral domain (not shown here), this perturbation appears as little "bumps" on the wings of the pulse spectrum. The amplification of the phase modulation corresponds to the Fourier components of the perturbation being rejected farther away from resonance. Numerical simulations have shown that the phase perturbations become amplitude and phase perturbations, which propagate at the normal group velocity of the medium, rather than the very low velocity v_e of the 2π pulse. When the 2π pulse has separated in time from the "noise" signal, it has reshaped into a lower energy sech pulse broadened by 0.7% in the particular example shown.

Pulse frequency evolution

Bloch's equations describe the transient response of the complex polarization for a two-level system. As mentioned in the introduction, the pulse frequency is no longer a conserved quantity in the presence of a transient polarization. The medium considered in this section can be represented by a collection of two-level systems, each with a homogeneous broadening of T_2^{-1}, with a frequency distribution represented by a function $g_{inh}(\omega_0' - \omega_{ih})$ (inhomogeneous broadening).

4.2. COHERENT INTERACTIONS WITH TWO-LEVEL SYSTEMS 167

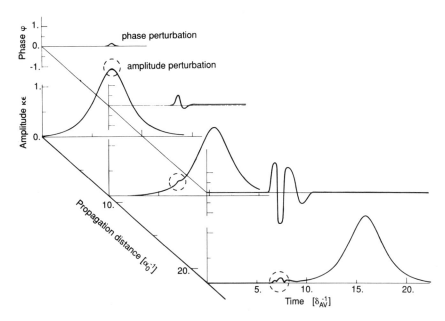

Figure 4.4: Propagation of a 2π pulse at resonance, showing the "filtering" of a small 90^0 out of phase Gaussian disturbance (shown in the dashed circles). The medium is inhomogeneously broadened with the line profile $g_{inh}(\Delta\omega) = (1/\sqrt{\pi})\exp[-(\Delta\omega/\delta_{AV})^2]$. The initial amplitude is $\kappa\mathcal{E}(t) = (2/1.5)\mathrm{sech}[(t-7)/1.5] + (0.1i/0.3\sqrt{\pi})\exp\{-[(t-7)/0.3]^2\}$, where the time t is expressed in units of δ_{AV}^{-1}. Three steps of propagation are shown. The amplitude perturbation is indicated by the dashed circle. The initial modulation is seen to be amplified in amplitude and phase. Because of its faster group velocity as compared to the envelope velocity of the 2π pulse, it separates in time from the latter. The amplification of the phase fluctuation indicates that the perturbation also separates in frequency from the main pulse (from [119]).

In this section, we will derive expressions from Bloch's equations that will allow us to establish quantitatively the evolution of the average carrier frequency of a pulsed signal as it propagates through the medium. We have seen in Chapter 1 that the frequency modulation $\dot\varphi$ has to be averaged in order to define a mean pulse frequency [Eq. (1.16)]. We will introduce the notation $\langle\dot\varphi\rangle$ for the average phase derivative (average deviation of the frequency from ω_ℓ):

$$\langle\dot\varphi\rangle = \frac{\int_{-\infty}^{\infty}\mathcal{E}^2\dot\varphi\, dt}{\int_{-\infty}^{\infty}\mathcal{E}^2 dt} = \frac{1}{2W}\left[\sqrt{\frac{\epsilon\epsilon_0}{\mu_0}}\int_{-\infty}^{\infty}\mathcal{E}^2\dot\varphi\, dt\right]. \qquad (4.28)$$

Multiplying both sides by W and taking the derivative we obtain:

$$\begin{aligned}\frac{d(W\langle\dot\varphi\rangle)}{dz} &= W\frac{d\langle\dot\varphi\rangle}{dz}+\langle\dot\varphi\rangle\frac{dW}{dz}\\ &= \frac{1}{2}\sqrt{\frac{\epsilon\epsilon_0}{\mu_0}}\frac{d}{dz}\int_{-\infty}^{\infty}\mathcal{E}^2\dot\varphi\,dt\end{aligned} \quad (4.29)$$

The last term of Eq. (4.29) can be directly calculated using the Maxwell-Bloch equations. In particular, we can insert the time derivative of Eq. (4.15) in this equation:

$$\begin{aligned}\frac{1}{2}\sqrt{\frac{\epsilon\epsilon_0}{\mu_0}}&\int_{-\infty}^{\infty}\left(\mathcal{E}^2\frac{\partial\dot\varphi}{\partial z}+\dot\varphi\frac{\partial\mathcal{E}^2}{\partial z}\right)dt\\ &= -\frac{\omega_\ell}{4}\int_{-\infty}^{\infty}dt\int_{0}^{\infty}d\omega_0'\,g_{inh}(\omega_0'-\omega_{ih})\left[\dot u\mathcal{E}-u\dot{\mathcal{E}}+2v\mathcal{E}\dot\varphi\right]\\ &= -\frac{\omega_\ell}{2}\int_{0}^{\infty}d\omega_0'\int_{-\infty}^{\infty}dt\,g_{inh}(\omega_0'-\omega_{ih})[\dot u\mathcal{E}+\dot\varphi v\mathcal{E}]\\ &= -\frac{\omega_\ell}{2}\int_{0}^{\infty}d\omega_0'\int_{-\infty}^{\infty}dt\,g_{inh}(\omega_0'-\omega_{ih})\left[(\omega_0'-\omega_\ell)v\mathcal{E}-\frac{u\mathcal{E}}{T_2}\right]\end{aligned}\quad(4.30)$$

where we have made use of the first Bloch equation (4.10). We have already derived an expression for the evolution of the pulse energy density W. Combining Eqs. (4.21),(4.29), and (4.30) yields the following expression for the evolution with propagation distance of the pulse carrier frequency:

$$\frac{d\langle\dot\varphi\rangle}{dz}=\frac{\omega_\ell}{2\kappa W}\int_0^\infty g_{inh}(\omega_0'-\omega_{ih})\left[\omega_0'-\omega_\ell-\langle\dot\varphi\rangle\right](w_\infty-w_0)\,d\omega_0'+\frac{2\langle k\rangle}{T_2}. \quad (4.31)$$

In analogy to the definition of the average frequency in Chapter 1 [cf. Eq. (1.16)], we have introduced the average contribution to the propagation vector due to the resonant dispersion of the two-level system:

$$\begin{aligned}\langle k\rangle &= \frac{\int_{-\infty}^{\infty}\mathcal{E}^2(\partial\varphi/\partial z)dt}{\int_{-\infty}^{\infty}\mathcal{E}^2\,dt}\\ &= \frac{\omega_\ell}{4W}\int_0^\infty d\omega_0'\int_{-\infty}^\infty dt\,g_{inh}(\omega_0'-\omega_{ih})u\mathcal{E}\end{aligned}\quad(4.32)$$

The polarization amplitude u — and hence the resonant contribution to the wave vector $\langle k\rangle$ — will shrink with time in presence of phase relaxation (finite T_2). The corresponding temporal modulation of the polarization is responsible for the second term of the right-hand side of Eq. (4.31). For very short pulses, however, $(\tau_p \ll T_2)$, this second term can be neglected.

4.2. COHERENT INTERACTIONS WITH TWO-LEVEL SYSTEMS

The frequency shift is proportional to the overlap integral of the lineshape $g_{inh}(w'_0 - w_{ih})$ with the (frequency dependent) change of the inversion $(w_\infty - w_0)$ times the detuning $(w'_0 - w_\ell - \langle \dot\varphi \rangle)$. The ratio of absorbed energy (which is proportional to $(w_\infty - w_0)$) to the pulse energy W is maximum in the weak pulse limit $(\theta \ll 1)$. Therefore, the frequency pushing as described by Eq. (4.31) is important in the weak pulse limit, and for narrow lines $[T_2 \to \infty;\ g_{inh}(w'_0 - w_{ih}) \approx \delta(0)]$.

Bloch's equations can be solved analytically in the weak, short pulse limit, i.e., for pulses that do not induce significant changes in population and have a duration short compared to the phase relaxation time T_2. The interaction equation (4.6) can be written in the integral form:

$$\tilde{Q}(t) = \int_{-\infty}^{t} \kappa \mathcal{E} w e^{-i[(w'_0 - w_\ell)t' - \varphi(t')]} dt' \qquad (4.33)$$

For weak pulses $(w \approx w_0)$ and the right hand side of Eq. (4.33) at $t = \infty$ is proportional to the Fourier transform of $\kappa \tilde{\mathcal{E}} w$. Thus we have:

$$|\tilde{Q}|^2 = u^2 + v^2 = \kappa^2 w_0^2 |\tilde{\mathcal{E}}(w'_0 - w_\ell)|^2$$
$$\approx -2w_0(w_\infty - w_0) \qquad (4.34)$$

where $\tilde{\mathcal{E}}(w'_0 - w_\ell)$ is the amplitude of the Fourier transform of the field envelope at the line frequency w'_0. The last equality results from the conservation of the length of the pseudo-polarization vector $(u^2 + v^2 + w^2 = w_0^2 = \text{constant})$. The approximation is made that the change in population is small: $w_\infty^2 = [w_0 + (w_\infty - w_0)]^2 \approx w_0^2 + 2w_0(w_\infty - w_0)$. Insertion of the approximation (4.34) into the energy evolution equation (4.21) leads to the expected result that the absorbed energy is proportional to the overlap of the pulse and line spectra:

$$\frac{dW}{dz} = \frac{\kappa w_\ell w_0}{4} \int_0^\infty |\tilde{\mathcal{E}}(w'_0 - w_\ell)|^2 g_{inh}(w'_0 - w_{ih}) dw'_0 \qquad (4.35)$$

A similar expression is found by inserting the approximation (4.34) into the frequency evolution equation (4.31):

$$\frac{d\langle\dot\varphi\rangle}{dz} = \frac{\kappa w_\ell w_0}{2W} \int_0^\infty (w'_0 - w_\ell - \langle\dot\varphi\rangle)|\tilde{\mathcal{E}}(w'_0 - w_\ell)|^2 g_{inh}(w'_0 - w_{ih}) dw'_0 \qquad (4.36)$$

Comparing expressions (4.35) and (4.36) leads to the conclusion that the frequency shift with distance is largest for sharp lines (compared to the pulse

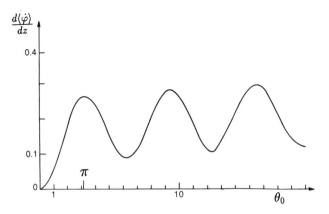

Figure 4.5: Frequency pushing with distance for an asymmetric pulse applied one linewidth off-resonance, as a function of initial pulse area. The medium is inhomogeneously broadened with the line profile $g_{inh}(\Delta\omega) = (1/\sqrt{\pi})\exp[-(\Delta\omega/\delta_{AV})^2]$. The pulse shape is given by $\kappa\mathcal{E}(t) = 1/[\exp(t/2)+\exp(t/0.8)]$. The frequency (time) is normalized to the (inverse of the) inhomogeneous broadening linewidth.

bandwidth), and pulses off resonance by approximately their bandwidth. The maximum possible shift is about half a pulse bandwidth per absorption length.

The initial frequency pushing with distance given by Eq. (4.36) has an interesting simple dependence on the initial pulse area, in the case of a sech^2 shaped pulse. It can be verified that the functional dependence is exactly:

$$\frac{d\langle\dot{\varphi}\rangle}{dz} = \frac{\kappa\omega_\ell w_0}{4\epsilon_0 cn}\frac{1-\cos\theta_0}{\theta_0^2}\Gamma_F(\omega_\ell), \tag{4.37}$$

where $\Gamma_F(\omega_\ell)$ is the spectral overlap function:

$$\Gamma_F(\omega_\ell) = \frac{\int_0^\infty \kappa^2\mathcal{E}^2(\omega_0' - \omega_\ell)g_{inh}(\omega_0' - \omega_{ih})d\omega_0'}{\int_{-\infty}^\infty \kappa^2\mathcal{E}(t)dt} \tag{4.38}$$

The above expression pertains to sech^2 shaped pulses. However, Fig. 4.5 illustrates that Eq. (4.37) still provides a reasonable approximation even for asymmetric pulse shapes.

The frequency shift with distance is a manifestation of the fact that the response time of matter is finite. We have already alluded to this in Chapter 1, when we wrote the polarization as a convolution integral, expressing that the instantaneous polarization is not an immediate function of the electric field of the light, but a function of the history of the field [Eq. (1.44)]. Such convolution expressions can be obtained directly by Fourier transformation of Bloch's equations [for instance Eq. (4.6)].

To appreciate physically why the carrier frequency should not be a conserved quantity, let us consider a square pulse sent through a near resonant absorbing medium at the average carrier frequency ω_ℓ where $\omega_\ell < \omega_{ih}$. High frequency components within the pulse spectrum are closer to the center of the inhomogeneous transition and, thus, are absorbed more than low frequency components. The result is a continuous shift of the pulse spectrum to lower frequencies. Essentially the same phenomenon has been discussed in Chapter 1 (Section 1.2.2). There, the cause was a frequency dependent imaginary part of the dielectric constant. It is left to a problem to find the connection between both descriptions. Another way to describe the frequency pushing is in the time domain. Initially — i.e. prior to the arrival of the light pulse — there were no induced dipoles, since there was no field present to induce them. The initial value of the susceptibility χ (or, using the notations of this chapter, the pseudo-polarization Q or u, v) is thus 0. As the leading edge of the pulse enters the medium, it excites the electronic dipoles, resulting in a change of χ and in a change of the refractive index, respectively. As we have seen before this results in a pulse chirp which can manifest itself in a shift of the average pulse frequency. Such an effect can be simply visualized with help of Fig. 4.6, which is a three dimensional representation of the electric field versus time and distance.

Since the frequency pushing with distance is a general manifestation of the response time of matter, it will occur whenever there is a transient in the polarization due to a change in applied field. Even continuous radiation will be frequency shifted if there are random fluctuations of the phase of the field. Analytical expressions can be derived for the shift of average frequency with distance for a phase fluctuating dc field [118].

Given a symmetric line shape, is the frequency pushing symmetric above and below resonance? There is a blue shift equal to the red shift, only within the framework of the slowly varying envelope approximation. Numerical simulations indicate that the material response is larger below resonance than above, leading to a larger red shift than blue shift. This may not come as a surprise, since the response of a system with a resonance at ω_0 will "follow" a low frequency ($\omega_\ell \ll \omega_0$) excitation, but has zero response at frequencies far beyond its resonance ($\omega_\ell \gg \omega_0$).

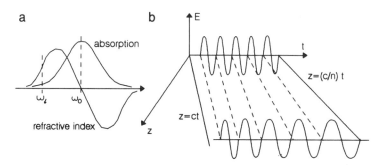

Figure 4.6: Sketch illustrating the evolution of the average frequency of a short pulse as it propagates through matter. Figure (a) shows the relative frequencies of the radiation (ω_ℓ) and material resonance (ω_0). These are however steady-state values. Prior to applying the field (and for the first optical cycle(s) of the radiation), the electric dipoles have not yet been induced, and the abscissa is the correct representation of both the absorption and dispersion. If a square pulse is applied, the value of the absorption and dispersion evolve with time from the initial condition (0) to the value on the corresponding curve at ω_ℓ. A representation of the evolution of the wave packet in time and space is shown in (b). The leading edge of the square pulse is not effected, since the dipoles have not yet been induced. The time varying index leads to a shift of the average pulse frequency to lower frequencies.

4.2.4 Steady-state pulses

A steady-state pulse is a pulse for which the envelope $\mathcal{E}(t-z/v_e)$ is conserved along propagation. The quantity v_e is the velocity of the pulse envelope. Usually it depends on properties of the medium and the pulse and is differs from the group and phase velocity. If the pulse envelope remains constant, various pulse parameters such as the pulse duration, energy and area should also be conserved.

In the case of inhomogenously broadened media, the area theorem applies. From Eq. (4.24), a necessary condition for the existence of a steady-state pulse is that $d\theta/dz = (\alpha_0/2)\sin\theta_0 = 0$. Obviously there are several values of the area which do not change with distance: they are $\theta_0 = 0, \pi, 2\pi, 3\pi, \ldots$ A quick glance at the graphical representation (Fig. 4.3) of Eq. (4.24) indicates that, in an absorbing medium ($\alpha < 0$) the areas $\theta_0 = 0, 2\pi, 4\pi \ldots$ are stable solutions, while for amplifying media, the stable areas correspond to uneven numbers of π. At resonance, steady-state pulses correspond to the π pulse. As mentioned previously, in the case of an absorber, only the area 2π corresponds to a stable stationary pulse. In the

4.2. COHERENT INTERACTIONS WITH TWO-LEVEL SYSTEMS

amplifier, however, even with constant area, the pulse energy will tend to grow to infinity unless balanced by a loss mechanism. We will use a scattering coefficient σ_s to describe such a loss mechanism in searching for steady state pulses in amplifiers. It is the same coefficient representing linear losses as the κ_1 introduced in Eq. (1.95) — the notation σ_s is chosen here to avoid confusion with the Rabi coefficient κ.

Transverse effects have been neglected in the search for steady-state solutions, and in the evolution equations derived in the previous sections. In free space, the plane wave approximation will only hold within the confocal parameter of a beam of finite size. Even within this limit, transverse effects in coherent interaction may contribute to self-focusing and defocusing of the beam [121]. As a rule of thumb, the plane wave approximation for self-induced transparency may be considered to hold for approximately five linear absorption lengths. There are however two important types of confinements for which the single dimensional approach remains valid over very long distances:

- optical fibers
- optical cavities

The search for steady state solutions in conditions of coherent interactions has more than purely academic interest. The main motivations for these studies relate to

- stability
- minimizing energy losses
- maximizing energy extraction

We have seen in the previous section that the 2π pulse in absorbing media acts as a filter "cleaning" the signal from amplitude and phase fluctuations. The steady-state pulse in an absorber has minimum energy loss, since it returns the absorbing two-level system to the ground state. In an amplifier or in a laser, steady-state pulses can be found that bring the two-level system from inversion to the ground state — and hence extract the maximum energy possible from the gain medium.

Steady-state pulses in amplifiers

For finite T_2, stable steady-state pulses do not exist in absorbing media. The loss of coherence leads to irreversible energy losses, causing to the collapse of an initial 2π pulse. In the case of an amplifier, however, steady state pulses can be found even in the presence of homogeneous broadening, because there is a gain mechanism to compensate for the losses. Since there is no mechanism to "push back" in time the leading edge of the pulse as in the case of "2π pulse propagation," the envelope velocity will generally be close to the phase velocity in an amplifier. In searching for steady state solutions off resonance, it is important to take into consideration the correct *dispersive* response, and the corresponding *phase velocity* v_p. It is this same near-resonant phase velocity — as opposed to the phase velocity in the host medium c/n — that is used in the definition of the slowly varying components of a propagating physical quantity $\tilde{X}$:

$$X = \frac{1}{2}\tilde{X}\exp[i\omega_\ell(t - \frac{z}{v_p})]. \tag{4.39}$$

Following the procedure of Ref. [122], we start directly from the second-order Maxwell equation for a linearly polarized plane wave:

$$\frac{\partial^2 E}{\partial z^2} - \frac{1}{c^2}\frac{\partial^2 E}{\partial t^2} = \mu_0 \left[\frac{\partial^2 P}{\partial t^2} + \sigma_s \frac{\partial E}{\partial t}\right]. \tag{4.40}$$

We are looking for solutions that satisfy the steady-state condition for the slowly varying envelopes, describing "form stable" pulse propagation. In a frame of reference moving with the (yet unknown) envelope velocity v_e, any function $\tilde{X}$ associated with the pulse (i.e., pulse envelope, polarization, medium inversion) should remain unchanged:

$$\frac{\partial}{\partial z}\tilde{X}(t - z/v_e, z) = 0. \tag{4.41}$$

It can be shown that, within the SVEA, $v_e \approx v_p$. Substituting the slowly varying field and "pseudo-polarization" amplitudes (4.6), (4.7) defined in the section on Maxwell–Bloch equations into Eq. (4.40), we find:

$$\beta\tilde{\mathcal{E}} + ig\tilde{Q} - i\gamma\tilde{\mathcal{E}} = 0 \tag{4.42}$$

where we have defined:

$$\beta = \omega_\ell^2\left(\frac{1}{v_p^2} - \frac{1}{c^2}\right) \tag{4.43}$$

4.2. COHERENT INTERACTIONS WITH TWO-LEVEL SYSTEMS

$$g = \mu_0 \omega_\ell^2 \tag{4.44}$$

$$\gamma = \mu_0 \omega_\ell \sigma_s. \tag{4.45}$$

Equation (4.42) expresses that, at steady-state, the "slippage" of the pulse envelope with respect to the wave (term β) results from a balance between gain (second term) and scattering losses (third term). According to Eq. (4.42), the steady-state pseudo-polarization $\tilde{Q} = Q \exp(i\vartheta)$ is proportional to the field $\mathcal{E}$:

$$\tilde{\mathcal{E}} = \tilde{A}\tilde{Q} \tag{4.46}$$

where

$$\tilde{A} = \frac{g\gamma - i\,g\beta}{\beta^2 + \gamma^2}. \tag{4.47}$$

Substituting in Eq. (4.6) for the time evolution of the pseudo-polarization yields:

$$\dot{\tilde{Q}} + i\dot{\vartheta}\tilde{Q} = -i(\omega_\ell - \omega_0)\tilde{Q} - \kappa w \tilde{A}\tilde{Q} - \frac{\tilde{Q}}{T_2}. \tag{4.48}$$

The real and imaginary parts of Eq. (4.48) and Bloch's equation (4.7) for the population difference w form a complete set:

$$\dot{Q} = -\frac{g\gamma\kappa}{\beta^2 + \gamma^2}wQ - \frac{Q}{T_2} \tag{4.49}$$

$$\dot{\vartheta} = -(\omega_\ell - \omega_0) + \frac{\kappa g\beta}{\beta^2 + \gamma^2}w \tag{4.50}$$

$$\dot{w} = \frac{\kappa g\gamma}{\beta^2 + \gamma^2}Q^2. \tag{4.51}$$

Taking the time derivative of the first of these equations [Eq. 4.49], and substituting $\dot{w}$ from the last Eq. (4.51), we find a (relatively) simple equation which is known [12] to have a sech solution for functions Q that tend to zero at both ends of the time scale:

$$Q\ddot{Q} - \dot{Q}^2 + \left[\frac{\kappa g\gamma}{\beta^2 + \gamma^2}\right]^2 Q^4 = 0 \tag{4.52}$$

Substituting the sech solution $\tilde{\mathcal{E}}(t) = \mathcal{E}_0 \text{sech}(t/\tau_s)\exp(i\varphi)$ into Eq. (4.46), and hence a form $\tilde{Q}(t) = Q_0 \text{sech}(t/\tau_s)\exp(i\vartheta)$ in Eq. (4.52), yields a relation between β and the field amplitude $\mathcal{E}_0$. If we choose the carrier frequency ω_ℓ to be the average pulse frequency, we impose that the average phase derivative over the pulse duration is zero ($\int_{-\infty}^{\infty} \dot{\varphi}|\mathcal{E}_0|^2 = 0$). Equation (4.49) provides an additional relation to determine the steady pulse parameters [pulse duration

		Steady State Pulse
Envelope	$\mathcal{E}$	$\mathcal{E}_0 \text{sech}(t/\tau_s) e^{i\varphi(t)}$
Field amplitude	$\mathcal{E}_0$	$[1/(\kappa\tau_s)]\sqrt{1+(\omega_\ell-\omega_0)^2 T_2^2}$
Pulse duration	τ_s	$T_2\left\{[\alpha_0/(\sigma_s\kappa^2\mathcal{E}_0^2\tau_s^2)]-1\right\}$
Average carrier frequency	ω_ℓ	ω_ℓ
Chirp	$\dot\varphi(t)$	$\tau_s^{-1}[T_2(\omega_\ell-\omega_0)]^2\tanh(t/\tau_s)/$
Pulse area	θ_0	$\pi\sqrt{1+(\omega_\ell-\omega_0)^2 T_2^2}$
Pulse energy density	W	$\mathcal{E}_0^2\tau_s\sqrt{\frac{\epsilon\epsilon_0}{\mu_0}}$

Table 4.1: Main parameters associated with steady state pulses in homogeneously broadened amplifiers. The pulse duration τ_s is defined in Table 1.1.

τ_s (time normalization factor of sech pulse shape, as defined in Table 1.1), phase φ, amplitude $\mathcal{E}_0$, and frequency ω_ℓ]. The relations yielding the steady state pulse parameters are summarized in Table 4.1.

The larger the amount off resonance, the larger the field amplitudes. A larger "power broadening" is required to extract the same energy from the gain medium off-resonance than on-resonance. This class of solutions include the "π pulse" solution of Ref. [123, 124] and the "$\pi\sqrt{2}$" solution derived in Ref. [118, 120]. These types of coherent steady state pulses are optimizing the energy extraction from the amplifier.

4.3 Multiphoton coherent interaction

4.3.1 Introduction

The high intensities associated with fs pulses lead to a nonlinear response of the real and imaginary parts of the polarization. The imaginary part of the nonlinear polarization is associated for instance with multiphoton transitions, and will exhibit a n-photon resonance when two levels of an atomic or molecular system can be connected by n optical quanta. As for single photon transition, there will be a linewidth and dephasing time associated with the higher order resonance. When the fs pulses are shorter than the multiphoton phase relaxation time, we are dealing with *coherent transient resonant multiphoton interaction*. This situation is as complex as its name, since we are

4.3. MULTIPHOTON COHERENT INTERACTION

cumulating the problems associated with nonlinear optics, coherent phenomena, transient absorption, transient dispersion, and propagation. In fact it is so complex that very few have addressed this problem either experimentally or theoretically. One might therefore wonder whether there is any benefit in even considering such situations?

Before we lay down the theoretical framework of multiphoton coherent resonant interaction, we show the potential benefits of fs coherent excitation with a few simple examples (an encouragement to the reader to proceed to the more tedious theoretical subsection).

There is a basic property of coherent transient interaction that is common to single- and multiphoton transitions. There is no longer a saturating intensity that tends to equalize the population of the levels connected by the radiation. Radiation coherently interacting with two-level systems can transfer all the population from one level into the other. If the two levels are initially inverted (amplifier), the total energy stored in the two-level system is transferred to the radiation ("π pulse" amplification). In the case of a system initially in the ground state (absorber), lossless transmission can be observed ("2π pulse" propagation). A typical application is harmonic generation in the presence of multiphoton resonances. The resonant enhancement leads to a better conversion efficiency, that saturates partly because of multiphoton absorption. Multiphoton coherent propagation effects can be exploited to achieve larger penetration depths — and hence better conversion.

Transitions always proceed via intermediate levels. In some cases, the effect of numerous intermediate levels far off-resonance can be combined into one or more "virtual" intermediate state(s). The multilevel system of equations then reduces to coupled equations involving only the two extreme upper and lower resonant levels. This approximation is typical of atomic systems where only a few levels are directly connected within the pulse bandwidth at low quantum numbers. In these systems, if two levels are connected by an n-photon transition, there is generally no intermediate resonance (of order $m < n$) within the pulse bandwidth.

The situation is different in molecules. There are generally so many levels around that one can describe the n-photon process as a "ladder" of single photon transitions. In the presence of a large number of levels — all with a different detuning with respect to the radiation — maintaining some degree of coherence requires a very particular excitation tailored to the transfer function of the system. Pulse shaping techniques have reached such a level of sophistication that it should be possible to provide a fs excitation with a shape designed to produce a pure inversion, or frequency selective excitation

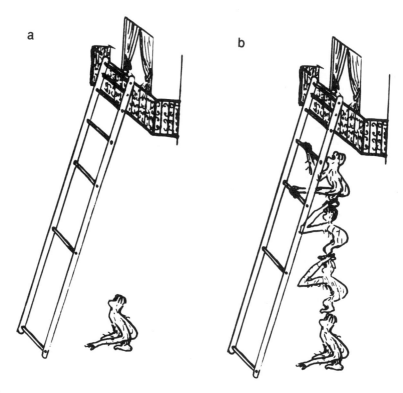

Figure 4.7: The problem of reaching a high level with an harmonic ladder (a), and the multiple-pulse multiphoton coherent solution (b)

of a particular level. The example of the multiphoton excitation of CH_3F by a *group of pulses* that will be considered in Section 4.3.2 shows that even in the case of such a complex anharmonic vibrational–rotational ladder, highly *selective* excitation can be achieved with *broadband* pulses. This complex problem has a solution nearly as simple as that of the unfortunate "Romeo" ape of Fig. 4.7(a) faced with an anharmonic ladder. In that analogy also, getting a group of friends to make a multi-ape coherent transition could succeed in getting the system in the upper state.

We will start in Section 4.3.2 with the more general (and complex) systems with a very large number of levels (such as molecules). In the following section we will proceed from this more general situation to the particular approximation that leads to the multiphoton analogue of Bloch's optical equations.

4.3.2 Multiphoton multilevel transitions

General formalism

The complex atomic/molecular system is represented by the unperturbed Hamiltonian H_0 to which corresponds a set of eigenstates ψ_k of energy $\hbar\omega_k$. In the presence of an electric field E, the state of the atomic/molecular system is described by the wavefunction ψ, a solution of the time dependent Schrödinger equation:

$$H\psi = i\hbar\frac{\partial \psi}{\partial t}, \qquad (4.53)$$

with the total Hamiltonian given by:

$$H = H_0 + H' = H_0 - p \cdot E(t) \qquad (4.54)$$

where p is the dipole moment. In the standard technique for solving time dependent problems, the wave function ψ is written as a linear combination of the basis functions ψ_k:

$$\psi(t) = \sum_k a_k(t)\psi_k. \qquad (4.55)$$

This expression for ψ is inserted in the time dependent Schrödinger equation (4.53). Taking into account the normalization conditions for the basis functions ψ_k, one finds the coefficients a_k have to satisfy the following set of differential equations:

$$\begin{aligned}\frac{da_k}{dt} &= -\frac{i}{\hbar}\hbar\omega_k a_k + \sum_j \frac{i}{\hbar}p_{k,j}\mathcal{E}(t)\cos[\omega_\ell t + \varphi(t)]a_j \\ &= -i\omega_k a_k + \sum_j \frac{i}{2\hbar}p_{k,j}[\tilde{\mathcal{E}}e^{i\omega_\ell t} + \tilde{\mathcal{E}}^* e^{-i\omega_\ell t}]a_j \end{aligned} \qquad (4.56)$$

where p_{kj} are the components of the dipole coupling matrix[2] for the transition $k \to j$, and a_k are the amplitudes of the eigenstates.

Apart from the basic assumption that the time scale is short enough for all phase and amplitude relaxation to be negligible, Eq. (4.56) is of a quite general nature, and it is ideally suited to numerical integration. It can be used to solve the most complex problem of light–matter interaction, since no assumptions have been made as to the transitions, resonances, detuning,

---

[2] We will use both the notation p_{ik} and $p_{i,k}$ depending on the complexity of the subscripts.

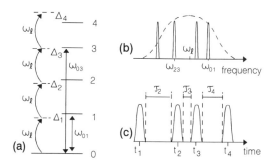

Figure 4.8: (a) Energy level diagram, (b) spectral representation, and (c) sequence of exciting pulses.

or degeneracies. However, the complete numerical treatment leaves little room for physical insight. We will therefore consider some simplified cases for which general trends emerge from the solution. Even though numerical analysis is still required, it seems possible to gain some intuition as to the response of the system.

As a first simplifying assumption, we will consider only the levels that can be connected directly by a single photon at ω_ℓ. The separations of the successive energy levels k from the ground state 0 are $\hbar\omega_{0k}$. We designate with an index k a state at an energy close to $\hbar k\omega_\ell$ (the ground state being thus labeled with the index 0). The important levels to consider will be those sketched in Fig. 4.8(a) for which the detuning:

$$\Delta_k = \omega_{0k} - k\omega_\ell \tag{4.57}$$

is at most comparable to the sum of the radiation bandwidth and the transition rate. Such a "near-resonance" condition is illustrated by the sketch of Fig 4.8(b) where the dashed line indicates the pulse spectrum overlapping the successive single-photon transitions. Consistent with this approximation, we replace the set of coefficients a_k, which have temporal variations at optical frequencies, by the "slowly varying" set of coefficients c_k, using the transformation:

$$a_k = e^{-ik\omega_\ell t} c_k. \tag{4.58}$$

This is the same type of transformation to a "rotating set of coordinates" as we applied to the set of density matrix equations (4.2) and (4.3) in the section on two-level systems.

Selecting from the sum in Eqs. (4.56) the particular pairs of levels that matches best the resonance condition for an incident field at frequency ω_ℓ,

4.3. MULTIPHOTON COHERENT INTERACTION

leads to the simpler form:

$$\frac{dc_k}{dt} = -i\Delta_k c_k + \frac{i}{2\hbar}p_{k-1,k}\tilde{\mathcal{E}}^*(t)c_{k-1} + \frac{i}{2\hbar}p_{k,k+1}\tilde{\mathcal{E}}(t)c_{k+1}. \quad (4.59)$$

The use of amplitudes c_k rather than elements of the density matrix is more convenient in dealing with multilevel systems. It is easier, however, to get a physical picture from a density matrix representation. The density matrix elements can be calculated from $\rho_{ii} = a_i a_i^*$ and $\rho_{ij} = \varrho_{ij}e^{i\omega_\ell t} = a_i a_j^* e^{i\omega_\ell t}$ for $i \neq j$. To verify that the two descriptions are equivalent, let us look at the equations for a two-level system which can be written in the form:

$$\frac{d}{dt}\begin{pmatrix} c_0 \\ c_1 \end{pmatrix} = \begin{pmatrix} 0 & i\frac{p_{01}\tilde{\mathcal{E}}}{2\hbar} \\ i\frac{p_{01}\tilde{\mathcal{E}}^*}{2\hbar} & -i\Delta_1 \end{pmatrix}\begin{pmatrix} c_0 \\ c_1 \end{pmatrix} \quad (4.60)$$

Multiplying by c_i^* ($i = 0, 1$), we find the following set of equations for the amplitudes $c_i^* c_j$ of the elements of the density matrix:

$$\dot{\rho}_{11} - \dot{\rho}_{00} = \frac{d}{dt}(c_1 c_1^* - c_0 c_0^*) = \frac{ip_{01}}{\hbar}[\tilde{\mathcal{E}}^*(c_0 c_1^*) - \tilde{\mathcal{E}}(c_0 c_1^*)^*] \quad (4.61)$$

$$\dot{\varrho}_{01} = \frac{d}{dt}(c_0 c_1^*) = i\Delta_1 c_0 c_1^* + \frac{ip_{01}}{2\hbar}\tilde{\mathcal{E}}(\rho_{11} - \rho_{00}). \quad (4.62)$$

The substitutions:

$$\kappa\tilde{\mathcal{E}} = \frac{p_{01}}{\hbar}\tilde{\mathcal{E}} \quad (4.63)$$

$$\frac{1}{2}\tilde{Q} = ic_0 c_1^* p_{01} \bar{N} \quad (4.64)$$

$$\rho_{11} - \rho_{00} = \frac{w}{p_{01}\bar{N}}, \quad (4.65)$$

lead identically to the equations of motion for the pseudo-polarization vector [Eqs. (4.6), (4.12)] derived earlier in this chapter. Consequently, some form of Bloch vector model can be used whenever the multilevel system can be reduced to a two-level system. It will be shown in the next subsection that such a simplification can be made even for multiphoton processes and n levels under certain conditions of detuning of the intermediate levels.

To simplify the notation, we will use complex Rabi frequencies to represent the electric field of the radiation:

$$\tilde{E}_{k+1} = \frac{i}{2\hbar}p_{k+1,k}\tilde{\mathcal{E}} = \frac{i}{2\hbar}p_{k+1,k}\tilde{\mathcal{E}}e^{\varphi_k}. \quad (4.66)$$

This particular notation is only used in the following sections of this chapter. The complex quantity $\tilde{E}_{k+1}$ represents the complex electric field envelope, expressed in units of frequency (the Rabi frequency for the transition $k \to k+1$). This definition involves an arbitrary choice of phase. In general, the initial cycles (near $t=0$) of the applied electromagnetic field are taken to have zero phase $[\varphi_k(t \approx 0) = 0]$. The definition (4.66) corresponds to transitions from level k to level $k+1$ in Eq. (4.59). The same definition applied to transitions from the lower level $[k-1 \to k$ in Eq. (4.59)] yields $\frac{i}{2\hbar}p_{k-1,k}\tilde{\mathcal{E}}^*(t)c_{k-1} = -\tilde{E}_{k-1}^*$. Let us consider the harmonic oscillator approximation, for which the dipole moments are proportional to the square root of the ratio of the indices of the successive levels [125]: $p_{k+1,k} \propto \sqrt{(k+1)/k}$. It results from this proportionality that the Rabi frequencies [defined by Eq. (4.66)] are proportional to the Rabi frequency of the first transition $0 \to 1$: $\tilde{E}_k = \sqrt{k}\,\tilde{E}_1$, where k is the level index. Let us assume in addition a constant anharmonicity: each successive transition frequency is reduced by the same frequency χ:

$$\begin{aligned}\omega_{0,j} - \omega_{0,j-1} &= \omega_{0,j-1} - \omega_{0,j-2} - \chi = \ldots \\ &= \omega_{0,1} - (j-1)\chi.\end{aligned} \quad (4.67)$$

As an illustrative example, let us consider a four level system excited by a single pulse of average frequency ω_ℓ. With the above mentioned assumptions and notations, the system of equations (4.56) reduces to:

$$\frac{d}{dt}\begin{pmatrix} c_0 \\ c_1 \\ c_2 \\ c_3 \end{pmatrix} = \begin{pmatrix} 0 & \tilde{E}_1 & 0 & 0 \\ -\tilde{E}_1^* & -i\Delta_1 & \tilde{E}_2 & 0 \\ 0 & -\tilde{E}_2^* & -i\Delta_2 & \tilde{E}_3 \\ 0 & 0 & -\tilde{E}_3^* & -i\Delta_3 \end{pmatrix}\begin{pmatrix} c_0 \\ c_1 \\ c_2 \\ c_3 \end{pmatrix}. \quad (4.68)$$

Equation (4.59) and the particular four level example Eq. (4.68) were established assuming the presence of levels nearly equally spaced (cf., Fig. 4.8), and a single frequency source of ultrashort pulses. We leave it as a problem at the end of this chapter to generalize this treatment to a polychromatic source of several pulses, matching stepwise transitions of a discrete level system, such as that of a simple atom. It can be shown that Eqs. (4.53) and (4.68) still apply, with an appropriate re-definition of the complex Rabi frequencies $\tilde{E}_i$ and detunings Δ_i.

4.3. MULTIPHOTON COHERENT INTERACTION

Climbing the ladder

One of the main interests in coherent interactions with ultrashort pulses is efficient creation of a population inversion. We will use the example of an anharmonic multilevel system to illustrate the high efficiency of optical pumping that can be achieved in conditions of coherent interactions.

The set of interaction equations (4.68) is particularly convenient for the study of the response of the multilevel system to a series of identical pulses (random sequence or regular train). Let us consider, for instance, the simple level structure of Fig. 4.8. We will asssume that only the first four transitions participate in the excitation process. The system is excited by the sequence of four ultrashort pulses of Fig. 4.8(c), with amplitude:

$$\tilde{\mathcal{E}}(t-t_1) + \tilde{\mathcal{E}}(t-t_2)e^{i\varphi_1} + \tilde{\mathcal{E}}(t-t_3)e^{i\varphi_2} + \tilde{\mathcal{E}}(t-t_4)e^{i\varphi_3} \qquad (4.69)$$

where t_i denotes the pulse delay. The interaction with a single pulse can be represented by a matrix that transforms any set of initial $c_i(t_0)$ into a set of coefficients $c_i(t_1^+)$ (where t_1^+ designates the time just following the pulse). The phase of each pulse corresponds to a rotation of that matrix. Between pulses, all the $\tilde{E}_i$ in Eq. (4.68) are zero, and the interaction matrix is diagonal. The solution of the system of equations (4.68) becomes in this case particularly simple. The eigenvalues are the successive detunings given by Eq. (4.57). For the four level system, the coefficients c_i after the first pulse will evolve according to:

$$\begin{aligned} c_0(t) &= c_0(t_1^+) \\ c_i(t) &= c_i(t_1^+)e^{-\Delta_i(t-t_1^+)} \quad \text{for } i = 1, 2, 3. \end{aligned} \qquad (4.70)$$

The c_i after a delay τ_2 [in Fig. 4.8(c)] are the initial conditions for the second pulse and so forth.

We have seen in the previous section how a "π pulse" excitation can completely invert a two-level system, while incoherent excitation or continuous irradiation (or irradiation with pulses longer than the phase relaxation time T_2) can at most equalize the population of the upper and lower states. Similarly, "coherent pumping" can also be used to complete inversion of multilevel systems. The situation is, however, more complex because of the plurality of Rabi frequencies and detunings involved. There is no simple analytical solution to this problem, which is to find a waveform $\tilde{\mathcal{E}}(t)$ which transforms the initial population state (1, 0, ..., 0) into a final state (0, 0, ..., 1). A practical method leading to the multilevel inversion is to excite the system with sequences of pulses, applied in a time short compared with the phase relaxation time. The general procedure is to optimize the relative phases and delays between pulses to maximize the transfers of population towards higher energy levels.

Before illustrating this concept by a few simple examples, let us examine first whether this problem should, in general, have a solution. Each pulse has set in motion a few near resonant harmonic oscillators in the molecular system, of amplitude given by the matrix elements $c_i c_j^*$. After passage of the first pulse, the matrix element $c_i c_j^*$ is left with a phase $\varphi_{ij}^{(1)}$, which, between pulses, will increase linearly with time as $(\Delta_j - \Delta_i)(t - t_1^+)$ [Eq. (4.59)]. Two parameters can be adjusted for the next (second) pulse:

- the interpulse spacing τ_2, to ensure that several matrix elements $c_i^* c_j$ are "caught" in phase by the next (second) pulse

- the phase of the next pulse, φ_2, equal or opposite to that of the various matrix elements that were in phase after the delay τ_2

For the sake of illustration, let us suppose that we want the second pulse to start in phase with $c_0 c_1^*, c_1 c_2^*$, and $c_2 c_3^*$. The common phase has to satisfy simultaneously the equations:

$$\begin{aligned}\varphi_{01}^{(1)} + \Delta_1 \tau_2 + m_1 \cdot 2\pi &= \varphi_2 \\ \varphi_{12}^{(1)} + (\Delta_2 - \Delta_1)\tau_2 + m_2 \cdot 2\pi &= \varphi_2 \\ \varphi_{23}^{(1)} + (\Delta_3 - \Delta_2)\tau_2 + m_3 \cdot 2\pi &= \varphi_2.\end{aligned} \quad (4.71)$$

The three equations (4.71) can be solved to determine τ_2, φ_2, m_1, m_2 and m_3. Since the m_i are integers, the system of equations (4.71) will, in general, have only approximate rather than exact solutions. Among the various approximate solutions within certain tolerance limits of φ_2, the one corresponding to the smallest delay τ_2 will be selected.

The procedure of selecting the proper delays and phases of the pulse train can be explained by means of a simple picture in the frequency domain. A sequence of equal pulses ($t_j - t_i$ is the delay between successive pulses) exhibits a modulated spectrum where the envelope is given by the single pulse spectrum. If the pulse delays and phases are chosen properly the modulated spectrum matches the transition profile of the energy ladder. Of course, finding the optimum pulse sequence is more complex than a simple spectral overlap of the optical excitation and the transition frequencies, because of the nonlinearity of the interaction.

4.3. MULTIPHOTON COHERENT INTERACTION

The various phases and elements ($c_i c_i^* = \rho_{ii}$, and $c_i c_j^* = \tilde{\varrho}_{ij} \exp[i\varphi_{ij}]$) of the density matrix can be represented, for convenience, at any stage during the excitation by the pulse sequence in the form of a matrix:

$$(\mathcal{M}) = \begin{pmatrix} \rho_{00} & 2|\tilde{\varrho}_{01}| & 2|\tilde{\varrho}_{02}| & 2|\tilde{\varrho}_{03}| \\ \frac{\varphi_{10}}{2\pi} & \rho_{11} & 2|\tilde{\varrho}_{12}| & 2|\tilde{\varrho}_{13}| \\ \frac{\varphi_{20}}{2\pi} & \frac{\varphi_{21}}{2\pi} & \rho_{22} & 2|\tilde{\varrho}_{23}| \\ \frac{\varphi_{30}}{2\pi} & \frac{\varphi_{31}}{2\pi} & \frac{\varphi_{32}}{2\pi} & \rho_{33} \end{pmatrix}.$$

Let us now apply the procedure outlined above to a specific multilevel system. The parameters chosen for this particular example are taken from the ν_3 C–F strech mode of vibration of the molecule CH_3F. For the first few levels of the vibrational ladder the transition frequencies and detunings are:

$$\frac{\omega_{01}}{2\pi} = 197.66 \text{ ps}^{-1} \quad \Delta_1 = 4.45 \text{ ps}^{-1}$$
$$\frac{\omega_{02}}{2\pi} = 392.34 \text{ ps}^{-1} \quad \Delta_2 = 5.92 \text{ ps}^{-1}$$
$$\frac{\omega_{03}}{2\pi} = 584.42 \text{ ps}^{-1} \quad \Delta_3 = 4.79 \text{ ps}^{-1}.$$

We assume that level 4 and higher energy levels are far off-resonance so that they can be neglected. Our intention is to show that indeed a suitable pulse sequence can lead to an almost complete inversion of the system (which is initially in the ground state). For the excitation of this model four level system, we use four Gaussian pulses of 1 ps FWHM at 1025 cm^{-1} (or an angular frequency of $\omega_\ell = 193.20$ ps^{-1}) [cf., Fig. 4.8(c)]. The pulse peak amplitudes (in terms of the complex Rabi frequency) are $|\tilde{E}_i(0)| = 2$ ps^{-1}. This pulse amplitude and duration, for a dipole moment of $p_{01} = 0.21$ Debye³, corresponds to a pulse energy density of 5 J/cm². The pulse phases and delays are:

$$\varphi_1 = \varphi_2 = 0 \qquad \varphi_3 = \varphi_4 = \pi/3$$
$$\tau_2 = \tau_3 = 1.8 \text{ ps} \qquad \tau_3 = \tau_4 = 0 \text{ ps}. \tag{4.72}$$

The successive density matrices, $(\mathcal{M}_1)$ at a time τ_2 after application of the first pulse and just before arrival of the second pulse, $(\mathcal{M}_2)$ at time τ_3 (following the second pulse) and $(\mathcal{M}_4)$ at time τ_4 (following the fourth pulse) are:

$$(\mathcal{M}_1) = \begin{pmatrix} 0.71 & 0.57 & 0.44 & 0.56 \\ 0.21 & 0.11 & 0.18 & 0.23 \\ 0.24 & 0.03 & 0.07 & 0.17 \\ -0.18 & 0.11 & 0.08 & 0.11 \end{pmatrix}$$

---

³1 Debye = 10^{-18} esu = $(1/3)\, 10^{-29}$ C· m

$$(\mathcal{M}_2) = \begin{pmatrix} \mathbf{0.46} & 0.12 & 0.49 & \mathbf{0.87} \\ 0.08 & 0.01 & 0.06 & 0.11 \\ 0.04 & -0.04 & 0.13 & 0.45 \\ 0.25 & 0.17 & 0.21 & \mathbf{0.40} \end{pmatrix}$$

$$(\mathcal{M}_4) = \begin{pmatrix} \mathbf{0.01} & 0.05 & 0.04 & 0.22 \\ -0.02 & \mathbf{0.05} & 0.07 & 0.42 \\ -0.10 & -0.08 & \mathbf{0.03} & 0.31 \\ -0.05 & -0.03 & 0.05 & \mathbf{0.91} \end{pmatrix}.$$

We note that after the first pulse, the diagonal terms of the density matrix indicate a distribution of 29% of the population among the three excited states. The population density in the middle of the pulse sequence (after excitation by the second pulse, see matrix $(\mathcal{M}_2)$) is particularly interesting, for having the population nearly equally distributed among the two extreme (ground and upper) states, and a maximum value for the off-diagonal matrix element $|\tilde{\varrho}_{03}|$. The subsequent pair of exciting pulses (pulse 3 and 4) is identical to the first one, except that it is phase–shifted by $\pi/3$. That second pulse sequence results in a 91% inversion of the four level system.

This type of excitation is the multiphoton analogue of a non-resonant "zero-area pulse excitation" [126] for single photon transitions, which consists of a sequence of two pulses π out of phase. We recall that, in the case of single photon transitions, referring to the "area theorem" Eq. (4.24), a signal consisting of successive equal positive and negative (phase $= \pi$) half will have a zero area θ_0, and conserve its zero area during propagation. The Fourier transform of such a waveform is zero at its resonance frequency, but has sidebands that can produce significant excitation of absorbing lines off-resonance.

In the preceding example of multi-step excitation, each of the two successive pair of pulses acts in an analogous way on the four level system. We have seen that a vector model can be constructed for two-level systems. It will be shown in the next section under which conditions such a model can be extended to a multiphoton resonance. In general, however, the sub-matrix $\begin{pmatrix} \rho_{00} & 2|\tilde{\varrho}_{0n}| \\ \ldots & \rho_{nn} \end{pmatrix}$ can lead to such a description at times τ_2, τ_3 and τ_4 [Fig. 4.8 (c)], but not *during* the application of the pulses, when all states are mixed by the electromagnetic field. The general strategy that emerged from a study of anharmonic systems [127] is to design a pulse (or a sequence of pulses)

4.3. MULTIPHOTON COHERENT INTERACTION

$\tilde{\mathcal{E}}_s(t)$ leading to a matrix $\begin{pmatrix} \rho_{00} & 2|\tilde{\varrho}_{0n}| \\ \cdots & \rho_{nn} \end{pmatrix} = \begin{pmatrix} 0.5 & 1 \\ \cdots & 0.5 \end{pmatrix}$. Repeating the same excitation, with a dephasing of π/n i.e., applying $\tilde{\mathcal{E}}(t)\exp(i\pi/n)$, generally leads to a good approximation of a pure inversion $\begin{pmatrix} 0 & 0 \\ \cdots & 1 \end{pmatrix}$, as in the case of the four-level system considered above.

The theory presented above applies to any physical system that can be modeled adequately by a set of isolated levels as sketched in Fig. 4.8. It can be generalized to a step-ladder excitation in atomic systems, where radiation pulses containing different frequencies $\omega_{\ell,i}$ matching the ladder of energy levels are applied.

A molecular system

The basic approximation made so far is that we have discrete, isolated levels (1,2,3,...) separated by about the photon energy of the exciting pulse(s). This, however does not apply to most molecular systems. In considering transitions of a real molecule, the model should be corrected to include all the vibrational–rotational transitions that fall within the excitation spectrum. That is, any level previously labelled by 1,2,3,..., corresponds now to a certain vibrational band which contains many rotational levels. We will show in this section that this additional element of complexity can be handled numerically, and that the general conclusion — namely that a total inversion can be achieved with properly shaped input signals — remains valid. Throughout this section we will neglect transitions between different electronic levels.

Depending on the molecular structure the actual vibration–rotational spectra can be very complex and its detailed derivation is beyond the scope of this book. The reader is referred to the Monograph by Herzberg [128] for example. In the case of molecular transitions, the wave function has to be expanded in a series of energy eigenfunctions corresponding to the quantum numbers v (vibrational quantum number) and J (rotational quantum number). A dipole allowed transition can only occur between two vibrational bands that satisfy the selection rules $\Delta J = \pm 1$ and $\Delta v = \pm 1$. Therefore the initial population of rotational states determines essentially the number of rotational levels that needs to be considered for each vibrational state during the excitation.

The definition of the detuning has to be extended to take into account the rotational level structure. Instead of Eq. (4.57), we define the detuning of the level labeled with the quantum number (v, J) as:

$$\Delta_{v,J} = \Delta_{v,0} + B\, J(J-1) \tag{4.73}$$

where

$$\Delta_{v,0} = \omega_{v,0} - v\omega_\ell \tag{4.74}$$

is the detuning of the molecule in the rotational ground state, and $\Delta_{0,0} = 0$. B is the rotational constant. To simplify the notations, we will use for the matrix elements of the dipole moment:

$$p_{v,+} = \langle v, J-1|p|J, v+1\rangle \tag{4.75}$$
$$p_{v,-} = \langle v, J+1|p|J, v+1\rangle. \tag{4.76}$$

The J dependence of the dipole moments $p_{v,+}$ and $p_{v,-}$ is given by [129, 130]:

$$p_{v,+} = p_v\sqrt{\frac{J+1}{2J+3}} \quad \text{and} \quad p_{v,-} = \sqrt{\frac{J}{2J-1}}. \tag{4.77}$$

For this simplified model of a vibrational–rotational level structure, the system of differential equations for the coefficients c_{kJ} takes a form similar to Eqs. (4.59). The first two equations are:

$$\begin{aligned}
\frac{d}{dt}c_{0J} &= -i\Delta_{0J}c_{0J} + \tilde{E}_1[p_{0,+}c_{1,J-1} + p_{0,-}c_{1,J+1}] \\
\frac{d}{dt}c_{1J} &= -\tilde{E}_1^*[p_{0,-}c_{0,J+1} + p_{0,+}c_{0,J-1}] - i\Delta_{1J}c_{1J} \\
&\quad - \tilde{E}_2[p_{1,+}c_{2,J-1} + p_{1,-}c_{2,J+1}].
\end{aligned} \tag{4.78}$$

As for the previous Eq. (4.56), the frame of reference is rotating at the angular velocity ω_ℓ.

The system of equations (4.78) can be solved numerically starting with an initial (Boltzmann) distribution for the population of the rotational levels in the vibrational state $(v = 0)$:

$$c_{0J}c_{0J}^* = A(J+1)e^{[-BJ(J+1)/k_BT]} \tag{4.79}$$

where k_B is Boltzmann's constant, T is the absolute temperature, and A a normalization factor (total population $= 1$). The numerical calculations were applied to the situation sketched in Fig. 4.9 and will be detailed below. A particular result is shown in Fig. 4.10.

4.3. MULTIPHOTON COHERENT INTERACTION

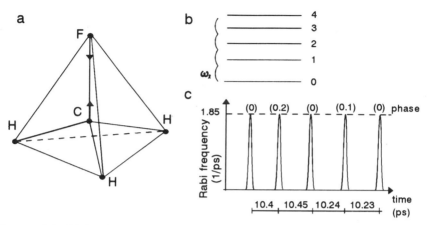

Figure 4.9: Multiphoton excitation of the CH_3F molecule. Sketch of the molecule identifying the C–F stretch mode (a), and corresponding energy level diagram (b). Optimized five pulse sequence leading to selective excitation of the fifth ($v = 4$) vibrational band (c). The absolute phase of each pulse (in radians) and the relative delays between pulses are indicated.

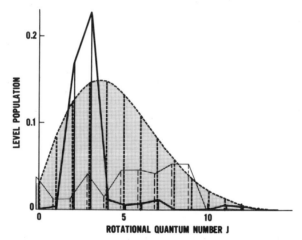

Figure 4.10: Selective pumping of the CF stretch mode of a CH_3F molecule by a sequence of five Gaussian pulses, each 1 ps FWHM and with a peak Rabi angular frequency of 1.85 ps^{-1}. With the first pulse taken as reference for phase and time, the successive pulses are at t = 10.4, 20.85, 31.09, and 41.32 ps, with a phase of 0.2, 0., 0.1, 0 radian, respectively. The ground state population versus rotational quantum number is represented before (dotted lines, shaded area) and after (thin dashed line) the five pulse excitation, as a function of rotational quantum number J. The heavy lines indicate the population distribution among the rotational lines after the four pulse excitation. The radiation wavelength is 9.7713 μm (from [127]).

Let us choose again as an example the excitation of the C–F stretch mode of the CH_3F molecule (Fig. 4.9), but now taking into account rotational levels. The transition frequencies are $\omega_{1,0}/(2\pi c) = 1048.6$ cm^{-1}, $\omega_{2,0}/(2\pi c) = 2081.4$ cm^{-1}, $\omega_{3,0}/(2\pi c) = 3100.4$ cm^{-1}, and $\omega_{4,0}/(2\pi c) = 4104.4$ cm^{-1}. The excitation consists of a sequence of five identical Gaussian pulses for which $1/\lambda_\ell = \omega_\ell/2\pi c = 1023.4$ cm^{-1}, slightly below exact four-photon resonance with the upper vibrational level (see Fig. 4.9). To compute the interaction with the train of pulses, a system of circa 50×4 coupled equations (4.78) has to be solved numerically.[4] As a result of the excitation by a first pulse, a complex distribution of populations is created among the rotational lines in each vibrational state. When a second identical pulse is applied, the modification of the population distribution is a function of the relative delays and phases of the two pulses.

The question of main interest — in fact the main purpose of dealing with the complexities of coherent interactions — is: can one find a distribution of pulses that excites *only a single rotation level in the upper vibrational state* (number 4)? The transfer function of this problem is too complex to solve the general inverse problem — i.e., find the pulse shape, intensity, phase, and spacing that would provide optimal selective excitation of the vibrational state $v = 4$. Numerical algorithms, however, exist to solve such problems in particular cases. With some intuition and luck, a trial and error method can, after a few iterations, lead to quite satisfactory solutions. The general guidelines are:

- each pulse should be short and intense enough to meet the bandwidth requirements of the anharmonicity. As indicated in Fig. 4.8(b), the spectrum of a single pulse (dashed line) should cover the frequencies of the successive transitions to be excited. We choose 1 ps pulses, of intensity corresponding to a peak Rabi frequency of $|\tilde{E}(0)_1| = 1.85 \, 10^{12}$ s^{-1} (notation defined by Eq. (4.66).

- The interpulse delay should be close to $\pi/2B$ (10.45 ps for CH_3F), which is one half of the period of the difference frequency between adjacent rotational lines.

Since there are two upward transition possible for any particular level with quantum numbers v, J ($v, J \to v+1, J+1$ and $v, J \to v+1, J-1$), we

---

[4]The exact number of equations needed depends on the number of rotational states initially populated, incremented by the number of additional states being accessed by stepwise $J \to J+1$ excitations.

can expect the initial Boltzman distribution of population in the ground vibrational state to be transposed as a broader distribution in the excited states (rotational line heating). The particular choice of interpulse delay ($\pi/2B$) should minimize this effect. Indeed, let us consider an isolated pair of vibrational levels resonant with the laser radiation, in a frame of reference rotating at the angular velocity corresponding to the (forbidden) $J \leftrightarrow J$ transition. A first pulse creates off-diagonal elements for the transitions $J \rightarrow J+1$ and $J \rightarrow J-1$. The precession after a delay $\pi/2B$ brings these two matrix elements exactly in opposite phase; hence, they cancel each other. The transitions to the next vibrational level will populate the sublevel J from a pure state at $J-1$ and $J+1$.

Systematic calculations of the excitation to the fifth level, for pulse sequences of various relative phases and delays, lead to the optimal five-pulse sequence sketched in Fig. 4.9(c). The population distribution created by the pulse sequence in the upper level ($v = 4$) is shown in Fig. 4.10 (heavy line), as a function of rotational quantum number. For comparison the Boltzmann distribution taken as initial condition for the ground state $v = 0$ (at a temperature of $T = 250°K$) is indicated by the light gray area under the dashed curve. The thin broken line shows the population distribution in the ground state after the pulse sequence. It is remarkable to note to which extent the choice of interpulse delay of $\pi/2B$ resulted in a cooling of the rotational temperature *in the excited state*. Indeed, the pulse sequence has essentially resulted in transferring most of the populations in the $J = 2$ and $J = 3$ rotational state of the upper vibrational level. A population inversion has been achieved for the vibrational levels (higher total population in $v = 4$ than in $v = 0$).

4.3.3 Simplifying a n-level system to a two-level transition

Taking into account the detailed level structure of a complex molecule, following the procedure of the previous section, is an arduous task. The atomic or molecular system can be assimilated to a two-level system (the upper and lower levels) if the intermediate levels are sufficiently off-resonance. To quantify the conditions under which multiphoton transitions — as opposed to the cascade of single photon transitions discussed in the previous sections — will be observed, two problems have to be solved. First, one has to determine the detuning $\Delta_n = \omega_{n0} - n\omega_\ell$ that provides the maximum amount of mixing (i.e., the largest amount of population exchange for a given optical field) between the extreme levels. Second, we have to determine the minimum detuning of the intermediate levels for which the perturbation of the intermediate levels is negligible. For simplicity of the analysis, we will aproximate the pulse by a square wave.

Two main assumptions are that (i) the detuning Δ_k of the intermediate level k is larger than the Rabi frequencies $\tilde{E}_k$ defined in the previous section, and that (ii) there is no resonance between any pair of intermediate levels (i.e., $\omega_{ij} \neq \omega_\ell$).

The following paragraph requires familiarity with properties of matrices, such as can be found in the textbook of Franklin [131]. It is not essential to the understanding of the remainder of this book.

We start from the time dependent Schrödinger equation (4.53) for the n-level system. The wave function $\psi(t)$ is expanded the usual way [cf. Eq. (4.55) in terms of the eigenfunctions ψ_k, corresponding to the eigenvalues $\hbar\omega_k$ of the unperturbed Hamiltonian H_0 [defined in Eq. (4.54)]. We proceed also to the same rotating wave approximation to represent the interaction by the set of differential equation put in matrix form in Eq. (4.68) (the latter equation is readily generalized to a $n-$level system).

Since we have assumed the field to be constant during the interaction (square pulse approximation), the coefficients $c_k(t)$ of this expansion can be found exactly through a calculation of the eigenvalues and eigenvectors of the matrix of this set of differential equations. The coefficients $c_k(t)$, solution of Eq. (4.68), are given by:

$$c_k(t) = \sum_{0}^{n-1} x_{kj} e^{\lambda_j t}, \qquad (4.80)$$

where x_{kj} are the eigenvectors corresponding to the eigenvalues λ_j of the interaction matrix. Because of the particular form of the interaction matrix $\mathcal{H}$ [the $n \times n$ square matrix in Eq. (4.68)], some recurrence relations exist to systematically search for the eigenvalues and eigenvectors. Indeed, the matrix $\mathcal{H}$ belongs to the class of tri-diagonal matrices, so-called because the only non-zero elements are of the type a_{ii}, $a_{i,i\pm 1}$ and $a_{i\pm 1,i}$. Properties of these matrices are given in Franklin [131], pages 251–253. Let Ψ_{n-1} be the determinant of the characteristic equation of the interaction matrix $\mathcal{H}$, and $\Psi_{k-1}(\lambda)$ be the determinant of the submatrix containing the first k rows and columns:

$$\Phi_{k-1}(\lambda) = (H_{k-1,k-1} - \lambda)\Phi_{k-2}(\lambda) - H_{k-1,k-2}H_{k-2,k-1}\Phi_{k-3}(\lambda). \qquad (4.81)$$

The characteristic equation $\Phi_{k-1}(\lambda) = 0$ has in general k solution which are the eigenvalues $\lambda_0, \lambda_1, \ldots \lambda_{k-1}$. There are in general k eigenvalues. To each

4.3. MULTIPHOTON COHERENT INTERACTION

solution λ_j corresponds a set of solutions or eigenvector of the eigenvalue equation:

$$\begin{pmatrix} H_{00}+\lambda & H_{01} & 0 & \cdots & \cdots \\ H_{10} & H_{11}+\lambda & H_{12} & 0 & \cdots \\ 0 & H_{21} & H_{22}+\lambda & H_{23} & \cdots \\ \cdots & 0 & \cdots & \cdots & \cdots \end{pmatrix} \begin{pmatrix} x_0 \\ \vdots \\ x_{k-1} \end{pmatrix} = 0. \quad (4.82)$$

Recurrence relations can be found in [131] for the eigenvectors of matrix $\mathcal{H}$:

$$\begin{aligned} x_{k-1} &= \Phi_{k-2}(\lambda); \\ x_v &= (-1)^{k-v} H_{v,v+1} \ldots H_{k-1,k} \Phi_{v-1}(\lambda), \end{aligned} \quad (4.83)$$

where v takes any value $< (k-1)$, and λ can take any of the λ_j values. The sum of the roots is equal to the sum of the diagonal elements $\sum_0^{k-1} H_{ii}$ and is the coefficient of the term of the characteristic equation in λ^{k-1} where k is the order of the submatrix. The product of the roots is equal to the value of the determinant, and is equal and opposite to the constant term of the characteristic equation.

These recurrence relations are sufficient to determine, for a given n-level system, the n roots λ_i of the characteristic equation, and the n^2 numbers $x_k(\lambda_i)$ that constitute the n eigenvectors of the characteristic equation, needed to determine the coefficient c_k according to Eq. (4.80). As stated earlier, it is desirable to be able to reduce the n-level system to an equivalent two-level system, between which the radiation induces multi-photon transitions. We outline the procedure of this reduction for a three-level system, leaving the generalization to a n-level as a problem.

In the particular case of the three-level system, the characteristic equation is:

$$\begin{aligned} \Phi_2(\lambda) &= \begin{vmatrix} \lambda & \tilde{E}_1 & 0 \\ -\tilde{E}_1^* & -i\Delta_1 + \lambda & \tilde{E}_2 \\ 0 & -\tilde{E}_2^* & -i\Delta_2 + \lambda \end{vmatrix} = 0 \\ &= \lambda^3 - i(\Delta_1 + \Delta_2)\lambda^2 - (\Delta_1\Delta_2 - |\tilde{E}_1|^2 - |\tilde{E}_2|^2)\lambda - i\Delta_2|\tilde{E}_1|^2. \end{aligned} \quad (4.84)$$

According to the recurrence expressions (4.83), the solutions $x_k(\lambda_i)$ ($k, i = 0, 1, 2$) should be proportional to:

$$x_0 = \tilde{E}_1 \tilde{E}_2 \quad (4.85)$$
$$x_1 = -\lambda \tilde{E}_2 \quad (4.86)$$
$$x_2 = |\tilde{E}_1|^2 - \lambda(i\Delta_1 - \lambda), \quad (4.87)$$

where λ takes any of the three values, solution, of the characteristic equation (4.84). Consistent with the original assumption that $\Delta_1 \gg \Delta_2$, $|\tilde{E}_1|$, $|\tilde{E}_2|$, one approximate solution to the characteristic equation (4.84) is $\lambda = i\Delta_1$. With a large detuning of the intermediate level, we will have thus one eigenvalue $\lambda_1 \approx i\Delta_1 \gg \lambda_0$, λ_2. Consistent with the absence of population transfer to and from level 1, The coefficient c_1 is proportional to $\exp(i\Delta_1 t)$. One of the eigenvalues (λ_1) is much larger than the two others, since according to Eq. (4.86), the values $x_1(\lambda_i)$ are proportional to λ.

Since our goal is to reduce the three-level system to an equivalent two-level system, the characteristic equation (reduced to second order) should have the same form as that of a two-level system, which can be seen by reducing the determinant in Eq. (4.84) to be:

$$\Phi_1(\lambda) = \lambda^2 - i\Delta\lambda + |\tilde{E}_1|^2, \qquad (4.88)$$

where $|\tilde{E}_1|$ is the Rabi frequency and Δ the detuning of the two-level system. For the similarity between the equation for the three level system and the two level model Eq. (4.88) to hold, the detuning of the upper state should be redefined as:

$$\Delta = \Delta_2 - \frac{|\tilde{E}_1|^2}{\Delta_1} - \frac{|\tilde{E}_2|^2}{\Delta_1}. \qquad (4.89)$$

The correction to the original detuning Δ_2, proportional to the square of the optical electric field, is a "Stark shift".

We note that in the case of the two-level system, the frequency of the Rabi cycling at exact resonance is determined by the difference between the two eigenvalues. "Exact resonance" in the case of our tri-level system implies [from Eq. (4.89)] that $\Delta_2 = [|\tilde{E}_1|^2 + |\tilde{E}_2|^2]/\Delta_1$. Substituting that value in the characteristic equation, and after some algebraic manipulations, one finds for the Rabi cycling frequency of the equivalent two-level system:

$$E_{\text{Rabi}} = \lambda_2 - \lambda_0 = \frac{|\tilde{E}_1||\tilde{E}_2|}{\Delta_1}. \qquad (4.90)$$

We conclude thus that the three level system, within the approximation of intermediate level far off-resonance, is equivalent to a two-level system. The resonance condition is modified by the "Stark shift" given by Eq. (4.89). A two-photon Rabi frequency can be defined, proportional to the product of the transition frequencies $\tilde{E}_1$ and $\tilde{E}_2$, hence proportional to the square of the electric field amplitude.

4.3. MULTIPHOTON COHERENT INTERACTION

The preceding considerations extend to an arbitrary number of levels. Quite generally, the multilevel system can, under certain conditions of large detuning of the intermediate levels, be reduced to a a two level system. The generalized Rabi frequency and the Stark shift can be incorporated into the density matrix equations of a two level system, or into the vector model.

For the n-level system, a key approximation is that the detuning of the intermediate levels be large compared with the Rabi frequencies $|\tilde{E}_k|$. This is a more complex condition than might appear at first glance: there should be no accidental resonance of order smaller than n for any pair of levels in the system being considered.

We refer the interested reader to [132] for a derivation of the generalized multilevel Rabi frequencies. One finds that the multilevel Rabi frequency E_{Rabi}, as in the case of the three level system, is the small difference of order $|\tilde{E}_k^{n-1}|/\Delta_j^{n-2}$ between two roots of the characteristic equation, and is

$$E_{\text{Rabi}} = 2\frac{\tilde{E}_1 \tilde{E}_2 \ldots \tilde{E}_{n-1}}{\Delta_1 \Delta_2 \ldots \Delta_{n-2}} = \kappa_{n-1}\tilde{\mathcal{E}}^{n-1}. \tag{4.91}$$

$\kappa_{n-1}\tilde{\mathcal{E}}^{n-1}$ is the generalized Rabi frequency, with the scale factor for the multilevel Rabi cycling given by:

$$\kappa_{n-1} = \frac{p_{01} \ldots p_{n-2,n-1}}{(\Delta_1 \ldots \Delta_{n-2})(2\hbar)^2}. \tag{4.92}$$

The Stark shift is the detuning $\delta\omega_s$ of the upper level Δ_{n-1} consistent with these solutions.

$$\delta\omega_s = \Delta_{n-1} = \frac{|\tilde{E}_1|^2}{\Delta_1} - \frac{|\tilde{E}_{n-1}|^2}{\Delta_{n-2}} \tag{4.93}$$

$$= \left[\left(\frac{p_{01}}{2\hbar\Delta_1}\right)^2 - \left(\frac{p_{n-1,n-2}}{2\hbar\Delta_{n-2}}\right)^2\right]|\mathcal{E}|^2. \tag{4.94}$$

Since the multilevel system is equivalent to a two-level system, a simple system of two density matrix equations should apply (or a modified Bloch vector model). Using the notation $\tilde{\varrho}_{ij} = c_i c_j^*$ for the off-diagonal matrix element in the rotating frame, we can write for the two-level system approximation:

$$\begin{aligned}\dot{\tilde{\varrho}}_{0,n-1} &= i(\Delta_{0,n-1} - \delta\omega_s) + i\kappa_{n-1}\tilde{\mathcal{E}}^{n-1}(\rho_{n-1,n-1} - \rho_{00}), \\ \dot{\rho}_{n-1,n-1} - \dot{\rho}_{00} &= 2\kappa_{n-1}\left[\tilde{\mathcal{E}}^{(n-1)*}\tilde{\varrho}_{0,n-1} + \tilde{\mathcal{E}}^{(n-1)}\tilde{\varrho}_{0,n-1}^*\right].\end{aligned} \tag{4.95}$$

The reduction of multiphoton interaction to a coherent model involving only the two extreme levels may seem like a coarse approximation. It has the value of making Bloch's vector model applicable to this complex problem. In particular, it has been shown to be valuable in designing pulse sequences leading to a complete population inversion in multilevel systems [127]. The validity of the two-level approximation has been tested numerically by computing the response of three to five level systems excited by sequences of pulses of different phase [132].

In the following subsection, we consider as an example of higher order two-level system four-photon resonance in mercury.

4.3.4 Four photon resonant coherent interaction

In an atomic system, the level density is much smaller than in molecules. Therefore, in the case of a multiphoton resonance, one will generally not find intermediate resonances between a pair of levels and the light field. The two-level approximation introduced above is generally appropriate. We will discuss in this section harmonic generation by multiphoton resonant coherent propagation in atomic vapors.

We have considered in Chapter 3 nonresonant nonlinear polarization, leading to self lensing, parametric generation and harmonic generation. The nonlinear susceptibility responsible for these effects can be enhanced by the proximity of a resonance. The presence of a resonance, on the other hand, complicates the interaction and introduces losses.

Coherent interaction processes take increasing importance the higher the order (n) of the nonlinear process[5] is. As we have seen in the previous sections, the transition rate between the two levels (a generalized Rabi frequency) is proportional to the nth power of the field envelope $\mathcal{E}$. Therefore, the parameter characteristic of the interaction is the pulse "area", proportional to the integral of the multiphoton Rabi frequency defined in Eq. (4.91), $\theta_n = \int \kappa_n \mathcal{E}^n(t) dt$. For a two photon transition, the area is proportional to the pulse energy. In the case of higher order processes, for the same pulse energy, shorter pulses have a larger area, hence favor the higher order nonlinear process. With the progress in pulse shaping techniques (see Chapter 7),

---

[5] In this subsection n is the *order* of the process, or the number of photons required to make the transition between the two levels, and should not be confused with the "n" representing the number of levels in the previous subsection.

4.3. MULTIPHOTON COHERENT INTERACTION

femtosecond pulses could be used to establish multiphoton coherences, manipulate population transfers, and generate with high efficiency higher order harmonics. This is a rather unexplored area of nonlinear optics, but extremely promising to generate efficiently short wavelength radiation, or store a large amount of energy in an atom or molecule.

Let us consider for instance the case of resonant n^{th} harmonic generation, where the fundamental radiation is in a kth ($k < n$) photon resonance with a particular atomic transition. At moderate powers, the conversion efficiency is orders of magnitude larger than in the non-resonance case, because of the intermediate resonance. However, as the power is increased, in an attempt to achieve higher conversion efficiencies, the k–photon absorption can deplete the fundamental radiation, before significant conversion is achieved. The energy of the fundamental radiation is transferred to the atomic system rather than to the harmonic radiation. A solution to this problem is to minimize the energy lost to the atomic system by the fundamental by making use of pulse shapes or sequences that propagate through the (multiphoton) resonant medium with minimum absorption losses.

This technique has been used successfully to increase the third harmonic conversion in the case of a two-photon resonant transition in Lithium vapor [133, 134, 135]. A first pulse excites (via a two-photon transition) the $4s$ level of lithium. For a second pulse, following the first one with a delay short compared to the phase relaxation time of the two photon transition, and sent through the same medium $\pi/2$ out of phase, two-photon stimulated emission occurs. The second pulse recovers the energy lost by the first one to the medium. If conditions of phase matching (for third harmonic generation) are not met, the peak intensity of the second pulse increases with propagation distance (two-photon stimulated emission) at the expense of the first pulse which is depleted. If conditions for phase matched third harmonic generation are met, however, energy is continuously transferred from the second pulse to the third harmonic field. As a result, instead of growing at the expense of the first pulse, the second pulse propagates with near constant peak intensity over a long distance. Efficient third harmonic generation occurs during that second pulse.

Among the multiphoton processes of various orders n, two photon resonant processes are somewhat unique, because, as long as the pulses are shorter than the inverse linewidth of the resonances, the result of the interaction is independent of the pulse duration and depends only on the pulse energy. This is mainly because, as we have seen before, the rate of population transfer (Rabi frequency) and level motion (Stark shift) are all proportional to the pulse intensity. For instance, if a 10 ps pulse of 1 mJ/cm^2 is a π pulse producing complete inversion, a 10 fs pulse of 1 mJ/cm^2, having a 10^3 times higher intensity, will produce exactly the same result.

The situation is quite different with higher order processes, as pointed out at the beginning of this section. We will therefore chose the example of a four-photon resonance to illustrate the potential benefits, as well as some of the difficulties associated with higher order resonances. Specifically, we will consider 4-photon coherent resonant excitation of mercury, as sketched in Fig. 4.11. If its intensity is sufficiently high, a fs pulse of amplitude $\tilde{\mathcal{E}}(t)$ at ω_ℓ (fundamental) will be attenuated through 4-photon absorption. The excited level (6^1D_2 in Fig. 4.11) can be seen as a "springboard" for third (frequency $3\omega_\ell$, field $\tilde{\mathcal{E}}_3$) and fifth (frequency $5\omega_\ell$, field $\tilde{\mathcal{E}}_5$) harmonic generation through Stokes and anti-Stokes Raman processes. The motivation for studying such a complex system is to generate efficiently third and/or fifth harmonic, using the four photon resonance to enhance the nonlinear susceptibility, and coherent propagation effects to minimize energy transfer (absorption) to the atomic system. We use this problem as an example to illustrate the various interesting phenomena that complicate significantly the study of higher order processes.

Since a resonant level has been excited by four-photon absorption (6^1D_2 in Fig. 4.11), generation of the third harmonic frequency is possible through a Raman process. Energy losses occur because of direct (four-photon) absorption, and subsequent absorption of another photon leading to (five-photon) ionization.

As discussed in the previous section, the system of interaction equations can be reduced to a system of three equations resembling Bloch's equations (cf., Eqs. (4.95)). One can define a fourth-order Rabi frequency proportional to the fourth power of the complex electric field amplitude $\tilde{\mathcal{E}}(t)$. The off-diagonal matrix element representing the four photon coherent resonant excitation is a quantity oscillating at the frequency $4\omega_\ell$, with a complex amplitude of the pseudo polarization $\tilde{Q}_4$. As shown in Appendix C, third and fifth harmonics $\tilde{\mathcal{E}}_3$ and $\tilde{\mathcal{E}}_5$ are generated by the combinations $\tilde{Q}_4\tilde{\mathcal{E}}^*$ and $\tilde{Q}_4\tilde{\mathcal{E}}$, respectively which appear as source terms in the wave equation for the electric field amplitudes. These coupling terms describe stimulated Raman processes. There is also a non-resonant contribution to the harmonic fields. In particular, the non-resonant third harmonic generation source term is proportional to the third power of the field [of the form $\chi^{(3)}\tilde{\mathcal{E}}^3(t)$] and may not be negligible compared to the Raman term $\tilde{Q}_4\tilde{\mathcal{E}}^*$. Combinations of these fields ($\tilde{\mathcal{E}}_1^4$, $\tilde{\mathcal{E}}_1\tilde{\mathcal{E}}_3$, $\tilde{\mathcal{E}}_1^*\tilde{\mathcal{E}}_5$ and $\tilde{\mathcal{E}}_1^2\tilde{\mathcal{E}}_5\tilde{\mathcal{E}}_3^*$) are the source terms in the interaction

4.3. MULTIPHOTON COHERENT INTERACTION

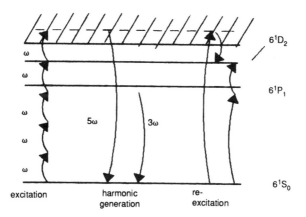

Figure 4.11: Energy levels of mercury and some of the possible excitation and harmonics generation mechanisms. The third and fifth harmonic can be generated by a Raman process involving the 6^1D_2 state for example. The main loss mechanisms are four-photon absorption and five-photon ionization.

equations (C.1) and (C.2) given in Appendix C.[6] The upper level of the transition is pumped by four photons at the fundamental frequency, but also by the sum of a third harmonic photon and a fundamental photon. If these two excitation processes have opposite phase, there is no more interaction between the resonant radiation and the medium (interaction quenching).

The complexity of the problem is due to the number of combinations of fields which give rise to an interaction of order 4 or lower. Second-order Stark shifts can sweep intermediate levels through resonance, creating "transient resonances."

The example of four-photon resonant transient coherent interaction in Hg vapor points to three puzzling effects typical of higher order systems:

1. interference between resonant and non-resonant harmonic generation

2. transient Stark shifts; and

3. interaction quenching.

All these effects are particularly dramatic when an intermediate level approaches resonance with an harmonic of the field. This is in particular the

---

[6]In Eqs. (C.1) and (C.2), the field amplitudes $\tilde{\mathcal{E}}_i$ are replaced by corresponding Rabi frequencies $\tilde{V}_i$ defined in Eq. (C.4).

case with the 6S–6D four photon resonance of Hg, where the 6P state approaches a three–photon resonant condition, as shown in Fig. 4.11. This near resonance accounts for the three above effects. First, the non-resonant susceptibility $\chi^{(3)}$ given in Eq. (C.8) is enhanced, resulting in a non-resonant contribution to the third harmonic, which can interfere with the resonant term in Eq. (C.9).

For the particular levels of Hg shown in Fig. 4.11, the four-photon area is:

$$\theta_4 = 0.0175 \frac{W^2}{\tau_p} \qquad (4.96)$$

where W is the energy density in J/cm^2 of a pulse of duration τ_p (in ps). This example shows indeed that higher order coherent interactions require fs pulses. An area of $\theta_4 = 0.7$ can be achieved with a 100 fs pulse of 2 J/cm^2 [200 μJ in a cross-section of $(0.1 \text{ mm})^2$].

Because of the proximity of the 6P level (indicated in Fig. 4.11), the main contribution to the Stark shift of the resonant transition is a shift of the upper 6D level:

$$\delta\omega_2 \approx \frac{1}{\sqrt{r_{04}}} \left[\frac{|p_{04}|^2}{(\omega_{34} - \omega_\ell)} \right] |\tilde{V}_1|^2, \qquad (4.97)$$

where the index 0, 3, and 4 have been given to the ground, the 6P and the 6D levels, respectively. Equation (4.97) is the Stark shift expression Eq. (C.5) in which the dominant term of the susceptibility [Eq. (C.6)] has been inserted. The symbol $|\tilde{V}_1(t)|$ represents the electric field in units of $s^{-1/4}$, defined through the four-photon Rabi frequency $\tilde{V}_1^4(t) = r_{04}\tilde{\mathcal{E}}_1^4(t)$. The various coefficients are defined in Appendix C.

As the field intensity of the fs pulse increases, the level 6D (level 4) is shifted in accordance with Eq. (4.97). The resulting change in the denominator of Eq. (4.97) implies a redefinition of the detuning $\delta\omega$. Thus, with increasing field amplitude, there is a nonlinear self-induced detuning effect, reducing the resonant coherent interaction with ultrashort pulses. This effect has been predicted within the framework of the approximation used in Refs. [134, 136]. It should be noted that in this particular example the field induced shifts are too large and too fast to be consistent with the adiabatic approximation used in the elaboration of the four photon resonant interaction equations (see Appendix C).

The last mentioned effect of interaction quenching is probably the most challenging – yet unsolved – dilemma of resonant multiphoton coherent inter-

4.3. MULTIPHOTON COHERENT INTERACTION

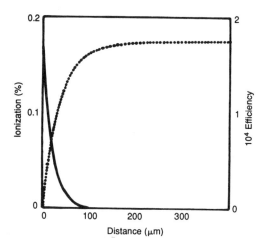

Figure 4.12: Five-photon ionization (in %; solid line) and third harmonic peak field conversion factor η (dashed line) versus propagation distance (in μm) in mercury vapor. The vapor pressure is 10 torr. The incident pulse shape is the Gaussian $\tilde{\mathcal{E}}_1(t) = \tilde{\mathcal{E}}_{10} \exp[-(t/\tau_G)^2]$, with $\tau_G = 3$ ps. The pulse energy density is 20 J/cm^2 (from [134]).

action. When intermediate resonances make the generation of an harmonic field particularly efficient, the source terms for the four-photon coherent excitation through the harmonic and through the fundamental are out of phase – hence cancel each other. In the density matrix equations for this interaction (as outlined in Appendix C), the "coherent excitation" is represented by an off-diagonal matrix element $\tilde{\rho}_{04}$ or its complex amplitude defined as $\tilde{Q}_4 = 2i\tilde{\rho}_{04} e^{-4i\omega_\ell t}$. The sequence of events that unfolds by solving the system of interaction equations (C.1) can be told through phenomenological arguments. As a strong pulse propagates through the medium, coherent excitation (of amplitude $\tilde{Q}_4$) is created through a source term proportional to the fourth power of the fundamental field $\tilde{\mathcal{E}}_1^4$, resulting also in some population of the level 4 (6^1D_2 in Fig. 4.11). The mixing of the four photon excitation and the fundamental leads to the generation of a third harmonic field ($\omega_{3\ell} = 3\omega_\ell$; $\tilde{\mathcal{E}}_3 \propto \tilde{Q}_4 \tilde{\mathcal{E}}_1^*$) and a fifth harmonic field ($\omega_{3\ell} = 5\omega_\ell$; $\tilde{\mathcal{E}}_5 \propto \tilde{Q}_4 \tilde{\mathcal{E}}_1$). Both the third and fifth harmonic field are themselves source terms for the coherent four photon excitation through second order processes: difference frequency generation $\omega_{5\ell} - \omega_\ell$ (source term for $\tilde{Q}_4 \propto \tilde{\mathcal{E}}_5 \tilde{\mathcal{E}}_1^*$) and sum frequency generation $\omega_{3\ell} + \omega_\ell$ (source term for $\tilde{Q}_4 \propto \tilde{\mathcal{E}}_3 \tilde{\mathcal{E}}_1$). The phase relation between the three source terms for the coherent excitation is such that they cancel each other. For all practical purposes, the radiation does not interact with the resonant system anymore. This effect is illustrated in the simulation of Fig. 4.12, where the generated third harmonic and the 5-photon ionization are plotted as a function of propagation distance. Because of the proximity of the 6^1P_1 level to a three-photon resonance condition, the third harmonic generation dominates the fifth harmonic in this particular example. The incident wavelength is at 566.7 nm, or 6 nm above the weak field resonance (560.7 nm) to compensate for the Stark shift. As the pulse propagates through the mercury vapor, the third harmonic generation is seen to saturate after a very short distance, while the ionization drops to zero. The third harmonic being generated with a phase opposite to that of the fundamental, the interaction vanishes, since the second order $(\omega_{3\ell} + \omega_\ell)$ and fourth order $(4\omega_\ell)$ mechanisms of four photon resonant excitation [Eq. (C.1)] cancel each other. This problem could be solved by adjusting the relative phase of the fundamental and third harmonic, and by using phased pulse sequences to control the relative phase of the coherent excitation (term of amplitude $\tilde{Q}_4$) and that of the radiation.

Multiphoton resonances reduce the energy requirements on higher order harmonic generation by enhancing the nonlinear susceptibilities. Energy losses associated with the inevitable multiphoton absorption associated with the resonance can be minimized by exploiting the property of reversibility of light matter energy transfers in coherent multiphoton resonant interactions with fs pulses. The above example however illustrates the complexity of multiphoton coherent effects. Mastering the theory and being able to generate the fs pulses shapes dictated by numerical simulations is the price to pay for high nonlinear conversion with low energy fs pulses.

These techniques are of increasing importance for the generation of femtosecond pulses in the UV, where one has to work with gases rather than crystals.

4.3.5 Miscellaneous applications

Because fs pulses can be shorter than dephasing times even in condensed matter, the reversibility of coherent interactions can be exploited in various applications.

All experiments using photons as a source of momentum can be made more efficient with ultrashort pulses. When an atom of mass M absorbs a

4.3. MULTIPHOTON COHERENT INTERACTION

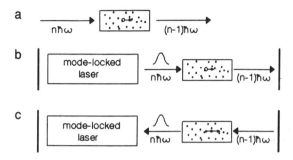

Figure 4.13: Momentum transfers between a collection of atoms and an optical pulse. In (a), a pulse of n photons loses one photon to the atomic system, providing a small recoil to the absorbing atom. (b) and (c) show two successive steps of intracavity interaction of a "π pulse" with an atomic beam. In a first passage, the atoms are inverted, resulting in a recoil to the right. The sketch assumes only one atom interacting with the beam. That atom absorbs one photon from the pulse, and is imparted a kinetic energy $Mv^2/2 = (\hbar\omega_\ell/2Mc^2)\hbar\omega_\ell$ (b). The reflected pulse recovers its energy by stimulated emission at the second passage, while giving an additional recoil to the right (c).

photon, the ratio of the recoil energy $Mv^2/2$ to the photon energy is the minute quantity $\hbar\omega_\ell/(2Mc^2) \leq 10^{-19}$.

To illustrate the latter point, let us consider the situation sketched in Fig. 4.13, where the intracavity beam path of a mode-locked laser crosses an atomic beam at a right angle. Let us assume that the laser is tuned to an absorbing transition of the atom, and the intracavity mode-locked pulse intensity is such as to be a "π pulse" for the atom. At the first passage of the pulse, m atoms will have been given a momentum Mv, while the pulse has lost m photons. If the number of photons in the pulse is $n \gg m$, the pulse reflected back by the mirror will still be a "π pulse," which will return the m atoms to the ground state, thereby restituting its original energy to the mode-locked pulse, and imparting an addition momentum Mv to the m atoms *in the same direction*. The net result is that the number of photons is conserved, but the atoms have been given a kinetic energy of $2Mv^2$. It is left as a problem at the end of this chapter to determine how the energy is conserved in this problem. A simple mechanical analogy for this problem is sketched in Fig. 4.14.

Many optical interactions that involve optical pumping can be performed more efficiently with ultrashort pulses. However, it may come as a surprise

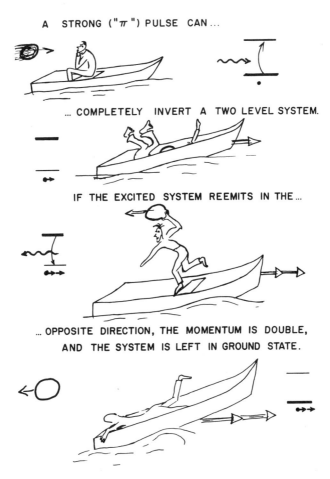

Figure 4.14: Momentum transfers between a short pulse and a two level system; the "mechanical" analogy.

that, with these ultrashort broad bandwidth pulses, *frequency selective* excitation is also possible. The frequency selectivity can be obtained from a combination of one or more pulses. For instance, it can easily be seen that a "zero-area pulse" has no Fourier component at its average frequency. Such a pulse can be designed to produce complete inversion off-resonance, and no excitation at resonance.

Similarly, a Gaussian pulse can be designed to be a "π pulse" at resonance, and a "2π pulse" off-resonance. The case of a square pulse is

4.3. MULTIPHOTON COHERENT INTERACTION

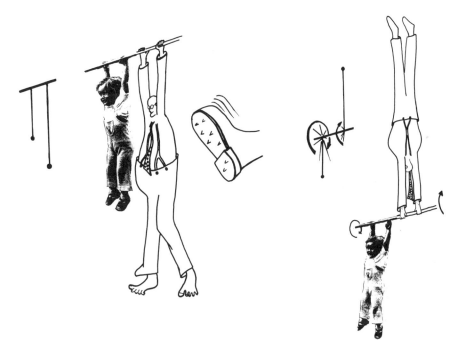

Figure 4.15: Selective optical excitation with a π pulse on resonance, and a 2π pulse off-resonance: the pendulum analogy.

the most obvious. For a square π pulse at resonance (ω_0) with a line to be excited $\theta_0 = \kappa \mathcal{E} \tau_p = \pi$. If we want to leave a line at frequency $\omega_1 = \omega_0 - \Delta\omega$ unexcited, the condition for the pulse duration and amplitude is $\tau_p \sqrt{\kappa^2 \mathcal{E}^2 + \Delta\omega^2} = 2\pi$ or $\kappa \mathcal{E} = \sqrt{3} \, \Delta\omega$ [as can be seen from the vector model sketched in Fig. 4.2 (a)]. These properties can be exploited for selective optical excitation, for instance, isotope separation [137, 138]. It can be shown [126] that, in the case of Doppler broadened transitions, the "π—2π" (π pulse on resonance; 2π pulse off-resonance) selective excitation scheme is more selective and efficient with Gaussian shaped than with square pulses. If the pulse duration and amplitude are appropriately chosen, the selected line can be completely inverted in a single shot, while the transition to be left undisturbed is completely returned to ground state. The selectivity is high, even though the pulse is short enough to interact with both lines. There is a simple mechanical analogy to this "π—2π" excitation. The analogue of the two absorbers at different frequencies are two pendulae of different lengths. The element to be selected is the longer pendulum (Fig. 4.15), while

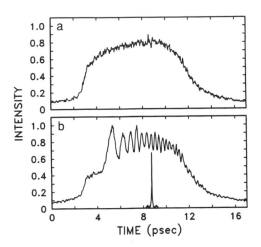

Figure 4.16: Transmission of a frequency swept pulse through a sodium cell. (a) Cross-correlation of the input pulse with a 22 fs probe pulse. (b) Cross-correlation of the output pulse with a 22 fs probe pulse. The modulation indicates that the frequency of the pulse transmitted through the fiber varies with time (chirp). The 22 fs probing pulse's autocorrelation (FWHM 34 fs) is also shown (from [140]).

the "unwanted" transition is the shorter toddler. The longer pendulum has a lower frequency. The potential energy is the analogue of the population difference (taking as reference the axis of the pendulum). For the same kick (light pulse) applied to both pendulae, it is possible to create a complete inversion for the selected transition, while the unwanted element has gone through a whole cycle and returned to the ground state.

Another application of coherent interaction involves the use of an atomic resonance as a local oscillator, to extract phase information of a chirped pulse [139] with frequency $\omega_\ell + \dot{\varphi}(t)$. The chirped pulse to be analyzed is sent through a cell containing sodium vapor with a sharp resonance at ω_0. The transmitted pulse interferes with the resonant re-radiation from the atomic line, resulting in a field component modulated at a frequency $\omega_\ell + \dot{\varphi}(t) - \omega_0$. This modulation carries the information about the chirp and is measured by cross-correlation with a shorter fs pulse. The result of such a measurement is illustrated in Fig. 4.16. The chirp was induced on a 5.4 ps pulse from a synchronously pumped dye laser by passage through a polarization preserving fiber (3 m long). The pulse bandwidth was increased by self phase modulation from 1 Å to 50 Å. A fraction of the original self-phase modulated pulse is sent through a two-stage compression (as described in Chapter 6) to create the 22 fs pulses [140] needed to cross-correlate the signal transmitted by the sodium cell. The upper trace (a) in Fig. 4.16 is the cross-correlation of the input to the sodium cell with the 22 fs pulse. The lower figure (b) shows the modulation induced in the sodium cell (optical thickness $a = 95$). Since the phase of the pulse changes by exactly π from one extremum to the next, a plot of the phase versus time can easily be made. Such a measurement is particularly useful to test the linearity of frequency chirping techniques.

4.4 Problems

1. Compare the population transfer when a two-level system is excited by (a) a step function dc field (zero carrier frequency) and (b) a step function electric field resonant with the transition frequency.

2. Find the steady state solution of Bloch's equations valid for monochromatic, cw incident radiation and a homogeneously broadened two-level system. Calculate the results of an absorption measurement for high incident field intensity (i.e., calculate the absorption versus wavelength and light intensity). The calculated width of the absorption profile will turn out to be a function of the incident intensity. Yet if you perform a pump probe experiment — i.e., you saturate the line with a field at frequency ω_1, and measure the absorption profile by tuning the frequency ω_ℓ of a weak probe beam — you find that the linewidth is $1/T_2$ independently of the intensity of the pump at ω_1. Explain.

3. Demonstrate the area theorem for a square pulse at resonance, in a medium with a square inhomogeneous broadening.

4. Show that a pulse with zero area has no Fourier component at its average frequency.

5. Show that the definitions (4.22) and (4.27) are equivalent for unchirped pulses at resonance.

6. Discuss the connection between the frequency pushing derived in Chapter 1 (Section 1.2.2) and $\langle \dot{\varphi} \rangle$ obtained in Chapter 4 (Section 4.3.3) for the weak pulse limit. Hint: Define a dielectric constant in terms of u, v, and w.

7. A short (weak) Gaussian pulse is sent through a resonant absorber with $T_2 \gg \tau_G$ (no inhomogeneous broadening). Being much narrower than the pulse spectrum, the absorbing line should act as a frequency filter. It is therefore a broadened pulse (in time) that should emerge from the absorber. Yet, according to the area theorem, the pulse area should decrease. Resolve this apparent contradiction in both frequency and time domains.

8. Calculate the initial frequency shift with distance $d\langle\dot\varphi\rangle/dz$ for a Gaussian pulse, off-resonance by $1/\tau_G$ with an absorbing (homogeneously broadened) transition with $T_2 = 100\tau_G$. Express your answer in terms of the linear attenuation.

9. Find an expression for $d\langle\dot\varphi^2\rangle/dz$ for a pulse propogating through an ensemble of two level systems. $\langle\dot\varphi^2\rangle$ is defined as $\int \dot\varphi^2 \mathcal{E}^2 dt / \int \mathcal{E}^2 dt$. *Hint*: Use a similar procedure as for the derivation of the expression for the frequency shift with distance, writing first an expression for the space derivative of $W\langle\dot\varphi^2\rangle$, and using Maxwell-Bloch's equations to evaluate each term of the right hand side.

10. Referring to Fig. 4.8(a), let us consider a stepwise excitation with a polychromatic pulse given by the sum $\tilde{\mathcal{E}}_1(t)e^{i\omega_{\ell,1}t} + \tilde{\mathcal{E}}_2(t)e^{i\omega_{\ell,2}t} + \tilde{\mathcal{E}}_3(t)e^{i\omega_{\ell,3}t} + \ldots$. The frequencies $\omega_{\ell,1}, \omega_{\ell,2}, \omega_{\ell,3} \ldots$, are nearly resonant with the successive transitions $(0 \to 1)$, $(1 \to 2)$, $(2 \to 3)$, $\ldots$. Derive Eq. (4.59) for this situation. *Hint*: Instead of Eq. (4.57), the detunings are now $\Delta_1 = \omega_{01} - \omega_{\ell,1}$, $\Delta_2 = \omega_{02} - (\omega_{\ell,1} + \omega_{\ell,2})$; $\Delta_3 = \omega_{03} - (\omega_{\ell,1} + \omega_{\ell,2} + \omega_{\ell,3})$, $\ldots$. The complex Rabi frequencies are to be defined as $\tilde{E}_{k+1} = \frac{i}{2\hbar}p_{k+1,k}\tilde{\mathcal{E}}_{k+1}$.

11. Consider the situation of Fig. 4.13. After having lost one photon, then gained one photon, the light pulse has the same original number n of photons. On the other hand, the atom has gained kinetic energy and momentum. Explain how the energy of the system atom and field is conserved in this situation.

Chapter 5

Ultrashort Sources

5.1 Introduction

5.1.1 Mode-locking

A mode-locked fs laser requires a broadband gain medium, which will sustain over 100,000 longitudinal modes in a typical laser cavity. The term "mode-locking" originates from the description of the laser in the frequency domain, where the emission is considered to be made up of the sum of the radiation of each of these (longitudinal) modes. In a free–running laser, the phases of a comb of equally spaced modes (frequency spacing Δ) can be a set of random numbers. The time domain transformation of such a frequency spectrum is an infinite series of identical bursts of incoherent light [Fig. 5.1(a)], spaced in time by $\tau_{RT} = 2\pi/\Delta$, which is the time needed to complete a cavity round-trip. Forcing all the modes to have equal phase — a procedure called "mode-locking" — implies in the time domain that all the waves of different frequency will add constructively at one point, resulting in a very intense and short burst of light [Fig. 5.1(b)]. The inverse Fourier transform of an infinite series of "δ functions" equally spaced in frequency, and *in phase*, is, in the time domain, an infinite train of "δ functions" or infinitely short pulses. The modes of a linear or ring laser cavity offer a good approximation to the infinite comb of delta functions in the frequency domain. An overwhelming proportion of ultrashort sources is based on some "mode locking" mechanism. The ratio of the laser cavity length to the pulse duration is a measure of the number of modes oscillating in phase. Typically, for a meter long laser producing a train of 100 fs pulses, there are over 100,000 modes contributing to the pulse bandwidth! If the comb of modes can be locked

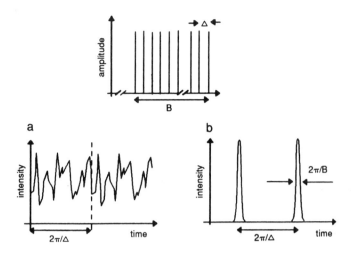

Figure 5.1: A set of equally spaced modes, and the inverse Fourier transform of that spectrum. In (a), the modes have a random phase distribution. In (b), all modes are "locked" to the same phase.

to a stabilized cavity, the laser is simultaneously a frequency standard and a fs probe. In general however, the comb of modes is allowed to drift in frequency (vibration of optical mounts and components make it generally impossible to prevent the cavity length to change on a wavelength scale).

In the time domain, it is often appropriate to view the mode-locked operation as the oscillation of one (sometimes two or three) single pulse(s) traveling back and forth in a linear cavity, or circulating in a ring. This is the description that we have tried to follow consistently in this book, and in particular in this chapter.

In the case of passive mode-locking, some intensity dependent loss or dispersion mechanism is used to favor operation of pulsed over continuous radiation. Another type of mode-locking mechanism is active: a coupling is introduced between cavity modes, "locking" them in phase. Between these two classes are "hybrid" and "doubly mode locked" lasers in which both mechanisms of mode locking are used. In parallel to this categorization in "active" and "passive" lasers, one can also classify the lasers as being modulated inside (the most common approach) or outside (usually in a coupled cavity) the laser.

We will begin our description of femtosecond lasers with a discussion of the various processes of pulse formation and compression.

5.1. INTRODUCTION

5.1.2 Elements of a fs laser

There are a few basic elements essential to a fs laser:

- a broadband ($\Delta\nu_g \gg 1$ THz) gain medium
- a laser cavity
- an output coupler
- a dispersive element
- a phase modulator
- a gain/loss process controlled by the pulse intensity or energy.

The items listed above refer more to a function than to physical elements. For instance, the gain rod in a Ti:sapphire laser can cumulate the functions of gain (source of energy), phase modulator (through the Kerr effect), loss modulation (through self lensing), and gain modulation. To reach femtosecond pulse durations, there is most often a dispersive mechanism of pulse compression present, with phase modulation to broaden the pulse bandwidth, and dispersion (positive or negative, dependent on the sign of the phase modulation) to eliminate the chirp and compress the pulse. It is the dispersion of the whole cavity that has to be factored in the calculation of pulse compression. Section 5.11 of this chapter is devoted to the analysis of typical fs cavities, including some mode analysis (an exception to the general rule we follow to analyze — whenever possible — the pulse formation in the time domain). It may come as a surprise that the modes of the cavity should never be exactly equally spaced. Ideally, the mode spacing should follow a quadratic frequency dependence, to match the chirp induced by the phase modulating element (if the chirp is linear).

Before turning to the details of fs pulse generation, let us discuss as an example two of the most widely used fs lasers: a ring dye laser, and a Ti:sapphire laser (Fig. 5.2). Both lasers are pumped typically by a cw Ar ion laser. In the dye laser, the ring configuration allows two counterpropagating trains of pulses to evolve in the cavity [141]. The gain medium is an organic dye in solution (for instance, Rh 6G in ethylene glycol), which, pumped through a nozzle, forms a thin (≈ 100 μm) jet stream. A similar section with a different dye (for instance, diethyloxadicarbocyanine iodide, or DODCI, in ethylene glycol) acts as saturable absorber. The two counter-propagating pulses meet in the saturable absorber (this is the configuration of minimum

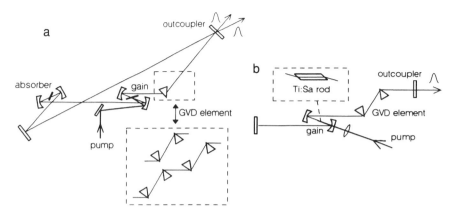

Figure 5.2: Sketch of a typical ring dye laser cavity (a) and a linear Ti:sapphire laser (b).

losses, as will be detailed in following sections). A prism sequence (one, two, or four prisms) allows for the tuning of the resonator group velocity dispersion. The pulse wavelength is determined by the spectral profiles of the gain and absorber dye. Limited tuning is achieved by changing the dye concentration. Pulses shorter than 25 fs have been observed at output powers generally not exceeding 10 mW with continuous (cw) pumping [142], and up to 60 mW with pulsed (mode-locked argon laser) pump [143].

The simplest cavity so far is that of the Ti:sapphire laser [Fig. 5.2(b)], which contains only the active element (the Ti:sapphire rod) and dispersive elements [144]. The technique is referred to as "self-mode-locking" [145], and led to the generation of pulses shorter than 15 fs [49, 146, 103, 147]. The output power is typically several hundred mW at pump powers of less than 10 W. Sometimes, to start the pulse evolution and maintain a stable pulse regime, a saturable absorber, an acousto-optic modulator, a wobbling end mirror, or synchronous pumping is used.

5.1.3 Pulse formation

The signal from any free-running oscillator originates from noise. This is not different with the radiation from a femtosecond laser. In continuously pumped lasers, the noise is due to spontaneous emission from the active medium. The evolution from noise to a regular train of pulses has been the object of numerous theories and computer simulations since the first

5.1. INTRODUCTION

mode-locked laser was operated (see, for instance, [148]). The pump power has to exceed a given threshold P_{th} for this transition from noise to pulsed operation to occur. This threshold is sometimes higher than the power required to sustain mode-locking: a mode-locked laser will not always restart if its operation has been interrupted. Such a hysteresis is sometimes observed with dye lasers, and is common with Ti:sapphire lasers. Generally, it is a loss or gain modulation that is at the origin of the pulse formation.

Emergence of a pulse from noise is only the first stage of a complex pulse evolution. Subsequently, the pulse — which may contain sub fs noise pulses and be as long as a fraction of the cavity round-trip time — will be submitted to several compression mechanisms that will bring it successively to the ps and fs range. These compression mechanisms are outlined in the next subsection. First, a loss (saturable absorption) and gain (synchronous pumping, gain saturation) mechanism will steepen the leading and trailing edges of the pulse, reducing its duration down to a few ps. Dispersive mechanisms — such as self-phase modulation and compression — take over from the ps to the fs range. There are mechanisms of pulse broadening that prevent pulse compression from proceeding indefinitely in the cavity. The most obvious and simple broadening arises from the bandwidth limitation of the cavity (bandwidth of the difference spectrum of gain minus losses). The bandwidth limit of the amplifier medium has been reached in some (glass lasers, Nd:YAG lasers), but not all, lasers. Other pulse width limitations arise from higher order dispersion of optical components and nonlinear effects (four wave mixing coupling in dye jets, two-photon absorption, Kerr effect, etc...).

The pulse evolution in a cw pumped laser leads generally to a "steady state," in which the pulse reproduces itself after an integer number of cavity round trips (ideally one). The pulse parameters are such that gain and loss, compression and broadening mechanisms, as well as shaping effects, balance each other.

5.1.4 Pulse compression mechanisms

We present here a qualitative overview of the various processes that can contribute to pulse shortening. The basic physics of light matter interaction underlying these processes has been discussed in detail in Chapters 3 and 4. Different compression mechanisms apply to the various ranges of pulse duration. A strong compression mechanism *in the femtosecond range* is required in a fs laser, but may not be sufficient to achieve stable operation. A more detailed description is given in the appropriate sections. We will use in the following a subscript a for the physical parameters relating to an absorber (if present), and a subscript g for the parameters relating to the gain medium. For instance, T_{1a} designates the energy relaxation time for the absorber, T_{1g} for the gain medium.

Gain saturation

We consider here only the case of small gain factors (as opposed to the more complex situation detailed in Chapter 6 on amplifiers). The leading edge of the pulse is amplified, until the saturation energy of the gain medium is reached [Fig. 5.3(a)]. In some condition, the width (FWHM) of the pulse is reduced. It is left as a problem at the end of this chapter to show that the fractional reduction in pulse width is of the order of the ratio of the energy density gained at the leading edge $\alpha_g d_g W_{sg}$ to the pulse energy density W.

Absorber saturation

As the pulse circulates in a cavity with saturable losses, its leading edge is eroded by linear absorption. If the pulse duration τ_p is longer than the energy relaxation time T_{1a} of the absorber, a fraction of the leading edge up to the saturation intensity I_{sa} is clipped. Once the intracavity pulse has been compressed down to a range (typically ps) where $\tau_p \ll T_{1a}$, the absorption saturates at the time t_s for which the saturation energy density W_{sa} has been reached: $W(t_s) = W_{sa} \int_{-\infty}^{t_s} I(t)dt$. The sketch of Fig. 5.3(b) illustrates the compression mechanism. There is an optimum ratio of pulse energy density to saturation energy density for which a maximum compression can be achieved. The energy lost at the leading edge is approximately $\alpha_a d_a W_{sa}$, where $|\alpha_a d_a|$ is the *linear* optical thickness of the absorber, assumed here $\ll 1$.

Phase modulation and dispersion

The underlying physical principles of phase modulation and dispersion have been introduced in Chapters 2 and 3. The most important self-phase modulation mechanisms in a fs laser are absorption and gain saturation (a resonant effect), and a non-resonant intensity dependent index of refraction. We will sometimes refer to the latter effect as the Kerr effect. Rigorously speaking, however, the optical Kerr effect is the birefringence induced by an optical

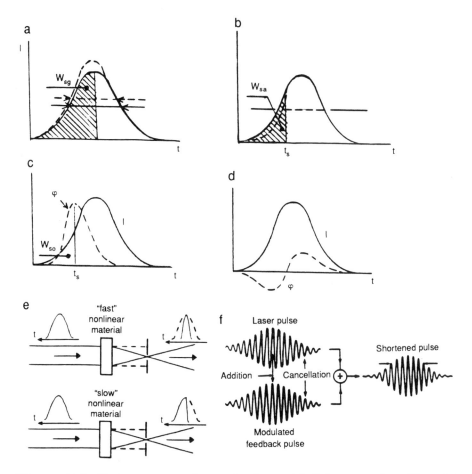

Figure 5.3: Various compression mechanisms. (a) Gain saturation: The original pulse and its pulse width are indicated by the solid line. The leading edge of the pulse is amplified, until the accumulated energy equals the saturation energy density W_{sg} at time t_s, as indicated by the dashed area. In the case of the figure, the pulse tail is not amplified, resulting in a shorter amplified pulse (dashed line). (b) Saturable absorption: The leading edge of the pulse is attenuated, until the saturation energy density W_{sa} is reached at t_s. (c) Frequency modulation (dotted line) due to saturation peaks at the time t_s when the saturation energy density W_{sa} is reached. For $t > t_s$, the pulse experiences a downchirp, if the carrier frequency of the pulse is smaller than the resonance frequency of the absorber. (d) Frequency modulation (dotted line) due to the Kerr effect. The central part of the pulse (intensity profile indicated by the solid line) experiences an upchirp. (e) Self-focusing by a fast and a slow nonlinearity combined with an aperture, leads to a compression by attenuating both leading and trailing edges. Only the trailing edge is trimmed by the slow nonlinearity. (f) Pulse compression by additive feedback. The output pulse is phase modulated and reflected into the cavity. The reinjected pulse front adds in phase, the pulse tail out of phase (courtesy E. Wintner).

field. However, a material with a large Kerr effect has a large intensity dependent index of refraction, which is the primary effect of interest here.

In a homogeneously broadened absorber, the whole absorption line is uniformly reduced by saturation. As a consequence of Kramers–Kronig relations, the dispersion line associated with the resonance is correspondingly reduced in amplitude. The saturable absorber of most fs dye lasers[1] has its peak absorption at a shorter wavelength than the laser wavelength. Therefore, with increasing saturation, the index of refraction will generally decrease. There will be a sweep of index of refraction during the pulse, resulting in a phase modulation. As detailed in Chapter 3, the time derivative of this phase modulation is a frequency modulation which peaks at the time t_s for which the energy accumulated by the pulse has approximately reached the saturation energy density W_{sa}. Ideally, this time should be in the leading edge of the pulse, so that the major part of the pulse experiences a monotonic downchirp [Fig. 5.3(c)]. The optimum condition for pulse compression by saturation mentioned above is also the optimum condition for frequency chirp, with the pulse energy density of the order of 6× the saturation energy density of the absorber. The advantage of having a downchirping mechanism dominate is that it takes simply the *positive* dispersion of glass to achieve intracavity pulse compression. From the expressions in Table 1.2, we can roughly estimate the maximum compression ratio for a Gaussian pulse with a downchirp on center $\ddot{\varphi}_0$, and a duration τ_G, to be $1/\sqrt{1 + (\ddot{\varphi}_0 \tau_G^2/2)^2}$.

The phase modulation can be either downchirp or upchirp. Downchirping occurs with saturable absorbers at relatively low intensity [Fig. 5.3(c)]. Upchirp occurs generally through the Kerr effect, and is proportional to *minus* the time derivative of the pulse intensity [Fig. 5.3(d)]. This chirp is dominant in solid state lasers (for instance, in the Ti:sapphire laser) or in fs dye lasers where the intensity is sufficiently high for the Kerr effect of the solvent to dominate. To compensate such an upchirp, a multiple (two or four) prism arrangement with large negative dispersion is generally used for intracavity pulse compression.

Self lensing effects

Being intensity dependent, the self phase modulation not only has a temporal dependence, but also is a function of the transverse coordinate. There will be a self-induced negative lens associated with the downchirp, i.e., self-

---

[1] A table of gain–absorber dyes combinations can be found in [149].

5.2. ROUND-TRIP MODEL OF A FS LASER

defocusing, and self-focusing associated with upchirp. This effect can be exploited to modulate the gain by increasing the overlap between the cavity mode and the pump pulse, and/or increasing the transmission through an aperture (both techniques are used in some fs Ti:sapphire lasers). With a fast nonlinearity, only the central peak of the temporal intensity profile will pass through the aperture, while with a slow nonlinearity, it is only the leading edge that is truncated by the combination aperture — self-lensing — as shown in Fig. 5.3(e). In some lasers, the self-defocusing is used in conjunction with an aperture to truncate the pulse tail. This method is used with high gain solid state lasers as discussed in Section 5.9.

Additive pulse mode-locking

A simple method to compress a pulse is to subtract some energy from the pulse wings and transfer it to the pulse center. Additive pulse mode-locking is an implementation of this idea, where the output pulse is reinjected into the laser, with a phase modulation such that the pulse center and leading edge add in phase with the intracavity pulse, and the pulse tail out of phase. In order to "reinject" a fraction of the output pulse, an auxiliary cavity has to be coupled to the main cavity [Fig. 5.3(f)].

5.2 Round-trip model of a fs laser

5.2.1 General

In all femtosecond lasers the pulse travels successively through the different resonator elements, each contributing to the pulse shaping in a particular manner. This model based on the block diagram of Fig. 5.4 is the basis for the most commonly used theoretical description of such lasers. Which elements need to be considered and in which order will depend on the type of laser to be modeled. Each block of the diagram of Fig. 5.4 represents a function rather than a physical element. For instance, the "saturable loss" in Fig. 5.4 can represent either a saturable absorber, as used in most dye lasers, or the intensity dependent loss encountered with passive negative feedback (see below) or Kerr lensing.

Computer simulations can be made starting from spontaneous emission noise, to calculate the evolution of the laser field. If we describe symbolically the action of each resonator element by an operator function T_i, the field

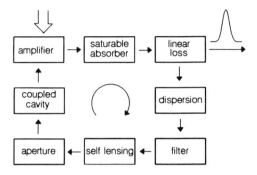

Figure 5.4: Schematic representation of a fs laser.

after one complete round trip can be written as

$$\tilde{\mathcal{E}}^{(n+1)}(t) = (T_N T_{N-1} \ldots T_2 T_1) \tilde{\mathcal{E}}^{(n)}(t), \qquad (5.1)$$

where we have numbered the resonator elements from 1 to N. The superscript indicates the number of round trips completed after starting the simulation (or switching on the laser). If the parameters of the laser elements are suitably chosen, the fields $\tilde{\mathcal{E}}^{(i)}$ will evolve towards a steady-state pulse, which reproduces itself after subsequent round trips, i.e., $\tilde{\mathcal{E}}^{(n+1)} = \tilde{\mathcal{E}}^{(n)}$ for n being large enough. In this stable mode-locking regime, pulse-to-pulse fluctuations are not an intrinsic property of the laser. Other possible solutions can possess periodic changes of one or several pulse parameters. For example, the envelope is stable, but the center frequency of the pulse fluctuates. Such a behavior can be observed, for example, in fs synchronously pumped dye lasers.

The advantage of the round trip model is the ease of incorporating various processes, leading to the modeling of virtually any laser. Even though the transient evolution towards steady state may take thousands of round trips, the modeling can generally be implemented with a personal computer.

5.2.2 Approximate analytical solutions

A number of approximate analytical procedures have been developed [150, 151, 152] to describe the steady state regime. The problem reduces to finding a complex pulse envelope $\tilde{\mathcal{E}}(t)$ that satisfies the steady-state condition:

$$\tilde{\mathcal{E}}(t+h) = \prod_{i=1}^{N} T_i \tilde{\mathcal{E}}(t). \qquad (5.2)$$

5.2. ROUND-TRIP MODEL OF A FS LASER

Equation (5.2) states that the pulse envelope reproduces itself after each round trip, except for a temporal translation h. The main challenge is to find appropriate operator functions for the different resonator elements that are amenable to an analytical evaluation of Eq. (5.2). A convenient approximation is to assume that the modification introduced by each resonator element is small, which allows one to terminate the expansion of the corresponding operator functions after a few orders. Another consequence of that approximation is that the order of the resonator elements is no longer relevant.

Some of the most frequently used operators representative of resonator elements are derived below. The transformation of the pulse envelope by a saturable loss/gain can be expressed as

$$\tilde{\mathcal{E}}_{out}(t) = \left\{ 1 + \frac{1}{2} a_a^{(0)} \tilde{L} \left[1 - \frac{W(t)}{W_{sa}} + \frac{1}{2} \left(\frac{W(t)}{W_{sa}} \right)^2 \right] \right\} \tilde{\mathcal{E}}_{in}(t). \tag{5.3}$$

Equation (5.3) is the rate equation approximation ($T_2 \to 0$) of Eq. (3.78). The expansion parameters are the small signal absorption gain coefficient and the ratio of pulse energy density to the saturation density of the transition. Equation (5.3) applies to a gain medium with the substitutions $a_a \to a_g$ and $W_{sa} \to W_{sg}$. The transfer function of a group velocity dispersion (GVD) element can be derived from Eq. (1.121) and reads

$$\tilde{\mathcal{E}}_{out}(t) = \left\{ 1 + ib_2 \frac{d^2}{dt^2} \right\} \tilde{\mathcal{E}}_{in}(t). \tag{5.4}$$

For this expansion to be valid, the dispersion parameter b_2 has to be much smaller than τ_p^2.

A corresponding expression for the action of a Kerr medium of length d is

$$\tilde{\mathcal{E}}_{out}(t) = \left\{ 1 - i \frac{k_\ell n_2 d}{n_0} |\tilde{\mathcal{E}}_{in}(t)|^2 \right\} \tilde{\mathcal{E}}_{in}(t) \tag{5.5}$$

which can easily be derived from Eq. (3.116).

A linear loss element, which for example represents the out-coupling mirror, can be modeled according to

$$\tilde{\mathcal{E}}_{out}(t) = \left\{ 1 - \frac{1}{2} \gamma \right\} \tilde{\mathcal{E}}_{in}(t) \tag{5.6}$$

where γ is the intensity loss coefficient (transmission coefficient).

Each resonator contains frequency selective elements which can be used to tune the frequency. Such elements are for example prisms, Lyot-filters, and mirrors with a certain spectral response. They ultimately restrict the bandwidth of the pulse in the laser. Let us assume a Lorentzian shape for the filter response in the frequency domain $\tilde{H} = [1+i(\Omega-\omega_\ell)/\Delta\omega_F]^{-1}$ where the FWHM $\Delta\omega_F$ is much broader than the pulse spectrum. After expansion up to second order and re-transformation to the time domain:

$$\tilde{\mathcal{E}}_{out}(t) = \left\{1 - \frac{2}{\Delta\omega_F}\frac{d}{dt} + \frac{4}{\Delta\omega_F^2}\frac{d^2}{dt^2}\right\}\tilde{\mathcal{E}}_{in}(t). \qquad (5.7)$$

If all passive elements are chosen to have an extremely broad frequency response, the finite transition profiles of the active media act as effective filters. This can be taken into account by using the operator defined in Eq. (3.78) for the media instead of Eq. (5.3) derived from the rate equations.

The operator describing the pulse change at each round-trip is obtained by multiplying the transfer functions of all elements of the cavity, neglecting products of small quantities. To evaluate the steady state (5.2), $\tilde{\mathcal{E}}(t+h)$ can be conveniently written as $(1 + h\frac{d}{dt} + \ldots)\tilde{\mathcal{E}}(t)$, leading to an integro-differential steady-state equation for the complex pulse envelope. A parametric approach is generally taken to solve the steady state equation. An analytical expression is chosen for the pulse amplitude and phase, depending on a number of parameters. This ansatz is substituted in the steady state equation, leading to a set of algebraic equations for the unknown pulse parameters. Several types of mode-locked fs lasers have been modeled by this approach [153, 154, 155, 156]. Changes in the beam profile due to the self lensing effect have been incorporated [157, 158]. The transverse dimension is included through a modification of the pulse matrices introduced in Chapter 2 to include the action of the various active resonator elements [158].

5.2.3 The continuous model

A slight variation of the previous approach describes the laser as an infinitely long medium, in which the resonator elements are uniformly distributed (Fig. 5.5). One looks for a stationary pulse, which is a shape-preserving signal propagating through this model medium. Such a pulse is called a soliton of first order or fundamental soliton. Pulses which reproduce after a certain periodicity length are labelled solitons of higher order. In an actual laser, the "higher order solitons" will reproduce after a given number of resonator round trips. We will discuss the soliton model and equations in more details in Section 5.8.

5.3. EVOLUTION OF THE PULSE ENERGY

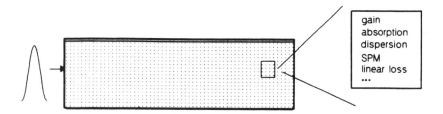

Figure 5.5: Representation of a fs laser as an infinitely long medium with the distributed properties of the cavity

5.3 Evolution of the pulse energy

In this section, we will look at simple "turn on" characteristics of a mode-locked laser. We will show that analytical approximations can be derived to describe the evolution of the energy of the pulse, as it cycles through the laser cavity. We will not be concerned in this section with the pulse shape evolution. Rather than the transfer functions for the pulse envelope defined in Eq. (5.1), we will derive transfer functions for the *pulse energy* only for arbitrary gain, losses, and energies. The simplified model used in this section is sketched in Fig. 5.6.

It is generally necessary to have an amplitude modulating element in the cavity to initiate the mode-locked train. In this section, we will assume that this function is performed by a passive mode locking or positive feedback element, which can be a saturable absorber, a nonlinear lens, or a nonlinear mirror. The passive mode locking element can most often be represented

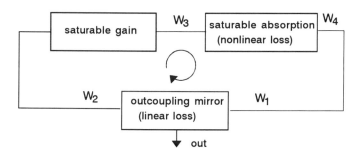

Figure 5.6: Simplified round-trip model for a passively mode-locked laser, showing the evolution of the pulse energy density.

by an intensity dependent intracavity loss. Larger losses at low intensity imply that the laser will have less gain – may even be below threshold – for low intensity cw radiation than for pulses with higher peak intensity. This leads to the emergence of a pulse out of the amplified spontaneous emission noise of the laser. Rather than concern ourselves with the primary process of formation of a precursor from random noise, let us assume a small pulse has emerged from noise, and follow its evolution into a fully shaped stable laser pulse.

In the time frame spanning the fraction of ns to 100 fs, one can generally make the approximation that the pulse duration is longer than the phase relaxation time T_2, and shorter than the energy relaxation time T_1 of intracavity gain and eventual absorbing media. We will establish a simple evolution equation for the pulse energy density $W = \int_{-\infty}^{\infty} I \, dt$.

In order to find an expression for the evolution of the pulse energy, let us make the approximation that the positive feedback element is controlled by the pulse energy W. At low energy, the losses are linear: $dW/dz = \alpha W$ where α (< 0) is the loss factor/unit length. The simplest positive feedback element converts the linear loss to a reduced constant loss at high energy, which can be written $dW/dz =$ constant $= \alpha W_s$. The simplest differential equation that combines the low and high energy limits is:

$$\frac{dW}{dz} = \alpha W_s [1 - e^{-W/W_s}]. \tag{5.8}$$

where W_s can now be interpreted as the characteristic energy at which the positive feedback mechanism turns on. In the case a laser with a saturable absorber of energy relaxation time T_1 larger than the pulse duration, Eq. (5.8) can be derived directly from the conservation law [Eq. (4.21)] and Bloch's equations (cf. Chapters 3 and 4).

Equation (5.8) can be integrated to yield the energy $\mathcal{W}_{out} = W_{out} A_a$ at the end of a loss element of thickness d_a, as a function of the input energy $\mathcal{W}_{in} = W_{in} A_a$. A_a is the cross section of the beam in this particular element. The transfer function $A(\mathcal{W}_{out}, \mathcal{W}_{in})$ that corresponds to the saturable nonlinear loss element, and transforms the pulse of input energy $\mathcal{W}_{in}$ into an output pulse of energy $\mathcal{W}_{out}$ is:

$$A(\mathcal{W}_{out}, \mathcal{W}_{in}) = A_a W_{sa} \ln \left[1 - e^{\alpha_a d_a} \left(1 - e^{\mathcal{W}_{in}/A_a W_{sa}} \right) \right], \tag{5.9}$$

Similarly, a transfer function $G(\mathcal{W}_{out}, \mathcal{W}_{in})$ corresponding to the gain medium can be defined, with a *positive* coefficient α_g and a saturation energy density

5.3. EVOLUTION OF THE PULSE ENERGY

$\mathcal{W}_{sg}$:

$$G(\mathcal{W}_{out}, \mathcal{W}_{in}) = A_g W_{sg} \ln\left[1 - e^{\alpha_g d_g}\left(1 - e^{\mathcal{W}_{in}/A_g W_{sg}}\right)\right]. \tag{5.10}$$

The dominant linear loss element is the output coupler, with (intensity) reflectivity $R = r^2$. The transfer function for that element is simply:

$$L(\mathcal{W}_{out}, \mathcal{W}_{in}) = R\mathcal{W}_{in} \tag{5.11}$$

and the energy of the output pulse is $(1-R)\mathcal{W}_{in}$. The evolution of the pulse energy in a single round-trip can simply be calculated from the product of all three transfer functions given by Eqs. (5.10), (5.9) and (5.11). For instance, if we consider a ring laser with the sequence mirror, gain, saturable loss, as in Fig. 5.6, the pulse energy $\mathcal{W}_4$ after the nonlinear loss (absorption) is given by the product $A(\mathcal{W}_4, \mathcal{W}_3)G(\mathcal{W}_3, \mathcal{W}_2)L(\mathcal{W}_2, \mathcal{W}_1)$.

Let us choose a reference point just after the saturable loss, where the beam cross section is A_a. It is left as a problem at the end of this chapter to establish algebraic relations for various permutations of the sequence loss–gain–nonlinear loss. For instance, the energy density W_4 at the end of series loss–gain–nonlinear loss is related to the energy density W_1 entering this sequence by:

$$1 + e^{-a_a}\left[e^{W_4/W_{sa}} - 1\right] = \left\{1 + e^{a_g}\left[e^{RW_1 A_a/(W_{sg} A_g)} - 1\right]\right\}^{\frac{W_{sg} A_g}{W_{sa} A_a}}. \tag{5.12}$$

For the small perturbations to develop, it is essential that the loss saturates before the gain. This condition corresponds to $\frac{W_{sg} A_g}{W_{sa} A_a} > 1$. Equation (5.12) can easily be solved numerically to find the pulse evolution from threshold to steady state. It is left as a problem at the end of this chapter to find the threshold condition, and the output energy. Solving Eq. (5.12) to first order in W_4/W_s leads to a linear round-trip equation for the energy $W_4 = RW_1 \exp(a_g + a_a)$. The cw threshold condition is thus $R = e^{-(a_a + a_g)}$. It can be shown from Eq. (5.12) that a threshold energy W_1 exists for which a pulse will be amplified in the cavity, even if $R < e^{-(a_a + a_g)}$. This threshold can be small provided $W_{sg} A_g / W_{sa} A_a$ is sufficiently large.

In high-power passively mode locked oscillators, both numbers a_a and a_g can be large, and the output coupling $(1 - R)$ can be as high as 50%. As a result, the *order of the elements* matters in the design of the laser, and in its performances. This point can be verified by solving directly equations for the cavity round-trip for different orders of the elements [Eq. 5.12] is one example] and cycling them repeatedly through the cavity given a small "noise pulse" as initial condition. With only the three components for the individual resonator elements considered here, there are effectively two distinct sequences possible. Let us assume we measure the intracavity power just after the gain section (the two possible sequences are then mirror–nonlinear loss–gain and nonlinear loss–mirror–gain). For the purpose of demonstration, let us make the simplifying assumption that the nonlinear loss is totally saturated. In full saturation, the input energy density W_{in} is related to the output energy density W_{out} by:

$$W_{out} = W_{in} - \alpha_a d_a W_{sa}, \qquad (5.13)$$

as can be seen from Eq. (5.8). The energies W_4 after single passage through the two different sequences of the same elements, for the same initial energy W_1 are given below. For the sequence mirror–nonlinear loss–gain:

$$\exp\left(\frac{W_4}{W_{sg}A_g}\right) = 1 - \left[1 - \exp\left(\frac{RW_1 - \alpha_a d_a W_{sa} A_a}{W_{sg}A_g}\right)\right] e^{\alpha_g d_g}. \qquad (5.14)$$

For the sequence nonlinear loss–mirror–gain:

$$\exp\left(\frac{W_4}{W_{sg}A_g}\right) = 1 - \left\{1 - \exp\left[R\left(\frac{W_1 - \alpha_a d_a W_{sa} A_a}{W_{sg}A_g}\right)\right]\right\} e^{\alpha_g d_g}. \qquad (5.15)$$

Let us take a numerical example for a high gain system such as the flashlamp pumped Nd:YAG laser considered at the end of this chapter, with $A_a W_{sa}/(A_g W_{sg}) = 0.1$, $\alpha_a d_a = -1.3$, $\alpha_g d_g = 1.5$, an output coupling of R = 0.8. For an initial energy $W_1/(A_g W_{sg}) = 1.$, we find:

- for the sequence mirror–nonlinear loss–gain $W_4/(A_g W_{sg}) = 1.66$,
- for the sequence nonlinear loss–mirror–gain $W_4/(A_g W_{sg}) = 1.71$,

The evolution of the pulse energy in the cavity can be calculated by repeated applications of products of operations such as

$$L(W_2, W_1) G(W_3, W_2) A(W_4, W_3)$$

for the sequence (nonlinear-loss element, gain, output coupler), starting from a minimum value of W_1 above threshold for pulsed operation [149], and recycling at each step the value of W_4 as new input energy W_1. Figure 5.7(a) shows the growth of intracavity pulse energy, as a function of the round-trip index i, for all possible element sequences. For a ratio of saturation energy

5.4. SPECIFIC PULSE SHAPING MECHANISMS

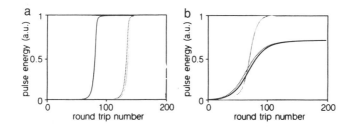

Figure 5.7: Intracavity pulse energy versus round-trip index. The solid, dashed and dotted line correspond to the sequence mirror—nonlinear loss—gain, nonlinear loss—gain—mirror, and nonlinear loss—mirror—gain, respectively. The initial pulse energy is 1% of the saturation energy in the gain medium. The linear gain coefficient is $\alpha_g d_g = 1.5$, and the output coupling $R = 0.8$. The other parameters for (a) are $A_a W_{sa}/(A_g W_{sg}) = 0.1$ and $\alpha_a d_a = -1.3$; for (b): $A_a W_{sa}/(A_g W_{sg}) = 0.8$ and $\alpha_a d_a = -1.2$.

densities closer to unity, as is often the case in cw mode-locked lasers, the influence of the order of the elements is even more pronounced, as can be seen in Fig. 5.7(b). For the latter figure, $A_a W_{sa}/A_g W_{sg} = 0.8$, the linear absorption is $\alpha_a d_a = -1.2$, the linear gain $\alpha_g d_g = 1.5$, and the output coupling corresponds to a mirror reflectivity of $R = 0.8$. Such large gain and saturable absorption are used in some hybridly mode-locked systems [143]. With a typical round-trip time of 12 ns, Fig. 5.7 gives the time required to reach the stable pulse energy. A semi-analytical treatment of a mode-locked laser for arbitrary pulse energies, gain, losses, and sequence of elements can be found in Ref. [4].

The simple model of energy evolution for the passively mode-locked laser illustrates the importance of the order of the elements – a fact confirmed by measurement on high gain lasers such as $Ti : Al_2O_3$ [144].

5.4 Specific pulse shaping mechanisms

5.4.1 Passive amplitude modulation

Most fs mode-locked lasers involve some intensity dependent loss mechanism. The examples discussed in this section are intracavity passive modulators. The case of coupled cavities is discussed in Section 5.10, "Additive pulse mode-locking". The typical passive mode-locking element favors pulsed over cw operation by reducing the cavity losses for high intensities. Since there should be at least one pulse/cavity round-trip time τ_{RT}, the recovery time τ_r

of the device should not exceed that time: $\tau_r \leq \tau_{RT}$. Within that constraint, there is still room for a distinction between "slow" and "fast" intensity dependent elements. A "slow" element — such as the saturable absorber of a fs laser — will recover in a time long compared with the pulse duration. A "fast" element — such as a Kerr lens — will have its time constant(s) even shorter than the fs pulse.

A popular passive mode locking element is the saturable absorber dye with a saturable energy density of the order of 1 mJ/cm^2, and an energy relaxation time ranging from 1 ps to several ns. Its functions, which are analyzed in the next few subsections, include pulse shaping by saturation, saturation induced phase modulation and self-lensing, Kerr effect induced chirp and self focusing.

Multiple quantum wells (MQW's) provide the substitute saturable absorber with the smaller saturation energy required to mode-lock semiconductor lasers [159]. Measurements performed at room temperature with cw radiation, and a 5 μm spot size [160] indicate a saturation intensity of less than 1 kW/cm^2 for MQW, against 10 kW/cm^2 for pure GaAs. It has been possible to achieve even more control upon the parameters of the saturable absorber (in particular its saturation intensity) by inserting a MQW in a Fabry–Perot used in antiresonance. Such a device is substituted to an end mirror of a mode-locked laser. Because of the antiresonance condition, the material inside the Fabry–Perot is subjected to a smaller field than the one present in the laser cavity, hence a better damage threshold and higher saturation intensity for the device than for the MQW used directly. Such a device has therefore been successfully applied to most cw mode-locked solid state lasers. A review of this topic work can be found in [161, 162].

Artificial saturable absorbers can be designed with nonlinear optics. For instance, Stankov [163, 164] demonstrated passive mode-locking in a Q-switched laser by means of a nonlinear mirror consisting of a second harmonic generating crystal and a dichroic mirror. Dispersion between the crystal and the dichroic mirror is adjusted so that the reflected second harmonic is converted back to the fundamental. The output coupler reflects only partially the fundamental, but all of the second harmonic, which is reconverted back into the fundamental. At high intensities, more second harmonic is generated, reflected back and reconverted to the intracavity fundamental. The losses are thus decreasing with intensity, just as is the case with a saturable absorber. The same principle has also been applied by Zhao and McGraw in a technique of parametric mode-locking [165], which can be viewed as a laser hybridly mode-locked by a nonlinear process. The third-order nonlinearity

5.4. SPECIFIC PULSE SHAPING MECHANISMS

of a crystal applied to sum and difference frequency generation is used in the mode-locking process. The electronic nonlinearity for harmonic generation responds in less than a few femtoseconds. However, because of the need to use long crystals to obtain sufficient conversion, the shortest pulse durations that can be obtained by this method are limited to the picosecond range by the phase matching bandwidth.

A last — rather complex — pulse shaping element is a two-photon absorbing semiconductor associated with a limiting aperture. This combination provides pulse compression through self defocusing by the two photon induced carriers. In addition, if a saturable absorber is used simultaneously in the cavity, the function of that absorber is optimized by limiting the intracavity pulse intensity to roughly the dye saturation intensity. A detailed description of this laser operation is given in Section 5.9.

5.4.2 Saturation

Let us now look at the influence of saturation on pulse shaping. We consider the effect of saturation for various time scales, starting with the steady state solutions of Bloch's equations (4.10)–(4.12), thereafter proceeding to the case of pulses shorter than the energy relaxation time of the saturable medium. It will be shown that saturation is enhanced in a standing wave configuration. We will also discuss the complex coupling between counter-propagating pulses resulting from the population grating created in such a standing wave configuration for pulses longer and shorter than T_1.

Pulses longer than T_1

In the initial phase of the pulse evolution, the changes in light intensity can be considered to be slow, even compared with the energy relaxation time of the absorber dye. We have derived in Chapters 3 and 4 a differential equation for the spatial derivative of the intensity in a saturable medium:

$$\frac{dI}{dz} = \frac{\alpha I}{1 + I/I_s} \qquad (5.16)$$

As defined previously, α is negative for an absorbing medium, and positive for an amplifier. The saturation intensity at ω_ℓ is related to the saturation intensity on line center I_{s0} by $I_s = I_{s0}(1 + \Delta_\ell^2 T_2^2)$.

The conditions for a pulse to develop in the cavity are sketched in Fig 5.8, which shows the intensity dependence of the gain and intracavity losses. It is

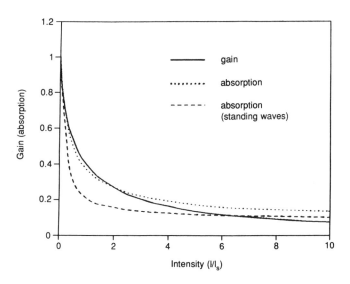

Figure 5.8: Intensity dependence of the intracavity gain and losses. The solid and dotted lines correspond to the gain and absorber medium, respectively. In the case of the dashed line, the absorber medium is saturated by standing waves (contacted dye cell, ring laser or antiresonant ring laser). The gain coefficient/round-trip is $\alpha_g d_g = 0.05/(1 + I/10)$. The total loss per round-trip is $1 - R - \exp(\alpha_a d_a)$ with $\alpha_a d_a = -0.1/(1 + 1.6I/20)$ and where $R = 0.9$ is the reflection coefficient of the output coupler. For the standing wave saturation: $\alpha_a d_a = -0.1/(1 + 4.8I/20)$.

only for very weak intensity, that the cavity is below threshold. Small fluctuations exceeding this intensity threshold corresponding to the first crossing point between the absorption and gain curves will be amplified, until the saturated gain becomes equal to the saturated losses (next crossing point of the gain and losses).

Standing wave saturation for $\tau_p \gg T_1$ For the two crossing points of the gain and loss curve in Fig. 5.8 to exist, it is necessary that the absorber saturates faster (at lower intensity) than the gain. The requirement for a large saturation cross-section can be relaxed by using standing wave saturation, or by focusing more tightly in the absorber than in the amplifier. The effective reduction of saturation in a standing wave configuration was first demonstrated and implemented by Ruddock and Bradley [166] by using a thin layer of absorber in close contact with an end-cavity mirror. The same idea was later implemented by using a ring laser configuration [141],

5.4. SPECIFIC PULSE SHAPING MECHANISMS

or a linear laser terminated by an antiresonant ring [167, 168]. The effective reduction of saturation intensity (for a slow absorber) simply results from the spatial averaging of the population grating induced by the beams in the medium. At the points of constructive interference (separated by $\lambda/2$), the local intensity is 4× the intensity of a single beam. At the nodes, it is zero. When a single beam of amplitude $\mathcal{E}_1$ is sent through the absorber, its attenuation is given by:

$$\frac{d(\mathcal{E}_1^2)}{dz} = \frac{\alpha(\mathcal{E}_1^2)}{1 + \frac{(\mathcal{E}_1^2)}{\mathcal{E}_s^2}}. \tag{5.17}$$

If a second beam of amplitude $\mathcal{E}_2 = \mathcal{E}_1$ is sent counterpropagating to the first, the medium is excited by the superposition of two interfering fields. The sum of the absorption of each beam in the presence of the other is:

$$2\frac{d(\mathcal{E}_1^2)}{dz} = \left\langle \frac{2\alpha(\mathcal{E}_1^2)(1 + \cos 2kz)}{1 + 2(\mathcal{E}_1^2)(1 + \cos 2kz)/\mathcal{E}_s^2} \right\rangle_z$$

$$\approx 2\alpha(\mathcal{E}_1^2) \left[\langle 1 + \cos 2kz \rangle_z - \frac{2(\mathcal{E}_1^2)}{\mathcal{E}_s^2} \langle (1 + \cos 2kz)^2 \rangle_z \right]$$

$$= 2\alpha(\mathcal{E}_1^2) \left(1 - 3\frac{(\mathcal{E}_1^2)}{\mathcal{E}_s^2} \right). \tag{5.18}$$

We have assumed a geometrically and optically thin absorber ($d_a \ll c\tau_p$, $|\alpha_a d_a| \ll 1$). The brackets $\langle \rangle_z$ indicate spatial averaging. The series expansion in Eqs. (5.18) is made under the assumption that the radiation amplitude $\mathcal{E}_1$ is small relative to the saturation field $\mathcal{E}_s$. Since symmetric illumination has been assumed, one can conclude from Eqs. (5.18) that the saturation law for each beam of intensity I_1, in presence of a counterpropagating beam of equal intensity, is:

$$\frac{dI_1}{dz} = \frac{\alpha I_1}{1 + 3I_1/I_s}, \tag{5.19}$$

instead of Eq. (5.16) for a single passage. Under the same assumption of optically and geometrically thin sample, for beams of unequal intensities I_1 and I_2, the result of averaging the spatial interferences of the counterpropagating beams is:

$$\frac{dI_1}{dz} = \frac{\alpha I_1}{1 + \beta I_1 + \theta I_2}. \tag{5.20}$$

In the above expression, $\beta = 1/I_s$ is called the self saturation coefficient, and θ the mutual saturation coefficient. For a purely homogeneously broadened line, $\theta = 2\beta$. In the limit of equal pulse intensities, Eq. (5.20) for homogeneous broadening reduces to the simple expression Eq. (5.19), or an effective saturation intensity three times smaller than for the single beam.

The reduced saturation intensity (reduced losses) for a pair of pulses crossing at a saturable absorber, as opposed to that for a single pulse, is the reason that the counterpropagating pulses meet at the absorber jet in a passively mode-locked ring laser. Noise pulses that meet coherently (with opposite k vector) in the absorber jet will be the first ones to exceed the lasing threshold. It is the minimum loss configuration for the counter-rotating intracavity pulses.

It has been demonstrated experimentally [169, 168] that the *duration* of a fs pulse is rather insensitive to the presence of a standing wave in the absorber. The reason is mainly that most often, in steady state, the dye is deeply saturated, and therefore a factor of three in saturation energy density merely affects a small fraction of the leading edge of the pulse. However, it was also demonstrated that the standing wave saturation does play an important role in the pulse evolution and stability of the laser. Indeed, the smaller the saturation energy density, the faster a weak pulse will emerge from noise. It has been observed experimentally that short pulse operation in a fs dye laser requires a pump power intensity threshold slightly higher than the basic lasing threshold. This particular threshold for short pulse operation was seen to increase from 1.7 to 3.5 W, if the absorber is located more than 400 μm away from the pulse crossing point [168].

In general, optimal pulse compression through saturable absorption requires that the saturation intensity be as large as compatible with the self starting requirement (built up of the oscillation from noise).

Pulses shorter than T_1

Let us consider next the more general situation when the pulse duration is comparable with- or shorter than- the energy relaxation time T_1. The main influence of a saturable absorber is to steepen the leading edge of the pulses circulating in the cavity. With femtosecond lasers, the pulse duration becomes inevitably much smaller than T_{1a}, and the absorber affects the pulse rise appreciably only up to the time t_s for which the accumulated energy density reaches the saturation energy density W_s:

$$\int_0^{t_s} I(t)dt = I_{sa}T_{1a} = W_{sa}. \qquad (5.21)$$

5.4. SPECIFIC PULSE SHAPING MECHANISMS

It is assumed that $W(t_s) = W_{sa}$ is reached at a time $t_s \ll T_{1a}$, where T_{1a} is the energy relaxation time of the absorbing transition. As the pulse intensity increases, the pulse duration shortens. Absorption saturation becomes less and less effective as a compression mechanism, because it affects a smaller and smaller fraction of the pulse leading edge. In some flashlamp pumped solid state lasers, the effectiveness of the saturation shortening mechanism is maintained by limiting the peak intensity to the saturation level. The limiting mechanism referred to as "negative feedback" is described in Section 5.9.

Trailing edge pulse shaping generally is the result of the gain depletion by stimulated emission. This mechanism becomes significant as the pulse energy approaches steady state (largest saturation of the gain).

The above considerations are sufficient to establish broad criteria for the gain and absorption in a passively mode locked laser. For the gain medium, a larger inversion at minimum cost of pump power calls for a long energy relaxation time T_{1g}. However, at moderate pumping, the gain recovery time (between pulses separated by the cavity round-trip time τ_{RT}) is equal to – or less than – T_{1g}. Therefore, operation of a passively mode locked laser with a single pulse/cavity round-trip requires that the cavity round-trip time does not exceed the gain lifetime T_{1g} by more than a factor 4 or 5. The effective lifetime of the gain is reduced by the pumping rate $\mathcal{R}$ [149]:

$$\frac{1}{T_{1g,\text{eff}}} = \frac{1}{T_{1g}} + \frac{2}{\mathcal{R}}. \tag{5.22}$$

Stable multiple pulse operation can be achieved at high pumping rate even in relatively short cavities [168]. Cross-correlation between successive pulses of the train show that the pulse separation is – within a pulse length – equal to an integer fraction of the cavity length.

In order for the pulse to grow in the cavity, it should saturate the absorber more than the gain. One will therefore generally select a saturable absorber with a larger or equal cross section as the gain medium (i.e., a smaller saturation energy density). In order to modulate effectively the cavity losses, the absorber should have a relaxation time T_{1a} shorter than the cavity round-trip time.

5.4.3 Induced grating effects for $\tau_p \ll T_1$

Two counterpropagating pulses meet coherently in the saturable absorber of a mode-locked ring laser. Each pulse scatters off the grating induced by the interaction of the standing wave with the nonlinear medium. Momentum conservation imposes that each pulse is scattered in the direction of the other pulse, resulting in a coupling between counterpropagating pulses. Saturation is only one of the coupling mechanism. Another one, which has been shown to dominate often [170], is two-photon resonant degenerate four wave mixing. We will study separately these two effects in the paragraphs that follow.

Saturation grating The saturable absorber is excited by two pulses of amplitude $\tilde{\mathcal{E}}_1(t)$ and $\tilde{\mathcal{E}}_2(t)$ propagating in the $+z$ and $-z$ directions, resulting in a total field:

$$\tilde{\mathcal{E}}_1 e^{i(\omega_\ell t - kz)} + \tilde{\mathcal{E}}_2 e^{i(\omega_\ell t + kz)}. \tag{5.23}$$

Substituting this field in the rate equation (4.18) results in source terms with spatial modulation $\exp(-2ikz)$ and $\exp(2ikz)$ for the time derivative of the population difference w. These terms in turn combine with the fields to produce still higher spatial frequencies which we shall neglect. To find a first order solution for the population differences, we decompose the population difference term w in a uniform component w_1, and a term w_2 representative of the amplitude of the induced grating:

$$w = w_1 + w_2 e^{2ikz} + w_2^* e^{-2ikz}, \tag{5.24}$$

where $|w_2| \ll |w_1|$, for the limitation of the expansion [Eq. (5.24)] to first order to be valid. Substituting into the rate equation (4.18) leads to a rate equation for each component w_1 and w_2:

$$\begin{aligned}
\dot{w}_1 &= -\left[\frac{w_1(\tilde{\mathcal{E}}_1^*\tilde{\mathcal{E}}_2 + \tilde{\mathcal{E}}_2^*\tilde{\mathcal{E}}_1) + w_2^*|\tilde{\mathcal{E}}_1|^2 + w_2|\tilde{\mathcal{E}}_2|^2}{\mathcal{E}_s^2} + (w_1 - w_0)\right]/T_1 \\
\dot{w}_2 &= -\left[\frac{w_2(\tilde{\mathcal{E}}_1^*\tilde{\mathcal{E}}_2 + \tilde{\mathcal{E}}_2^*\tilde{\mathcal{E}}_1) + w_1|\tilde{\mathcal{E}}_1|^2}{\mathcal{E}_s^2} + w_2\right]/T_1.
\end{aligned} \tag{5.25}$$

Within the rate equation approximation, the source term for Maxwell's propagation equation is $v = w\mathcal{E}/(1 + \Delta_\ell^2 T_2^2)$ [steady state solution of Eqs. (4.10) and (4.11)]. Grouping the terms according to the propagation direction, results in the coupled equations for the field amplitudes:

$$\begin{aligned}
\left(\frac{\partial}{\partial z} + \frac{1}{v_g}\frac{\partial}{\partial t}\right)\tilde{\mathcal{E}}_1 &= A_1\tilde{\mathcal{E}}_1 + A_2\tilde{\mathcal{E}}_2 \\
\left(\frac{\partial}{\partial z} - \frac{1}{v_g}\frac{\partial}{\partial t}\right)\tilde{\mathcal{E}}_2 &= A_1\tilde{\mathcal{E}}_2 + A_2^*\tilde{\mathcal{E}}_1,
\end{aligned} \tag{5.26}$$

5.4. SPECIFIC PULSE SHAPING MECHANISMS

with

$$A_i = \frac{\mu_0 \omega_\ell c^2}{2} \kappa T_2 w_i. \tag{5.27}$$

Consistently with the first order approximation used in writing Eq. (5.24), $|A_2| \ll |A_1|$. If the geometrical thickness of the absorber jet can be considered to be much smaller than the pulse length $c\tau_p$, one can replace each of the Eqs. (5.26) by a finite difference for the pulse in their retarded frame of reference (the left side is replaced by $\Delta \tilde{\mathcal{E}}_i / \Delta z$). For small absorption coefficients $|\alpha_a d_a| \ll 1$ and weak saturation, one can find analytical solutions for the amplitudes $\tilde{\mathcal{E}}_1$ and $\tilde{\mathcal{E}}_2$ [61]. For example,

$$\tilde{\mathcal{E}}_1(t, z = d_a) = \tilde{\mathcal{E}}_1(t, 0) \left\{ 1 - \frac{1}{2}\alpha_a d_a \left[1 - 3\frac{W_1(t,0)}{W_{sa}} + 5\frac{W_1^2(t,0)}{W_{sa}^2} \right] \right\}. \tag{5.28}$$

The factor 3 in the linear saturation term is expected from our earlier discussion.

If the absorber thickness is not negligible compared to the pulse length, the coupled Eqs. (5.26) have to be solved in each successive "time slice" Δt within the absorber. Instead of a retarded time frame of reference, it is then more convenient to use at each instant moving spatial coordinates, i.e. the transformation $z^\pm = z \pm v_g t$, and $t^\pm = t$. In these moving coordinates, the left hand side of each of Eqs. (5.26) is replaced by $\partial \tilde{\mathcal{E}}_1 / \partial t^+$, and $\partial \tilde{\mathcal{E}}_2 / \partial t^-$, respectively [149].

Phase-conjugated coupling The coupling between counterpropagating fields is generally larger than that given by A_2 in Eqs. (5.26). In the case of several saturable absorber dyes such as DODCI, there exist a two photon resonance corresponding to the laser wavelength, enhanced by the presence of the near resonant single photon absorption [170]. A detailed analysis of this interaction can be made using the multiphoton coherent interaction model outlined in Chapter 4. The overlapping counterpropagating pulses create a two-photon excitation in the medium, or an off-diagonal element with an amplitude $\tilde{\varrho}_{02}$ given by Eq.(4.95). The mechanism by which the medium reacts on each pulse is a two-photon stimulated emission. The polarization interfering with the pulse $\tilde{\mathcal{E}}_1$ is proportional to $i\tilde{\varrho}_{02}(t)\tilde{\mathcal{E}}_1^*(t)$. The mutual coupling is thus of the form:

$$\begin{aligned} \frac{\partial \tilde{\mathcal{E}}_1}{\partial t^+} &= B_2 \tilde{\mathcal{E}}_2^* \\ \frac{\partial \tilde{\mathcal{E}}_2}{\partial t^-} &= B_2 \tilde{\mathcal{E}}_1^*, \end{aligned} \tag{5.29}$$

where t^+, t^- are the time coordinates in the moving spatial coordinates defined in the previous paragraph. The magnitude of the coupling term B_2 can be estimated by using a steady-state approximation for the two-photon interaction equations (4.95), assuming exact two-photon resonance, and neglecting the Stark shift. One finds:

$$B_2 = -\frac{\mu\omega_\ell c^2}{2} \frac{\kappa^2(\tilde{\mathcal{E}}_1^*\tilde{\mathcal{E}}_2 + \tilde{\mathcal{E}}_2^*\tilde{\mathcal{E}}_1)}{\Delta_1^2}\kappa w_0 T_2^{(2)}, \qquad (5.30)$$

where $T_2^{(2)}$ is the phase relaxation time of the two-photon interaction, and Δ_1 is the detuning of the intermediate (S_1-) state. We note that this mutual coupling is of the same order (with respect to the field amplitudes) as the mutual coupling due to the saturation induced grating discussed in the previous paragraph. Therefore, the conclusions reached in Section 5.4.2 – in particular the effective reduction of saturation energy density – remain valid. Comparing the magnitude of the mutual coupling due to two-photon interaction given by Eq. (5.30) with that due to mutual saturation given by A_2 in Eq. (5.27), we find for the ratio

$$\frac{B_2}{A_2} \approx \frac{T_2^{(2)}}{T_2}\frac{1}{\Delta_1^2 T_1 T_2}. \qquad (5.31)$$

A phase conjugated coupling dominates when this ratio is $\gg 1$, as is the case for DODCI as saturable absorber [170]. The consequence of a phase conjugated coupling is that the frequencies of the longitudinal modes for the two directions of propagation in the cavity are not coupled by the saturable absorber. Let us assume that the clockwise circulating pulses $\tilde{\mathcal{E}}_1$ are being given a phase shift $+\delta/2$, and the counterclockwise circulating pulses $\tilde{\mathcal{E}}_2$ a phase shift $-\delta/2$ at each round-trip. Being phase conjugated, the two-photon resonant coupling from beam $\tilde{\mathcal{E}}_2$ into $\tilde{\mathcal{E}}_1$ will have the phase $-(-\delta/2)$, hence be in phase with $\tilde{\mathcal{E}}_1$.

Consequences of phase-conjugated coupling We have seen in the previous paragraph that the counterpropagating pulses are coupled *in amplitude but not in phase* at the saturable absorber of the ring dye laser. The amplitude coupling ensures that the ratio of the intensities of the counterpropagating pulses is stable (and close to unity) against any external perturbation. Interference measurements between the counterpropagating pulse trains [171, 172] have shown that, indeed, the saturable absorber does not couple the phases of the two counterpropagating beams. Since the same

5.4. SPECIFIC PULSE SHAPING MECHANISMS

comb of modes applies to the counter-propagating trains of pulses, very small nonreciprocal effects can be observed as a beat frequency between the two outputs of the laser [173]. As will be discussed in Chapter 12, even though the laser bandwidth is of the order of 10^{13} Hz, beat notes — representative of the difference between longitudinal mode spacing for both directions — as small as tens of Hz have been measured.

5.4.4 Phase modulation and dispersion

Pulse compression through chirping and dispersion

We have discussed so far direct amplitude modulation of the intracavity field through intensity dependent gain or losses. In this section we analyze compression through purely dispersive effects, i.e., the compression mechanism does not affect the pulse energy. It is believed that a dispersive mechanism alone cannot effectively mode-lock a laser. Of course, a "purely dispersive" laser does not exist: there is always a saturable gain mechanism, and there are also time dependent losses due to the lensing effects associated with self-phase modulation which are discussed in the following section.

As the circulating pulse shortens, its bandwidth increases, and the index of refraction of the various cavity components can no longer be considered to be constant over the frequency range of the pulse. Of particular importance is the *total* dispersion of the cavity. An average index n_{av} can be easily measured for any cavity of perimeter P, by measuring the longitudinal mode spacing. Such a measurement [174] is typically performed with a frequency counter, which records the cavity round-trip time (which is the inverse of the longitudinal mode spacing $\Delta\nu$) as a function of wavelength $\lambda_\ell = 2\pi c/\omega_\ell$, of the laser operating with long pulses. The average index is:

$$n_{av}(\omega_\ell) = \left(\frac{c}{P}\frac{1}{\Delta\nu}\right) = \left(\frac{\text{mode spacing of empty cavity}}{\text{measured mode spacing}}\right). \quad (5.32)$$

As we have seen in Chapters 1 and 2, the factor affecting the pulse broadening/round trip is the group velocity dispersion. It is possible also to exploit group velocity dispersion for pulse compression, by phase modulating the pulse at each round-trip. For instance, if the pulse is given a downchirp at each round-trip and the cavity has a positive group velocity dispersion, the leading edge of a pulse is slowed down with respect to the trailing edge, at each cavity round-trip, resulting in pulse compression. The same properties hold for upchirp with negative group velocity dispersion. Phase modulation leading to spectral broadening will generally be achieved with an intensity dependent mechanism. Since the pulse intensity has a radial as well as a temporal intensity profile, wavefront modulation resulting in self lensing will generally be a corollary of temporal (passive) phase modulation.

An example of a "downchirping" element is the saturable absorber. In most practical situations involving passive mode locking with a saturable absorber dye, the laser radiation is applied *below resonance* with the absorber. The contribution of the absorbing resonance to the index of refraction is positive on the long wavelength side. Consequently, saturation of the absorption will also imply a reduction of the index of refraction below resonance (cf. Fig. 3.3). A decreasing index with saturation leads to a *downchirp* of the pulse and *self-defocusing* of the beam. Both processes – analyzed quantitatively in the next subsection – can simultaneously affect the mode locking process.

Sources of downchirped pulses

A pulse with instantaneous frequency decreasing with local time is called a downchirped pulse. Such a pulse has thus its higher frequency components at its leading edge. In a normally dispersive medium such as glass in the visible spectral region (group velocity decreasing with frequency), the pulse front will propagate slower than the pulse tail, resulting in pulse compression.

It has been shown in Chapter 3 that saturation of an absorber (amplifier) below (above) resonance can result in a downchirp. Exact expressions were derived for the instantaneous frequency versus time induced by propagation through an optically thick [Eqs (3.66) and (3.67)] or thin [Eq. (3.68)] saturable absorber or amplifier. This chirp is due to the change in population differences of the two-level system traversed by the pulse near resonance. The medium saturation is a function of the intensity, itself a function of both time and radial coordinate. The temporal dependence of the intensity leads through saturation to chirp. The radial dependence leads to self lensing effects that will be studied in the next section. To introduce an expression for the temporal and spatial variation of the phase to be used in the next section on self-lensing, and as an application of Bloch's equations of Chapter 4, we will rederive the saturation induced chirp [Eq. (3.68)] of Chapter 3.

We have shown in Chapter 4 that the propagation equations for the field amplitude $\mathcal{E}$ and phase φ are related to the v and u components of the pseudo-polarization vector $\vec{\mathcal{P}}$. The rate equation approximation essentially

5.4. SPECIFIC PULSE SHAPING MECHANISMS

consists of solving the two first Bloch equations (4.10) and (4.11) for steady state. The same approximation can be applied to Maxwell's equation (4.15) for the phase evolution, substituting the steady state solution for u:

$$\begin{aligned}\frac{\partial \varphi(t,r)}{\partial z} &= -\frac{\mu_0 \omega_\ell c}{2n} \frac{u}{\mathcal{E}} \\ &= -\frac{\mu_0 \omega_\ell c}{2n} \frac{\Delta_\ell T_2 w \kappa T_2}{1 + \Delta_\ell^2 T_2^2} \\ &= \frac{\sigma_{01}^{(0)} \Delta_\ell T_2}{1 + \Delta_\ell^2 T_2^2} \Delta\gamma(t,r),\end{aligned} \quad (5.33)$$

where $\sigma_{01}^{(0)}$ is the absorption cross section at the center of the line of the absorbing dye, and $\Delta\gamma(t,r)$ is the radial and temporal distribution of population density. In Eq. (5.33), we have explicitly written the dependence on the radial coordinate r to emphasize the origin of lensing effects, and assumed cylindrical symmetry. For a small optical thickness of the dye jet, the propagation equation can be linearized. Integrating the rate equation for the population difference, we find that passage through a thickness d of a medium (dye) leads to a space-time phase distortion:

$$\varphi(t,r) = a\frac{\Delta_\ell T_2}{1 + \Delta_\ell^2 T_2^2} \exp\left\{-\int_{-\infty}^{t} \frac{I(t',r)dt'}{W_s}\right\}, \quad (5.34)$$

where $a = \bar{N}\Delta\gamma(t=-\infty)\sigma_{01}^{(0)}d$ is the linear attenuation coefficient at the line center. The frequency modulation obtained by taking the time derivative of Eq. (5.34) is identical to Eq. (3.68), which is the thin medium approximation of Eq. (3.66) or (3.67). Expressions were derived in Chapter 3 for the frequency modulation as illustrated by Figs. 3.6 and 3.7.

The frequency modulation is proportional to the time derivative of the integral factor in Eq. (5.34), and plotted for three values of the pulse energy density in Fig. 5.9. As we have seen in Chapter 1, a linear downchirp is desirable, in order to be able to compress the pulse with a quadratic positive dispersion. As can be inferred from Fig. 5.9, there is an optimum ratio of the pulse energy density to the saturation energy density. For too low a pulse energy (dotted line), the chirp is small, and as much up as down. For too high a pulse energy (dashed line), most of the phase modulation takes place before the main portion of the pulse. For a pulse energy density equal to the saturation energy density, there is a nearly uniform downchirp for the central portion of the pulse (indicated by two vertical lines on the temporal profile of the pulse shown on the left Fig. 5.9).

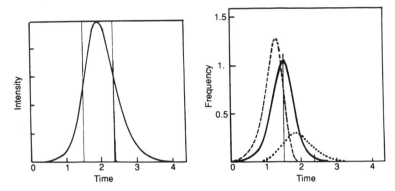

Figure 5.9: Intensity versus time, and saturation induced chirp. The exciting pulse (left) has the temporal profile $I(t) = I_0 \left(\exp(-t/\tau_r) + \exp(t/\tau_f)\right)^{-2}$ with $\tau_r = 0.44\tau_p$ and $\tau_f = 0.73\tau_p$, where τ_p is the pulse duration (FWHM). The instantaneous frequency (right) corresponds to $a_a = -0.1$ in Eq. (5.34), and a detuning of $\Delta_\ell = \tau_p^{-1}$. The pulse energies are 0.5× (dotted line), 1× (solid line) and 10× (dashed line) the saturation energy in the absorber.

Sources of upchirped pulses

The mechanism described above can also generate upchirped pulses (for instance, gain saturation below resonance, or absorption saturation above resonance). The largest effect, however, is mostly from the Kerr effect. It is the main modulation mechanism in solid state lasers — as, for instance, in the Ti:Al$_2$O$_3$ laser — because of the relatively long gain section and large value of $\bar{n}_2$ (cf. Table 6.2 in Chapter 6). It is even the main mechanism for upchirp in dye lasers with tight focusing in the absorber jet [169, 168].

Intracavity compression of an upchirped pulse requires a negative dispersion, as can be provided by the angular dispersion of a four prism sequence [175], as discussed in Chapter 2. In dye lasers, the upchirp can be increased by using additives with large Kerr constant such as 2-methyl-4-nitroaniline (MNA) to the saturable absorber solution [168, 176].

The variation of the phase factor $\varphi(t,r)$ due to propagation through a Kerr medium of thickness d and nonlinear index $\bar{n}_2$ is:

$$\varphi(t,r) = -\frac{2\pi\bar{n}_2 d}{\lambda_\ell} I(t,r) \qquad (5.35)$$

The different types of chirp are illustrated in Fig. 5.10. Neither the downchirp due to absorber dye saturation below resonance, nor the Kerr effect–induced chirp can be considered to be linear chirps. Therefore, com-

5.4. SPECIFIC PULSE SHAPING MECHANISMS

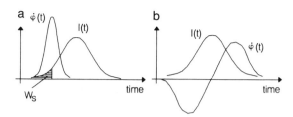

Figure 5.10: Comparison of the frequency modulation or chirp induced by saturation below resonance (a) and by Kerr effect (b). The frequency modulation (a) peaks at a time such that the integrated intensity equals the saturation energy density W_s. The Kerr effect induced phase modulation is proportional to minus the time derivative of the intensity.

plete compensation of the chirp is not possible with a finite number of linear optical elements.

The chirp induced by saturation of an absorbing two-level system has only two components: upchirp for the leading edge, downchirp near the middle of the pulse if the interaction is below resonance. The upchirped portion of the leading edge will speed ahead of the pulse and be eliminated by linear absorption, if there is a positive quadratic intracavity dispersion to compress the downchirped part of the pulse. In the case of a Kerr induced chirp the uncompressed portion *trailing* the main pulse will not be eliminated by saturable absorption. However, because of gain depletion at the end of the pulse, it may remain a small perturbation.

5.4.5 Self-lensing

Importance of self-lensing

Self-lensing does take place in nearly all ultrashort pulse mode-locked lasers. There is often a dispersive pulse shaping mechanism in the cavity, as discussed in the previous section. Some loss modulation is generally associated with the dispersive mechanism. Temporal self-phase modulation implies necessarily a spatial modulation of the wavefront, hence self-lensing. As a result of self-lensing, the size of the cavity modes is modified, leading to an increase or reduction of losses because

1. there is a change in transmission through an aperture, or
2. there is a change in spatial overlap between the cavity mode and the pump beam.

Another aspect of self-lensing that has received little attention is its influence on the mode spacing. Self-lensing affects the *transverse mode spacing*. Indeed, one could imagine a situation with an extreme self-lensing modifying a cavity from confocal to plane parallel. In that drastic situation, the mode spacing is changed by a factor 2! As pointed out by Kafka [177], mode-locking does not require a laser to be purely single transverse mode. One can construct cavities where the transverse mode spacing is equal to the longitudinal mode spacing (see Section 5.11). The advantage is a larger mode volume — hence a more efficient extraction of the gain.

Self-lensing is important in solid state lasers — in particular the Ti:sapphire laser — because the Kerr effect takes place over 1 cm or more (the full length of the laser crystal). The self lensing effect can also be due to off-resonance saturation.

In general the nonlinear interaction leads to a phase shift which, to a first approximation, can be assumed to be quadratic in the radial coordinate $\Delta\varphi(r) \approx Br^2$. The radial phase variation of a spherical wave of curvature R is $\varphi(r) = -k_\ell r^2/2R$. Given a quadratic radial phase variation $B(t)r^2$, the beam will focus in a distance f_{nl} given by:

$$f_{nl} = \frac{k_\ell}{2B(t)}. \tag{5.36}$$

Kerr lensing

Let us assume a Kerr medium of thickness d small compared to either the Raleigh range ρ_0 or the induced focal length f_{nl} (to be defined). A Gaussian beam is incident with its beam waist w_0 at the sample input. The phase shift due to the nonlinear index is:

$$\begin{aligned}\Delta\varphi(r,t) &= -k_\ell d \frac{\bar{n}_2 I(r,t)}{n_0} \\ &\approx -k_\ell d \frac{\bar{n}_2 I(t)}{n_0}\left(1 - \frac{2r^2}{w_0^2}\right),\end{aligned} \tag{5.37}$$

where $I(t) = I(r = 0, t)$ is the intensity on axis. From Eq. (5.37), we obtain $B =\approx 2\bar{n}_2 k_\ell I(t)d/(n_0 w_0^2)$. Substituting in Eq. (5.36), we find for the Kerr–induced lensing:

$$f_{nl} = \frac{n_0 \pi w_0^4}{8\bar{n}_2 P d}, \tag{5.38}$$

where $P = \pi w_0^2 I(t)/2$ is the power.

5.4. SPECIFIC PULSE SHAPING MECHANISMS

It is convenient to use the ABCD matrix approach [178] to evaluate the intensity dependent losses introduced by Kerr lensing in a particular cavity [179]. The ABCD matrix of the resonator is calculated starting from a reference plane at the position of the Kerr medium. Let $\mathcal{M}_0 = \begin{pmatrix} A_0 & B_0 \\ C_0 & D_0 \end{pmatrix}$ be the ABCD matrix for low intensity (negligible Kerr effect). At high intensity, the nonlinear lensing effect modifies this matrix as follows:

$$\mathcal{M} = \begin{pmatrix} 1 & 0 \\ -\frac{1}{f_{nl}} & 1 \end{pmatrix} \begin{pmatrix} A_0 & B_0 \\ C_0 & D_0 \end{pmatrix} = \begin{pmatrix} A & B \\ C & D \end{pmatrix}$$

$$= \mathcal{M}_0 + \begin{pmatrix} 0 & 0 \\ -\frac{A_0}{f_{nl}} & -\frac{B_0}{f_{nl}} \end{pmatrix}, \tag{5.39}$$

where f_{nl} is a time dependent quantity. A Gaussian beam is uniquely characterized by its complex beam parameter $\tilde{q}$ (cf. Chapter 1). The inverse complex radius of curvature of the beam, at the location of the Kerr medium, $\tilde{s} = 1/\tilde{q} = 1/R - i\lambda_\ell/(\pi w^2)$ is a solution (eigenmode) of the cavity round-trip equation:

$$\tilde{s} = \tilde{s}_0 + \delta \tilde{s} = \frac{C + D\tilde{s}}{A + B\tilde{s}} \tag{5.40}$$

where $\tilde{s}_0$ is the eigenmode in the absence of Kerr lensing. Differentiating Eq. (5.40) after multiplication of both sides by the denominator, and keeping in mind that in the particular case of Kerr lensing the variation of the matrix elements A and B is null [Eq. (5.39)]:

$$\begin{aligned}
\delta \tilde{s} &= \frac{\delta C + \tilde{s}_0 \delta D}{A_0 + 2B_0 \tilde{s} - D_0} \\
&= \frac{1}{f_{nl}} \left[\frac{-(A_0 + B_0 \tilde{s}_0)}{A_0 + 2B_0 \tilde{s}_0 - D_0} \right] \tag{5.41} \\
&= \frac{8\bar{n}_2 P \pi d |\mathrm{Im}(\tilde{s}_0)|^2}{n_0 \lambda_\ell^2} \left[\frac{-(A_0 + B_0 \tilde{s}_0)}{A_0 + 2B_0 s_0 - D_0} \right] \tag{5.42}
\end{aligned}$$

where the substitutions $\delta C = -A_0/f_{nl}$, $\delta D = -B_0/f_{nl}$, and $w_0^2 = \lambda_\ell/[\pi |\mathrm{Im}(\tilde{s}_0)|]$ have been made.

The change in the complex beam parameter $\delta \tilde{s}$ implies that the lensing effect results in a change in beam size at any location in the cavity. Typically, an aperture is used at a particular location of the cavity, where, ideally, the self lensing results in the largest reduction in beam size. Let

$$\mathcal{M}_m = \begin{pmatrix} A_m & B_m \\ C_m & D_m \end{pmatrix}$$
be the ABCD matrix that connects the reference point of the cavity (location of the Kerr lens) to the position of the aperture. The complex beam parameter at the aperture is:

$$\tilde{s}_m = \frac{C_m + D_m \tilde{s}}{A_m + B_m \tilde{s}}. \tag{5.43}$$

The relative change in beam size at the aperture $\delta w_m / w_m$ is related to the change in inverse complex radius of curvature $\delta \tilde{s}_m$:

$$\frac{\delta w_m}{w_m} = -\frac{1}{2} \frac{\text{Im}(\delta \tilde{s}_m)}{\text{Im}(\tilde{s}_m)}. \tag{5.44}$$

The change in beam parameter at the aperture $\delta \tilde{s}_m$ can be inferred from the change in beam parameter $\delta \tilde{s}$ at the point of reference:

$$\begin{aligned}
\delta \tilde{s}_m &= \frac{[(A_m + B_m \tilde{s}_0) D_m - (C_m + D_m \tilde{s}_0) B_m]}{(A_m + B_m \tilde{s}_0)^2} \delta \tilde{s} \\
&= \frac{\delta \tilde{s}}{(A_m + B_m \tilde{s}_0)^2} \tag{5.45} \\
&= \left[\frac{-(A_0 + B_0 \tilde{s}_0)}{(A_0 + 2B_0 s_0 - D_0)(A_m + B_m \tilde{s}_0)^2} \right] \frac{8 \bar{n}_2 P \pi d |\text{Im}(\tilde{s}_0)|^2}{n_0 \lambda_\ell^2}. \tag{5.46}
\end{aligned}$$

The last Eq. (5.46) contains all the information necessary to estimate the effect of Kerr–induced lensing on a cavity. As a coarse estimate, one could define an effective diameter w_a for the aperture, such that the transmission factor be $P_2/P_1 \approx (w_a/w_m)^2$, where P_2 and P_1 are the power of the pulse respectively after and before the aperture. It follows that:

$$\begin{aligned}
\Delta P &= P_2 - P_1 = -P_1 \left(1 - \frac{w_a^2}{w_m^2}\right) \\
&= -P_1 \left[1 - \frac{w_a^2}{w_{m0}^2}\left(1 - 2\frac{\delta w_m}{w_{m0}^2}\right)\right] \\
&\approx \frac{aP_1}{1 + \frac{P_1}{P_s}}, \tag{5.47}
\end{aligned}$$

where $a = 1 - (w_a/w_{m0})^2$ is the power loss coefficient (low power limit) at the aperture, w_{m0} is the beam size at zero power at the aperture ($w_m = w_{m0} + \delta w_m$), and

$$P_s = \left(\frac{w_a}{w_{m0}}\right)^2 \text{Im}(\tilde{s}_M) \left[1 - \left(\frac{w_a}{w_{m0}}\right)^2\right] \frac{P}{\text{Im}(\delta \tilde{s}_m)} \tag{5.48}$$

5.4. SPECIFIC PULSE SHAPING MECHANISMS

is the equivalent "saturation power" of the Kerr lens–aperture combination which can be calculated by substituting expression (5.46) for δw_m. The constants were named a and P_s in the last Eq. (5.47) to draw attention on the analogy with saturation. To first order, the action of Kerr lensing combined with an aperture on the pulse power is similar to the action of a fast saturable absorber on the pulse intensity. As in the saturable absorber, pulse compression results from the stronger absorption on the pulse leading and trailing edges than at the pulse peak.

Saturation lensing

An expression (5.34) has been derived earlier in this section for the space-time phase variation due to saturation. Even though the gain or absorbing medium can be very thin, as is the case in dye lasers, self-lensing can be important [157]. The reason is that the use of focusing mirrors introduces astigmatism, hence a distinct sagittal and tangential focus which cannot be co-located with the dye jet. To understand why the saturation induced lensing effect can be important in a dye laser, let us consider as an example a system of two curved mirrors separated by twice the focal length of 15 mm, on which the beam incidence is 10^0. For a beam diameter of 2 mm on the mirrors, the Raleigh range at the beam waist is roughly 50 μm, while the distance between sagittal and tangential foci is 450 μm. Because the jet will always be at least four Raleigh ranges away from one focus, self-lensing will have an impact on the cavity configuration.

The procedure to calculate the loss modulation due to saturation self-lensing is essentially the same as the one presented in the previous section for Kerr lensing. As for the Kerr effect, a quadratic approximation can be made for the radial phase dependence of Eq. (5.34) [i.e., a phase of the form $B(t)r^2$]. The coefficient $B(t)$ of the phase variation is given by:

$$B(t) \approx a \frac{\Delta_\ell T_2}{1 + \Delta_\ell^2 T_2^2} \left(\int_{-\infty}^{t} \frac{I(t')dt'}{W_s} \right) \frac{1}{w_0^2}. \tag{5.49}$$

Inserting B into Eq. (5.36) yields the temporal dependence of the self-lensing focal distance due to saturation. Following the same procedure as in the previous section, this nonlinear focal distance is inserted in Eqs. (5.42) and (5.45) to determine the changes in complex parameter $\delta\tilde{s}$ and $\delta\tilde{s}_m$, which lead to the change in beam size at a particular point of the cavity due to self lensing.

As pointed out earlier, the impact of self-lensing would be minimal in dye lasers, if it were not for the large astigmatism of these cavities. The beam waist in a dye jet can be as small as $w_0 = 5$ μm [149]. Short curvature mirrors are generally used to focus the beam. The angle of incidence θ_i on the mirrors should be large enough for the incident and returning beams to clear the focal region with the saturable absorber. An angle of 8^0 is a practical minimum (slightly more than twice the half diffraction angle). If mirrors of 5 cm radius of curvature are used, the distance between sagittal and tangential foci is $R[(1/\cos\theta_i) - \cos\theta_i] = 500$ μm. This distance is much larger than the Raleigh range of $\rho_0 = 125$ μm (for a wavelength of 620 nm and a beam diameter of 2 mm). The negative lensing calculated above corresponds to a focal distance of - 10 mm. Calculations show [157] that such a lensing has an appreciable influence on the stability of a cavity, by affecting differently the ABCD round-trip matrix for the sagittal and tangential plane of the cavity.

5.5 Initiation of mode-locking

In the case of a laser mode-locked passively with a saturable absorber, the oscillation builds up gradually from noise fluctuations saturating the absorber. The starting mechanism is not obvious in the case of the laser mode-locked by a dispersive process, particularly when the gain medium has a long lifetime. Some lasers where a dispersive pulse compression mechanism dominates require an external starting mechanism; others are "self starting."

The dynamic gain saturation appears to play a determinant role both in pulse formation and starting mechanism [180, 181]. We follow the approach of Wang [182] to establish a self-starting criterion based on dynamic gain saturation. The laser is assumed to contain a saturable gain and absorption element, and a mirror of reflectivity R. To define this concept of dynamic gain saturation, let us consider a cw laser, at equilibrium with a photon flux F, saturating the gain to the value $a_g = \sigma_g \Delta\gamma_s d_g$ where σ_g is the emission cross-section, and $\Delta\gamma_s$ the (saturated) inversion density. The total *net gain* of the laser is $G = \exp[a_g + a_a + \ln R]$. Considering first only the gain medium, a small photon flux perturbation $\Delta F(t)$ will result in the change in gain: It can easily be seen that the gain response ΔG is:

$$\Delta G = \Delta a_g G = -\sigma_g G \int_{-\infty}^{t} \Delta F(t) dt. \qquad (5.50)$$

5.5. INITIATION OF MODE-LOCKING

Equation (5.50) is written under the assumption that the recovery time of the gain is much slower than the duration of the fluctuation $\Delta F(t)$. It results in a decreased gain as a small perturbation $\Delta F(t)$ is applied on top of a cw laser oscillation. In the case of *slow* saturable absorbers, we could set $a_a = 0$ and define a_g as the *net gain* of the complete laser. Saturation of an intracavity absorber will then cause an increase Δa_g of this net gain. We will consider in the following a laser with a slowly recovering gain medium characterized by the gain a_g, and a fast mode-locking mechanism characterized by a_a.

In general, the effect of most passive mode-locking mechanisms (Kerr lensing for instance) can be characterized as producing an instantaneous *increase* in net gain upon application of a perturbation $\Delta F(t)$ [see, for instance, Eq. (5.48)]:

$$\Delta G \approx \Delta a_a = b\Delta F(t). \tag{5.51}$$

The proportionality constant b is characteristic of the mode-locking mechanisms applicable to the laser being considered (fast saturable absorption, Kerr lensing, additive pulse mode-locking). In the case of additive pulse mode-locking, for example, the constant b can be calculated from Eq. (5.74) to be $4\pi^2 \hbar c n_2 L(1-r^2)/\lambda_\ell^2$ [182], where $r^2 = R$.

A fluctuation in photon flux $\Delta F(t)$ will result in a total change in gain ΔG given by the sum of Eqs. (5.50) and (5.51), which will in turn increase or decrease the average intensity. If the overlap of the change in gain with the fluctuation $\int \Delta G \Delta F dt$ is positive, the perturbation ΔF will grow. This is the "self-starting" condition for the mode-locked laser, obtained by substituting the expressions for ΔG given by of Eqs. (5.50) and (5.51) into the overlap integral $\int \Delta G \Delta F dt$:

$$b \int_{-\infty}^{\infty} \Delta F^2 dt - \sigma_g G \int_{-\infty}^{\infty} \Delta F(t) dt \int_{-\infty}^{t} \Delta F(t') dt' > 0. \tag{5.52}$$

Assuming the perturbation is a pulse $I(t)$ of finite duration τ_p, we can rewrite Eq. (5.52) for the total length of the fluctuation:

$$\frac{b}{G} > \sigma_g \frac{\left[\int_{-\infty}^{\infty} \Delta F dt\right]^2}{\int_{-\infty}^{\infty} \Delta F^2 dt} = \sigma_g \frac{\left[\int_{-\infty}^{\infty} I(t) dt\right]^2}{\int_{-\infty}^{\infty} I^2(t) dt} \tag{5.53}$$

If the perturbation is a square pulse of intensity I_0 and duration τ_p, it can easily be seen that the right-hand side of Eq. (5.53) is equal to $\sigma_g \tau_p$. For any particular shape of perturbation, we can write for the ratio $[\int I(t) dt]^2/$

$\int I^2(t)dt = \beta\tau_p$ where β is a shape dependent factor. The self starting condition reduces now to the simple inequality:

$$\frac{b}{G} > \sigma_g \beta \tau_p. \tag{5.54}$$

The saturated gain factor G (of the order of one) does not vary much from one mode-locked laser to another. The parameter b is determined by the mode-locking mechanism. A weak Kerr lensing effect may be sufficient to start a laser with a relatively small gain cross section such as Ti:sapphire ($\sigma_g = 2.7 \; 10^{-19}$ cm^2). The same Kerr lensing may not be sufficient for self-starting a dye laser with a gain cross section of the order of 10^{-16} cm^2 [182].

The gain dependence of the self-starting condition Eq. (5.54) suggests that mode-locking should start right at threshold. Generally however, a laser will have a first threshold for continuous operation, and another threshold for mode-locked operation. For the self-mode-locked Ti:sapphire laser, the first threshold may be of the order of 100 mW, the second threshold (for mode-locked operation) higher by a factor 2 or 3, and optimum (shortest pulse) operation may require more than 1.5 W pump power.

A different approach is that of Krausz et al. [183] which consists of introducing phenomenologically a decay time τ_f in the equation for the growth of a perturbation $\Delta F(t)$. We notice from Eq. (5.53) that the growth of the perturbation due to the nonlinear (mode-locking) element is proportional to $b\Delta F^2$. With such a driving term, the simplest equation to describe the growth and decay of a perturbation is:

$$\frac{\partial}{\partial t}\Delta F = \frac{b}{\tau_{RT}}\Delta F^2 - \frac{\Delta F}{\tau_f}, \tag{5.55}$$

where b is the coefficient introduced before to describe the compression mechanism, and a decay time τ_f is introduced to account phenomenologically for the influence of the gain medium. It was suggested that τ_f can be extracted from a measurement of the spectral width of the beat note of the two longitudinal modes with the highest (net) gain. If $\Delta\nu_{rf}$ is the linewidth of that beat note, $\tau_f = \frac{1}{\pi}\frac{1}{\Delta\nu_{rf}}$. The homogeneous contribution to the linewidth $\Delta\nu_{rf}$ assimilated to the mechanism of dynamical gain saturation cited above, and an inhomogeneous contribution to spatial hole burning resulting in a distribution in axial mode spacing. The basic conclusion of such a study [183] is that, to improve the starting characteristics of the laser, one could try to reduce the inhomogeneous contribution to $\Delta\nu_{rf}$ by eliminating the spatial hole burning. The latter creates a population grating in the gain medium. This grating could be erased by moving the gain medium. Another option to eliminate spatial hole burning is to use a ring, unidirectional laser [184].

5.6 Passively mode-locked lasers

5.6.1 Generalities

The two essential components in the cavity of a passively mode-locked laser are a gain medium and a "positive feedback element." By the latter it is meant a device or material that has losses decreasing with intensity or energy.

Analytical and numerical approaches have been used to study the pulse evolution and/or determine the steady-state pulse. In general, analytical models of the laser replace the localized elements by a continuous distribution throughout the cavity. The assumption has to be made that the pulse evolution during any round-trip is infinitesimally small. It is well known that finite difference equations can have oscillatory solution that do not have their counterpart in the corresponding system of differential equations. These particular oscillatory solutions [149], as well as the influence of the order of the elements, are lost in the infinitesimal models. The analytical approach has been used extensively in the past. Its main advantage is that it gives a better identification of the respective role of the various physical processes involved in the pulse formation.

Numerical codes have been developed that attempt to include all physical phenomena affecting pulse shape and duration [185, 186]. Unfortunately, the shear number of these mechanisms makes it difficult to reach a physical understanding of the pulse generation process, or even identify the essential parameters. Therefore, the most popular approach is to construct a simplified analytical model on a selected mechanism. Historically, progress in fs pulse generation have been made by focusing on specific processes. Such specific processes have been listed and analyzed in Section 5.4. For instance, we have shown that a positive feedback element will sharpen the leading edge of a pulse because of the progressive decrease in attenuation. If the positive feedback is very fast compared to the pulse duration (Kerr lensing, or saturable absorber with very short lifetime), both the pulse leading edge and trailing edge will be sharpened.

Gain saturation also plays a role in the pulse evolution. In the case of a laser with saturable absorber, as pointed out by Haus [152], there is a net gain inside the cavity after the absorber has bleached. Before the absorber recovers (within a time T_{1a}), there is a net gain for any disturbance that follow the pulse. The same equations used to describe saturation in an absorber apply to the gain medium. Let us define a time t'_s as the time at which the saturation energy density has been reached in the gain medium:$W(t'_s) = W_{sg}$. By proper adjustment of the focusing conditions, saturation in the gain medium will be reached at a time $t'_s > t_s$. As it is being amplified, the pulse will deplete the gain medium itself, leaving a net negative gain in the cavity, in which small perturbations will be damped. It is essential for the operation of a passively mode-locked laser that the gain saturates at a higher energy than the characteristic energy of the positive feedback, to ensure a time window $t_s < t < t'_s$ during which pulse amplification can occur. In that time window, the losses are saturated, but the gain has not yet saturated. This time window will narrow as t'_s recedes (with pulse amplification), and the pulse leading edge steepens. A detailed analytical and numerical treatment of the pulse evolution due to saturable gain and absorption has been made by New [150, 151].

5.6.2 Study of the passively mode-locked dye laser with the round-trip model

So far we have discussed essential features of passive mode-locking by considering certain isolated aspects rather than their complex interplay. This allowed us to establish a physical picture of the various mechanisms affecting the pulse evolution without extensive mathematical means. However, in order to obtain more quantitative results, we need to consider the simultaneous effect of all resonator elements. The round-trip model leads to a powerful method to analyze mode-locked operations, as illustrated in the following example of passive mode-locking.

Analytical solutions

Despite their limited application range, analytical approaches have found interest because of their ease of physical interpretation. Let us consider a laser that contains a saturable absorber, an amplifier, an outcoupling mirror, and a frequency filter. The product of the corresponding transfer functions Eqs. (5.3), (5.6), and (5.7) yields for the steady state condition:

$$\left\{ G(W) - \left(\frac{2}{\Delta\omega_F} + h\right)\frac{d}{dt} + \left[\left(\frac{2}{\Delta\omega_F}\right)^2 - \frac{1}{2}h^2\right]\frac{d^2}{dt^2} \right\} \mathcal{E}(t) = 0 \quad (5.56)$$

5.6. PASSIVELY MODE-LOCKED LASERS

where

$$G(F) = \frac{1}{2}(a_{gi} + a_a - \gamma) - \frac{1}{2}\frac{W(t)}{W_{sg}}(m\,a_a + a_{gi}) + \frac{m^2 a_a}{4}\left(\frac{W(t)}{W_{sg}}\right)^2 \quad (5.57)$$

is the time dependent net gain and

$$a_{gi} = a_g\left[1 - \frac{W}{W_{sg}}(e^{\tau_{RT}/T_{1g}} - 1)^{-1}\right] = a_g[1 - \Delta_u] \quad (5.58)$$

is the gain coefficient at the leading edge of the pulse. As previously, we use the notation $a_j = \alpha_j d_j$ for the optical thickness of an amplifying ($j = g$) or absorbing ($j = a$) medium). The parameter $m = W_{sg}A_g/W_{sa}A_a$ is the ratio of saturation energy in the gain and absorber medium. As noted earlier, for instance in Section 5.4.2, this quantity should be multiplied by a factor ranging from one to three if standing wave saturation takes place in the absorber. Provided that radiation is exactly resonant with the active media (i.e., no chirping), Haus [152] found a closed form solution with the ansatz $\mathcal{E} \propto \sqrt{W/W_{sg}}\text{sech}(1.76t/\tau_p)$ where τ_p is the pulse duration and W is the pulse energy density. These two pulse parameters can be determined by substituting the ansatz into the steady state condition Eq. (5.56) and solving the resulting system of algebraic equations. This approach is also suitable for more complex problems. If, for example, we allow for a detuning ($\Delta_g, \Delta_a \neq 0$) of the laser frequency with respect to the maxima of the absorption/gain profile (i.e., chirping), we obtain a similar steady state condition as before but with complex coefficients $\tilde{a}_{a,g}$, cf. Chapter 3. As shown by Kühlke et al. [187] an appropriate ansatz is now:

$$\tilde{\mathcal{E}} = \mathcal{E}e^{i\phi} \quad (5.59)$$

where $\mathcal{E} \propto \sqrt{W/W_{sg}}\text{sech}(1.76t/\tau_p)$ and $d\phi/dt = b\tanh(1.76t/\tau_p)$. The solution for the system of algebraic equations gives for the pulse duration:

$$\Delta\omega_F \tau_p = \frac{4W_{sg}}{mW}\sqrt{\frac{2}{5|a_a|}} \quad (5.60)$$

and for the pulse energy:

$$\frac{W}{W_{sg}} = A + \sqrt{A^2 - \frac{a_g + a_a - \gamma}{15m^2 a_a/8}} \quad (5.61)$$

where $A = [3ma_a + a_g(1 + 2\Delta\omega_g T_{2g})]/(7.5m^2 a_a)$. The gain and absorption coefficients $a_{a,g}$ in the above expressions are the real part of $\tilde{a}_{a,g}$. For the chirp parameter we find

$$b = \frac{1.76}{\tau_p}\left[\frac{3}{2}\left(\frac{\omega_\ell - \omega_a}{\Delta\omega_a}\right) + \text{sign}(\omega_\ell - \omega_a)\sqrt{\left(\frac{\omega_\ell - \omega_a}{\Delta\omega_a}\right)^2 + 2}\right]. \quad (5.62)$$

From the relation (5.60) for τ_p we expect shorter pulses with increasing saturation and pulse energy. This dependence is consistent with the mechanism of erosion of the leading edge of the pulse by saturable absorption discussed earlier in this chapter. For given parameters of the absorber/amplifier, the pulse duration is limited by the bandwidth of the laser which is represented here in $\Delta\omega_F$. The sign of the chirp parameter is determined by the relative position of the laser frequency and resonance frequency of the absorber (saturation of the absorber is the only chirp generating mechanism in this simplified model). The magnitude of the chirp is inversely proportional to the pulse duration. This is not surprising since τ_p determines the slope of the change in index of refraction.

Stable pulses can only be observed in a limited range of absorption, gain and linear loss. This "stability range" can be included in the analytical model by requiring the net gain to be negative in the leading and trailing edges of the pulse. Both quantities are controlled by the pulse energy, and the ratio of the cavity round-trip time to the lifetime of the gain in steady state. The stability condition prevents the growth of any fluctuations (spontaneous emission) between two successive pulses. Figure 5.11 shows as an example the behavior of pulse duration and energy versus gain, as well as the corresponding stability range.

Most passively mode-locked dye lasers contain no extra frequency selective element. Following a simple model [187] the laser frequency is determined by the maximum net gain which can be obtained by subtracting the weighted gain and absorption profiles. This model correctly accounts for the experimentally observed shift in laser wavelength resulting from a change in absorber concentration [188].

Numerous variations on this analytical laser model exist, based either on more complex cavities or concentrating on a particular aspect of the laser. The incorporation of coherent effects, i.e., the consideration of transfer functions of the media according to Eq. (3.78), avoids the necessity to incorporate a fictitious bandwidth limiting element and relates the stationary pulse durations to the transition width of amplifier and absorber [154]. It also explains

5.6. PASSIVELY MODE-LOCKED LASERS

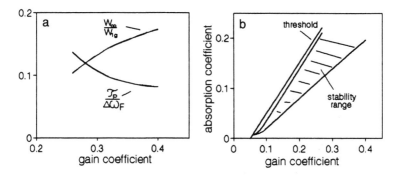

Figure 5.11: (a) Normalized pulse energy and pulse duration vs small signal gain. (b) Stability ranges. Laser parameter: $|a_a| = 0.2, \gamma = 0.05, \tau_{RT}/T_{1g} = 0.8, m = 4,$, standing wave saturation.

that the generation of a phase modulation and its compensation by a GVD element is an additional shortening factor. Further effects of dispersion and phase modulation were discussed in [155] and in the models of Haus and Silberberg [156], and Martinez et al. [189]. Other theories discuss intracavity second harmonic generation in a passively mode locked fs dye laser (Zhang et al. [190]) and the "self-starting" problem (Haus and Ippen [191]).

Numerical simulation

Although the analytical models give satisfactory results as far as essential parameter dependencies are concerned, more accurate models are often needed for a more detailed comparison with the experiment. Another advantage of numerical codes is that they can provide information not only on the stationary pulse regime but also on the pulse evolution and the temporal laser response to external perturbations.

In order to study the switch-on dynamics one starts from noise. The noise bandwidth is roughly given by the width of the fluorescence curve of the amplifier while its magnitude corresponds to the light emitted spontaneously into the solid angle defined by the cavity modes. In most cases the particular noise features vanish after few round trips and the final results are independent of the field originally injected. Figure 5.12 shows as an example the development of the pulse envelope, instantaneous frequency, and energy as a function of round trips completed after the switch-on of the laser. Obviously these parameters become stationary after several hundred round trips, which amounts to several microseconds.

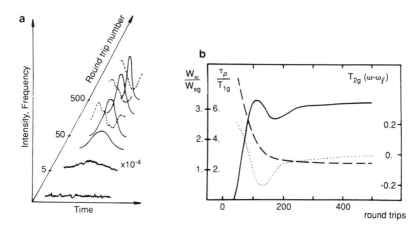

Figure 5.12: (a) Evolution of pulse envelope (solid line) and instantaneous frequency (dashed line) after switch on of the laser, and (b) corresponding steady-state pulse energy (solid line), pulse duration (dashed line), and frequency (dotted line). The active media were described by the density matrix equations introduced in Chapter 3. All pulse parameters including the average frequency develop as a result of the interplay of resonator elements. No extra frequency selective element was necessary to limit the pulse duration. (From [192].)

A convenient method to increase the available pulse energy at the expense of repetition rate is cavity dumping, which was also successfully applied in fs passively mode locked dye lasers at a frequency of 3 MHz [193]. The cavity dumper introduces a periodical perturbation for the steady state pulse regime. Can the fs laser respond to an external modulation in such a way that a near quasi steady-state is established and maintained? Questions such as this one can be directly answered by computer simulation. An example is depicted in Fig. 5.13. The cavity is dumped with an efficiency of 70% after every 30th round trip ($\sim$ 30 MHz).

As explained in previous sections chirp generation and compensation can be an essential pulse shortening mechanism in fs lasers. There have been attempts to increase the chirp production by adding a medium with large n_2 to the dye solution [168, 176]. Figure 5.13(b) shows the behavior of the pulse duration as obtained from a numerical simulation [185]. For a given GVD value of the cavity, the pulses become shorter until they reach the boundary of the stability range.[2] The shortening of stable pulses "saturates" at high

---

[2] At small nonlinearities the boundaries of the stability range are defined as previously.

5.6. PASSIVELY MODE-LOCKED LASERS

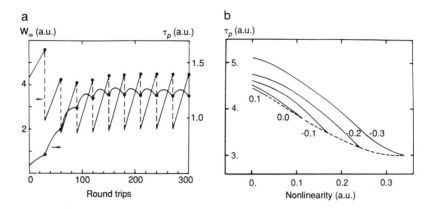

Figure 5.13: (a) Behavior of pulse energy and duration in a cavity dumped, passively mode-locked dye laser. The cavity dumper was turned on at round-trip number zero. The bullets indicate the switching of the cavity dumper. After a response time of about 150 round trips, a new quasi-stationary state occurs, which delivers a stable output at 1/30th of the original repetition rate but of about ten times larger energy (out-coupling losses ~ 70%) (adapted from [192]). (b) Normalized pulse duration as function of the nonlinearity introduced by an additional n_2 element for different values of the cavity GVD (adapted from [185]).

nonlinearities. The reason is the behavior of the boundary of the stability range for larger nonlinearities. Additional instabilities arise from the fact that the chirp is nonlinear and the GVD basically can only compensate the linear part. The result is pulse compression in the pulse center and pulse broadening in the wings. The wings can eventually split into satellite pulses. Stable single pulse operation is not found under these conditions. Therefore, in lasers with a large n_2 nonlinearity such as the Ti:sapphire, the third- and (if necessary) higher-order order dispersion are adjusted to achieve an additional pulse shortening.

What ultimately limits the pulse duration? Neglecting linear dispersive effects, the finite response time of the nonlinear interaction remains as bandwidth limiting factor. The "slowest" components in this respect are the amplifier and absorber because the interaction with the pulse is near resonance. Thus the finite width of the transition profiles limit the pulse bandwidth that can be handled by them. At a high pumping rate, however,

---

However, their relevance can now be proved directly if we add to the pulse a small amount of noise (spontaneous emission) every time it passes through the amplifier. Outside the stability range this additional noise prevents a stable pulse from developing.

the normalized field strength becomes of the order of the Rabi frequencies (see Chapter 4) of the media. In this case stationary pulses were found that approach π pulses in the amplifier and 2π pulses in the absorber [194]. Their spectra can, in principle, be (much) broader than the transition profiles.

Other numerical models concentrate on the explanation of periodic pulse trains [186], the comparison of amplitude and phase shaping [195], bandwidth limitations [196], and the evolution of the pulse average frequency, as well as the role of an absorber isomer [194].

5.7 Synchronous and hybrid mode-locking

5.7.1 Synchronous pumping, stability, noise

A simple method to generate short pulses is to excite the gain medium at a repetition rate synchronized with the cavity mode spacing. This can be done by using a pump that emits pulses at the round trip rate of the cavity to be pumped. One of the main advantages of synchronous mode-locking is that a much broader range of gain media can be used than in the case of passive mode-locking. For instance, laser dyes such as styryl 8, 9, and 14, which have too short a lifetime to be practical in cw operation, are quite efficient when pumped with short pulses.

Ideally, the gain medium in a synchronously pumped laser should have a short lifetime, so that the duration of the inversion is not larger than that of the pump pulse. Synchronous pumping is sometimes used in situations that do not meet this criterion, just as starting mechanism. This is the case in some Ti:sapphire lasers, where the gain medium has a longer lifetime than the cavity round-trip time, and therefore synchronous pumping results in only a very small modulation of the gain. The small modulation of the gain coefficient $\alpha_g(t)$ is sufficient to start the pulse formation and compression mechanism by dispersion and self-phase-modulation. The initial small gain modulation grows because of gain saturation by the modulated intracavity radiation, resulting in a shortening of the function $\alpha_g(t)$, and ultimately ultrashort pulses.

The simple considerations that follow, neglecting the influence of saturation, show the importance of cavity synchronism. If the laser cavity is slightly longer than required for exact synchronism with the pump radiation, stimulated emission and amplified spontaneous emission will constantly accumulate at the leading edge of the pulse, resulting in pulse durations that could be even longer than the pump pulse. Therefore, to avoid this situation, the cavity length should be slightly shorter than that required for exact syn-

5.7. SYNCHRONOUS AND HYBRID MODE-LOCKING

chronism with the pump radiation. Let us assume first perfect synchronism. The net gain factor per round-trip is

$$G(t) = e^{[\alpha_g(t)d_g - L]}, \tag{5.63}$$

where L is the natural logarithm of the loss per cavity round-trip. After n round-trips, the initial spontaneous emission of intensity I_{sp} has been amplified sufficiently to saturate the gain α_g, and thus the pulse intensity is approximately $I(t) \approx I_{sp} \times \left\{ e^{[\alpha_{g0}(t)d_g - L]} \right\}^n = I_{sp} \times [G_0(t)]^n$. The pulse is thus $\sqrt{n}$ times narrower than the unsaturated gain function $G_0(t)$.

For a cavity shorter than required for exact synchronism, in a frame of reference synchronous with the pulsed gain $\alpha_g(t)$, the intracavity intensity of the jth round-trip is related to the previous one by:

$$I_j(t) = I_{j-1}(t + \delta) e^{[\alpha_g(t)d_g - L]}. \tag{5.64}$$

where δ is the mismatch between cavity round-trip time and the pump pulse spacing. The net gain for the circulating pulse $e^{[\alpha_g(t)d_g - L]}$ exists in the cavity for a time $n\delta$ only, as can be seen from Fig. 5.14. The laser oscillation will

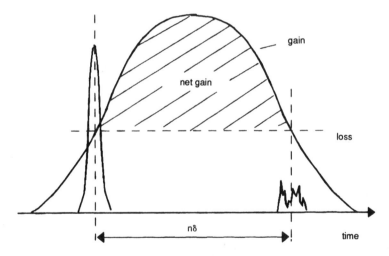

Figure 5.14: Net gain (gain minus loss) temporal profile as it appears at each periodicity of the pump pulse. If the round-trip time of the laser cavity is slightly shorter (by δ) than the pump periodicity, radiation emitted at the right edge of the gain profile will reappear shifted to the left by that amount δ at each successive round-trip. A pulse will experience gain for a maximum of n passages, given by the ratio of the duration of the net gain to the mismatch δ.

start from a small noise burst $I_{sp}(t)$. The intracavity pulse after n round trips can be approximated by:

$$I(t) = I_{sp}(t + n\delta) \left[e^{(\alpha_{av} d_g - L)} \right]^n, \tag{5.65}$$

where $I_{sp}(t + n\delta)$ is the spontaneous emission noise present in the cavity in the time interval $(n-1)\delta \to n\delta$, and $\alpha_{av} = \frac{1}{n\delta} \int \alpha_g(t) dt$ is a gain coefficient averaged over the n round-trips.

These simple considerations indicate that in the absence of any spectral filtering mechanism and neglecting the distortion of the gain curve $\alpha_g(t)$ by saturation, the pulse should be roughly $\sqrt{n}$ times shorter than the duration of the gain window. The mismatch δ is an essential parameter of the operation of a synchronously mode-locked laser. The shape of the autocorrelation (see Chapter 8) is typically a double sided exponential, which — as pointed out by Van Stryland [197] — is a signature for a possible random distribution of pulse duration in the train. The interferometric autocorrelation also indicates a random (Gaussian) distribution of pulse frequencies [198]. These fluctuations in pulse duration and frequency have also been observed in theoretical simulations by New and Catherall [199] and Stamm [200].

Gain saturation — neglected in the elementary model above — does play an essential role in pulse shaping and compression for synchronously pumped dye lasers. We refer to a paper by Nekhaenko et al. [201] for a detailed review of the various theories of synchronous pumping. In a typical synchronously pumped dye laser, the net gain (at each round-trip) is "terminated" by depletion at each passage of the circulating pulse. This shortening of the gain period result in a dye laser pulse much shorter than the pump pulse. This mechanism was analyzed in detail by Frigo et al. [202]. It has been verified experimentally [203] that the shortest pulse duration is approximately $\tau_p \approx \sqrt{\tau_{pump} T_{2g}}$. Numerical simulations have been made to relate the number of round-trips required to reach steady state to the single-pass gain [204].

5.7.2 Stabilization

Regenerative feedback As we have seen at the beginning of the previous section, the laser cavity should never be longer than the length corresponding to exact synchronism with the pump radiation in order to generate pulses shorter than the pump pulse. This implies strict stability criteria for the pump cavity, its mode-locking electronics, and the laser cavity (invar or quartz rods are generally used). Considerations of thermal expansion of

5.7. SYNCHRONOUS AND HYBRID MODE-LOCKING

the support material and typical cavity lengths clearly shows the need for thermal stability. Indeed, the thermal expansion coefficient of most rigid materials for the laser support exceeds $10^{-5}/°C$. Since the cavity length approaches typically 2 m, even a temperature drift of 0.5 °C would bring the laser out of its stability range. However, since it is the *relative* synchronism of the laser cavity with its pump source that is to be maintained, a simpler and efficient technique is to use the noise (longitudinal mode beating) of the laser itself, to drive the modulator of the pump laser [205]. This technique, sometimes called "regenerative feedback," has been applied to some commercial synchronously pumped mode-locked lasers.

Seeding Even if somewhat oversimplified, the representation of Fig. 5.14 gives a clue to an important source of noise in the synchronously pumped dye laser. The seed $I_{sp}(t)$ has a complex electric field amplitude $\tilde{\epsilon}(t)$ with random phase. As pointed out in [206] and in [200] it is this spontaneous emission source that is at the origin of the noise of the laser. Could the noise be reduced by adding to $\tilde{\epsilon}$ a minimum fraction $\eta E(t)$ of the laser output, just large enough so that the phase of $\eta E(t) + \tilde{\epsilon}(t)$ is equal to the phase of the output fields $E(t)$ (which essentially implies $\eta E(t) \gg \tilde{\epsilon}$. Both calculation and experiment have demonstrated a dramatic noise reduction by seeding the cavity with a small fraction of the pulse *in advance* of the main pulse [207]. The emphasis here is on small: only a fraction of the order of 10^{-7} (not exceeding 10^{-5}) of the output power should be re-injected. A possible implementation would consist of reflecting back a fraction of the output pulse delayed by slightly less than a cavity round-trip. This amounts to a weakly coupled external cavity. A much simpler implementation demonstrated by Peter *et al.* [207] consists in inserting a thin glass plate (microscope cover for instance) in front of the output mirror (Fig. 5.15). The amount of light re-injected is adjusted by translating the glass plate in front of the beam. The timing of the re-injected signal is determined by the thickness of the plate.

5.7.3 Hybrid mode-locking

Combining passive and synchronous mode-locking

Synchronous pumping alone can be considered as a good source of ps rather than fs pulses. For dye lasers, the disadvantages of this technique, as com-

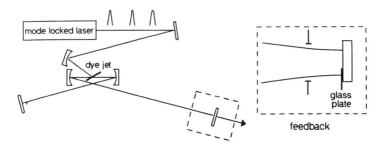

Figure 5.15: Typical synchronously pumped dye laser. The length of the dye laser cavity has to be matched to the repetition rate of the pump pulses. The noise in a synchronously pumped laser can be reduced by re-injection of a portion of the output *ahead* of the main intracavity pulse. A thin glass plate on the output mirror intercepts and reflects part of the beam into the cavity, with the desired advance. The fraction of energy reflected (of the order of 10^{-6}) is determined by the overlap of the aperture and the glass plate (adapted from [207]).

pared to passive mode-locking, are:

- a longer pulse duration
- larger amplitude and phase noise
- the duration of the pulses of the train are often randomly distributed [197]
- when attempting to achieve the shortest pulse durations, the pulse frequencies are randomly distributed [198]

One solution to these problems is to combine the techniques of passive and active mode-locking in a hybrid system [208, 209]. Depending on the concentration of the absorber dye, the hybrid mode-locked laser is either a synchronously mode-locked laser perturbed by the addition of saturable absorption or a passively mode-locked laser pumped synchronously. The distinction is obvious to the user. The laser with a very dilute absorber will have the noise characteristics and cavity length sensitivity typical of synchronously pumped lasers, but a shorter pulse duration. The laser with a concentrated saturable absorber shows intensity autocorrelation traces identical to those of the passively mode-locked laser [168]. The sensitivity of the laser to cavity detuning decreases. The reduction in noise can be explained as being related to the additional timing mismatch introduced by the absorber, which partially compensates the pulse advancing influence of the gain and spontaneous emission [185].

5.7. SYNCHRONOUS AND HYBRID MODE-LOCKING

The hybrid mode-locked operation leads to pulse durations as short as those of passively mode-locked lasers, with higher output powers. Kubota et al. [143] reported 29 fs pulses from a hybridly mode-locked "Kiton red" dye laser, at an average power of 60 mW at 76 MHz (peak power of 27 kW). The saturable absorber used was a mixture of diethyloxadicarbocyanine iodide (DODCI) and diethylquinoloxacarbocyanine iodide (DQOCI).

Hybrid mode-locking of dye lasers has extended the palette of wavelength hitherto available through passive mode-locking, making it possible to cover a broad spectral range spanning from covering the visible from the UV to the near infrared. A list of dye combinations is given in Table 5.1. Except when noted, the laser cavity is linear, with the absorber and the gain media at

Gain dye	Absorber[1]	λ_ℓ nm	range	τ_{pmin} fs	at λ_ℓ nm	Remarks
Disodium fluorescein	RhB	535	575	450	545	
Rh 110	RhB	545	585	250	560	
Rh6G	DODCI	574	611	300	603	
Rh6G	DODCI			110	620	Ring laser
Rh6G	DODCI			60	620	Antiresonant ring
Kiton red S	DQOCI			29	615	
Rh B	Oxazine 720	616	658	190	650	
SRh101	DQTCI	652	682	55	675	Pumping with
	DCCI	652	694	240	650	doubled Nd:YAG
Pyridine 1 [2]	DDI			103	695	
Rhodamine 700	DOTCI	710	718	470	713	
Pyridine 2	DDI,DOTCI			263	733	
Rhodamine 700	HITCI	770	781	550	776	
LDS-751	HITCI	790	810	100		
Styryl 8	HITCI			70	800	
Styryl 9 [3]	IR 140 [4]	840	880	65	865	Ring laser
Styryl 14	DaQTeC			228	974	

Table 5.1: Femtosecond pulse generation by hybrid mode-locking of dye lasers pumped by an argon ion laser, except as indicated (from [149]).
[1] See Appendix D for abbreviations.
[2] Solvent: propylene carbonate and ethylene glycol.
[3] Solvent: propylene carbonate and ethylene glycol.
[4] in benzylalcohol.

opposite ends. Another frequently used configuration is noted "antiresonant ring". The saturable absorber jet is located near the pulse crossing point of a small auxiliary cavity, in which the main pulse is split into two halves, which are recombined in a standing wave configuration in the absorber [167, 168]. The ring laser appears only once in Table 5.1 [210], because of the difficulty of adjusting the cavity length independently of all other parameters.

Femtosecond pulse generation mechanism

Unless the pump pulses are very short (i.e., $\ll$ 50 ps), synchronous pumping leads typically to ps rather than fs pulses. In the presence of a concentrated saturable absorber, the pulse generation mechanism in hybrid mode-locking is the same as in passive mode-locking. One can roughly estimate that the combined effect of saturation and gain modulation brings the pulse duration down to 0.5 ps. At this level of pulse duration, however, dispersion and phase modulation become the dominant pulse shaping mechanisms. Depending on the particular laser, either the gain medium or the saturable absorber (dye solution or multiple quantum well) will be the main (self) phase modulator. The dispersion (positive or negative) has to be adjusted for optimum pulse duration, which is not an easy task since the cavity length has to be kept in synchronism with the pump rate. The most commonly used technique is to adjust the dispersion by translating one intracavity prism, and correct the cavity length by translating one end mirror. As pointed out in [211], it is possible to select the direction of translation of a dispersive prism, such that only the intracavity quadratic dispersion is modified, without any change in group velocity (or cavity length).

5.8 Solitons in femtosecond lasers

We have alluded in Section 5.2.3 to a laser model where the resonator elements are replaced by a uniform medium (cf. Fig. 5.5). Essentially, instead of having a pulse circulating in the cavity, we are considering the propagation of a pulse, in plane wave approximation, through an inverted medium of infinite extent. The resulting differential equation is analyzed for a shape-preserving solution, called soliton.

The existence of solitons is related to a particular structure of the propagation equations through the composite medium. No exact soliton solutions have been found in any model incorporating most of the resonator elements. However, several subsystems have been found to lead to soliton solutions.

5.8. SOLITONS IN FEMTOSECOND LASERS

For instance, considering *only* the amplifier and an absorber as cavity elements, soliton solutions can be found [194] which are related to the π and 2π pulse propagation, as detailed in Chapter 4. Another subsystem that has been used considers the laser to consist only of a GVD and Kerr medium. It is this latter model, where absorption and gain are assumed to balance each other exactly, that will be discussed here in more detail.

The elegance of the theory is at the expense of simplifying assumptions that are not quite compatible with a pulse *formation* mechanism. Since the only compression mechanism assumed in the model presented below is purely dispersive, there is no intensity dependent mechanism that could amplify the noise fluctuations of the laser in order to start the pulse operation. Even though the Kerr effect is taken into account, one assumes that the self-focusing associated with it has no influence on the cavity parameters.

The ring laser model reduces to a product of two operations, as sketched in Fig. 5.16: phase modulation in the time domain (upper part of the figure) and dispersion in the frequency domain. Both operations are combined into a single equation (see problem at the end of this chapter) shown in a square box in the middle of Fig. 5.16.

This equation is the nonlinear Schrödinger equation, which is analyzed in the section on pulse propagation in fibers (Chapter 6). The solutions of this equation are detailed in an article by Zakharov and Shabat [212], using the inverse scattering method [213]. The problem is reduced to a search for eigenvalues of coupled differential equations. The soliton is the eigenfunction associated with that eigenvalue. The order of the soliton is the number of poles associated with that solution. A soliton of order 1 is a sech-shaped pulse. It exhibits a stable pulse shape, propagating without distortion. Solitons of order n are periodic solutions with n characteristic frequencies. Periodic evolution of the pulse train has been observed in some dye lasers [214, 215]. Salin *et al.* [216] interpret the periodicity in pulse evolution of a fs laser as a manifestation of a soliton of order larger than 1. In the case of a dye laser, however, the nonlinear Schrödinger equation is only a crude approximation of the complex pulse evolution. In the case of lasers with large gain, such that the continuity approximation is no longer valid, periodic oscillation of the pulse energy can also be observed [217, 149]. Despite the oversimplifications of this soliton model, it appears to describe certain features of the stationary operation of a fs Ti:sapphire laser [218, 48]. For instance, the unchirped pulse that returns identical to itself after each round-trip is associated with the soliton of order 1. With some minor changes in alignment, a periodicity is observed in the pulse train. If this periodicity

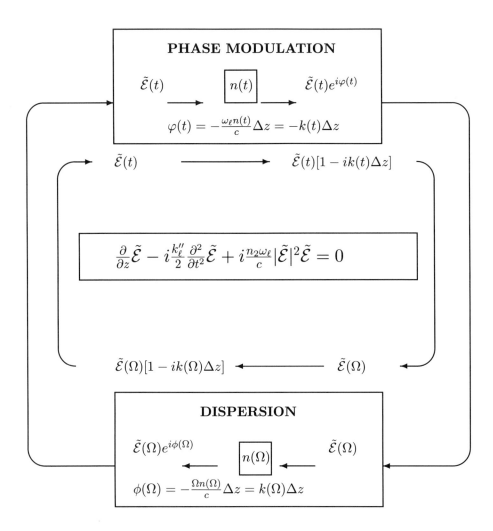

Figure 5.16: Model used to describe a "soliton" laser. Only infinitesimal self-phase modulation (top part) and dispersion (bottom part of the figure) are applied on the pulse circulating in the cavity. Δk_k is the nonlinear contribution to the wave vector $\omega_\ell n_2 |\tilde{\mathcal{E}}|^2/c$. The combined operation results in the equation written in the center of the picture for which soliton solutions are known (nonlinear Schrödinger equation [212]). A steady state in which gain and loss compensate is assumed.

contains m frequencies, it is often possible to represent the pulse by a soliton of order $n = m - 1$.

5.9 Negative feedback

5.9.1 Introduction

We have seen that the pulse formation — in passively mode-locked lasers — is associated with a positive feedback element (Kerr lensing, saturable absorber) which enhances positive intensity fluctuations (generally through a decrease of losses with increasing intensity). While a positive feedback leads to pulse formation, it is inherently an unstable process, since intensity fluctuations are amplified. Therefore, it is desirable, in particular in high-power lasers, to have a *negative* feedback element that sets in at higher intensities than the positive element. In addition to its stabilizing function, negative feedback can also play an important role in pulse compression. We will study here the passive negative feedback technique applied to flashlamp pumped solid state lasers. While these lasers are not specifically femtosecond lasers, the remarkable pulse to pulse stability has made these sources a primary pump for fs generation through nonlinear processes.

In the case of negative feedback applied to solid state lasers, the train of pulses is extended by limiting the energy of the intracavity pulse, thereby limiting the gain depletion by stimulated emission at each round-trip. Extension of the pulse train implies that the intracavity pulse makes a much larger number of round-trips. Therefore, intracavity pulse compression can be implemented more effectively. It should be noted that, in most lasers, the pulses do not become as short as the limit set by the gain bandwidth, because of the various pulse broadening and intracavity filtering mechanisms. The mode-locked laser with passive negative feedback is an exception to that general trend, in the sense that the pulse compression mechanisms are so efficient that the minimum pulse duration corresponding to the bandwidth limit of the gain medium is routinely obtained. Pulses of 10, 5, and less than 1 ps have been generated with this technique with Nd:YAG, Nd:YAP and Nd:glass lasers, respectively. More importantly for the fs field, the pulse-to-pulse reproducibility (better than 0.2% [219]) makes these lasers ideal pump sources for synchronous or hybrid mode-locking. The flashlamp pumped solid state laser with negative feedback provides much higher energy/pulse, at lower pulse duration, than the cw mode-locked laser used conventionally as pump for fs systems. The use of negative feedback to effectively pump a fs dye laser was demonstrated by Angel *et al.* [204].

5.9.2 Feedback mechanisms

Electronic feedback

A typical flashlamp pumped, mode-locked Nd laser generates a train of only five to ten pulses of all different intensities. In the first implementation of "negative feedback," an electronic feedback loop *increases* the cavity losses if the pulse energy exceeds a well defined value. Martinez and Spinelli [220] proposed to use an electro-optic modulator to actively limit the intracavity energy in a passively mode locked glass laser. They demonstrated that the pulse train could be extended. A fast high voltage electronics led to the generation of μs pulse trains in a passively mode locked glass lasers [221] and in hybrid Nd:glass lasers [222].

These electronic feedback systems are expensive and sophisticated. There is a minimum response time of one cavity roundtrip before the feedback can react [222]. As a result, the pulse train envelope does not have the stability required for the generation of fs pulses. We will therefore analyze in some detail passive negative feedback.

Passive negative feedback

A passive feedback system can provide immediate response — i.e., on the time scale of the pulse rather than on the time scale of the cavity round-trip. We will here restrict our description to the Nd laser using a semiconductor (GaAs) for passive negative feedback, because this type of laser has been used successfully for pumping femtosecond sources.

A typical laser using passive negative feedback generally includes an acousto-optic modulator for active mode-locking, a saturable absorber dye cell for Q-switching and mode-locking, and an energy limiter. An energy limiter that can be used for passive negative feedback is illustrated in Fig. 5.17. A two-photon absorber (typically GaAs) is located near a cavity end mirror. After double passage through this sample, the beam is defocused by a lens originating from the free carriers generated through two-photon absorption. The defocused portion of the beam is truncated by an aperture. Self-defocusing in the semiconducting two-photon absorber sets in at a power level that should be close to the saturation intensity in the saturable absorber

5.9. NEGATIVE FEEDBACK

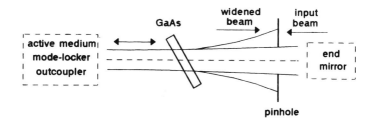

Figure 5.17: Passive negative feedback is typically achieved by inserting in the cavity an energy limiter, which can consist of a GaAs plate (acting as two-photon absorber and subsequent defocusing element) and an aperture.

dye. Because the pulse intensity is close to the pulse saturation intensity, there is optimal pulse compression at the pulse leading edge by saturable absorption. Because of self defocusing by the two-photon generated carriers, the pulse trailing edge is clipped off, resulting in further pulse compression and energy loss.

The stabilization and compression of the individual pulses result from a delicate balance of numerous physical mechanisms. It is remarkable that pulse compression can be achieved by a two-photon process, even though two-photon absorption is known to lead generally to pulse stretching [223, 224]. A simple overview of the various physical mechanisms involved is given below. Details of the experimental implementation and theoretical analysis can be found in the literature cited in the review [225].

The semiconductor used for passive negative feedback has typically a small linear absorption (α = -0.04 cm^{-1} for GaAs) and a two photon absorption coefficient (β = 26 cm/GW at 1.06 μm). In the retarded frame of reference, the depletion of the beam of intensity $I = I(x,y,z)$ due to the combined action of one- and two- photon absorption far below saturation is given by:

$$\frac{dI}{dz} = \alpha I - \beta I^2. \tag{5.66}$$

Equation (5.66) can be directly integrated to yield the intensity versus distance z in the sample:

$$I(z,t) = \frac{I(0,t)e^{\alpha z}}{1 - \beta I(0,t)[1 - e^{\alpha z}]/\alpha}. \tag{5.67}$$

The average transverse distribution of free carrier density $N(x, y, t)$ for the sample thickness d is:

$$N(x, y, t) = \frac{1}{d} \int_0^d \left[\frac{\beta I^2}{2\hbar\omega_\ell} - \frac{\alpha I}{\hbar\omega_\ell} \right] dz. \tag{5.68}$$

There are several contributions to the change in index δn that are controlled by the free carrier density $N(x, y, t)$. The classical Drude model [11] gives the contribution:

$$\delta n_d = -\frac{n_0 e^2}{2m^* \epsilon_0 \omega_\ell^2} N(x, y, t), \tag{5.69}$$

where m^*, e is the effective mass and charge, respectively of the electrons in the conduction band. We refer to the appropriate literature for additional contributions to δn such as the interband contribution [226], and an additional electronic contribution [227, 228]. All these contributions are generally expressed as the product of a factor and δn_d. Therefore, the total phase shift after transmission through the sample of thickness d can be calculated after substitution of Eq. (5.68) into the expression for the change in index Eq. (5.69). It takes the form:

$$\delta\phi(x, y, t) = C d \, \delta n_d. \tag{5.70}$$

The proportionality constant C is approximately 10 for GaAs. The change in amplitude of the field due to single- and two-photon absorption can be calculated from Eq. (5.67). Knowing the amplitude and phase of the field exiting the sample, the free space propagation of the beam can be calculated through application of the diffraction integral Eq. (1.134). The effect of the pinhole is obtained by multiplication by an aperture function. The result of such a calculation is depicted in Fig. 5.18, showing the temporal profile of the pulse transmitted through the open aperture and closed aperture. The energy loss is negligible for the pulse transmitted through the open aperture. The pulse tail is truncated by self defocusing in Fig. 5.18(b), resulting in significant pulse compression. For this technique to effectively compress pulses, it is essential that the two-photon absorption losses be negligible compared with the energy loss due to finite aperturing of the defocused beam.

Numerical techniques have been developed to study the operation of lasers with "Passive Negative Feedback" [225]. The rational for using numerical techniques is that, after nonlinear absorption and refraction, the

5.9. NEGATIVE FEEDBACK

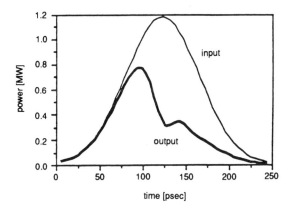

Figure 5.18: Power versus time of a ps pulse transmitted through GaAs and an aperture. The curve labelled "input" corresponds as well to the input intensity as to the output intensity (negligible absorption). However, the pulse tail can be eliminated by an aperture (curve labelled "output"), resulting in pulse compression.

beam is no longer Gaussian. Often, the phase modulation $\delta\phi(x, y, t)$ is induced in an optically thin medium. Let $\mathcal{E}(x, y, z = 0, t) \exp[i\delta\phi(x, y, t)] = \tilde{\mathcal{E}}(x, y, z = 0, t)$ be the amplitude distribution of the field leaving the nonlinear element. Further propagation of the beam *in free space* is computed by integration of the wave equation:

$$\frac{\partial \tilde{\mathcal{E}}}{\partial z} = -\frac{i}{2k_\ell}\left(\frac{\partial^2}{\partial x^2} + \frac{\partial^2}{\partial y^2}\right)\tilde{\mathcal{E}} \qquad (5.71)$$

which can be done by taking the Fourier transform along the transverse spatial coordinates, as detailed in Chapter 1 [cf. Eq. (1.134)].

The pulse evolution through the entire laser cavity can be calculated with a succession of localized elements (lenses, jets, phase modulators, mirrors, *etc.*) modifying the field amplitude and/or phase, followed by stretches of free space propagation. As any numerical simulation, this technique is limited to solving particular problems, rather than providing general solutions. The theory explain most of the features of the operation of negative feedback, but not the particular details of the components (selection criteria for the semiconductor, or the need for an acousto-optic modulator in addition to the saturable absorber).

In summary, all successfull implementations of passive negative feedback included the following elements: (a) a high gain solid state laser,

(b) hybrid mode-locking (acousto-optic modulator and saturable absorber), (c) an intracavity GaAs sample, and (d) a beam limiting aperture. Typical results [219] show a *rectangular* pulse train of 100 to 200 *identical* pulses; sa pulse duration of 10 ps in Nd:YAG; 10 μJ per pulse, and a shot to shot stability better than 0.5%, measured over 10,000 pulses.

5.10 Additive pulse mode-locking

5.10.1 Generalities

There has been in the late 1980s a resurrection of interest in developing additive pulse mode-locking (APML), a technique involving coupled cavities. One of the basic ideas — to establish the mode coupling outside the main laser resonator — has been suggested in 1965 by Foster *et al.* [229] and applied to mode-locking a He-Ne laser [230]. In that earlier implementation, an acousto-optic modulator is used to modulate the laser output at half the inter-mode spacing of the laser. The frequency shifted beam is reflected back through the modulator, resulting in a first-order diffracted beam which is shifted in frequency by the total mode spacing, and re-injected into the laser cavity through the output mirror. The output mirror of the laser forms, with the mirror used to re-inject the modulated radiation, a cavity with the same mode spacing as the main laser cavity. If the laser is close to threshold, a small extracavity modulation fed back into the main cavity can be sufficient to lock the longitudinal modes.

The recent APML methods are passive implementations of this older technique. In the purely dispersive version, pulses from the coupled cavity are given some phase modulation, such that the first half of the pulse fed back into the laser adds in phase with the intracavity pulse, while the second half has opposite phase [231]. At each round-trip, the externally injected pulse thus contributes to compress the intracavity pulse, by adding a contribution to the leading edge and subtracting a certain amount from the trailing edge, as sketched in Fig. 5.19. This technique has first been applied to shortening pulses generated through other mode-locking mechanisms. A reduction in pulse duration by as much as two orders of magnitudes was demonstrated with color-center lasers [232, 233, 234, 235] and with Ti:sapphire lasers [236].

It was subsequently realized that the mechanism of pulse addition through a nonlinear coupled cavity is sufficient to passively mode-lock a laser. This technique has been successfully demonstrated in a Ti:sapphire laser [237], Nd:YAG [238, 239], Nd:YLF [240, 241], Nd:glass [242], and KCl color-center

5.10. ADDITIVE PULSE MODE-LOCKING

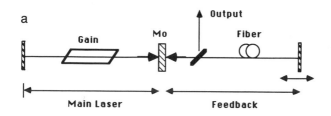

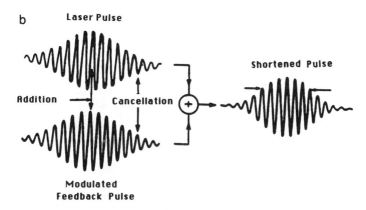

Figure 5.19: A typical additive-pulse mode locked laser (a). At the output mirror, the pulse of the main cavity [(b), top left] adds coherently to the pulse of the auxiliary cavity [(b), bottom left], to result in a shortened pulse [(b), right] (courtesy E. Wintner).

lasers [243]. A detailed description of the coherent addition of pulses from the main laser and the extended cavity which takes place in the additive-pulse mode locking has been summarized by Ippen *et al.* [244].

Coherent field addition is only one aspect of the coupled cavity mode-locked laser. The nonlinearity from the coupled cavity can be for example an amplitude modulation, as in the "soliton" laser [245], or a resonant nonlinear reflectivity via a quantum well material [159].

5.10.2 Analysis of APML

Analysis of APML [234, 244] has shown that the coupling between a laser and an external nonlinear cavity can be modeled as an intensity-dependent reflectivity of the laser end mirror. Let r be the (field amplitude) reflection coefficient of the output mirror. The radiation transmitted through that mirror into the auxiliary (external) cavity returns to the main cavity having experienced a field amplitude loss γ ($\gamma < 1$) and a total phase shift $-[\phi + \Phi(t)]$. The nonlinear phase shift $\Phi(t)$ induced by the nonlinearity is conventionally chosen to be zero at the pulse peak [244], so that the linear phase shift ϕ includes a bias due to the peak nonlinear phase shift. Therefore, if $r\tilde{\mathcal{E}}(t)$ is the field reflected at the mirror, the field transmitted through the output mirror, the auxiliary cavity (loss γ) and transmitted a second time through the output mirror is $(1 - r^2)\gamma e^{-i\phi}\tilde{\mathcal{E}}(t)e^{-i\Phi(t)}$. If d is the length of the nonlinear medium, and assuming a $\bar{n}_2$ nonlinearity, according to the definition (3.110):

$$\Phi(t) = \frac{2\pi\bar{n}_2}{\lambda_\ell} \left[I_{ax}(t) - I_{ax}(0)\right] d \qquad (5.72)$$

where $I_{ax}(t)$ is the intensity of the field in the auxiliary cavity. We determine the total reflection by adding the contribution of the re-injected field from the auxiliary cavity to the field reflection r of the output mirror, which leads to a time dependent complex "reflection coefficient" $\tilde{\Gamma}$:

$$\tilde{\Gamma}(t) = r + \gamma(1 - r^2)e^{-i\phi}[1 - i\Phi(t)]. \qquad (5.73)$$

In Eq. (5.73), it has been assumed that Φ is small, allowing us to substitute for the phase factor $e^{-i\Phi}$ its first order expansion. There is a differential reflectivity for different parts of the pulses. If one set $\phi = -\pi/2$, then $|\tilde{\Gamma}|$ has a maximum value at the pulse center where $\Phi = 0$, and smaller values at the wings:

$$\Gamma = r + \gamma(1 - r^2)\Phi(t) + i[\gamma(1 - r^2)]. \qquad (5.74)$$

The reflection is thus decreasing when Φ becomes negative in the wings of the pulse, which is the "coherent field subtraction" sketched in Fig. 5.19. The compression factor is determined by the ratio of $\gamma(1 - r^2)$ to r, which can be related to the ratio of energy in the auxiliary cavity to that in the main cavity [$\gamma(1 - r^2)$ is the maximum amount of energy that can be subtracted from the pulse in the main cavity at each round-trip].

This dynamic reflectivity can be adjusted for pulse shortening at each reflection, until a steady-state balance is achieved between the pulse shortening and pulse spreading due to bandwidth limitation and dispersion.

5.11 Femtosecond laser cavities

Resonators are an essential part of any laser. Therefore, most of this section has mainly a general tutorial purpose. We will review first the mode spec-

5.11. FEMTOSECOND LASER CAVITIES

trum of a laser cavity. "Mode-locked" operation requires a well defined mode structure. It is generally understood that the longitudinal modes are locked in phase. A transverse mode structure will generally contribute to amplitude noise (at frequencies corresponding to the differences between mode frequencies). Most fs lasers operate in a single TEM_{00} transverse mode. We will see however that some multiple transverse mode lasers have the same longitudinal mode structure as the fundamental TEM_{00}.

The next subsection reviews standard ABCD matrix calculations of the stability of laser resonators. Most fs laser cavities have at least one beam waist (for instance, one for the gain medium, and one for an eventual passive mode-locking element). Since mirrors are used as focusing elements, astigmatism complicates significantly the calculation and design (optical positioning of the elements) of the resonator.

5.11.1 Cavity modes

A beam with an electric field amplitude having a Gaussian radial dependence is uniquely defined by its complex $\tilde{q}$ parameter defined by Eq. (1.133). The phase variation on axis ($r = 0$) of the beam is determined by the phase angle $\theta(z) = \arctan z/\rho_0$ according to Eq. (1.130).

The parameter of the fundamental Gaussian beam that can reproduce itself in a cavity can be determined by the standard technique of ABCD matrices [14]. Let A, B, C, and D be the elements of the 2×2 matrix obtained by making the product of all ABCD matrices, starting from a point P, along a cavity round-trip. The complex $\tilde{q}$ parameter of the Gaussian beam at the point P is given by:

$$\frac{1}{\tilde{q}} = -\frac{A - D}{2B} - i\frac{\sqrt{1 - (\frac{A+D}{2})^2}}{B}. \tag{5.75}$$

Identifying with the definition of the $\tilde{q}$ parameter Eq. (1.133) leads to the beam characteristics at point P:

$$R = -\frac{2B}{A - D} \tag{5.76}$$

$$\frac{\pi w^2}{\lambda_\ell} = \frac{B}{\sqrt{1 - (\frac{A+D}{2})^2}}. \tag{5.77}$$

The "modes" of a cavity are determined by the condition that, after a round-trip from point P to point P, the total phase variation be a multiple of

2π. For the fundamental TEM$_{00}$ mode, and a simple cavity consisting of two concave mirrors at distance d_1 and d_2 from the beam waist, the phase variation for the half round-trip should be a multiple of π:

$$\begin{aligned} l\pi &= k(d_1 + d_2) + \theta(d_1) - \theta(d_2) \\ &= k(d_1 + d_2) + \Delta\theta \\ &= k(d_1 + d_2) + \arctan\left(\frac{d_1}{\rho_0}\right) + \arctan\left(\frac{d_2}{\rho_0}\right) \end{aligned} \quad (5.78)$$

The longitudinal mode spacing are determined by the condition $k(d_1 + d_2) + \Delta\theta = l\pi$, requiring an integer number of wavelengths in the round-trip cavity length $2(d_1 + d_2)$, or, in the frequency domain, a mode spacing of $\Delta\nu = c/[2(d_1 + d_2)]$, if $d_{1,2} \ll \rho_0$.

The Fourier transform of the output of a continuously mode-locked laser is a "comb" of frequency spikes that match the longitudinal mode structure represented by Eq. (5.78). Apertures are often inserted in the cavity of mode-locked lasers to avoid the complication introduced by a transverse mode-structure. In some lasers, a saturable absorber acts as an aperture. The small cross-section of the inverted region can also act as mode-limiting apertures (as is usually the case in argon laser pumped lasers).

The fundamental TEM$_{00}$ Gaussian beam is not necessarily the only existing mode in a fs laser. Let us consider the influence that the transverse mode structure can have on the operation of a mode-locked laser. If the laser is multimode, the electric field can be expressed as an expansion of Hermite–Gaussian modes (see, for instance, [14]):

$$E(x, y, z, t) = \sum_n \sum_m c_{nm} u_n(x) u_m(y) e^{i(\omega t - kz)} + c.c., \quad (5.79)$$

where

$$\begin{aligned} u_n(x) &= \left(\frac{2}{\pi}\right)^{\frac{1}{4}} \sqrt{\frac{\exp[i(2n+1)\theta(z)]}{2^n n! w(z)}} \\ &\quad \times H_n\left(\frac{\sqrt{2}x}{w(z)}\right) \exp\left\{-\left[i\frac{kx^2}{2R(z)} - \frac{x^2}{w^2(z)}\right]\right\}, \end{aligned} \quad (5.80)$$

where H_n are Hermite–Gaussian polynomials, and a similar expression for $u_m(y)$. m, n are the transverse mode indices.

5.11. FEMTOSECOND LASER CAVITIES

For the simple cavity consisting of two curved mirrors introduced above, the spacing between transverse modes is:

$$\Delta \nu = \frac{c}{2(d_1 + d_2)}(\Delta n + \Delta m)\left[\arctan\left(\frac{d_1}{\rho_0}\right) + \arctan\left(\frac{d_2}{\rho_0}\right)\right] \quad (5.81)$$

where Δn and Δm are the difference in transverse mode numbers. If the various transverse modes oscillate independently with random phase, the output of the laser will have a noise component corresponding to the beating of the various traverse modes. This noise component will be periodic if the mode spacing is equal in the two transverse directions. In some cases [3], the transverse modes can be locked in phase, resulting in a periodic spatial scanning of the beam. The noise contribution corresponding to the transverse mode beating will be low frequency if $d_i \leq \rho_0$.

For our simple model two-mirror cavity, the longest possible stable cavity is the concentric one for which the mirror radii $R_1 = d_1$ and $R_2 = d_2$. For that stability limit, $d_i \gg \rho_0$, and we can see from Eq. (5.81) that the transverse mode separation becomes equal to the longitudinal mode separation. As a result, the laser has the same well defined set of modes that can be locked in phase as a single transverse mode laser. Unfortunately, the concentric cavity is only marginally stable. It is however possible, by introducing one or more additional focusing elements in the cavity, to have an additional phase shift for transverse modes equal to a multiple of π. Any beam passing through a focus experiences a phase shift close to π (an effect known as "Guoy phase shift" [14]). A cavity such as sketched in Fig. 5.20 will have its transverse modes coincident with the longitudinal modes, while satisfying the stability criterion [177]. A multiple transverse mode operation has the advantage of a larger mode volume, hence much more efficient energy extraction from the gain medium than in TEM_{00} operation. The parameters of the cavity sketched in Fig. 5.20 correspond to the Ti:sapphire laser first reported to "self-mode-lock" by Spence *et al.* [246]. Because mode-locking was suppressed by forcing single mode operation with an aperture, it was suggested by Kafka [177] that multiple transverse mode operation may also play a role in the starting mechanism of the Ti:sapphire laser.

It is important to realize that the degeneracy of transverse modes necessary for this multimode mode-locking is only possible with cavities without astigmatism. In cavities folded with focusing mirrors, the transverse mode spacing is generally different in the sagittal and tangential planes. As shown in the following section, astigmatism compensation can easily be made in a Ti:sapphire laser cavity such as that of Fig. 5.20.

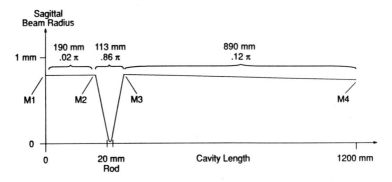

Figure 5.20: Beam size $w(z)$ in a linear, four-mirror cavity, computed for a laser with a 20 mm long Ti:sapphire rod between focusing mirrors M_2 and M_3. The phase shift $\theta(z)$ on axis between elements is indicated. Because of the Guoy phase shift of 0.86π between M_2 and M_4, a total phase shift of π/(half round-trip) can easily be achieved with that cavity (from [177]).

5.11.2 The geometrical cavity

Generalities

Calculation of the exact stability diagram, position and size of the beam waists, is a tedious but essential task in the design of fs lasers. Since it is not essentially different from the design of any laser, we will refer to the appropriate literature for details. A few general criteria to consider in the design of the laser cavity are:

- minimal losses

- a flexibility of varying the ratio of beam waists in various resonator elements in large proportions. The saturation level in the gain medium will be a factor in determining the intracavity power. In the case of Kerr lensing, it is important to identify the location where the beam size variation due to self-lensing will be maximum (positioning of an aperture). In the case of passive mode-locking, the ratio of beam waists in the absorber/amplifier is important, since this is the parameter that determines the relative saturation of the gain and absorber.

- in the case of cw Ar-laser pumped lasers, a small spot size in the amplifier is desirable for efficient pumping and heat removal.

5.11. FEMTOSECOND LASER CAVITIES

- a round spot is desirable in the passive mode-locking element (for instance, the amplifier rod in the case of a Ti:sapphire laser, or the saturable absorber jet in a dye laser) in order to have the most uniform possible wavefront across the beam, since it is a region of the cavity which contributes to the phase modulation

Mirrors or lenses can be used to create beam waists in a laser cavity. Because of the requirement of minimum losses, one will generally choose reflective optics over lenses.

Astigmatism and its compensation

It is not always an easy task to create a waist of minimal size with off-axis reflective optics. Indeed, let us consider the typical focusing geometry sketched in Fig. 5.21. The smaller the focal spot in A, the larger the diameter w of the incident beam on the mirrors, hence the larger the clearance angle θ required to have the focal point fall outside of the incident beam cross section. However, the astigmatism caused by a large angle of incidence θ will make it impossible to obtain the desired small focal spot with a cylindrically symmetric Gaussian beam incident from the left.

Since a tight focusing is required in the nonlinear elements of a mode-locked laser, there is clearly a need for minimizing or reducing the astig-

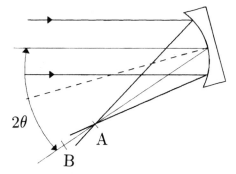

Figure 5.21: Off-axis focusing of a Gaussian beam leading to astigmatism. In the plane of the figure the focal distance of the mirror is $(R/2)\cos\theta$, where R is the radius of curvature of the mirror. The first focalization is therefore a line perpendicular to the plane of the figure originating from A. In the orthogonal plane the focal distance of the mirror is $(R/2)/\cos\theta$. There will therefore be a focal line in the plane of the figure at B.

matism. There are some exceptions. A large astigmatism may sometimes be desirable in the gain medium. This is the case when it is desirable to take maximum advantage of a self-lensing effect, as mentioned earlier in this chapter. Another example of such a need is to match the elongated shape of the gain profile of semiconductor lasers.

Let us choose as transverse coordinate y for the plane of incidence (the plane of the figure in Fig. 5.21), and x for the orthogonal direction. The two focal lines corresponding respectively to the plane of the figure and to the orthogonal plane are:

$$f_x = \frac{f}{\cos\theta}$$
$$f_y = f\cos\theta, \qquad (5.82)$$

where θ is the angle of incidence and $f = R/2$ is the focal length of the mirror.

Other elements in a cavity, such as Brewster plates, also have astigmatic properties that can limit the performance of the system. The gain medium of a Ti:sapphire laser or of a dye laser is generally a plane-parallel element put at Brewster angle. Kogelnik et al. [247] have shown under which condition the astigmatism can be compensated by such elements.

Let us consider the propagation of a Gaussian beam through a plate of thickness d put at Brewster angle directly next to the beam waist (size w_0) at $z = 0$. Upon entering the medium, the beam waist takes the values:

$$w_{0x} = w_0$$
$$w_{0y} = w_0 \frac{\cos\theta_r}{\cos\theta_B} = w_0 \frac{\sin\theta_B}{\cos\theta_B} = nw_0, \qquad (5.83)$$

where $\theta_B = \arctan n$ is the Brewster angle, and θ_r is the refracted angle. To a thickness d, there corresponds a propagation distance

$$\chi = \frac{d}{\cos\theta_r} = d\frac{\sqrt{1+n^2}}{n}. \qquad (5.84)$$

Applying the propagation law (1.129) for the beam waist across the thickness d yields:

$$w_x = w_0\sqrt{1 + \left(\frac{\lambda_\ell \chi}{n\pi w_0^2}\right)^2} = w_0\sqrt{1 + \left(\frac{\lambda_\ell}{\pi w_0^2}\frac{d\sqrt{1+n^2}}{n^2}\right)^2} \qquad (5.85)$$

$$w_y = nw_0\sqrt{1 + \left(\frac{\lambda_\ell \chi}{n^3\pi w_0^2}\right)^2} = nw_0\sqrt{1 + \left(\frac{\lambda_\ell}{\pi w_0^2}\frac{d\sqrt{1+n^2}}{n^4}\right)^2}. \qquad (5.86)$$

5.11. FEMTOSECOND LASER CAVITIES

Equations (5.85) and (5.86) show that the effective propagation distances for a beam propagating through a Brewster parallel plate are:

$$d_x = d\frac{\sqrt{1+n^2}}{n^2} \tag{5.87}$$

$$d_y = d\frac{\sqrt{1+n^2}}{n^4} \tag{5.88}$$

The beam issued from the waist w_0 will be again collimated by a mirror of radius R at an angle θ if the difference between the two distances d_x and d_y compensates the difference in focal distances f_x and f_y. The condition is:

$$f\left(\frac{1}{\cos\theta} - \cos\theta\right) = d\frac{\sqrt{1+n^2}}{n^2}\left(1 - \frac{1}{n^2}\right) \tag{5.89}$$

In the case of a typical Ti:sapphire laser, the crystal is inserted near the focal point between two curved mirrors. Expressing that the astigmatism of the two curved mirrors has to be compensated by the crystal of total thickness d leads to the following condition for a "Z" configuration:[3]

$$\frac{2d}{R}\frac{\sqrt{n^4-1}\sqrt{n^2-1}}{n^4} = \frac{\sin^2\theta}{\cos\theta}. \tag{5.90}$$

For a 9 mm long Ti:sapphire crystal inserted between two mirrors of 10 cm curvature, Eq. (5.90) indicates astigmatism compensation at an angle of 9.5°.

In the case of dye lasers, we find that for a typical jet thickness of 200 μm and a mirror curvature of 5 cm, the compensated angle is less than 2.5°. Compensation is impossible for tight focusing in a saturable absorber jet of typical thickness of 50 μm. In situations where there are two or more waists in the cavity, it is possible to find conditions where the astigmatism in one part of the cavity compensates for the astigmatism in another part. Calculations show that large angles can actually result in astigmatism compensation, and, even with angles of incidence on the focusing mirrors exceeding 10° , large stability ranges have been found [168]. Resonators for dye lasers have been designed to provide a round and minimum size spot in the absorber jet, using the cavity geometry (relative location of the components) to compensate the astigmatism, without a need for inserting additional "compensating

---

[3]The beam line — input beam, beam reflected at mirror 1, beam reflected at mirror 2, output beam — describes a "Z".

elements." Returning to Fig. 5.21, the beam incident from the left will focus first on a line originating from A, perpendicular to the plane of the figure, next in a line at B in the plane of the figure. However, if the beam incident from the left is collimated in the plane of the figure, but convergent in the orthogonal direction, the focal line B will recede towards the focus A. The incident beam parameters can be adjusted such as to create a tight round focal spot.

With the availability of symbolic manipulation languages, multiplication of all the matrices representing elements and propagation in free space in the cavity can be performed analytically. The result will be an ABCD matrix for the plane of the cavity, and another for the orthogonal plane.

In order to assess the importance of astigmatism, let us consider a simple ring cavity with two beam waists. We assume that one beam waist is formed by two lenses of 15 mm focal distance, spaced by 30 mm. The other waist is formed by two mirrors of focal distance $f = 25$ mm separated by a distance d. The distances between the two waists are 1 m and 3 m in a 4 m perimeter ring cavity. From the expression for the ABCD matrix for this cavity one finds that the stability range is $f < d < 1.01f$. If astigmatism due to an angle of incidence θ on the two mirrors of $f = 25$ mm is taken into account, there will be a different stability condition corresponding to each of the two focal distances $f_x = f/\cos\theta$ and $f_y = f\cos\theta$. The cavity is stable if the two stability ranges overlap. For an angle $\theta \geq 5.7^0$, this cavity is no longer stable. Other degrees of freedom, and the astigmatism of other portions of the cavity, have to be taken into account to find stability regions.

Linear cavity

As an example, let us consider the basic linear cavity terminated by an antiresonant ring, as sketched in the lower part of Fig. 5.22. The detailed expression for the ABCD matrix is given in Appendix B. For each value of the angle of incidence θ on the folding mirror, and distance d between mirrors of the amplifier section, the stability limits z_{min} and z_{max} are plotted in Fig. 5.22. Only solutions of Eq. (B.1) for which the spot size on any folding mirror does not exceed 3 mm are considered as stable. The folding angle at the antiresonant ring is taken to be 8^0. The radii of curvature of the mirrors at the amplifier and absorber sections are respectively 5 and 3 cm.

A cross section of the three dimensional stability diagram corresponding to the angle $\theta = 16^o$ is shown in Fig. 5.23. The spot sizes in the absorber and amplifier are indicated. The minimum beam waist in the absorber cor-

5.11. FEMTOSECOND LASER CAVITIES

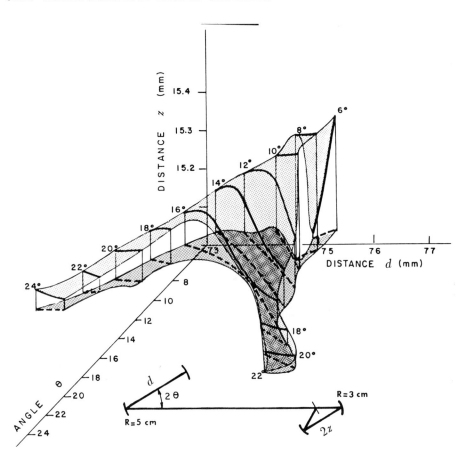

Figure 5.22: Three-dimensional stability diagram for the antiresonant ring linear cavity with mirrors of curvature 5 cm at the amplifier, and 3 cm at the absorber jet. The cavity is stable in the region between the upper and lower surface. The angle of incidence at the folding mirror of the antiresonant ring is 8°. The distance between absorber and gain is 1 m (from ref. [168]).

responds to a round spot with $w_H = w_V = w_a = 1.2$ μm (at the wavelength of 620 nm). The inter-mirror spacings are $d = 75.8$ mm and $2z = 30.06$ mm.

Detailed calculations of the stability range and the beam parameters have been published for two sets of focusing optics ($R_1 = R_2 = 7.5$ cm and 5 cm; and $R_3 = R_4 = 5$ cm and 3 cm) [168]. In general, the stability diagram (plots of the stable regions in the θ, d, z space) shows forbidden ranges between stable domains at large angles. The practical consequence is that

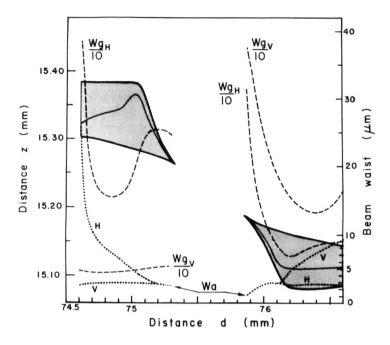

Figure 5.23: Cross section through the stability diagram for the antiresonant ring linear laser, for $\theta = 16°$. The stability range is indicated in the shaded region. The subscripts H and V refer to the calculations in the plane of the figure (xz) and the orthogonal plane (yz), respectively. The radii of the beam waist at the absorber w_a, for which the absolute value of the difference $w_{aH} - w_{aV}$ is minimized (and even zero for the larger values of d), are plotted as a function of the distance d between mirrors of the amplifier section. The corresponding radii of the beam waists in the amplifier section are also shown (dashed lines). The values of the distances d and z corresponding to the plotted spot sizes are given by the thin solid line in the stability diagram (shaded region) (from ref. [149]).

the laser cannot be tuned continuously from one stable region to the other (for instance by increasing the distance d and optimizing simultaneously the other parameters).

An important goal of these cavity calculations is to determine the ratio of focal spots at two locations of the cavity. We have seen, for instance that the ratio of focal spots in the amplifier to the absorber sections of a passively mode-locked laser with a saturable absorber is an essential parameter in the operation of the laser. The example of Fig. 5.23 illustrates particularly well this point, since the ratio of the focal spots of the amplifier to the absorber can exceed 50,000. The beam waist in the absorber being a monotonous function of the mirror spacing d, it is possible to adjust this critical ratio as required for a particular combination of amplifier and absorber dyes.

Complete calculation of the cavity parameters would be only an academic exercise, if it were not possible to accurately align the cavity in predetermined geometry. A systematic procedure to pre-align a cavity with the parameters optimized by the computer calculations is given in Appendix B. It should be noted that the calculations of the cavity of Fig. B.1 can be applied to other linear cavities. For instance, the antiresonant ring section can be replaced by two curved mirrors, or a curved mirror and a flat mirror. The results of calculations applied to more complex linear cavities commonly used for hybrid mode-locking and to ring lasers are also given in Appendix B.

5.12 Generation of synchronized wavelengths

5.12.1 Nonlinear mixing

Parametric sources

The active element is here a nonlinear crystal oriented for phase matched down frequency conversion of a pump pulse. Optical parametric oscillators (OPO's) offer the possibility to generate synchronous multiple wavelength fs pulses. The basic principles of this second order nonlinear process were explained in Chapter 3.

We shall organize this section according to the mode of pumping. The various options leading to tunable fs pulse generation are:

1. direct pumping of a nonlinear crystal with a fs pulse laser

2. insertion of a nonlinear crystal in the cavity of a fs pump laser

3. synchronous pumping of a crystal inserted in a cavity resonant for at least one of the generated wavelengths (singly resonant cavity).

Traveling wave optical parametric generators (OPG)

In this approach, the radiation has to start from noise (a condition often labelled superfluorescence), and, consequently, a very high gain has to be achieved. Direct pumping has been used to generate picosecond pulses from OPG's pumped by temporally compressed pulses from Nd:YAG lasers [248,

249]. The obvious disadvantage of direct pumping a crystal is that it requires a pump power close to the damage threshold to achieve superfluorescence.

The main advantage of a parametric source is its broad tuning range. The signal wavelength can be determined either by angle tuning, temperature tuning, or pump wavelength tuning. While fixed frequency pump sources are the most common, it can be advantageous to use a tunable pump laser to maintain the condition of 90^0 phase matching [10, 65] throughout the tuning range. A source tunable in the 1 μm to 3 μm range was demonstrated, using KTiOPO$_4$ (KTP) as nonlinear crystal, pumped by a tunable Ti:sapphire laser. Within the constraint of 90^0 phase matching in KTP, the signal wavelength is related to the pump wavelength λ_p (in μm) by [250]:

$$\lambda_s = -1.5646\lambda_p^4 + 3.771\lambda_p^3 - 1.2079\lambda_p^2 - 0.834\lambda_p + 1.2963.$$

The crystal KTiOAsO$_4$ (KTA), an isomorph of KTP, has an even broader tuning range, because its transparency in the IR extends to 5 μm. In:KTA is a KTA crystal doped with a trivalent oxide of In ($\approx$ 0.2 % in weight) used to improve the formation of large single crystals. The tuning curve for In:KTA pumped at 850 nm by a Ti:sapphire laser is shown in Fig. 5.24. Details on the properties of this crystal as an OPO can be found in Ref. [251].

Intracavity pumped OPO

The nonlinear crystal in a singly resonant cavity is inserted in the cavity of a continuously pumped fs dye laser [253], or a Ti:sapphire laser [254, 255]. This parametric oscillator is essentially the same as the conventional synchronously pumped parametric oscillator described in the following subsection. The main difference is that, in order to minimize the perturbation of the pump laser, the OPO cannot be used in the regime of large depletion (which will be referred to as "giant pulse operation") Such a design [256, 255, 252] offers fs pulses at a very high repetition rate (typically 80 MHZ), but low energy/pulse (at most 1 nJ). A typical cavity is shown in Fig. 5.25. The pump laser shown here is a ring dye laser, with four dispersion compensating prisms, a gain section G, and an absorber section A. Any cw modelocked fs laser can be used. The nonlinear crystal (KTP in this particular example) is located at a beam waist of both the pump cavity, and a cavity resonant for the signal or idler. A piezoelectric element allows for fine adjustment of the length of the latter cavity, to achieve exact synchronism between pump and signal (or idler). A short interaction length — which implies generally tight focusing — is used to prevent pulse broadening due to a too narrow

5.12. GENERATION OF SYNCHRONIZED WAVELENGTHS

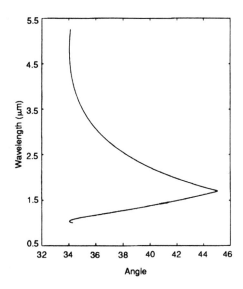

Figure 5.24: Tuning curve for an In:KTA crystal used as OPO with a 850 nm Ti:sapphire pump laser and a non-collinear angle between the pump and signal waves of 2.8° (from [252]).

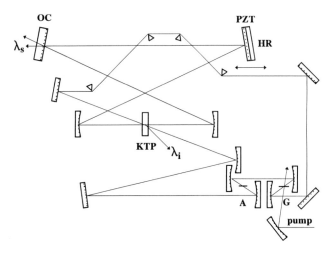

Figure 5.25: Intracavity optical parametric oscillator (adapted from [87]). A piezoelectric transducer is used to match the cavity length of the OPO to the length of the laser cavity.

phase matching bandwidth. Because of the high repetition rate, the material has to withstand a local thermal loading (high average power density) in addition to having a high peak field damage threshold.

Synchronously pumped parametric oscillators

The crystal is inserted in a resonator for one (or both) of the generated wavelengths of the signal, λ_s, and idler, λ_i. It is pumped by a pulse train whose repetition frequency matches those of the OPO cavitiy. The oscillation has a longer effective gain pathlength to emerge from noise — the crystal length times the effective number of round trips. The cavity for a synchronously pumped optical parametric oscillator can be singly or doubly resonant. In the latter case, both the idler and signal beams are part of a resonant interferometer. A doubly resonant configuration is not practical for fs pulse generation. Indeed, because of the different dispersion in nonlinear crystals for signal and idler, the OPO output spectrum has cluster structure. The output of a doubly resonant fs OPO is unstable, because the phase relationship between the three waves (pump, idler and signal) cannot be maintained over the spectrum of a fs pulse. Generally, only the signal or the idler is incorporated in a resonant cavity.

In a linear cavity, the signal pulse is attenuated at each return passage through the crystal. This passage through an unpumped crystal can be avoided either through "double pumping the crystal" [86] (Fig. 5.26), or

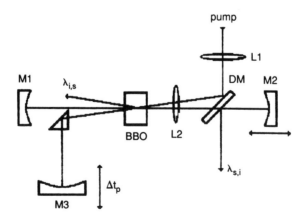

Figure 5.26: Sketch of a synchronously pumped linear OPO cavity (mirror M_1 and M_2) with double pumping (adapted from [86]).

5.12. GENERATION OF SYNCHRONIZED WAVELENGTHS

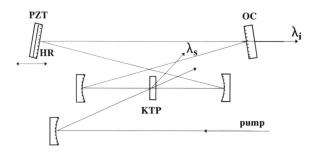

Figure 5.27: Singly resonant (for the idler wave) ring cavity of an OPO (adapted from [257]).

using a ring configuration (Fig. 5.27), as used in [257]. In the latter work, a KTP OPO in a ring cavity was pumped externally by a continuously mode-locked Ti:sapphire laser. The OPO signal was a continuous beam of 185 mW average power (pump power 800 mW) of 62 fs pulses at 1.3 μm (repetition rate 76 MHz). The signal pulses were not transform limited. Amplification of the chirp of the pump pulse — as noted in Section 3 — could be one factor contributing to the large bandwidth of the output pulses. As noted by the authors [257], self phase modulation could also be an important factor, in view of the very large peak intensity of the pulses in the KTP crystal (> 40 GW/cm^2).

Compression mechanisms The parametric gain being practically instantaneous on the time scale of the pump pulse, the dynamics of the synchronously pumped OPO are totally different from that of population inversion lasers. For instance, the depletion of the inversion that plays pivotal role in the compression of synchronously pumped dye laser pulses does not exist in parametric oscillators. There is no "saturable loss" as with saturable absorbers. Because of the nonlinearity of the conversion process, one expects the generation of signal or idler pulses only slightly shorter than the pump pulse duration. For large conversion efficiencies however, it is possible to find conditions in which re-conversion of the pump radiation at the end of the idler leads to large compression of the generated wave, as shown in the next paragraph.

Compression in "nonlinear" OPO Let us consider the OPO ring cavity sketched in Fig. 5.27 in which the idler is resonant. In the analysis of such a parametric oscillator, one has to take into consideration the relative group velocities of pump, idler and signal, as well as the cavity length detuning. The latter determines the time of arrival of the idler with respect to the pump at each round-trip. After a sufficient number of round-trips, the idler has reached a power level such that, during the passage of the idler through the crystal, the sign of the energy exchanges between pump and idler reverses. This regime is sometimes referred to as "nonlinear" parametric oscillation, sometimes as "giant pulse operation" [258, 259]. A steady state situation under which up to 20-fold pulse compression can be achieved is sketched in Fig. 5.28. With a proper cavity length mismatch (lengthening), the idler pulse reaches the crystal at the trailing edge of the pump pulse. The faster idler traverses through the pump pulse from its trailing to leading edge as it propagates through the crystal. The main compression mechanism starts once a sufficiently high intensity has been reached at the peak of the idler. Depletion of the pump causes the latter to have a steeper trailing edge (Fig. 5.28).

It should be noted that this mode of pulse compression requires an accurate control of pump pulse duration and intensity. Hence the importance of passive negative feedback, which has made possible the practical implementation of this technique [260, 261]. In a singly resonant cavity, such as used in [257] (see Fig. 5.27), a KTP OPO is synchronously pumped by 10 ps pulses from a frequency doubled Nd:YAG laser stabilized by passive negative feedback. Figure 5.29 shows the autocorrelations of signal and idler pulses. The 0.6 ps autocorrelation width corresponds to a (sech^2) pulse duration of 390 fs. The autocorrelation of the signal (Fig. 5.29(b)) has the triangular shape typical of a square pulse. This is to be expected, since the idler can be

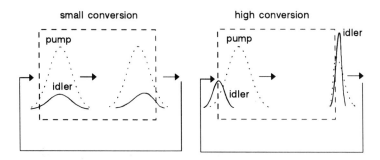

Figure 5.28: Successive relative positions of the pump, idler and crystal as they propagate through the crystal, in the condition of "giant pulse" compression.

5.12. GENERATION OF SYNCHRONIZED WAVELENGTHS

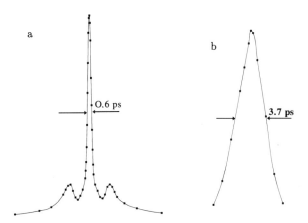

Figure 5.29: Autocorrelation of the output of a synchronously pumped OPO. The cavity length is longer by 0.105 mm than the length corresponding to maximum output. The pump pulse energy is 2.5 times above threshold. The idler wavelength is 1.28 μm. The resonant idler (a) has an autocorrelation width of 0.6 ps, which corresponds to a (sech^2) pulse duration of 390 fs. The signal (b) has a triangular autocorrelation typical of a square pulse.

approximated by a δ-function as compared to the other waves. The signal is continuously generated for the transit time of the idler through the crystal. The duration τ_s of the signal pulse depends on the crystal length l and the group velocity mismatch between idler and signal pulses $\tau_s = l[1/v_i - 1/v_s]$ where $v_{i,s}$ are idler and signal pulse velocities and range from 5 to 5.6 ps in this particular example.

The mechanism outlined above is independent of phase modulation and dispersion considerations that could also be exploited for further compression. Indeed, as noted in Chapter 3, a slightly chirped pump pulse may, under certain conditions, generate an idler and signal with larger and opposite chirp. The strongly chirped signal or idler can easily be compressed by dispersive elements. Using the natural chirp of frequency doubled pulses from a Nd:glass laser, pulses of 50 fs at 920 nm were obtained by this technique [88].

Harmonic generation

High conversion efficiency in harmonic generation is mostly associated with a narrow bandwidth. For instance, efficient second harmonic generation requires that the fundamental and the second harmonic fields remain in phase as they propagate through the length of the crystal. This condition of index matching or "phase matching" can be realized at a particular wavelength. For a given crystal length, the phase matching bandwidth is the range of fundamental frequencies over which the fundamental and second harmonic remain in phase. The second harmonic pulse will generally be broader than the fundamental pulse if the latter is shorter than the inverse of the phase matching bandwidth. There is an important exception to this rule, which involves second harmonic generation type II, where the fundamental radiation is split in two pulses orthogonally polarized as ordinary (o) and extraordinary (e) waves. For particular group velocities v_e and v_o for the extraordinary and ordinary fundamental waves, the second harmonic can be much shorter than the inverse of the phase matching bandwidth. The excess bandwidth is provided by amplitude modulation resulting from successive energy transfers between the fundamental and second harmonic. A theoretical investigation of this mode of compression has been made by Wang and Dragila [70] and Stabinis and Valiulis [81]. Experiments have confirmed a four-fold compression of pulses of 1 ps duration (or less) [262, 263]. More recently, up to 30- fold compression of 11 ps pulses from a Nd:YAG laser stabilized by passive negative feedback was demonstrated [82]. Frequency doubling crystals for type II SHG are used, in which two fundamental pulses of orthogonal polarization are launched with a well defined relative delay. In a frame of reference moving with the group velocity of the second harmonic v_{g2}, these two fundamental pulses at v_e and v_o move in opposite direction with nearly equal velocities.

Energy conversion to the second harmonic occurs in the time interval corresponding to the pulse overlap. Some re-conversion at the peak of the second harmonic occurs, resulting in compression of the harmonic, and transfer of energy to the fastest component of the fundamental. The theory [82] predicts a 40-fold compression in a 4 cm long KDP crystal at a pump intensity of 4.6 GW/cm^2, and a 50-fold compression in a 5 cm long deuterated KDP crystal (designated either as DKDP or KD*P), at a pump intensity of 2.4 GW/cm^2. The optimum pre-delay of the e-polarized fundamental pulse is between 7 and 8 ps.

The most important parameters in selecting an appropriate crystal are the group lengths in the frame of reference moving with the second harmonic: $L_e = \tau_p[1/v_e - 1/v_2]$ and $L_o = \tau_p[1/v_o - 1/v_2]$ where τ_p is the pump pulse duration. Other important parameters are the walk-off between the three waves, and the damage threshold. The theory outlined in Ref. [81]

5.12. GENERATION OF SYNCHRONIZED WAVELENGTHS

shows that the maximum compression occurs for a specific ratio between L_o and L_e, $m = L_o/L_e = -1$. A deviation from that ideal value leads to a decrease in compression ratio and a decrease in conversion efficiency through re-conversion to the fundamental. However, both the energy conversion and compression ratio can be optimized by adjusting the ratio of the fundamental intensities to $I_o/I_e \approx m$. The values are $m = 1.32$ and $m = 1.37$ for KDP and DKDP, respectively.

A typical experimental set-up for second harmonic generation and compression through this method is shown in Fig. 5.30. The expanded beam from a Nd:YAG laser (pulse duration 12 ps) amplified to 20 mJ is split into an e and o wave by a polarizing beam splitter. The intensity ratio of the o to e beams is controlled by a $\lambda/2$ plate after the last amplifier stage (not shown). The two waves are recombined at a second polarizing beam splitter after having been given a relative delay τ_d. The duration (FWHM of the autocorrelation in ps) and energy of the second harmonic pulse are plotted in Fig. 5.31 as a function of delay. The peak conversion efficiency of 50% is achieved by delaying the e polarized fundamental by approximately 2 ps. The dotted line indicates the autocorrelation width on beam center rather than averaged over the beam cross section (solid curve). A minimum pulse duration on beam center of 500 fs is measured for a 9 ps delay of the e wave.

Since the beam intensity should exceed 3 GW/cm^2 over a cross section of at least 1 cm, this type of compression is particularly attractive for doubling large energy (over 50 mJ/pulse) Nd:YAG ps laser sources. The fundamental laser pulse should also be free of any phase modulation. Because of the sensitivity of compression to the pulse intensity, accurate control and reproducibility of the duration and profile of the fundamental pulses are required.

5.12.2 Other techniques

There are other techniques to achieve multiple-wavelength femtosecond laser emission. One possibility is to cascade several lasers synchronously pumping each other [264], a complex setup which has low efficiency. A recent example of cascade operation involves the Ti:sapphire laser synchronously pumping a dye laser [265].

Multiple wavelength operation is easily achievable within a laser cavity, provided the bandwidth of the gain medium is sufficiently large. This is in particular the case of the Ti:sapphire laser, for which various configurations for two-color fs operation have been demonstrated. Simultaneous operation

290   CHAPTER 5.   ULTRASHORT SOURCES

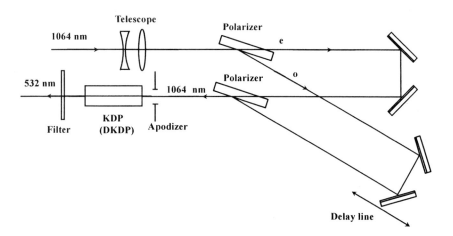

Figure 5.30: Typical experimental set-up for second harmonic generation and compression through group velocity mismatch in type II second harmonic generation.

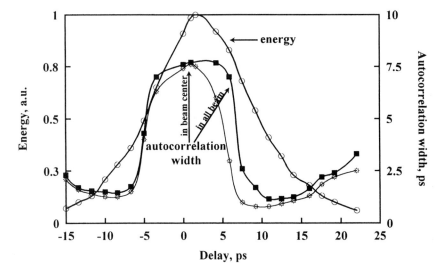

Figure 5.31: Energy (left scale) and autocorrelation width (right scale) on beam center (thin line) or averaged over the beam cross section (solid line) as a function of delay of the e wave.

of two *synchronized* pulse trains is generally desired. One synchronization mechanism exploited by Dykaar and Darack [266] is cross-phase modulation via the Kerr effect, in a region of the gain medium where the pulses are spatially and temporally close enough to interact. The pump beam (an Ar^+ laser) is split to excite two adjacent volumes of the Ti:sapphire crystal. These two adjacent gain volumes are associated with two cavities tuned to different wavelengths. Up to the largest wavelength offset observed of about 20 nm, the two pulse trains appear perfectly synchronous, as indicated by a cross-correlation width of 70 fs.

Another technique is to use the traditional prism sequence needed to adjust the group velocity dispersion as a spectral device to select two spectral regions in the cavity. Barros and Becker [267] inserted a double slit between the prism pair and the end cavity mirror, to select two center frequencies. Dual wavelength spacing as large as 60–75nm with less than 20 fs time jitter was achieved. A third technique applies to even larger wavelengths spacing. Evans *et al.* [268] used two prism pairs (sharing of the first prism) and two output couplers. They obtained a wavelength separation as large as 90 nm and less than 70 fs time jitter between two pulse sequences.

As noted by Kafka *et al.* [269] and Luo *et al.* [270], dual wavelength pulse trains do not need to be synchronized to serve a useful purpose.

5.13 Miniature dye lasers

It seems somewhat paradoxal to be using large (of the order of meters) resonators to generate pulses of spatial extension in the micron range. The narrow bandwidth associated with each longitudinal mode of the long cavity contributes to the amplitude stability of the pulse train, while the fixed mode spacing ensures that the pulses are generated at a regular time interval. The large resonator serves an essential purpose when a sequence of pulses — rather than a single pulse — is needed. Emission of a short pulse by the long resonator laser requires — as we have seen at the beginning of this chapter — a coherent superposition of these modes with fixed phase relation. If, however, only a single pulse is needed, there is no need for more than one longitudinal mode within the gain profile. Ultrashort pulses are generated in small cavity lasers through resonator Q-switching and/or gain switching. Aside from gain bandwidth limitations, the pulse duration is set by the spectral width of the longitudinal mode, and hence the resonator lifetime. The latter in turn is limited by the resonator round trip time $2L/c$. Ideally, the laser cavity should have a free spectral range exceeding the gain bandwidth.

Two methods of short pulse generation use either ultrashort cavities (Fabry–Perot dye cells of thickness in the micron range) or no traditional cavity at all (distributed feedback lasers). Another type of miniature laser is the integrated circuit semiconductor laser, which will be described in Section 5.14.

5.13.1 Distributed feedback dye lasers

The fs distributed feedback laser is an example of combined gain and cavity Q-switching. Gain switching is achieved by pumping with a ps pulse. Q-switching is accomplished by creating a transient resonator.

In a standard Fabry–Perot resonator, the counterpropagating beams create a standing wave pattern in the gain medium. The gain is only extracted at the antinodes of this pattern. Rather than having the feedback from the emission into the gain with localized mirrors, in the *distributed feedback* laser, the inversion is only *produced* at the antinodes of a standing wave pattern. The gain pattern is produced by interfering two pump beams in the gain medium, hence creating a periodic modulation of the gain and refractive index. The feedback is thus distributed and time dependent. The central wavelength of the emission spectrum is given by $2n\lambda_a$, where λ_a is the period of the population grating. This follows from the Bragg condition which must be satisfied for positive feedback. The wavelength selectivity $\Delta\lambda/\lambda$ is simply the inverse of the number of gain lines in the pattern. If 1 mm wide pump beams are used to make a pattern for 600 nm, $\Delta\lambda/\lambda \approx 1/1700 = \Delta\nu/\nu$. The bandwidth of this grating is in the THz range, broad enough to support the generation of subpicosecond pulses.

It may seem at first that a perfectly coherent source is required to pump such a laser. In principle, even white light can be used, if the grating is formed by imaging a holographic grating into the gain medium with the pump radiation. As shown in Fig. 5.32, in a perfectly symmetric configuration, the pump beam diffracted into the +1 order is made to interfere with the −1 order in a dye cell [271, 272]. A detailed background review of various experimental arrangements for picosecond distributed feedback lasers can be found in a publication by Bor and Müller [273].

As a result of the excitation by a single pump pulse, a series of pulses within the pump pulse duration can be generated — each pulse being terminated by gain depletion. There is, however, a range of pump pulse power and duration leading to the generation of only one single pulse, with a duration which can be much shorter than the pump pulse. Quenching of the emission

5.13. MINIATURE DYE LASERS

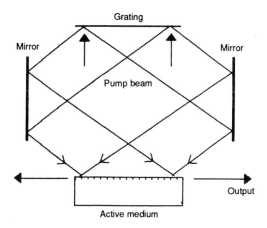

Figure 5.32: Creating an inversion grating with a source of low coherence, using a holographic grating (from [272]).

can eventually be achieved by an additional laser pulse sent through the pumped region, which terminates the gain [274]. For fs pulse generation, it is essential to create a short pumped region. In the configuration of Szatmari and Schäfer [275], for example, the pump pulse images a holographic transmission grating through a microscope objective into the active medium.

5.13.2 Short cavity lasers

Another approach to single short pulse generation is the short cavity laser with a low cavity Q. In a typical "short cavity" laser [276], the wavelength is tuned by adjusting the thickness of the dye cell in a 3 to 5 μm range with a transducer bending slightly the back mirror of the cavity. With a round-trip time of the order of only 10 fs, it is obvious that the pulse duration will not be longer than that of a ps pump pulse. As with the distributed feedback laser, the dynamics of pump depletion can result in pulses considerably shorter than the pump pulses. The basic operational principles of this laser can be found in [277]. Technical details are given in [276].

A cascade of short cavity dye lasers can be used to convert ns pump pulses (Nd:YAG, excimer lasers) to sub-picosecond dye laser pulses, as demonstrated by Hung and Meyer [278]. Starting with 40 mJ, 6 ns pulses from a Q-switched Nd:YAG laser, in a first step, they produced 500 fs, 200 μJ dye laser pulses. In a second step they generated a white light continuum from these dye laser pulses and amplified a selected spectral region. After compression, they obtained pulses of less than 100 fs duration, 50 µJ energy, tunable between 540 and 650 nm.

Using an excimer laser, Szatmari and Schäfer [275] produced 500 fs pulses, tunable from 400 to 760 nm, in a cascade of distributed feedback and short cavity dye lasers. After self-phase modulation and recompression, pulses as short as 30 fs in a spectral range from 425 to 650 nm were obtained [279]. Frequency doubling and amplification in another excimer gain module allowed the generation of subpicosecond UV pulses.

These miniature devices are becoming an economical alternative to much larger, and more expensive oscillator-amplifiers systems.

5.14 Semiconductor lasers

Semiconductor lasers are the next obvious candidate for fs pulse generation, because of their large bandwidth. A lower limit estimate for the bandwidth of a diode laser is $k_B T$ (where k_B is the Boltzmann constant and T the temperature in °K), which at room temperature is $(1/40)$ eV, corresponding to a 15 nm bandwidth at 850 nm, or a minimum pulse duration of 50 fs. The main advantage of semiconductor lasers is that they can be directly electrically pumped. In the conventional diode laser, the gain medium is a narrow inverted region of a p–n junction. We refer to a publication of Vasil'ev [280] for a detailed tutorial review on short pulse generation with diode lasers. We will mainly concentrate here on problems associated with fs pulse generation in external and internal cavity (integrated) semiconductor lasers. The main technical challenges associated with laser diodes result from the small cross-section of the active region (typically 1µm by tens of µm), the large index of refraction of the material ($2.5 < n < 3.5$, typically) and the large nonlinearities of semiconductors.

The cleaved facets of a laser diode form a Fabry–Perot resonator with a mode spacing of the order of 1.5 THz. Two options are thus conceivable for the development of fs lasers: integrate the diode with a waveguide in the semiconductor, to construct fs lasers of THz repetition rates, or attempt to "neutralize" the Fabry–Perot of the chip, and couple the gain medium to an external cavity. We will consider first the latter approach.

5.14.1 External cavity semiconductor laser

Because of the high refractive index of the semiconductor, it is difficult to eliminate the Fabry–Perot resonances of the short resonator made by the

5.14. SEMICONDUCTOR LASERS

cleaved facets of the crystal. Antireflection coatings have to be of exceptionally high quality. Even though reflectivities as low as 10^{-4} can be achieved, a good quality antireflection coating remains a technical challenge. A solution to this problem is the angled-stripe semiconductor laser [281], which has the gain channel making an angle of typically 5° with the normal to the facets (Fig. 5.33). Because of that angle, the Fabry–Perot resonance of the crystal can easily be decoupled from that of the external cavity. The laser can be operated in an external cavity with a standard quality antireflection coating.

Femtosecond pulse operation in a semiconductor laser with an external cavity is similar to that of a dye laser. Best results so far were obtained in hybrid operation, using radio-frequency current modulation for gain modulation (synchronous pumping), and a saturable absorber. The low intracavity power of the external cavity semiconductor laser — as compared to the dye laser — makes the use of conventional saturable absorbers (i.e., dyes, bulk semiconductors) impractical. It has been necessary to develop absorbing structures with a very low saturation energy density. These are the multiple quantum well (MQW) absorbers, which are analyzed in the subsection that follows. The laser diode is modulated at the cavity frequency (0.5 W RF power applied via a bias tee [282]). Modulation of the index of refraction is associated with the gain depletion and the saturation of the MQW. Since the gain depletion results in an increase of the index, a negative dispersion line appears appropriate. At present, however, bandwidth limited operation has not been achieved directly from a mode-locked semiconductor laser.

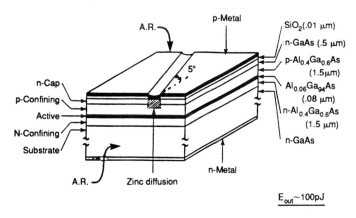

Figure 5.33: Structure of an angled stripe semiconductor laser (from [281]).

An external dispersion line with gratings resulted in pulse durations of 200 fs [283].

The exact phase modulation mechanism of this laser is complex. The index of refraction of the diode is a function of temperature and free carrier density, itself a function of current, bias light intensity, etc. As with other high-gain solid state lasers, changes in the pulse parameters can be as large as 50% from one element to the next [282].

Multiple quantum well absorber

The very high density of states of a dye solution or of a band–band transition of a semiconductor results in a high saturation intensity (or energy density). The absorption often used in multiple quantum wells is the room temperature exciton resonance near the band edge. This excitonic absorption saturates at 10 times lower intensities than the GaAs band-to-band transition. The reduction in saturation for GaAs/GaAlAs MQW samples, as compared to GaAs, has been measured with cw lasers focused onto a 5 μm spot size [160].

The lower absorption saturation intensity in MQW is attributed to screening of optically excited carriers [160]. The absorption recovers at the carrier recombination rate of $T_1 \approx 30$ ns, which is too long for high repetition rate GaAs lasers. Several methods are available to reduce the effective absorption recovery rate:

1. tight focusing,

2. proton bombardment, and

3. low-temperature growth.

With tight focusing, the carriers diffuse laterally out of the interaction region faster than they recombine. The effective relaxation time of the nonlinear absorption τ_A is:

$$\frac{1}{\tau_A} = \frac{1}{T_1} + \frac{1}{\tau_d}, \qquad (5.91)$$

where T_1 is the carrier recombination time, and τ_d the diffusion time. A simple two dimensional diffusion equation yields the value for the diffusion constant:

$$\tau_d = \frac{w_0^2}{8D}, \qquad (5.92)$$

5.14. SEMICONDUCTOR LASERS

where w_0 is the Gaussian beam waist, and $D = 13$ cm^2/s is the diffusion constant in the plane of the MQW layers [284]. A 2 μm spot size corresponds to a diffusion time $\tau_d \approx 400$ ps.

A reduction of the carrier lifetime can be achieved by creating additional relaxation channels through the insertion of defects. The most commonly used method is proton bombardment. To ensure long-term stability, the proton bombardment is generally followed by a gentle annealing (10 min at 300°C). Shallow traps may otherwise anneal spontaneously, causing increases in absorption recovery time. For a typical MQW sample consisting of 80 periods of 102 Å GaAs and 101 Å Ga$_{0.71}$Al$_{0.29}$As, bombardment with 200 keV protons at a level of 10^{12}/cm^2 and 10^{13}/cm^2 resulted in recovery times of 560 ps and 150 ps, respectively [160]. Structures with shorter wells (70 to 80 Å) separated by 100 Å barriers yield broader absorption bands [282], with the same recovery time of 150 ps after a 10^{13}/cm^2 proton bombardment and annealing.

Another technique to introduce defects is to grow the semiconductor at very low substrate temperature. The experimental data of Gupta *et al.* [285] for As-based III–V semiconductors indicate that molecular beam epitaxial (MBE) growth at very low substrate temperatures leads to an excess As incorporation in the epilayers, resulting in a large density of deep-level defects. Therefore, the carrier lifetime is dramatically reduced by the trapping and subsequent recombination of carriers at these centers. Figure 5.34 shows a plot of the carrier lifetime versus MBE growth temperature. The measurement is performed by focusing a 100 fs pump pulse onto a 20–30 μm spot on the semiconductor. A 10 times attenuated (as compared to the pump) probe pulse is focused into a 10 μm island within the pumped region. Both pump and probe are at 620 nm. The reflectance of the probe is measured as a function of probe delay (insert in Fig. 5.34). The carrier lifetime is defined as the initial decay $(1/e)$ of the reflectance versus delay curve.

The properties of semiconducting materials relevant to fast optical switching are listed in Table 5.2.

5.14.2 Integrated circuit fs lasers

Instead of trying to couple the laser to a standard laser cavity, one can integrate the semiconductor into a waveguide cavity. Such devices ranging in length from 0.25 mm to 2 mm have been constructed and demonstrated for example by Chen et al. [286]. The end mirrors of the cavity are — as in a conventional diode laser — the cleaved faces of the crystal (InP) used

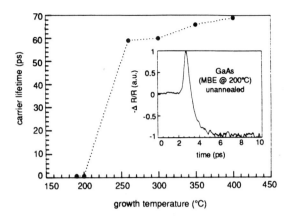

Figure 5.34: Carrier lifetime of GaAs versus MBE growth temperature. The inset shows a reflectance curve for a 200°C grown unannealed sample (from [285]).

Material	Carrier lifetime (ps)	Mobility (cm^2/Vs)
Cr-doped GaAs	50-100	1000
Ion implanted InP	2-4	200
Ion-damaged Si-on-sapphire	0.6	30
Amorphous silicon	0.8-20	1
MOCVD CdTe	0.45	180
GaAs (MBE, 200°C)	0.3	150
In$_{0.42}$Al$_{0.48}$As (MBE, 150°C)	0.4	5

Table 5.2: Semiconducting materials with carrier lifetimes and mobilities (from [285]).

as substrate. Waveguiding is provided by graded index confining InGaAsP layers. Gain and saturable absorber media consist of multiple quantum wells of InGaAs. The amount of gain and saturable absorption are controlled by the current flowing through these parts of the device (reverse bias for the absorber). As shown in the sketch of Fig. 5.35, the saturable absorber is located at the center of symmetry of the device, sandwiched between two gain regions. This configuration is analogous to that of the ring dye laser, in which the two counterpropagating pulses meet coherently in the absorber jet. In the case of this symmetric linear cavity, the laser operation of minimum

5.14. SEMICONDUCTOR LASERS

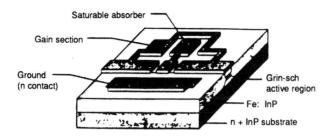

Figure 5.35: Layout of an integrated semiconductor fs laser (from [286]).

losses will correspond to two circulating pulses overlapping as standing waves in the saturable absorber.

These devices are pumped continuously and are thus the solid-state equivalent of the passively mode-locked dye lasers. The sizes and orders of magnitude are, however, significantly different, While the average output power is only slightly inferior to that of a dye laser (1 mW), at the much higher repetition rate (up to 350 GHz), the pulse energy is only in the fW range! For these ultrashort cavity lengths, there are only a handful of modes sustained by the gain bandwidth.

5.14.3 Semi-integrated circuit fs lasers

Total integration as shown above results in a very high duty cycle, at the expense of a lower energy per pulse. On can seek a compromise between the discrete elements semiconductor laser, as discussed in Section 5.14.1 and the total integrated laser of Section 5.14.2 above. For instance, the integration of the gain and saturable absorber of the integrated laser in [286] can be maintained in a single element coupled to an external cavity. Such a design has been successfully tested by Lin and Tang [287]. The absorber consists of a 10 μm island in middle of the gain region, with an electrical contact, isolated from the gain structure by two 10 μm shallow etched regions (without any electrical contact). The absorption — as in the case of the totally integrated laser of Section 5.14.2 — can be controlled through the bias of the central contact. To prevent lasing action of the 330 μm long gain module, the end facets — after cleavage — are etched (chemically assisted ion beam etching) at 10° from the cleaved plane. The autocorrelation of the pulses has a width of approximately 700 fs [287].

5.15 Fiber lasers

5.15.1 Introduction

In most lasers discussed so far, the radiation is a free propagating wave in the gain or other element of the cavity. The gain length is limited by the volume that can be pumped. The length of a nonlinear interaction is also limited by the Raleigh range (ρ_0). By confining the wave in a waveguide, it is possible to have arbitrarily long gain media, and nonlinear effects over arbitrarily long distances. A fiber is an ideal waveguide for this purpose. Its losses can be as small as a few dB/km. Yet the pulse confinement is such that substantial phase modulation can be achieved over distances ranging from cm to m. The fiber is particularly attractive in the wavelength range of negative dispersion (beyond 1.3 μm), since the combination of phase modulation and dispersion can lead to pulse (soliton) compression (see Chapter 7). The gain can be provided by Stimulated Raman Scattering (SRS) in the fiber material, Such "Raman soliton lasers" are reviewed in the next subsection. In the following subsection, we will consider the case of doped fibers, where the gain medium is of the same type as in conventional glass lasers.

5.15.2 Raman soliton fiber lasers

Stimulated Raman scattering (SRS) is associated with intense pulse propagation in optical fibers. A review of this topic can be found in [289] for example. The broad Raman gain profile for the Stokes pulse extends up to the frequency of the pump pulse. An overlap region exists because of the broad pump pulse spectrum. The lower frequency components of the pulse can experience gain at the expense of attenuation of the higher frequency components. In addition, the amplification of spontaneously scattered light is possible. Either process leads to the formation of a Stokes pulse which separates from the pump pulse after the walk-off distance due to GVD. These processes can be utilized for femtosecond Raman soliton generation in fibers, and fiber lasers [290, 291, 288]. An implementation of this idea is shown in Fig. 5.36. The pulses from a cw modelocked Nd:YAG laser (100 MHz, 100 ps, 1.32 μm) are coupled through a beam splitter BS into a ring laser containing an optical fiber. The fiber was tailored to have a negative dispersion for $\lambda_\ell > 1.46$ μm. While traveling through the fiber the pump pulses at 1.319 μm produce Stokes pulses at $\lambda_1 = 1.41$ μm. This first Stokes pulse in turn can act as pump source for the generation of a second Stokes pulse ($\lambda_2 = 1.495$ μm), which is in the dispersion region that enables soliton

5.15. FIBER LASERS

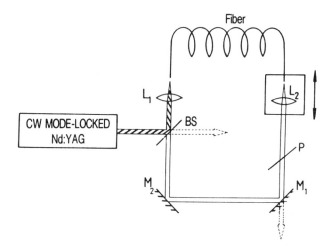

Figure 5.36: Experimental configuration of a synchronously pumped fiber ring laser (from [288]).

formation. Of course, for efficient synchronous pumping, the length of the ring laser had to be matched to the repetition rate of the pump. Second Stokes pulses as short as 200 fs were obtained.

5.15.3 Doped fiber lasers

Fibers can be doped with any of the rare earth ions used for glass lasers. Whether pumped through the fiber end, or transversely, these amplifying media can have an exceptionally large optical thickness ($a_g = \alpha_g d_g \gg 1$). An initial demonstration of this device was made by Duling [292, 293]. Passively mode locked rare earth doped fiber lasers have since evolved into compact, convenient, and reliable sources of pulses shorter than 100 fs. Two gain media have been demonstrated: Nd^{3+} operating at 1050 nm and Er^{3+} at 1550 nm. These lasers may include bulk optic components, e.g., mirror cavities, dispersion compensating prisms, or saturable absorbers, or they may be all-fiber, using a variety of pigtailed optical components and fused tapered couplers for output and pumping.

As compared to conventional solid state lasers, fibers have the advantage of a very large surface to volume ratio (hence efficient cooling is possible). The specific advantages of the single mode fiber geometry over bulk solid state (rare earth) media for mode locking are:

- Efficient conversion of the pump to the signal wavelength. Erbium, for example, is a three-level system and the tight mode confinement of the pump in a fiber allows for efficient depopulation of the ground state and thus high efficiency.

- Nonradiative ion–ion transitions which deplete the upper laser level are minimized. Such interactions are especially egregious in silica because of its high phonon energy and because the trivalent dopants do not mix well into the tetravalent silica matrix, tending instead to form strongly interacting clusters at the high concentrations necessary for practical bulk glass lasers [294]. The confinement of both the laser and pump modes allows the gain dopant to be distributed along greater lengths of fiber at lower concentration, obviating the need for high concentrations and so eliminating the interactions cited above.

- Diode laser pumping is practicable (due in large part to the previous two points). Single-mode laser diodes have been developed at 980 nm and 1480 nm for erbium fiber amplifiers in telecommunications applications. The four-level structure of neodymium allows for pumping even by multimode lasers, such as high-power laser diode arrays, by using fibers designed to guide the pump light in the cladding [295].

- Tight mode confinement and long propagation lengths maximize the self-phase modulation by the weak nonlinear index of silica ($\bar{n}_2 = 3 \; 10^{-16}$ cm^2/W).

Configurations in which the Kerr nonlinearity is converted into an artificial saturable absorption led to fs pulse generation with fiber lasers. The two main approaches that have been realized so far are

1. nonlinear loop mirrors [296]; and

2. nonlinear polarization rotation [297].

The pulse input to the nonlinear element (a fiber Sagnac interferometer for the approach 1, a birefringent fiber for 2) is split into two unequal pulses (propagating in opposite directions around the loop, or on the orthogonal birefringence axes). The two propagating components accumulate a nonlinear phase via self-phase modulation. After propagation (around the loop, or to the end of the birefringent fiber), the pulses are recombined. In the

5.15. FIBER LASERS

loop mirror, the variation of the accumulated differential phase across the combined pulse will cause different portions of the pulse to be reflected back toward the input or transmitted through the opposite port in proportion to the intensity. As a function of intensity, the transmission function of this nonlinear mirror is sinusoidal. Thus, the loop mirror behaves as a fast saturable absorber from low intensity up to the first transmission maximum.

In a system using nonlinear polarization rotation, the differential accumulated phase yields an intensity dependent state of polarization across the pulse. This polarization state is then converted into an intensity dependent transmission by inserting a polarizer at the output of the birefringent element, oriented, for example, to transmit the high intensity central portion of the pulse and reject the wings. Both approaches are of course closely related to additive pulse mode locking. The most studied fiber laser implementation of the nonlinear mirror is the figure-eight laser [292], so named for the schematic layout of its component fibers (see Fig. 5.37), with a nonlinear amplifying loop mirror [298]. In the latter, a 50% splitter is used at the input, and the imbalance between the opposite propagation directions is obtained by placing the gain in the interferometer loop offset from the center. The figure-eight optical circuit is completed by connecting the two ports of the loop through an isolator.

Fiber lasers operating on the 1050 nm transition of Nd^{3+} in silica require bulk optic elements (prism sequences) for compensating the substantial normal dispersion (30 ps/nm· km) of the gain fiber at the operating wavelength, and so are generally constructed as a bulk optic external cavity around the gain fiber. Passive mode locking is obtained via nonlinear polarization rotation in the gain fiber, and the Brewster angled prisms serve as the polarizer. Pulses as short as 100 fs have been demonstrated [299].

Femtosecond fiber lasers operating in the 1530–1570 nm gain band of erbium are of obvious interest for their potential application in telecommunications. This wavelength range is in the low loss window of silica fibers, and such a source is obviously compatible with erbium fiber amplifiers. Of particular interest also is the anomalous dispersion exhibited by silica at this wavelength. The precise value of the dispersion may be tailored through the exact fiber design. This implies that a mode-locked laser with Er gain may be constructed entirely from fiber, with no need for dispersion compensating prisms as in the Nd fiber lasers or most other ultrafast sources. Indeed, subpicosecond erbium lasers have been demonstrated with all-fiber figure-eight, linear, and ring cavities, using both nonlinear mirrors and nonlinear

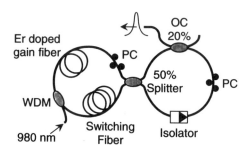

Figure 5.37: Schematic representation of the figure eight laser. The pump radiation at 980 nm is injected via the directional coupler WDM in the gain fiber (europium doped). Let us follow a signal as it emerges from the isolator, passes through a polarization controller PC (essentially a succession of $\lambda/4$—$\lambda/2$—$\lambda/4$ elements), and is coupled out (20%). Equal fractions of the remaining circulating signal are equally split into the two directions in the gain fiber (left). Both counter-rotating signals return to the splitter between the two loops with approximately the same gain (2 to 3 dB), but having experienced a different intensity dependent phase shift (the counterclockwise pulse being more intense, having been amplified before entering the switching fiber) in the switching fiber. As a result of this phase shift, it is the most intense fraction of the signal that is coupled back in the counterclockwise direction in the right loop.

polarization rotation. In addition, systems using semiconductor saturable absorbers have recently been demonstrated.

While soliton-like models have been used to describe a number of ultrafast laser systems as shown earlier in this chapter, the nonlinear dynamics of soliton propagation plays a more direct role in the erbium fiber laser than is seen in any other. The generated pulses are typically transform limited $sech^2$ in intensity profile – the shape expected from the soliton solution of the nonlinear Schrödinger equation. The average intracavity energy per pulse corresponds reasonably well to the energy of a soliton of the same length propagating in fiber with dispersion equal to the average cavity dispersion.

It has been demonstrated that the minimum pulse length obtainable in erbium fiber lasers is approximately proportional to the total dispersion inside the cavity [300]. This is to some degree surprising: As the pulse propagation is soliton-like, the fiber dispersion is continuously balanced by the self-phase modulation of the fiber. In principle, solitons of any length will form as long as the amplitude of the input pulse exceeds the threshold value of Eq. (7.35) (cf. soliton description in Chapter 7). However, the coupling of energy into the dispersive wave increases exponentially as the pulse shortens, thus limiting the minimum obtainable pulse width [301]. This loss becomes important only when the cavity length is of the order of the characteristic soliton length defined in Eq. (7.37) in Chapter 7. This is also why dispersive wave dynamics do not play an important role in other mode-locked lasers which can be described by a soliton model: In such systems the soliton length corresponds to many cavity round trips, much longer than the cavity lifetime of the dispersive wave. To obtain very short pulses, then, it is necessary to minimize the total cavity dispersion, either by using dispersion shifted fiber components, or very short cavities, or by including lengths of dispersion compensating fiber specially designed to have normal dispersion at 1550 nm. Such approaches have yielded pulse widths of less than 100 fs [302]. With such short pulses, third order dispersion plays an important role in limiting the pulse width and may impose a nonlinear chirp on the pulse [303].

5.16 Problems

1. Consider the saturable gain medium of a mode-locked laser. Show that the fractional reduction in pulse width per passage of the pulse through the saturable gain medium is of the order of the ratio of the energy density gained at the leading edge $\alpha_g d_g W_{sg}$ to the input pulse energy density W. Show that in the case of strong saturation, the upper limit for the gain contribution to the fractional reduction in pulse width/round-trip is simply the total loss factor L for the cavity.

2. By simple energy conservation arguments, find (in the approximation $T_1 \to \infty$) an expression for the energy gained (lost) per unit distance for an amplifier (absorber) $\frac{dW}{dz}$ as function of the change in population difference, and of the photon energy. Introduce into that expression the linear gain (absorption) coefficient, and combine with the rate equation to derive the simple evolution equation (5.8) for the pulse energy density.

3. Derive an evolution equation for the pulse energy in a mode-locked laser ring cavity consisting of (a) the sequence output mirror (reflectivity r — gain — saturable absorber; (b) the sequence gain — mirror — absorber; and (c) the sequence absorber — gain — mirror. Equation (5.12) is the solution for case (a).

4. Solve Eq. (5.12) in the approximation $W_1/W_{sa} \ll 1$ and $W_1/W_{sg} \ll 1$. Show that, even for a laser below threshold for cw operation ($R = r^2 < \exp[-(a_a + a_g)]$, two solutions can be found for W_1. The first solution is the minimum intracavity energy required to start mode-locked oscillation. The second solution is the steady state intracavity energy. Discuss the stability of both solutions. Under which condition can the first solution be small compared to the steady state intracavity energy?

5. Derive the equation for the soliton laser, following the procedure sketched in Fig. 5.16.

6. Calculate the spectrum of a continuously mode-locked Ti:sapphire laser emitting a continuous train of identical Gaussian pulses, 40 fs FWHM each, at a repetition rate of 80 MHz. The laser cavity is linear, with two prisms separated by 60 cm $[(dn/d\lambda)^2 = 10^{-9}$ nm$^{-2}]$. Neglecting all other contributions to the dispersion, calculate the longitudinal mode spectrum of the cavity. Is the latter spectrum identical to the Fourier amplitude (squared) of the train of pulses? If not, explain the difference(s).

7. Consider a typical passively mode-locked ring laser cavity. Assume as an initial condition that a pulse is generated every 0.99 ns (in each direction). The cavity round-trip time is 12 ns, and the distance between the absorber and gain is 4 ns×c. How will the pulses be distributed in the cavity after a large number of round-trips? Verify the answer by a simple numerical computation, using the following model to find the evolution of the pulse energy $\mathcal{W}$. Assume that the gain coefficient a_g recovers between pulses according to $da_g/dt = \mathcal{R}$, with the pumping rate $\mathcal{R} = 0.15$ ns^{-1}. Assume for simplicity that the losses (loss factor $e^L = 0.7$) are localized at the gain medium [i.e. at each passage through the gain, the pulse energy is multiplied by $(e^{a_g}e^L)$]. Following amplification of the pulse of energy $\mathcal{W}$, the depleted gain is $e^{a_g} \approx 1 + a_{g0}\exp(-\mathcal{W}/W_{sg})$, where W_{sg} is the saturation energy density and a_{g0} is the coefficient of the small signal gain. For each clockwise propagating pulse of energy $\mathcal{W}^{(+)}$, the energy loss $\Delta \mathcal{W}^{(+)}$ upon passage through the saturable absorber is: $\Delta \mathcal{W}^{(+)} = -A\mathcal{W}^{(+)}[1 - B\mathcal{W}^{(+)}/W_{sg}(1 + 2\mathcal{W}^{(-)}/W_{sg})]$. A similar expression holds for the counterpropagating pulse passing the absorber, with an interchange of the indices (+) and (-). Use for instance

5.16. PROBLEMS

$A = 0.03$ and $B = 0.02$, and initial pulse energies W/W_{sg} ranging from 0.45 to 0.55.

8. Three mode-locked lasers generate pulses of 40 fs duration at the wavelengths of 620, 700 and 780 nm, respectively. Find under which condition the output of these lasers could be combined to provide a train of much shorter pulses. What are the practical problems to be solved? Assuming perfect technology, how short would these pulses be?

9. Derive expression (5.3) for the pulse duration in a passively mode-locked dye laser. For simplicity assume that both saturable absorber and amplifier are traversed at resonance.

10. In short cavity and distributed feedback lasers short pulse generation is possible in the single (longitudinal) mode regime. At first sight this seems to be in contradiction to the necessity of a broad laser spectrum. Make a quantitative estimate of the essential laser parameters required for the generation of a 500 fs pulse.

11. Calculate the proportionality constant b between flux and gain fluctuations defined in Eq. (5.51), for the case of additive pulse mode-locking. [Refer to Eq. (5.74).]

Chapter 6

Femtosecond Pulse Amplification

6.1 Introduction

As discussed in the previous section, femtosecond pulse oscillators typically generate pulse trains with repetition rates of about 100 MHz at mean output powers which range from several mW (passively mode-locked dye laser) to several hundred mW (Ti:sapphire laser). Corresponding pulse energies are between several tens of pJ and several nJ. Femtosecond pulses with larger energies are needed for a variety of practical applications. Therefore a number of different amplifier configurations have been developed (for a review see also [304, 305, 306]). These amplifiers differ in the repetition rate and energy gain factor that can be achieved, ranging from 0.1 Hz to several MHz, and from 10 to 10^{10}, respectively. Both parameters cannot be chosen independently of each other. Instead, in present amplifiers the product of repetition rate and pulse energy usually does not exceed several hundred mW, i.e., it remains in the order of magnitude of the mean output power of the oscillator. The power of a single amplified pulse, however, can be in the terawatt range [307, 308]. It is not necessary for many applications to reach this power level. The specific function of the fs pulse will dictate a compromise between single pulse energy and repetition rate. It should also be noted that it is mostly the *intensity* of the focused pulse that matters, rather than the pulse power. Therefore, a clean beam profile providing the possibility of diffraction limited focusing is desired, eventually at the expense of a reduction in pulse energy. For many applications in spectroscopy, one

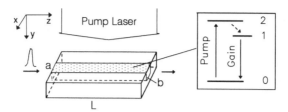

Figure 6.1: Light pulse amplification.

needs to generate a white light continuum. Typical threshold intensities that have to be reached for this purpose are on the order of 10^{12} W/cm^2.

The basic design principles of amplifiers have already been established for ps and ns pulse amplification. The pulses to be increased in energy are sent through a medium which provides the required gain factor (Fig. 6.1). However, on a femtosecond time scale new design methods are required to (i) keep the pulse duration short and (ii) circumvent undesired nonlinear effects caused by the extremely high intensities of amplified fs pulses. A popular technique to circumvent the problems associated with high peak powers is to use dispersive elements to stretch the pulse duration to the ps scale, prior to amplification.

Development of high repetition rate devices has been spurred by the dramatic progress in opto-electronics and optical materials (in particular, solid state laser materials) in the last decade.

Femtosecond pulse amplification is a complex issue because of the interplay of linear and nonlinear optical processes. The basic physical phenomena relevant to fs amplification are discussed individually in the next sections.

6.2 Fundamentals

6.2.1 Gain factor and saturation

It is usually desired to optimize the amplifier to achieve the highest possible gain coefficient for a given pump energy. In many cases of fs pulse amplification the gain medium is pumped transversely by the pulse of a pump laser. To simplify our discussion let us assume that the pump inverts uniformly the part of the gain medium (Fig. 6.1) which is traversed by the pulse to be amplified (single pulse). To achieve this uniform inversion with transverse optical pumping, we have to choose a certain concentration $\bar{N}$ of the

6.2. FUNDAMENTALS

(amplifying) particles which absorb the pump light and a certain focusing of the pump. The focusing not only determines the transverse dimensions of the pumped volume, $\Delta x = a$, but also controls the saturation coefficient s_p and the depth $\Delta y = b$ of the inverted region. The saturation parameter $s_p = W_{p0}/W_s$ was defined as the ratio of incident pump pulse density and saturation energy density [see Eq. (3.58)].

In practice, the pump energy is set by equipment availability and other experimental considerations. Therefore, to change s_p, we have to change the focusing. Note that here the total number of excited particles corresponding to the number of absorbed pump photons remains constant. To illustrate the effect of the pump focusing let us determine the depth distribution of the gain coefficient for various pump conditions. For simplicity, we assume a three-level system for the amplifier where $|0\rangle \to |2\rangle$ is the pump transition and $|1\rangle \to |0\rangle$ is the amplifying transition. The relaxation between $|2\rangle$ and $|1\rangle$ is to be much shorter than the pumping rate. Starting from the rate equations for a two-level system Eqs. (3.50) and (3.51), it can easily be shown that the system of rate equations for the photon flux density of the pump pulse F_p and the occupation number densities $\gamma_i = \tilde{N}\rho_{ii}$ ($i = 0, 1, 2$) reads now:

$$\frac{\partial}{\partial t}\gamma_0(y,t) = -\sigma_{02}\gamma_0(y,t)F_p(y,t) \tag{6.1}$$

$$\frac{\partial}{\partial y}F_p(y,t) = -\sigma_{02}\gamma_0(y,t)F_p(y,t) \tag{6.2}$$

and

$$\gamma_1(y,t) = \bar{N} - \gamma_0(y,t) \tag{6.3}$$

where σ_{02} is the interaction (absorption) cross-section of the transition $|0\rangle \to |2\rangle$. The coefficient of the small signal gain, a_g, is proportional to the occupation number difference of levels $|1\rangle$ and $|0\rangle$:

$$a_g = \sigma_{10}(\gamma_1 - \gamma_0)L = \sigma_{10}\Delta\gamma_{10}L \tag{6.4}$$

where L is the amplifier length. With the initial conditions $\gamma_0(y,0) = \gamma_0^{(e)}(y) = \bar{N}$ (all particles are in the ground state) we find from Eqs. (6.1), (6.2), and (6.3) for the inversion density $\Delta\gamma_{10}$ after interaction with the pump:

$$\Delta\gamma_{10}(y) = \bar{N}\left\{1 - \frac{2}{1 - e^a(1 - \exp s_p)}\right\} \tag{6.5}$$

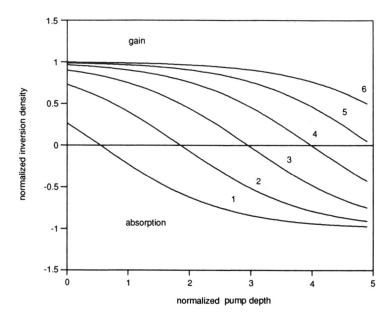

Figure 6.2: Inversion density $\Delta\gamma_{10}$ as a function of normalized depth y/ℓ_a, where $\ell_a = |\sigma_{02}\bar{N}|^{-1}$, for different saturation parameters $s_p = W_{p0}/W_s$.

where $a = -\sigma_{02}\bar{N}y$ is the coefficient of the small signal absorption for the pump. Figure 6.2 shows some examples of the population inversion distribution for different intensities of the pump pulse.

In the limit of zero saturation the penetration depth is roughly given by the absorption length $\ell_a = |\sigma_{02}\bar{N}|^{-1}$ defined as the propagation length at which the pulse intensity is $1/e$ of its original value. If the pump density is large enough to saturate the transition $0 \to 2$ the penetration depth becomes larger and, moreover, a region of almost constant inversion (gain) is built.

Given a uniformly pumped volume, the system needs to be optimized for maximum energy amplification of the signal pulse. Using Eq. (3.56) the energy gain factor achieved at the end of the amplifier can be written as:[1]

$$\begin{aligned} G_e = \frac{W(L)}{W_0} &= \frac{\hbar\omega_\ell}{2\sigma_{10}W_0}\ln\left[1 - e^{a_g}\left(1 - e^{2\sigma_{10}W_0}\right)\right] \\ &= \frac{1}{s}\ln\left[1 - e^{a_g}(1 - e^s)\right]. \end{aligned} \quad (6.6)$$

---

[1] Note that for the amplification, the relations found for the two-level system hold if we assume that during the amplification process no other transitions occur. This is justified in most practical situations.

6.2. FUNDAMENTALS

Medium	λ_ℓ [μm]	$\Delta\lambda$ [nm]	σ_{10} [cm^2]	Life time [s]	Typical pump
organic dyes	0.3...1	≥ 50	$\geq 10^{-16}$	$10^{-8}...10^{-12}$	laser
color centers	1...4	≈ 200	$\geq 10^{-16}$	$\leq 10^{-6}$	laser
XeCl	0.308	1.5	7×10^{-16}	$\approx 10^{-8}$	discharge
XeF	0.351	≤ 2	3×10^{-16}	$\approx 10^{-8}$	discharge
KrF	0.249	≈ 2	3×10^{-16}	$\leq 10^{-8}$	discharge
ArF	0.193	≈ 2	3×10^{-16}	$\leq 10^{-8}$	discharge
alexandrite	≈ 0.75	≈ 100	7×10^{-21}	2.6×10^{-4}	flashlamp
Cr:LiSAF	≈ 0.83	≈ 250	5×10^{-20}	6×10^{-5}	flashlamp
Ti:sapphire	≈ 0.78	≈ 400	3×10^{-19}	3×10^{-6}	laser
Nd:glass	1.05	≈ 21	3×10^{-20}	3×10^{-4}	flashlamp

Table 6.1: Optical parameters of gain media

Figure 6.3(a) shows this gain factor (on a logarithmic scale) as a function of s for different values of the small signal gain a_g. The saturation parameter s can be controlled by adjusting the cross-section of the pulse to be amplified. As expected from Eq. (6.6), as long as saturation is negligible, the energy gain varies exponentially with a_g (linear slope for the logarithm of the gain G_e versus gain coefficient). The total gain is drastically reduced by saturation. It should be noted, however, that for the sake of high energy extraction from the system, the amplifier has to be operated near saturation. Active media with larger saturation energy densities (smaller gain cross-sections) are therefore clearly favored if very high pulse energies are to be reached. Table 6.1 shows some important parameters of gain media used in fs pulse amplification. The small σ_{10} (large saturation energy density W_s) in connection with the long energy storage time ($\sim$ fluorescence life time) make solid state materials mostly attractive for high-energy amplification.

How close to saturation should an amplifier operate? If "chirped pulse amplification" is used (cf. section at the end of this chapter), it is essential that the amplifier operates in the *linear* regime. In other cases, it is advantageous to have at least one stage of amplification totally saturated. The reason is that the saturated output of an amplifier is relatively insensitive to fluctuations of pulse energy. A quantitative assessment of the relative fluctuations of the amplified pulses $\Delta W(L)/W(L)$ in terms of the input fluctuations $\Delta W_0/W_0$ can be found by differentiating Eq. (6.6) and defining a

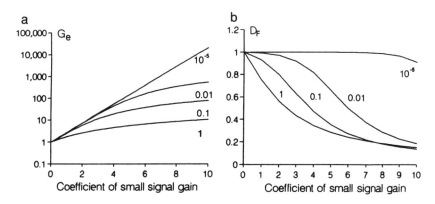

Figure 6.3: (a) Energy gain factor G_e and (b) damping of input fluctuations D_F as function of the coefficient of the small signal gain, a_g, for different saturation parameters $s' = 2s$.

damping factor:

$$D_F = \frac{\Delta W(L)/W(L)}{\Delta W_0/W_0} = \frac{e^{a_g} e^s}{[1 - e^{a_g}(1-e^s)]} \frac{1}{G_e}. \quad (6.7)$$

Obviously for large saturation the output can be expected to be smoothed by a factor of G_e^{-1} (Fig. 6.3b). This noise reduction is at the expense of a reduction of the amplification factor.

From the preceding discussion we may want to optimize the amplifier geometry as follows. From the given pump energy and the known absorption cross-section we can estimate the focusing conditions for the pump pulse to achieve a uniformly pumped volume. For maximum amplification, the cross-section of the signal beam has to be matched to this inverted region. If the saturation is too large or too small with respect to the overall design criteria, readjustment of either signal or pump focusing can correct the error. However, there are a number of additional effects that need to be considered in designing the amplification geometry, which are discussed below.

6.2.2 Shaping in amplifiers

Saturation

Saturation has a direct and indirect pulse shaping influence. The direct impact of saturation arises from the time dependent amplification. As the gain saturates, the dispersion associated with the amplifying transition changes,

6.2. FUNDAMENTALS

resulting in a phase modulation of the pulse. While the phase modulation does not affect the pulse envelope directly, it does modify the propagation of the pulse through the dispersive components of the amplifier (glass, solvent, isolators).

The mathematical framework to deal with the effect of saturation on both the pulse shape and its phase has been given in Chapters 3 and 4. We present here a few examples to illustrate the importance of these shaping mechanisms in specific configurations.

The change in the pulse intensity profile resulting from saturation — excluding dispersive effects — is found by evaluating Eq. (3.54), describing the intensity of a pulse at the output of an absorbing or amplifying medium as a function of the integrated intensity at the input $W_0(t) = \int_\infty^t I_0(t')dt'$:

$$I(z,t) = I_0(t) \frac{e^{W_0(t)/W_s}}{e^{-a_g} - 1 + e^{W_0(t)/W_s}}. \qquad (6.8)$$

Figure 6.4 shows the normalized shape of the amplified pulse for different input pulse shapes and saturation. As expected, saturation in the amplification process favors the leading edge of the pulse. Thus the pulse center shifts toward earlier times whereby the actual change in pulse shape depends critically on the shape of the input pulse. In particular the wings of the amplified pulses are a sensitive function of the initial slope. To minimize pulse broadening or even to obtain shortening during amplification, it is recommended to have pulses with a steep leading edge. For reasons to be discussed later amplifiers usually consist of several stages isolated by saturable absorbers. These absorbers can also serve to steepen the pulses.

If the pulse is detuned from resonance, we have seen in Chapter 3 how saturation can also result in a chirp. This effect will be discussed in Section 6.3 on nonlinear refractive index effects.

Group velocity dispersion

While being amplified the pulses travel through a certain length of material and are thus influenced by dispersion. In the case of *linear gain*, the pulse shaping is only due to group velocity dispersion. Shortest amplified pulses will be obtained either by sending up- (down-)chirped pulses through the amplifier if its net GVD is positive (negative). Alternatively (the only appropriate procedure if the input pulses were bandwidth limited) the broadened and chirped pulse at the end of the amplifier can be sent through a dispersive device, for example a prism or grating sequence, for recompression.

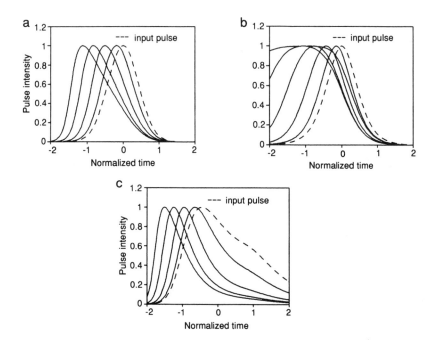

Figure 6.4: Behavior of the pulse shape in an amplifier for different shapes of the input pulse and different saturation parameters $s = W_0/W_s$ for a small signal gain $e^{a_g} = 10^4$. $2s$ varies from 10^{-4} to 1 (increment factor 10) in the order of increasing shift of the pulse maximum. The initial pulse shapes are a Gaussian $I(t) \propto \exp(-2t^2)$ (a) (t being the normalized time); a sech pulse $I(t) \propto \text{sech}^2(t)$ (b); and an asymmetric pulse, Gaussian in the wings $\exp(-t^2) + \frac{1}{2}\exp[-(t-1)^2]$ (c).

For unchirped input pulses, the magnitude of the broadening that occurs depends on the length of the amplifier and the dispersive length (for the pulse being amplified) defined in Chapter 1 [cf. Eq. (1.84)]. Some dispersion parameters for typical materials of fs amplifiers are shown in Table 6.2. Group velocity dispersion can generally be neglected in gases.

Gain narrowing

In the preceding discussion we assumed the bandwidth of the gain medium to be larger than the spectral width of the pulse to be amplified. Depending on the active medium (see Table 6.1) this assumption becomes questionable when the spectral width of the input pulse approaches a certain value. Now

6.2. FUNDAMENTALS

Material	λ_ℓ [μm]	k_ℓ'' [fs^2/cm]	L_D [cm]	$\bar{n}_2$ [cm^2/W]
water	0.6	480	21	0.67×10^{-16}
methanol	0.6	400	25	
benzene	0.6	1700	6	8.8×10^{-15}
ethylene glycol	0.6	840	12	3×10^{-16}
fused silica	0.6	590	17	$^1 3 \times 10^{-16}$
SF10	0.6	2530	4	$^2 1.3 \times 10^{-15}$
SF14	0.6	3900	2.5	
phosphate glass	1.06	330	30	1×10^{-15}
sapphire	0.78	610	16	8×10^{-16}
diamond	0.6	1131	8.8	6.7×10^{-15}
diamond	0.25	3542	2.8	-8×10^{-15}
air (1 atm)	1.06	0.0007	1.4×10^7	2.8×10^{-19}
air (1 atm)	0.6	0.0075	1.3×10^6	3.3×10^{-19}
air (1 atm)	0.25	0.96	1.05×10^4	7.8×10^{-19}

Table 6.2: Optical parameters for typical materials used in fs pulse amplifiers. The dispersion length L_D is given for a pulse duration of 100 fs. $^1 \lambda_\ell = 1.06$ μm, $^2 \lambda_\ell = 1.06$ μm, SF6.

we have to take into account that different spectral components of the pulse experience different gain. Since typical gain curves of active media have a finite bandwidth, the amplification is accompanied by a narrowing of the pulse spectrum. Thus, in the linear regime (no saturation), an unchirped pulse broadens while being amplified. This behavior can easily be verified assuming an unchirped Gaussian pulse at the amplifier input, with a field spectrum varying as $\tilde{\mathcal{E}}_0(\Omega) = A_0 \exp[-(\Omega \tau_G/2)^2]$ [cf. Eq. (1.35)], and a small signal gain $G(\Omega) = e^{a_g(\Omega)}$ where $a_g = a_0 \exp[-(\Omega T_g)^2/2]$. For simplicity we expand the Gaussian distribution and use $a_g \simeq a_0[1 - (\Omega T_g)^2/2]$. The spectral field amplitude behind the amplifier, neglecting saturation, is:

$$\begin{aligned}
\tilde{\mathcal{E}}(\Omega) &= \tilde{\mathcal{E}}_0(\Omega) e^{a_g(\Omega)/2} \\
&\simeq A_0 e^{-(\Omega \tau_G)^2/4} e^{a_0/2[1-(\Omega T_g)^2/2]} \\
&= A_0 e^{a_0/2} e^{-\Omega^2(\tau_G^2 + a_0 T_g^2)/4}
\end{aligned} \quad (6.9)$$

where τ_G is a measure of the input pulse duration $\tau_p \simeq 1.18\tau_G$ (see Table 1.1) and $\Delta\omega_g \simeq 2.36/T_g$ is the FWHM of the gain curve. As can be seen from Eq. (6.9) the spectrum of the amplified pulse becomes narrower; the FWHM is given by $\simeq 2.36/\sqrt{\tau_G^2 + a_0 T_g^2}$. The corresponding pulse duration at the amplifier output is

$$\tau_p' \simeq \tau_p\sqrt{\tau_G^2 + a_0(T_g/\tau_p)^2}. \tag{6.10}$$

If saturation occurs, the whole set of density matrix and Maxwell's equations has to be analyzed, as indicated in Chapters 3 and 4, to describe the behavior of the pulse on passing through the amplifier.

The situation is somewhat different in inhomogeneously broadened amplifiers if they are operated in the saturation regime. Roughly speaking, since field components which see the highest gain saturate the corresponding transitions first, those amplified independently by the wings of the gain curve can also reach the saturation level if the amplifier is sufficiently long. Therefore a net gain which is almost constant over a region exceeding the FWHM of the small signal gain can be reached and thus correspondingly shorter pulses can be amplified. This was demonstrated by Glownia et al. [309] and Szatmari et al. [310] who succeeded in amplifying 150–200 fs pulses in XeCl.

6.2.3 Amplified spontaneous emission (ASE)

So far we have neglected one severe problem in (fs) pulse amplification, namely amplified spontaneous emission (ASE), which mainly results from the pump pulses being much longer than the fs pulses to be amplified. As a consequence of the medium being inverted before (and after) the actual amplification process, spontaneous emission traveling through the pumped volume can continuously be amplified and can therefore reach high energies. ASE reduces the available gain and decreases the ratio of signal (amplified fs pulse) to background (ASE), or even can cause lasing of the amplifier, preventing amplification of the seed pulse. For these reasons the evolution of ASE and its suppression has to be considered thoroughly in constructing fs pulse amplifiers. Here we shall illustrate the essential effects on basis of a simple model [304] (illustrated in Fig. 6.5), and compare the small signal gain (for the signal pulse) with and without ASE. For simplicity, let us assume that the ASE starts at $z = 0$ and propagates toward the exit of the amplifier while being amplified. The photon flux of the ASE is thus given by:

$$F_{ASE}(z,t) = F_{ASE}(0)\exp\left[\int_0^z \sigma_{ASE}\left(\gamma_1(z',t) - \gamma_0(z',t)\right)dz'\right] \tag{6.11}$$

6.2. FUNDAMENTALS

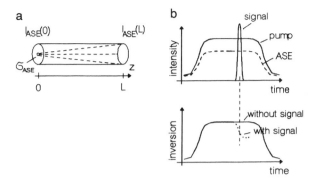

Figure 6.5: (a) Geometry of unidirectional ASE evolution (b) Temporal behavior of pump pulse, ASE, signal pulse and inversion (from [304]).

where σ_{ASE} is the emission cross-section which, with reference to Fig. 6.1 describes the transition between levels $|1\rangle$ and $|0\rangle$. With strong pumping the ASE will follow the pump intensity almost instantaneously, and after a certain time a stationary state is reached in which the population numbers do not change. This means that additional pump photons are transferred exclusively to ASE while leaving the population inversion unchanged. Under these conditions the rate equations for the occupation numbers read:

$$\frac{d\gamma_0(z,t)}{dt} = -\sigma_{02}\gamma_0(z)F_p(t) + \sigma_{10}\gamma_1(z)F_{ASE}(z,t) = 0 \qquad (6.12)$$

and

$$\gamma_0(z) + \gamma_1(z) = \bar{N}. \qquad (6.13)$$

Combination of Eq. (6.11) with Eqs. (6.12) and (6.13) yields an integral equation for the gain coefficient $a(z)$ for a signal pulse which has propagated a length z in the amplifier:

$$a(z) = \int_0^z \sigma_{10}\bar{N}\frac{F_p/F_{ASE}(0) - e^{a(z')}}{F_p/F_{ASE}(0) + e^{a(z')}}dz'. \qquad (6.14)$$

In the absence of ASE the gain coefficient is:

$$a = \sigma_{10}\bar{N}z. \qquad (6.15)$$

The actual gain under presence of ASE is reduced to $G_a = \exp[a(z)]$ from the larger small-signal gain in the ideal condition (without ASE) of $G_i =$

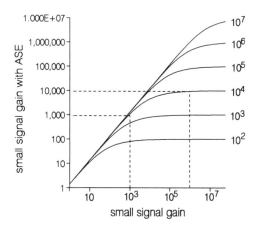

Figure 6.6: Comparison of small signal gain with and without ASE for different values of the normalized pump intensity $F_p/F_{ASE}(0)$.

$\exp(\sigma_{10}\bar{N}z)$. The ASE at $z = 0$ can be estimated from:

$$F_{ASE}(0) = \frac{\eta_F \Delta\Omega \hbar\omega_{ASE}}{4\sigma_{ASE}T_{10}} \quad (6.16)$$

where $\Delta\Omega = d^2/4L^2$ is the solid angle spanning the exit area of the amplifier from the entrance, T_{10} is the fluorescence life time, and η_F is the fluorescence quantum yield. Figure 6.6 shows the result of a numerical evaluation of Eq. (6.14). Note that a change in small signal gain at constant F_p can be achieved by changing either the amplifier length or the concentration $\bar{N}$. As can be seen at high small signal gain the ASE drastically reduces the gain available to the signal pulse. In this region a substantial part of the pump energy contributes to the build-up of ASE.

One solution to the problem of gain reduction due to ASE is the segmentation of the amplifier in multiple stages. To understand the nature of this improvement, let us compare a single- and a two-stage amplifier. With a normalized pump power $F_p/F_{ASE}(0) = 10^4$ and $G_i = 10^6$, we expect a small signal amplification of about 10^4 in the single-stage amplifier (Fig. 6.6). In contrast, we obtain a gain of about 10^3 in one cell and thus 10^6 in the whole device when we pump two cells by the same intensity, and each has half the length of the original cell. Another advantage of multi-stage amplifiers is the possibility to place filters between the individual stages and thus to reduce further the influence of ASE. If saturable absorbers are used this leads also to a favorable steepening of the leading pulse edge. Moreover, in multi-stage arrangements the beam size and pump power can be adjusted to control the saturation, taking into account the increasing pulse energy. For a more quantitative discussion of the interplay of ASE and signal pulse amplification as well as for the amplifier design, see, for example, [311, 312].

6.3 Nonlinear refractive index effects

6.3.1 General

As discussed already in previous chapters, the propagation and amplification of an intense pulse will induce changes of index of refraction in the traversed medium. As a result, the optical pathlength through the amplifier varies along the beam and pulse profile, leading possibly to self-phase modulation (SPM) and self-lensing. The origins of this pulse induced change in refractive index can be

(i) saturation in combination with off-resonant amplification (absorption), or

(ii) nonresonant nonlinear–refractive index effects in the host material.

The nonlinearity is somewhat more complex in semiconductor amplifiers, since it is related to the dependence of the index of refraction on the carrier density (which is a function of current, light intensity, and wavelength). The nonlinearities are nevertheless large and can contribute to significant spectral broadening in semiconductor amplifiers [313].

While SPM leads to changes in the pulse spectrum, self lensing modifies the beam profile. Being caused by the same change in index, both effects occur simultaneously, unless the intensity of the input beam does not vary transversely to the propagation direction. Such a "flat" beam profile can be obtained by expanding the beam and filtering out the central part to almost constant intensity.

It will often be desirable to exploit SPM in the amplifier chain for pulse compression, while self focusing should be avoided. There are a number of successful attempts to achieve spectral broadening (to be exploited in subsequent pulse compression) through SPM in a dye amplifiers [314, 315] and semiconductor amplifiers [313]. We will elaborate on this technique towards the end of this chapter. At the same time, self focusing should be avoided, because it leads to instabilities in the beam parameters such as filamentation or even to material damage.

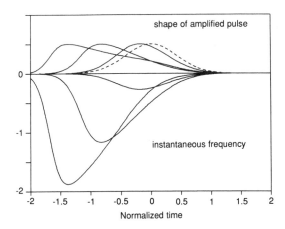

Figure 6.7: Chirp due to amplification of a Gaussian input pulse for different saturation parameters s ($s = 0.0001; 0.01; 1$ in the order of increasing chirp). The shape of the amplified pulse is also indicated. The other parameters are detuning $\Delta_\ell = 0.5$, small signal gain $e^a = 10^4$.

We proceed with some estimates of the self-phase modulation that occurs in amplifiers. We have derived in Chapter 3 an expression [Eq. (3.67)] for the change in instantaneous frequency with time due to gain depletion:

$$\delta\omega(t) = -\frac{(\omega_\ell - \omega_{10})T_2}{2} \frac{e^{-a} - 1}{e^{-a} - 1 + e^{W(t)/W_s}} \frac{I(t)}{W_s}. \tag{6.17}$$

The contribution from the nonlinear refractive index $\bar{n}_2$ of the host material is given by:

$$\delta\omega(t) = -k_\ell \bar{n}_2 \int_0^z \frac{\partial}{\partial t} I(z', t) dz' \tag{6.18}$$

The principal functional behavior of this frequency modulation is the same as that shown in Fig. 5.10 for intracavity chirping. The nonresonant refractive index change always results in up-chirp at the pulse center while the sign of the chirp due to gain saturation depends on the sign of the detuning $\Delta_\ell = (\omega_\ell - \omega_{10})$. The corresponding refractive index variation transverse to the propagation direction can lead to self-focusing as well as to self-defocusing. Chirp due to saturation may play a role in dye as well as in solid state amplifiers if they are highly saturated. Figure 6.7 illustrates the pulse shaping and chirping that arises for high values of the saturation parameter ($s \geq 0.0001$ at the amplifier input and $e^a = 10^4$).

6.3. NONLINEAR REFRACTIVE INDEX EFFECTS

In amplifier chains operating up to high saturation parameters, saturable absorbers will generally be used to isolate stages of high gain, reducing the effect of ASE. In addition the saturable absorber can counteract the broadening effect of the amplifier due to:

- pulse shaping (steepening of the leading edge) by saturable absorption and thus elimination of subsequent broadening by the saturable gain (cf. Fig. 6.4);

- pulse compression due to the combination of downchirping (if the pulse has a longer wavelength than that of the peak of the absorption band) by the absorber and propagation in a gain medium of normal (linear) dispersion.

6.3.2 Self-focusing

The situation is more complex if we consider self-focusing effects which have been introduced in Chapter 3. In this section, to obtain some order-of-magnitude estimations, we will use the relations derived for cw Gaussian beams. If the instantaneous peak power of the amplified pulse exceeds the critical power for self-focusing defined in Eq. (3.132)

$$P_{cr} = \frac{(1.22\lambda_\ell)^2 \pi}{32 n_0 \bar{n}_2}, \tag{6.19}$$

particular attention has to be given to the beam profile. Even weak ripples in the transverse beam profile may get strongly amplified, and lead to breaking-up of the beam in filaments. The critical size of these beam fluctuations, w_{cr}, below which a beam of intensity I (and power $P > P_{cr}$) becomes unstable against transverse intensity irregularities can be estimated from:

$$P_{cr} = \frac{\pi w_{cr}^2}{2} I. \tag{6.20}$$

If a smooth transverse beam profile is used, the amplifier may be operated above the critical power, provided the optical path through the amplifier L does not exceed the *self focusing length* L_{SF} [cf. Eq. (3.134)]:

$$L_{SF}(t) = \frac{0.5\rho_0}{\sqrt{P/P_{cr} - 1}}, \tag{6.21}$$

where $P = P(t)$ refers to the instantaneous power on axis of the Gaussian beam and the beam waist is at the sample input.

Values of $\bar{n}_2$ are listed for various transparent materials in Table 6.2.[2] To a typical value of $\bar{n}_2 \approx 5 \times 10^{-16}$ cm^2/W at 0.6 μm corresponds a critical power of $P_{cr} \approx 700$ kW, or only 70 nJ for a 100 fs pulse. Much higher energies are readily obtained in fs amplifiers. To estimate the self-focusing length, let us consider a saturated amplifier. The pulse energy $W = P\tau_p$ is of the order of the saturation energy density $\hbar\omega_\ell/2\sigma_{01}$ times the beam area $S \approx 0.5\pi w_0^2$. For a dye amplifier operating around 600 nm ($\sigma_{01} \approx 10^{-16}$ cm^2), the saturation energy density is of the order of 3 mJ/cm^2; hence the peak power for a 100 fs pulse in a beam of 1 cm^2 cross section is 3×10^{10} W/cm^2. Inserting this peak power in Eq.(6.21) leads to a self-focusing length of about 1 m.

Self-focusing is therefore generally not a problem in dye amplifiers, because the gain medium saturates before L_{SF} is reduced to dimensions of the order of the amplifier. Solid state media have a much lower cross-section σ_{01}, hence a much higher saturation energy density. For instance, if a Ti:sapphire laser amplifier ($\sigma_{01} = 3 \times 10^{-19}$ cm^2, or a saturation energy density of 0.66 J/cm^2) were driven to full saturation as in the previous example, the peak power would be 0.66×10^{13} W/cm^2. At $\lambda = 1$ μm, the corresponding self-focusing length is only 4 cm.

The smaller the interaction cross-section the shorter is L_{SF} and thus the more critical is self-focusing in a saturated amplifier. This problem has been a major obstacle in the construction of very high power amplifier sources. The solution is to stretch the pulse *prior to amplification*, to reduce its peak power, and recompress it thereafter. This solution, known as chirped pulse amplification [308], is outlined in Section 6.3.4.

6.3.3 Thermal noise

As the efficiency of an amplifier medium never approaches 100%, part of the pump energy is wasted in heat. In amplifiers as well as in lasers, thermally induced changes in index of refraction will result from a non-uniform heating. In most materials, all nonlinear lensing mechanisms are dwarfed by the thermal effects. A "z-scan" experiment [318] can be performed to appreciate the size of this nonlinearity, for example.

Average power levels of a few mW are sufficient to detect a thermal nonlinear index. The problem of thermal lensing is much more severe in

---

[2]The values are often expressed in Gaussian units (such as for instance the values for air [316]), or have to be derived from values of the third order susceptibility. A detailed discussion of the conversion factors can be found in [317].

6.3. NONLINEAR REFRACTIVE INDEX EFFECTS

amplifiers than in lasers, because of the larger pump energies and larger volumes involved. For instance, the heat dissipated by the pump can easily be carried away by the transverse flow in a typical dye laser jet, with a spot size of the order of μm. The larger cross-section of the amplifier calls generally for the use of cells. Non-uniform heating results in convection, turbulence, and a random noise in the beam profile and amplification. In the case of dye laser amplifiers, a simple but very effective solution consists of using *aqueous dye solutions* cooled near $4^\circ C$, since $(dn/dT)|_{4^\circ C} = 0$ at that temperature (or $11.7^\circ C$ for heavy water). Pulse-to-pulse fluctuations in the output of a Cu vapor laser pumped amplifier have been considerably reduced by this technique [319].

In the case of solid state amplifiers, a careful design of a cylindrically symmetric pump (and cooling) geometry is required to prevent thermal lensing from causing beam distortion.

6.3.4 Combined pulse amplification and chirping

The preceding section has established that self focusing sets a limit to the maximum power that can be extracted from an amplifier chain. Within that limit, self-phase modulation and subsequent compression can be combined with pulse amplification [315, 313]. In the implementation of [315], the pulse is self-phase modulated through the nonlinear index of the solvent in the last stage of the amplifier, for the purpose of subsequent pulse compression (Fig. 6.8). If the gain medium is not used at resonance, saturation can result in an even larger phase modulation, which can be calculated with Eq. (6.17).

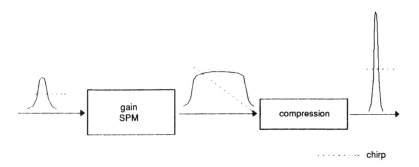

Figure 6.8: Block diagram showing the combination of pulse amplification, self-phase modulation, and compression.

With the constraints set above for the absence of self-focusing, the pulse propagation equation (1.142) is basically one-dimensional:

$$\frac{\partial}{\partial z}\tilde{\mathcal{E}} = \frac{1}{2}ik_\ell'' \frac{\partial^2}{\partial t^2}\tilde{\mathcal{E}} + \mathcal{B}_1 + \mathcal{B}_2; \qquad (6.22)$$

where

$$\mathcal{B}_1 = -\frac{\omega_\ell^2 \mu_0}{2k_\ell}\mathcal{P}_{gain}(t,z) \qquad (6.23)$$

and

$$\mathcal{B}_2 = -ik_\ell n_\ell n_2 |\tilde{\mathcal{E}}|^2 \tilde{\mathcal{E}} \qquad (6.24)$$

are nonlinear source terms (nonlinear polarization) responsible for the time dependent gain and the nonlinear refractive index, respectively. The polarization for the time dependent gain can be determined with Eq. (3.38), or with Eq. (3.42) if the rate equation approximation can be applied. We have outlined in Chapter 1 the basic procedure to study the propagation of a given input pulse through such an amplifier. Saturated amplification and group velocity dispersion will affect mainly the pulse temporal amplitude, while self-phase modulation will affect the pulse spectrum.

Figure 6.9(a) shows an example of pulse shape evolution, in amplitude and phase, through such an amplifier. The broadening and the development of a time dependent frequency can be clearly seen. As for the case of optical fibers, the interplay of SPM and group velocity dispersion leads to an almost linear chirp at the pulse center. The temporal broadening increases with the input pulse energy as a result of stronger saturation and larger SPM leading to a larger impact of group velocity dispersion [Fig. 6.9(b)]. It is also evident that the chirped and amplified pulses can be compressed in a quadratic compressor following the amplifier. Detailed numerical and experimental studies [315] show that an overall pulse compression by a factor of two is feasible for typical parameters of dye amplifiers.

6.3.5 Chirped pulse amplification

As mentioned earlier, the smaller the gain cross-section σ_{10}, the larger the saturation energy density $\hbar\omega_\ell/2\sigma_{10}$, which is a measure of the largest energy density that can be extracted from an amplifier. Since the maximum peak power is limited by self-focusing effects (the amplifier length has to be smaller than the self-focusing length), one logical solution is to limit the power by stretching the pulse in time. Dispersion lines with either positive or negative

6.3. NONLINEAR REFRACTIVE INDEX EFFECTS

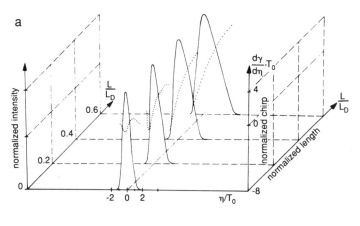

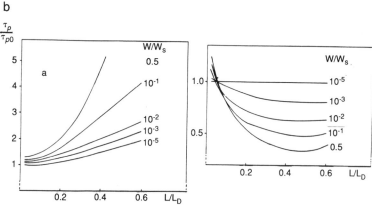

Figure 6.9: (a)Evolution of pulse shape and chirp in an amplifier with group velocity dispersion and SPM, (b) Broadening of pulses in an amplifier with group velocity dispersion and SPM (left) and pulse duration normalized to that of the input pulse after an optimum quadratic compressor (right). The initial pulse shape is Gaussian, applied at resonance with the gain medium. Parameters: small signal gain 2×10^4, $\tau_p 0 = 100 fs$, $k_\ell'' = 6 \times 10^{-26} s^2 m^{-1}$, $n_2 = 4 \times 10^{-23} m^2 V^{-2}$, $\sigma_{10} = 10^{-16} cm^2$. (adapted from ref. [315])

CHAPTER 6. FEMTOSECOND PULSE AMPLIFICATION

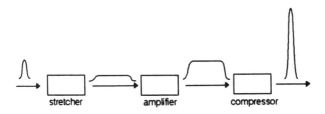

Figure 6.10: Block diagram of chirped pulse amplification.

group velocity dispersion can be made with combinations of gratings and lenses (cf. Chapter 2). The concept of "chirped pulse amplification" (CPA), as introduced by Mourou et al. [308], is to stretch (and chirp) a fs pulse from an oscillator up to 10,000 times with a dispersion line, increase the energy by *linear amplification*, and thereafter recompress the pulse to the original pulse duration and shape with the conjugate dispersion line (dispersion line with opposite group velocity dispersion). A similar concept was utilized earlier by Damm et al. [320] in order to use a fiber-grating compressor for the compression of high-power ps pulses from a Nd:glass laser. A block diagram illustrating the concept of CPA is shown in Fig. 6.10. Stretching of a pulse up to 10,000 times can be achieved with a combination of gratings and a telescope, as discussed in Chapter 2. Such a combination of linear elements does not modify the original pulse spectrum. For the amplification to be truly *linear*, two essential conditions have to be met by the amplifier:

- the amplifier bandwidth exceeds that of the pulse to be amplified; and
- the amplifier is not saturated.

It is only if these two conditions are exactly met that the original pulse duration can be restored by the conjugated dispersion line. It is not unusual to operate an amplifier in the wings of its gain profile, where the first condition is best met. For instance, Ti:sapphire, with its peak amplification factor close to 800 nm, is used as an amplifier for 1.06 μm, because of its flat gain profile in that wavelength range. A pulse energy of 1 mJ has been obtained in such a Ti:sapphire amplifier chain, corresponding to a gain of 10^7. Further *linear* amplification with Ti:sapphire requires rods of too large a diameter to be economical. With Nd:glass as a gain medium pulse energies as large as 20 J were obtained [308]. Because of the bandwidth limitation in the last 10^4 factor of amplification, the recompressed pulse has a duration of 400 fs, a fivefold stretch from the original 80 fs.

6.4 Amplifier design

6.4.1 Gain media and pump pulses

Parameters of gain media crucial for the amplification of fs pulses are

- The interaction cross-section. For a given amplifier volume (inverted volume) this parameter determines the small signal gain and the maximum possible energy per unit area that can be extracted from the system. The latter is limited by gain saturation.

- The energy-storage time of the active medium. If there is no ASE this time is determined by the life time of the upper laser level T_{10} and indicates (a) how long gain is available after pump pulse excitation for $\tau_{pump} < (<) T_g$ or (b) how fast a stationary gain is reached if $\tau_{pump} > (>) T_g$. The corresponding "response" time can be significantly shorter if ASE occurs, it is then roughly given by:

$$T_{ASE} = \frac{\hbar \omega_{ASE}}{I_{ASE}(L) \sigma_{ASE}}. \tag{6.25}$$

When short pump pulses are used, the latter quantity provides a measure of the maximum jitter allowable between pump and signal pulse without perturbing the reproducibility of amplification.

- The width of the gain profile $\Delta \omega_g = (\omega_g / \lambda_g) \Delta \lambda_g$ and the nature of the line broadening. The minimum pulse duration (or maximum pulse bandwidth) that can be maintained by the amplification process is of the order of $2\pi / \Delta \omega_g$ (or $\Delta \omega_g$).

These parameters were given in Table 6.1 for typical materials used as gain media in fs amplifiers. The wavelength and bandwidth of the seed pulse dictates the choice the gain medium. Presently, it is only in the near infrared that a selection can be made among various types of gain media, dyes, and solid state materials for fs pulse amplification. At pulse durations of the order of 10^{-14} s gain narrowing effects of single dyes dominate [306] leading to pulse broadening. These difficulties can be overcome by using a mixture of several dyes with different transition frequencies providing optimum amplification for a broad input spectrum [321]. The achievable energies with dye amplifiers are on the order of 1 mJ. This value is determined by the saturation energy density and the dye volume that can be uniformly pumped with available pump lasers.

Laser	Pulse energy [mJ]	Duration [ns]	Rep. rate [Hz]	λ [nm]
Ar$^+$(cav.dump.)	10^{-3}	15	3×10^6	514
Nd:YAG				
Q-switched	300	5	10	532
regen.ampl.	2	0.07	10^3	532
diode pumped	0.05	10	800	532
Nd:YLF (Q-sw.)	10	400	10^4	523.5
copper vapor	2	15	5000	510, 578
excimer	100	20	10	308

Table 6.3: Typical parameters of pump lasers for fs pulse amplifiers.

Shorter pulses and higher energies can, in principle, be extracted from certain solid state amplifiers. With the additional advantage of compactness, such systems are attractive candidates for producing pulses in the TW and PW range. These systems are presently limited to the red and near infrared spectral range.

At certain wavelengths in the UV (see table 6.1) excimer gases can be used for fs pulse amplification (see for example, [309, 310, 322, 323, 324, 325]). The interaction cross-section being similar to that of dyes, the saturation energy density of excimer gain media is also of the order of millijoules/cm^2. Much larger pulse energies however — ranging from tenths of joules to the joule range — can be extracted, because the active volume that can be pumped is much larger than in dye amplifiers. As compared to solid state or liquid materials, another advantage of excimer gases is the smaller susceptibilities associated with undesired nonlinear effects (such as self-focusing). Unfortunately, the relatively narrow gain bandwidth of excimers limits the shortest pulse duration that can be amplified and the tunability.

Essential pulse parameters, such as the achievable energy range and repetition rate, that can be reached are determined by the pump laser of the amplifier. Table 6.3 summarizes data on lasers that have successfully been used for pumping fs amplifiers. Usually these pump lasers have to be synchronized to the high repetition rate oscillators for reproducible amplification. On a nanosecond time scale this synchronization can be achieved electronically. With picosecond pump pulses, satisfactory synchronism requires generally that the pump pulses for the femtosecond oscillator and amplifier be derived from a single master oscillator.

6.4. AMPLIFIER DESIGN

6.4.2 Amplifier configurations

Usually the amplifier is expected to satisfy certain requirements for the output radiation which can be achieved by a suitable design and choice of the components. Table 6.4 shows some examples. Different applications of amplified pulses have different requirements, and subsequently various amplifier configurations have been developed. In particular, trade-off between pulse energy and repetition rate will call for a particular choice of amplifier design and pump source. A feature common to nearly all femtosecond amplifiers is that they are terminated by a linear optical element to recompress the pulses.

Multi-stage amplifiers

Low–repetition rate systems (< 500 Hz) used for high-gain amplification consist mostly of several stages traversed in sequence by the signal pulse. A typical configuration is sketched in Fig. 6.11. This concept, introduced by Fork *et al.* [326] for the amplification of fs pulses in a dye amplifier, has the following advantages. (i) Each stage can be adjusted separately for maxi-

Requirements	Realization
(a) clean beam profile	homogeneously inverted gain region, no self-focusing, proper (linear) optical design
(b) high peak power	same as above, with chirped pulse amplification (CPA)
(c) high energy amplification	high pump power, amplification reaches saturation level
(d) low background	ASE suppression through spatial and/or spectral filtering, filtering through saturable absorption
(e) certain repetition rate	repetition rate of pump, suitable gain medium
(f) no temporal broadening	group velocity dispersion adjustment
(g) no spectral narrowing	gain medium with broad bandwidth

Table 6.4: Design requirements of a fs pulse amplifier.

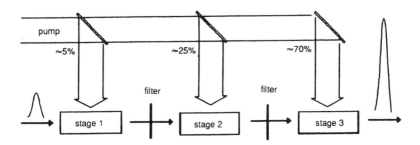

Figure 6.11: Sketch of a multi-stage amplifier.

mum gain, considering the particular signal pulse energy at that stage. The splitting of the pump energy among the various stages has to be optimized, as well as the pump focalization to match the volume to be pumped. Typically, only a few percent of the pump pulse is tightly focused into the first stage, resulting in a gain of several thousands. More than 50% of the pump energy is reserved for the last stage (to pump a much larger volume) resulting in a gain factor of about 10. (ii) The unavoidable ASE can be suppressed with filters inserted between successive stages. Ideally, these filters are linear attenuators for the ASE, but are saturated by the signal pulse. Of course, the filter remains "open" after passage of the signal pulse for a time interval given by the energy relaxation time, and subsequently ASE within this temporal range cannot be suppressed. Edge filters, such as semiconductors and semiconductor doped glasses, can be used for ASE reduction whenever the ASE and the signal pulse are spectrally separated. Finally, since the beam characteristics of ASE and signal pulse are quite different, a spatial filter (for example a pinhole in the focal plane between two lenses) can enhance the signal-to-ASE ratio. Typical pump lasers for multi-stage dye amplifiers are Q-switched Nd:YAG lasers [326] and excimer lasers [327, 328] with pulse durations of about 5 ns and 20 ns, respectively. Typically the repetition rates are below 100 Hz, and maximum pulse energies of a few millijoules have been reported. In the last decade many modifications of the setup shown in Fig. 6.11 have been made. For example, multiple passes through one and the same stage to extract more energy or/and to use smaller pump lasers were implemented. To increase the homogeneity of the gain region longitudinal pumping is frequently used in the last amplifier stage(s).

An example for a fs multi-stage amplifier pumped by a XeCl excimer laser is shown in Fig. 6.12. The excimer laser, consisting of two separate discharge channels, serves to pump the dye cells and to amplify the frequency doubled

6.4. AMPLIFIER DESIGN

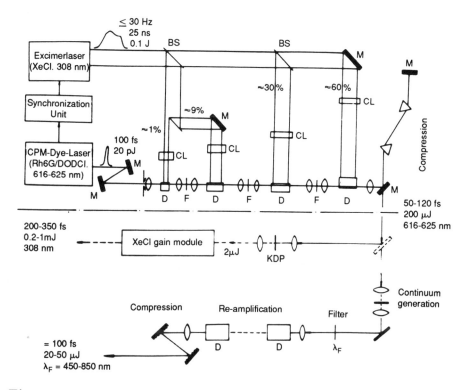

Figure 6.12: A fs dye amplifier pumped by an excimer laser. The amplification stages are decoupled by saturable absorbers (semiconductor doped glasses) or k-space filters. The prism sequence serves to compress the phase modulated, amplified pulses. To extend the wavelength range of available fs pulses, the frequency doubled output can be amplified in the second discharge channel of the excimer laser. Another option is to amplify the spectrally filtered white light continuum. (From [324].)

fs output at 308 nm. Another option is to generate a fs white light continuum and to amplify a certain spectral component. With the UV pump pulses and different dyes a wavelength range from the NIR to the UV can be covered.

6.4.3 Single-stage, multi-pass amplifiers

For larger repetition rates one has to use pump lasers working at higher frequencies. Since the mean output power of table-top pump lasers cannot be increased arbitrarily, higher repetition rates are achieved at the expense of energy per pulse. To obtain still reasonable gain factors one has to increase

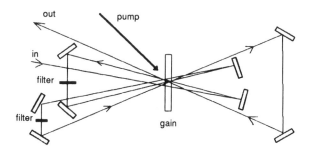

Figure 6.13: Sketch of a single-stage, multi-pass amplifier (after [99]).

the efficiency of converting pump energy into signal energy as compared to the configurations described above. In this respect the basic disadvantage of single-pass amplifiers is that only a fraction of the energy pumped into the gain media is used for the actual amplification. The main reason is that the pump process is often much longer than the recovery time of the gain medium; hence, a considerable part of the pump energy is converted into ASE. The overall efficiency can be enhanced by sending the pulse to be amplified several times through the amplifier. The time interval between successive passages should be of the order of the recovery time of the gain medium T_g. The number of passages should not exceed the ratio of pump pulse duration to T_g.

In a first attempt to amplify femtosecond pulses at high repetition rates, Downer et al. [193] used a cavity-dumped Ar^+-laser to pump a dye amplifier. Despite the high repetition rate (3 MHz), this approach did not find broad application because of the relatively small net gain (∼100) resulting from the low energy of the pump pulses. The use of copper-vapor lasers, working at repetition rates from 5 to 15 kHz, turned out to be a more practicable concept to pump dye amplifiers [329]. Knox et al. [99] used such a laser to pump a single dye jet and to amplify 100 fs pulses to microjoule energies. The dye jet was passed six times to match the pump pulse duration (∼25 ns) where the reported small signal gain per pass was five to eight. The disadvantage of such a configuration, as sketched in Fig. 6.13, is its complexity and large number of optical elements. Two saturable absorbers were implemented to suppress ASE. Higher output powers could be reached using a dye cell for the gain medium [330]. Other concepts distinguish themselves by a minimum number of optical components and simplicity of adjustment [331, 319]. The gain medium is inserted in a resonator-like structure. A convenient structure

6.4. AMPLIFIER DESIGN

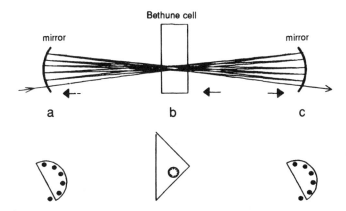

Figure 6.14: Schematic diagram of a multi-pass (copper vapor laser pumped) amplifier and view on the beam geometry at the focusing mirrors and the Bethune cell (adapted from [319]).

for uniform transverse illumination of a cylindrical volume is the "Bethune cell" [332]. The volume to be pumped is inserted in a total reflection prism, at a location such that adjacent sections of the pump beam are reflected to all four quadrants of the cylinder (Fig. 6.14). In the arrangement of Ref. [319], the beam to be amplified is sent 11 times through the gain cell. A series of apertures on a circular pattern (Fig. 6.14) are a guide for the alignment and prevent oscillation in the stable resonator configuration. The latter being close to concentric, the first 11 paths are focused to a small beam waist in the gain medium. This amplifier is intentionally operating at saturation for the last few passes, in order to reduce its sensitivity to fluctuations of the input. The beam is sent back for two more passes through the center of the 2 mm diameter amplifying cell, providing unsaturated amplification to 15 μJ.

Copper vapor laser pumped amplifiers have also been successfully applied to generate powerful femtosecond pulses in the near infrared [333].

6.4.4 Regenerative amplifiers

As discussed previously, a broad gain bandwidth and high saturation energy density make some solid state materials (cf. Table 6.1) prime candidates for the generation of powerful femtosecond light pulses. A large energy storage time (10^{-6} s) is generally associated with the small gain cross-section of these amplifying media. Regenerative amplification is the most efficient method to transfer efficiently energy to a fs pulse from an amplifier with a long storage

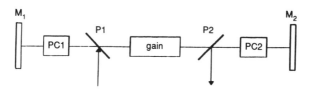

Figure 6.15: Principle of regenerative amplification

time. The concept of regenerative amplification is illustrated in Fig. 6.15. The gain medium is placed in a resonator built by the 100% mirrors M_1 and M_2. After the seed pulse is coupled into the resonator through a polarizer P_1, the Pockels cell PC1 is switched to rotate the polarization of the seed pulse and Q-switches the resonator. The pulse circulates in the resonator and is continuously amplified. After a certain number of round trips (determined by the energy storage time and/or the time needed to reach saturation) a quarter-wave voltage is applied to the cavity-dumping Pockels cell PC2 and the amplified pulse is coupled out by reflection from the second polarizer P2.

Regenerative amplifiers were originally developed to amplify the output of cw mode-locked solid state (e.g., Nd:YAG, Nd:YLF) lasers at repetition rates up to 2 kHz and energies up to the millijoule level (e.g., [334, 335, 336]), to obtain ps pulses at microjoule energies. These pulses served as pump for dye amplifiers [337]. Since the pump pulse duration is on the order of 100 ps the amplification process can be much more efficient than with ns pump lasers. Synchronization between pump and fs signal is achieved by pumping the dye laser synchronously with the same master oscillator used to provide the seed pulses for the regenerative amplifier. More recently regenerative solid state amplifiers have been used to amplify fs light pulses directly in alexandrite [338] and Ti:sapphire [339] utilizing chirped pulse amplification. Using a Q-switched Nd:YLF laser as pump for the Ti:sapphire crystal in the regenerative amplifier a repetition rate as high as 7 kHz could be reached [340].

A multi-terawatt, 30 fs, Ti:sapphire laser system based on a combination of a regenerative amplifier and a multi-pass amplifier, operating at 10 Hz, was reported by Barty *et al.* [47]. A sketch of the system is shown in Fig. 6.16. The 20 fs (5 nJ, 800 nm) pulses from a modelocked Ti:sapphire laser are stretched to 300 ps. Amplification in a 14-pass regenerative amplifier (50 mJ pump pulse at 532 nm) yielded 9 mJ output pulses. A 4-pass amplifier (235 mJ pump pulse at 532 nm) increases the pulse energy to 125 mJ. Finally, after re-compression, 30 fs pulses were obtained. To reduce the effect of gain-

6.4. AMPLIFIER DESIGN

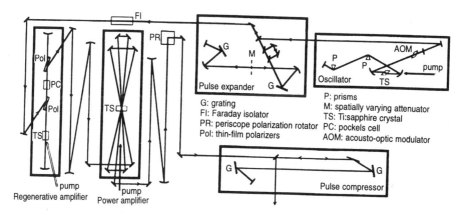

Figure 6.16: Multi-terawatt fs laser system (from [47]).

narrowing the spectrum of the pulse prior to amplification was flattened. This was accomplished by spectral filtering (element M in Fig. 6.16, see also Chapter 7).

6.4.5 Traveling wave amplification

Amplified stimulated emission limits strongly the overall efficiency of the amplification process, in particular at shorter wavelengths where the ratio of spontaneous emission to stimulated emission is larger. One of the causes of a large ASE to signal ratio is that the duration of the pump — and hence that of the gain — generally exceeds that of the pulse to be amplified by several orders of magnitude. Considerably higher conversion from pump energy into signal pulse energy is therefore expected when using femtosecond pump pulses. It may seem ludicrous to use a powerful fs light pulse, that is, a fs pulse already amplified, to amplify another fs pulse. However, such schemes offer the prospect of efficient frequency conversion with continuous tunability. A first implementation is the traveling wave amplifier (TWA) introduced by Polland et al. [341] and Bor et al. [342] in a transverse pumping arrangement with ps light pulses. The TWA technique was successfully extended to the fs time scale by Hebling and Kuhl [343]. A theoretical analysis of TWA can be found in [342, 344, 345]. An example of implementation of TWA is sketched in Fig. 6.17. The active medium can be a dye cell. The tilt of the pulse front of the pump with respect to the propagation direction of the signal pulse is chosen so as to invert the gain medium in synchronism

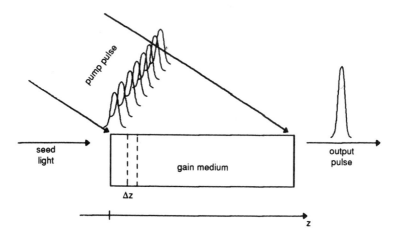

Figure 6.17: Transverse traveling wave amplification. A fs pulse, a spectrally filtered fs continuum or suitable (quasi) cw radiation (for example, ASE) can serve as seed light.

with the propagating signal light. This "traveling wave amplification" leaves practically no time for ASE in front of the signal pulse to develop.

It would seem that a tunable fs source — such as that provided by a spectrally filtered white light continuum — is required as seed pulse to produce fs light pulses at new frequencies by TWA. Such a sophisticated seed is fortunately not required, because the output pulse duration of the TWA amplifier does not depend on the input pulse duration. Hebling and Kuhl [346], for example, simply selected a certain spectral component from the ASE excited in the active medium as amplifier input. With three different dyes and pump pulses at 620 nm a spectral range between 660 nm and 785 nm could continuously be covered. In the absence of external seeding, the ASE excited in the very first part of the amplifier serves as seed signal. The pulse evolution is similar to that in a synchronously pumped laser: The propagation from one TWA amplifier slice to the next compares to successive passages through the gain jet in a synchronously pumped laser. In view of the short pump pulse, the output will also be a fs light pulse with a mean wavelength given roughly by the maximum of the net gain profile of the amplifier. In a synchronously pumped laser the pulse evolution starts from noise (spontaneous emission) and a certain number of round trips are needed for the pulse to form (cf. Chapter 5). In the TWA the number of round trips translates into a minimum number of "slices," which corresponds to a minimum

6.4. AMPLIFIER DESIGN

amplification length. This analogy leads to conclude that the output pulse duration is shorter than the pump pulse and is a sensitive function of the tilt angle (the latter determines the timing mismatch of signal and pump pulse at any given location in the amplifier).

The general approach introduced in Chapter 1 can be applied to a quantitative study of TWA. The complex electric field amplitude of the signal pulse at position $z + \Delta z$, $\tilde{\mathcal{E}}_s(t, z + \Delta z)$ is related to the amplitude at z through:

$$\tilde{\mathcal{E}}_s(t, z + \Delta z) = \tilde{\mathcal{E}}_s(t, z) + \delta_g \tilde{\mathcal{E}}_s(t, z) + \delta_{nl} \tilde{\mathcal{E}}_s(t, z) + \delta_{k''} \tilde{\mathcal{E}}_s(t, z) \quad (6.26)$$

where Δz is the slice width, and $\delta_g \tilde{\mathcal{E}}_s$, $\delta_{nl} \tilde{\mathcal{E}}_s$, and $\delta_{k''} \tilde{\mathcal{E}}_s$ describe the amplitude change due to gain, nonlinear refractive index effects, and GVD, respectively. These changes can easily be calculated by means of Eqs. (1.146), (6.8), and (3.116). Neglecting GVD, for the photon flux density and phase of the signal pulse we find:

$$F_s(t, z + \Delta z) = F_s(t, z) + |\tilde{L}(\omega_{10} - \omega_s)|^2 \sigma_{10}^{(0)} (\gamma_1 - \gamma_0) F_s(t, z) \Delta z \quad (6.27)$$

$$\phi_s(t, z + \Delta z) = \phi_s(t, z) - \frac{1}{2} \text{Im}[\tilde{L}(\omega_{10} - \omega_s)] \sigma_{10}^{(0)} (\gamma_1 - \gamma_0) \Delta z \quad (6.28)$$

$$- i \left(\frac{2\omega_m}{c^2 n_0 \epsilon_0} \right) n_2 [F_s(t, z) + F_p(t, z)] \Delta z \quad (6.29)$$

where, for the sake of simplicity, we have introduced a mean frequency of signal and pump pulse, ω_m, and the linear refractive index n_0 at this frequency. F_p denotes the photon flux density of the pump pulse. As indicated in these equations, the phase modulation originates from (near) resonant (saturation) and non-resonant (host medium) contributions to the changes in index of refraction. The inversion density $\Delta \gamma_{10}(t, z) = \gamma_1 - \gamma_0$ is obtained by solving a system of rate equations for the population numbers in each slice. Assuming a three-level system as shown in Fig. 6.1:

$$\frac{d}{dt} \gamma_0(t, z) = -\sigma_{02} \gamma_0 F_p(t, z) + \sigma_{10}^{(0)} |\tilde{L}(\omega_{10} - \omega_s)|^2 \gamma_1 F_s(t, z) \quad (6.30)$$

$$\frac{d}{dt} \gamma_1(t, z) = -\sigma_{10}^{(0)} |\tilde{L}(\omega_{10} - \omega_s)|^2 \gamma_1 F_s(t, z) + \frac{\gamma_2}{T_{21}} \quad (6.31)$$

$$\bar{N} = \gamma_0 + \gamma_1 + \gamma_2. \quad (6.32)$$

Because of the short pump pulse duration, we cannot neglect the population in level 2 and have to consider a nonzero relaxation time T_{21}. Starting either with the photon flux of a small seed pulse or with cw light at $z = 0$ the successive application of Eq. (6.26) yields the signal pulse for an amplifier

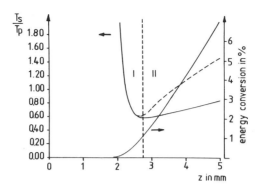

Figure 6.18: Ratio of signal pulse to pump pulse duration and energy conversion as function of the propagation length for typical parameters of a dye amplifier. The dashed line describes the behavior of the pulse duration when group velocity dispersion plays a role (from [345]).

length z. Figure 6.18 shows corresponding results for the evolution of pulse duration and pulse energy as a function of the propagation length [345]. Region I is characterized by a decrease of signal pulse duration and an almost exponential increase in energy. This is due to the fast rise of the gain (short pump pulse duration) resulting in a rapid build-up of a steep leading edge of the signal pulse, and to the fact that the gain is essentially unsaturated until the pumping process stops. The latter is responsible for the shaping of the trailing edge of the signal pulse. With the onset of saturation (region II) the pulse duration increases, a tendency which becomes more pronounced with significant group velocity dispersion.

6.5 Problems

1. An important parameter of a gain medium is the energy storage time. Consider a one-stage amplifier of 5 mm length transversely pumped by a pulse of 20 ns duration and 5 mJ energy. A (fs) pulse with an energy of 100 pJ is to be amplified. The gain medium consists of a three-level system (cf. Fig. 6.1). The relaxation time from level 2 to level 1 is assumed to be extremely fast. Calculate and compare the energy amplification achievable in a single-pass configuration for a gain medium with a lifetime of the upper gain level, T_{10}, of (a) 100 ps, (b) 1 μs. For simplification you may assume a rectangular temporal and spatial profile for both pump and pulse to be amplified. Perform

6.5. PROBLEMS

your calculation for a beam size of $50 \times 50 \ \mu m^2$. Assume homogeneous gain and equal cross-sections for absorption and amplification, $\sigma \approx 10^{-17} \ cm^2$.

2. Explain the different effect of gain saturation on the shaping of the wings of a Gaussian and a sech pulse (cf. Fig. 6.4).

3. By means of Fig. 6.6 design a three-stage dye amplifier to amplify the output of a fs dye laser (100 fs, 100 pJ) to > 0.5 mJ. The second harmonic of a Nd:Yag (50 mJ, 10 ns) is to be used as pump. For the absorption and emission cross-section use a value of $10^{-16} \ cm^2$. Specify the split of the pump energy among the three stages and make an estimate for the focusing conditions.

4. Let us consider a 2 cm long cuvette filled with a solution of Rhodamine 6G. The dye is pumped longitudinally by a Nd:Yag laser beam of uniform intensity $I = 10 \ MW/cm^2$. The dye has an absorption coefficient of 5 cm^{-1} at the pump laser wavelength. The absorption cross-section of the solution is $5 \ 10^{-16} \ cm^2$. Approximate the dye solution by a three-level system, with the upper level being common to the pump and lasing transition. The pump transition is from the ground state to the upper level. The lasing transition (cross-section $\sigma_g = 5 \times 10^{-17} \ cm^2$) is from the upper level (lifetime of 2.5 ns) to an intermediate level which relaxes to the ground state with a characteristic energy relaxation time of 1 ps. Find the gain distribution $\alpha_g(z)$ along the propagation direction (z) of the pump beam. A 50 fs pulse, with an energy density of 100 $\mu J/cm^2$ is sent through the medium along the same z direction. Calculate the energy of the pulse exciting the gain cell. Solve the problem for a 100 $\mu J/cm^2$ pulse sent in the *opposite* direction $(-z)$. How do the results differ? Assume next that the pump beam diameter decreases linearly from 1 cm at the cell entrance down to 1 mm at $z = 2$ cm. The pump power at the cell entrance is 1 MW. The beam to be amplified has the same geometry, the initial pulse energy being 5 μJ. Find the amplified pulse energy for co- and counterpropagating pump pulse and pulse to be amplified.

Chapter 7

Pulse Shaping

On a fs time scale, many interactions depend on the particular temporal shape of the waveform being applied. For many applications it is desirable and necessary to modify the pulses from the source in a well-defined manner. A compression of the intensity profile leads to shorter pulses and higher peak powers. Closely spaced femtosecond pulses with controllable phase relations are needed for coherent multi-photon excitation and the selective excitation of, for example, certain molecular vibrations as detailed in Chapter 4.

The distortion of the complex pulse envelope caused by most linear and nonlinear optical processes has been discussed in previous chapters. In this chapter we shall concentrate on techniques applied to compress or shape pulses in amplitude and phase. (for a comprehensive review on pulse compression, see also [289]). While shaping of ns and ps pulses can advantageously be achieved by electronically driven pulse shapers, such as electro-optic modulators [347], all-optical techniques have to be applied for fs pulse shaping. Dispersion leads to pulse shortening or lengthening depending on the input chirp. Saturable absorption tends to steepen the leading edge of the pulse.

7.1 Pulse compression

7.1.1 General

Optical pulse compression is the optical analogue of a well established technique for the shaping of radar pulses [348]. Its implementation into Optics was in the late 1960's [349, 39, 350, 351] for compression of ps pulses.

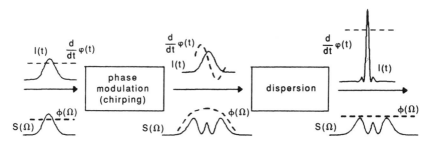

Figure 7.1: Sketch of optical pulse compression through self-phase modulation and chirp compensation.

Optical pulse compression is generally achieved by the two-step process sketched in Fig. 7.1. Let us assume a bandwidth limited input pulse. In the first step a phase modulation $\varphi(t)$ is impressed on the pulse which for example can be obtained by self-phase modulation in a nonlinear refractive index material. The transformed pulse is spectrally broadened, or, in the time domain, a chirped pulse. The temporal intensity profile $I(t)$ (or $|\tilde{\mathcal{E}}(t)|^2$) is generally unchanged by this first step, which modifies only the phase function of the pulse. As we have seen in Chapter 3, for a purely dispersive nonlinearity (n_2 real, no two-photon absorption), self-phase modulation in an n_2 medium does not change the pulse shape. Only the spectral phase is modified by the nonlinear interaction.

The second step can be seen as the Fourier transform analogue of the first one: The phase function $\phi(\Omega)$ of the pulse spectral field $\tilde{\mathcal{E}}(\Omega)$ is modified, without affecting the spectral intensity ($\propto |\tilde{\mathcal{E}}(\Omega)|^2$). The spectrally broadened, nonbandwidth limited pulse is transformed into a bandwidth limited pulse. This process is sometimes referred to as "chirp compensation." Since the spectrum does not change in the second step, the new pulse has to be shorter than the input pulse. The compression factor K_c is given roughly by the ratio of the spectral width before ($\Delta\omega_{in}$) and after ($\Delta\omega_{out}$) the nonlinear element (step 1):

$$K_c \sim \frac{\Delta\omega_{out}}{\Delta\omega_{in}} \tag{7.1}$$

To analyze the second step, the amplitude at the output of the phase modulator is written in the frequency domain:

$$\tilde{\mathcal{E}}(\Omega) = |\tilde{\mathcal{E}}(\Omega)|e^{i\Phi(\Omega)} \tag{7.2}$$

7.1. PULSE COMPRESSION

As discussed in Chapter 1, the action of a linear optical element is described by its optical transfer function $\tilde{H}(\Omega) = R(\Omega)e^{-i\Psi(\Omega)}$. The amplitude of the pulse transmitted by such an element is the inverse Fourier transform of $\tilde{H}(\Omega)\tilde{\mathcal{E}}(\Omega)$:

$$\tilde{\mathcal{E}}(t) = \frac{1}{2\pi}\int_{-\infty}^{\infty} R(\Omega)|\tilde{\mathcal{E}}(\Omega)|e^{i[\Phi(\Omega)-\Psi(\Omega)]}e^{-i\Omega t}d\Omega \tag{7.3}$$

Let us consider a rather common situation where the amplitude response R is constant in the spectral range of interest. The peak amplitude will be highest if all spectral components add up in phase, which occurs if $\Phi(\Omega) = \Psi(\Omega)$. The linear element with the corresponding phase factor $\Psi(\Omega)$ is called an ideal compressor. Its phase response matches exactly the spectral phase of the chirped pulse. It was shown in Chapter 1 that the output pulse remains unchanged if the difference $(\Phi - \Psi)$ is a nonzero constant or a linear function of Ω. Possible techniques to synthesize the ideal compressor through spectral "filtering" [352] are presented in Section 7.2.

If a pulse is linearly chirped, its spectral phase varies quadratically with frequency. An ideal compressor is then simply an element with adjustable GVD, for example a prism or grating sequence, provided higher order dispersion can be neglected. A piece of glass of suitable length can compress a linearly chirped pulse, as discussed earlier. If the spectral phase of the pulse deviates from a parabola by a term $b_3\Omega^3$, we can use two different linear elements in series to construct the corresponding phase response, provided their ratio of third- and second-order phase response is different. This can be a grating and a prism pair, which allows us to tune the overall GVD and third-order dispersion independently. To illustrate this procedure let us formally write the phase of the transfer function of a prism (P) and a grating (G) sequence as:

$$\Psi_{P,G} = L_{P,G}\left(b_2^{(P,G)}\Omega^2 + b_3^{(P,G)}\Omega^3\right) \tag{7.4}$$

where $L_{P,G}$ is the prism (grating) separation and $b_i^{(P,G)}$ are device constants (cf. Chapter 2). To fit a spectral phase

$$\Phi = a_2\Omega^2 + a_3\Omega^3 \tag{7.5}$$

we need to solve a simple system of algebraic equations to find the adjustment for the prism and grating separation

$$a_2 = L_P b_2^{(P)} + L_G b_2^{(G)} \tag{7.6}$$

$$a_3 = L_P b_3^{(P)} + L_G b_3^{(G)} \tag{7.7}$$

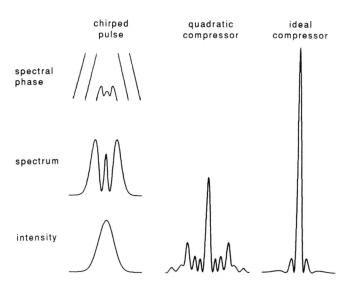

Figure 7.2: Gaussian pulse chirped in a nonlinear index medium and after compression. Note: The spectral phase is shown in the interval $(-\pi/2, +\pi/2)$.

A third element would be needed to compensate for the next term in the Taylor expansion of $\Phi(\Omega)$. Only one GVD element will be required if the phase behavior of the pulse to be compressed is sufficiently simple. However, even though the second-order dispersion dominates, higher order dispersion terms are unavoidable in all linear elements. These terms become more important as broader pulse spectra have to be handled. If pure GVD is desired, two elements are needed to eliminate third order dispersion.

As we have seen in Chapter 3, focusing a pulse in a medium with a nonlinear index leads to a spectral broadening where the frequency is a complicated function of time. A comparison of the compression of such a pulse in an ideal and a quadratic compressor is made in Fig. 7.2. Owing to the strong nonlinear behavior of the frequency modulation (upchirp in the center, downchirp in the wings), cf. Fig. 3.13, the spectral phase is far from being a parabola. Hence, a quadratic compressor cannot compensate the chirp very well. We can, for example, adjust for compression in the pulse center but then encounter broadening of the wings. Much better results can be obtained using an ideal compressor. However, the compressed pulse exhibits also satellites in this "ideal" case, because of the particular shape imparted to the spectrum $|\tilde{\mathcal{E}}(\Omega)|^2$ by self phase modulation. Since all frequency components are

7.1. PULSE COMPRESSION

in phase after passage through the "ideal compressor," the temporal shape is that with the highest peak power, corresponding to the particular pulse spectrum.

For some applications, it may be more important to achieve a smooth, narrow temporal profile, rather than a maximum peak intensity. In this section we shall focus mostly on simple and practical devices, such as quadratic compressors.

The drawback of using bulk materials as a nonlinear medium is the strong nonlinear behavior of the frequency modulation and the poor pulse quality after compression. Moreover, as discussed in Chapter 3 and 5, SPM is associated with self-lensing if the transverse beam profile is not uniform. Therefore, the achievable spectral broadening is limited.

An approach other than Gaussian optics is needed to achieve larger frequency modulations. Of these, the most widely used for pulse compression is the optical single-mode fiber. As discussed next, it leads to the production of almost linearly chirped output pulses [353].

7.1.2 The fiber compressor

Pulse propagation in single-mode fibers

An optical field can travel in single-mode fibers over long distances while remaining confined to a few microns. This is a typical property of a guided wave; see, for example, [354]. Figure 7.3 summarizes some properties of single-mode fibers made from fused silica. To support only a single-mode, the core radius must satisfy the relation $R < \lambda_\ell/(\pi\sqrt{n - n_c^2})$. Modes of higher order are generally undesired in guiding ultrashort light pulses be-

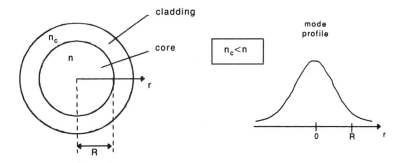

Figure 7.3: Cross-section of a single-mode fiber and corresponding mode profile.

cause of mode dispersion. The core diameters are therefore not larger than a few microns in the VIS. This implies a limit for the pulse power to avoid material damage. Typical damage intensities are in the order of $\mu J/\mu m^2$. With the focus on chirping for pulse compression, the only nonlinearity to be considered is the nonlinear refractive index effect. To describe fiber propagation under such conditions we have to use the three-dimensional wave equation supplemented by a nonlinear source term according to Eq. (3.113). A perturbative approach [13] is used to study nonlinear pulse propagation in the fiber. Assuming linear polarization the electric field in the fiber can be written:

$$E(x,y,z,t) = \frac{1}{2}\tilde{u}(x,y)\tilde{\mathcal{E}}(t,z)e^{i(\omega_\ell t - k_\ell z)} + c.c. \tag{7.8}$$

The quantity $\tilde{u}$ determines the mode profile which, to first order, is not affected by the nonlinear index change. Using the same procedure as in Chapters 1 and 3 to deal with the dispersion and the nonlinearity, respectively, the equation for the complex envelope is found to be:

$$\frac{\partial}{\partial z}\tilde{\mathcal{E}} - i\frac{k_\ell''}{2}\frac{\partial^2}{\partial t^2}\tilde{\mathcal{E}} = -i\gamma|\tilde{\mathcal{E}}|^2\tilde{\mathcal{E}}, \tag{7.9}$$

where $\gamma = n_2^{eff}\omega_\ell/c$. The differences as compared to the bulk medium can be roughly explained in terms of the refractive index, which we now write:

$$n_{NL}^{eff} = n_{eff} + n_2^{eff}|\tilde{\mathcal{E}}|^2. \tag{7.10}$$

The notation *effective* is to indicate that

(i) the refractive index which determines the dispersion is given not only by the material properties of the fiber core, but also by that of the cladding, as well as by the core shape and size;

(ii) the action of the nonlinearity must be averaged over the fiber cross-section which means $n_2^{eff} = n_2/A_{eff}$ where the effective fiber area is given by $A_{eff} = \left(\int |u|^2 dA\right)^2 / \int |u|^4 dA$.

Also, the propagation constant k_ℓ'' differs slightly from its value in the bulk material, a small deviation that only becomes important in the vicinity of the "zero-dispersion" wavelength λ_D where $k_\ell'' = 0$ (Fig. 7.4). This zero-dispersion wavelength can be shifted by suitable dopants and shaping of the core cross-section. The effective area, which depends on the fiber geometry

7.1. PULSE COMPRESSION

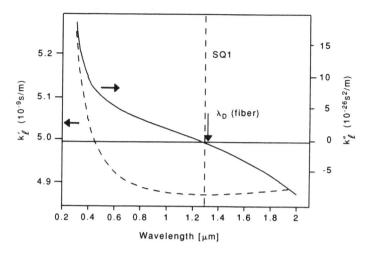

Figure 7.4: Dispersion parameters k'_ℓ and k''_ℓ of fused silica. For comparison, the zero-dispersion wavelength λ_D of a typical single-mode fiber made from SQ1 is also shown. Frequently, the dispersion of fibers is expressed in terms of $D = dk'_\ell/d\lambda = -(2\pi c/\lambda^2)k''_\ell$ in units of ps/nm·km. It describes the group delay in ps per nm wavelength difference and per km propagation length.

and the refractive indices n, n_c, can be somewhat smaller or larger than the core cross section. Its value ranges from 10 to 25 μm^2 in the visible and can be larger in the infrared because of the usually larger core radii.

It is sometimes useful to express Eq. (7.10) in terms of normalized quantities, as for example in [355]. Dimensionless coordinates which are particularly convenient for the description of soliton propagation in the spectral region where $k''_\ell < 0$ (see the next section) are $s = t/t_c$ and $\xi = z/z_c$, the two normalization constants satisfying the relation:

$$k''_\ell = -t_c^2/z_c. \tag{7.11}$$

Using a normalized amplitude $\hat{u} = \sqrt{\gamma z_c}\tilde{\mathcal{E}}$, the propagation equation becomes:

$$\frac{\partial}{\partial \xi}\hat{u} + i\frac{1}{2}\frac{\partial^2}{\partial s^2}\hat{u} = -i\gamma z_c|\hat{u}|^2\hat{u}. \tag{7.12}$$

This equation governing the pulse propagation in fibers with GVD and a nonlinear refractive index resembles the Schrödinger equation known from quantum mechanics. This analogy becomes most obvious after associating the nonlinear term with a potential, and interchanging position and time coordinates in Eq. (7.12). For this reason, this equation is often called the nonlinear Schrödinger equation (NLSE).

The propagation of ultrashort pulses in single-mode fibers is affected by dispersion and an "n_2" (often referred to as "Kerr type") nonlinearity. These effects were studied independently from each other in Chapters 1 and 2. For their characterization we introduced a dispersion length $L_D = \tau_{p0}^2/k_\ell''$ and a nonlinear interaction length $L_{NL} = (\gamma|\tilde{\mathcal{E}}_{0m}|^2)^{-1}$. Both quantities contain material parameters and properties of the input pulse. In terms of the two characteristic lengths, the limiting cases in which one effect dominates are valid for propagation lengths $L \approx L_D \ll L_{NL}$ and for $L \approx L_{NL} \ll L_D$, respectively. It is the intermediate situation characterized by the interplay of GVD and n_2 effect which shall be of interest now.

The behavior of pulses propagating through single-mode fibers is substantially different in the spectral range where $k_\ell'' > 0$ and $k_\ell'' < 0$. For wavelengths $\lambda_\ell < \lambda_D$ envelope and phase shaping appropriate for subsequent compression is achieved. For longer wavelengths $\lambda_\ell > \lambda_D$, "soliton" shaping may occur.

Compression of pulses chirped in the normal dispersion regime ($k_\ell'' > 0$)

Grisckkowsky and Balant [353] recognized the possibility of shaping optical pulses in single-mode fibers for subsequent pulse compression. To obtain the characteristics of a pulse as it travels through an optical single-mode fiber, we need to solve Eq. (7.10). The general case can only be solved numerically, for instance through the procedure outlined in Chapter 1. An example of such a calculation is shown in Fig. 7.5.

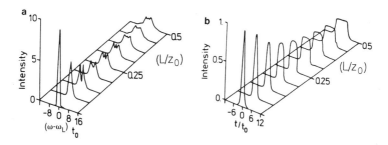

Figure 7.5: Development of spectrum and envelope of a pulse propagating through an optical fiber in the normal dispersion regime ($z_0 = 0.5L_D$). (a) Spectral intensity versus frequency. (b) Intensity versus normalized time (from [356]).

7.1. PULSE COMPRESSION

These results can easily be interpreted as follows. Group velocity dispersion broadens the initially unchirped pulse. Because $k''_\ell > 0$, longer wavelength components travel faster and accumulate along the leading edge of the pulse. Self-phase modulation produces new frequency components where longer (shorter) wavelength components arise in the leading (trailing) edge. These new frequency components induce an even faster pulse broadening. Owing to group velocity dispersion and the fact that new frequency components are preferably produced in the pulse edges, where the derivative of the intensity is large, the frequency develops an almost linear behavior over the main part of the pulse and the envelope approaches a rectangular shape. The broadening is associated with a decrease of the pulse intensity, and thus the self-phase modulation becomes less important for larger propagation lengths. Eventually a regime is reached where the spectrum remains almost unchanged and the fiber acts like an element with linear group velocity dispersion only.

Before discussing some quantitative results of the numerical evaluation of the nonlinear Schrödinger equation, we shall analyze the fiber propagation by means of a simple, heuristic model, describing temporal and spectral broadening under the simultaneous action of group velocity dispersion and a nonlinearity. We will establish two ordinary differential equations. For simplicity and to exploit previously derived results, we make the approximation that the pulse is linearly chirped and Gaussian over the entire propagation length:

$$\tilde{\mathcal{E}}(z,t) = \mathcal{E}_m(z) e^{-[1+ia(z)](t/\tau_G)^2} \tag{7.13}$$

with

$$\tau_G(z) = \sqrt{2\ln 2}\ \tau_p(z) \tag{7.14}$$

and

$$\tau_p(z) \Delta\omega_p(z) = 4\ln 2 \sqrt{1 + a^2(z)} \tag{7.15}$$

$$\mathcal{E}_m^2(z)\tau_p(z) = \mathcal{E}_m(0)^2 \tau_p(0) = \mathcal{E}_{0m}^2 \tau_{p0}. \tag{7.16}$$

The latter relations simply follow from the pulse duration bandwidth product, cf. Eq. (1.39), and from the requirement of energy conservation. From Eqs. (1.79) and (7.15) we find for the change in pulse duration:

$$\frac{d}{dz}\tau_p(z) = \frac{4\ln 2}{\tau_p(z)} k''_\ell \sqrt{\frac{\tau_p^2(z)\Delta\omega_p^2(z)}{(4\ln 2)^2} - 1 + \frac{\Delta\omega_p^2(z) k''^2_\ell}{\tau_p(z)} z}. \tag{7.17}$$

Next we need an equation for the change of the pulse spectrum due to self-phase modulation. Let us estimate the chirp coefficient as $a \approx 0.5 \delta\omega_p \tau_p/(4\ln 2)$

with $\delta\omega_p\tau_p$ given by Eq. (3.122). We introduced the factor 0.5 here to account for the fact that the actual chirp from self-phase modulation is not monotonous over the entire pulse. Together with Eqs. (7.15) and (7.16) this chirp parameter a yields for the change of the spectrum

$$\frac{d}{dz}\Delta\omega_p(z) = \frac{\ln 2}{\tau_p^3(z)}\left(\frac{\tau_{p0}}{L_{NL}}\right)^2 \frac{z}{\sqrt{1 + [\tau_{p0}/4\tau_p(z)]^2(z/L_{NL})^2}} \quad (7.18)$$

where $L_{NL} = (\gamma\tilde{\mathcal{E}}_{0m}^2)^{-1}$ as introduced earlier. In normalized quantities $\alpha = \tau_p(z)/\tau_{p0}$ for the temporal broadening, $\beta = \Delta\omega_p(z)/\Delta\omega_{p0}$ [where, from Eq. (7.15), $\Delta\omega_0 = 4\ln 2/\tau_{p0}$] for the spectral broadening, and $\xi = z/L_D$, the system of differential equations can be written as

$$\frac{d}{d\xi}\alpha = \frac{4\ln 2}{\alpha}\sqrt{\alpha^2\beta^2 - 1} + (4\ln 2)^2\frac{\beta^2}{\alpha}\xi \overset{\xi\gg 1}{\to} 4\ln 2\beta\left(1 + \frac{4\ln 2\beta}{\alpha}\xi\right), \quad (7.19)$$

$$\frac{d}{d\xi}\beta = \frac{1}{4\alpha^3}\xi\left(\frac{L_D}{L_{NL}}\right)\left[1 + \frac{1}{4}\left(\frac{\xi}{\alpha}\right)^2\left(\frac{L_D}{L_{NL}}\right)^2\right]^{-\frac{1}{2}} \overset{\xi\gg 1}{\to} \frac{L_D}{L_{NL}}\frac{1}{\alpha^2}, \quad (7.20)$$

with the initial conditions $\alpha(\xi = 0) = 1$ and $\beta(\xi = 0) = 1$. It is interesting to note here that the parameters of the input pulse and the fiber enter this equation only as L_D/L_{NL} if we measure the propagation length in units of L_D. This set of ordinary differential equations can easily be integrated numerically. The results for $L_D/L_{NL} = 1600$ are depicted in Fig. 7.6.

For fused silica ($k_\ell'' \approx 6 \times 10^{-26}$ s^2/m, $\bar{n}_2 \approx 3.2 \times 10^{-16}$ cm^2/W), an effective fiber area of 10 μm^2 and 500 fs input pulses at 600 nm, $L_D/L_{NL} = 1600$ corresponds to a peak power of 12 kW where $L_D \approx 4.2$ m and $L_{NL} \approx 2.6$ mm. The figure illustrates the properties discussed previously, in particular the "saturation" of the spectral broadening and the linear behavior of the temporal broadening for large propagation length. In our example the spectral broadening reaches a value of about 20, which sets a limit to the compression factor.

For the purpose of pulse compression, a large spectral broadening is desirable. Figure 7.6 suggests that very long fibers are not essential, since most of the spectral broadening occurs within a finite length L_F (which, however, is still larger than L_{NL}). To obtain an approximate relationship between the maximum spectral broadening $\bar{\beta}$ and L_F at which a certain percentage m of $\bar{\beta}$ is achieved, one can proceed as follows. Equations (7.19) and (7.20) are solved asymptotically in a perturbative approach. Substituting $\beta = \bar{\beta}$ into

7.1. PULSE COMPRESSION

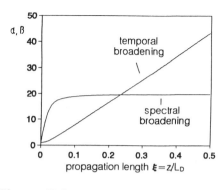
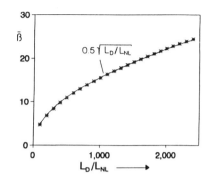

Figure 7.6: Temporal broadening α and spectral broadening β for pulse propagation in single-mode fibers as function of the normalized propagation length $\xi = z/L_D$ for $L_D/L_{NL} = 1600$ (left) and asymptotic spectral broadening, $\bar{\beta}$, versus L_D/L_{NL} (right).

Eq. (7.19) gives as solution for the temporal broadening $\alpha = 2(\sqrt{5}+1)\ln 2\bar{\beta}\xi$. This solution can be inserted into Eq. (7.20), which yields after integration from ξ_F to ∞:

$$\bar{\beta} - \beta(\xi_F) \approx \frac{L_D}{L_{NL}} \frac{1}{20\bar{\beta}} \frac{1}{\xi_F}. \tag{7.21}$$

If we chose $\xi_F = L_F/L_D$ so that at this length $m\%$ of the maximum spectral broadening occurs, we obtain:

$$\bar{\beta}^3 L_F \left(\frac{L_{NL}}{L_D^2}\right) \approx \frac{1}{20(1 - m/100)}. \tag{7.22}$$

The numerical evaluation of Eqs. (7.19) and (7.20) shown in Fig. 7.6 revealed that $\bar{\beta}$ varies as:

$$K_c \approx \bar{\beta} \approx 0.5 \sqrt{\frac{L_D}{L_{NL}}}. \tag{7.23}$$

$\bar{\beta}$ is also a rough measure of the compression that can be achieved. To satisfy relation (7.22), the fiber length at which a certain spectral broadening can be expected must be proportional to:

$$L_F \propto \sqrt{L_D L_{NL}}. \tag{7.24}$$

In our example, the propagation length at which 95% of the maximum broadening occurs is $L_F \approx 2.9\sqrt{L_D L_{NL}} \approx 0.1 z/L_D \approx 43$ cm, which is in good agreement with Fig. 7.6.

Using the inverse scattering technique [357], Meinel found an approximate analytical solution for the pulse after a long propagation length L [358]:

$$\tilde{\mathcal{E}}(t) = \begin{cases} \mathcal{E}_m e^{ia(t/\tau_p)^2} & |t| \leq \tau_p/2 \\ 0 & |t| > 0 \end{cases} \tag{7.25}$$

where

$$a \approx 0.7 \frac{\tau_p}{\tau_{p0}} \sqrt{\frac{L_D}{L_{NL}}} \tag{7.26}$$

and

$$\tau_p \approx 2.9 \frac{L}{\sqrt{L_D L_{NL}}} \tau_{p0}. \tag{7.27}$$

A linear element must be found for optimum pulse compression. For this particular pulse, in order to produce a chirp free output, the b_2 parameter of a quadratic compressor should be chosen as [359]:

$$b_2 \approx \frac{\tau_p^2}{4a(1 + 22.5/\tau_p^2)}. \tag{7.28}$$

For the actual analysis of the compression step one can favorably use the Poisson integral Eq. (1.66) to calculate the pulse behind the linear element and to determine its duration. The compression factor is found to be [358]:

$$K_c \approx 0.5 \sqrt{\frac{L_D}{L_{NL}}}. \tag{7.29}$$

As mentioned earlier the requirements for producing chirp-free output pulses of the shortest achievable duration and best quality can not be satisfied simultaneously. Tomlinson et al. [356] solved the nonlinear Schrödinger equation numerically and varied b_2 in the compression step to obtain pulses with the highest peak intensity. They found this to be a reasonable compromise between pulse duration and pulse quality (small satellites). Figure 7.7 shows this optimum compression factor as a function of the fiber length for various values of $\sqrt{L_D/L_{NL}}$. Depending on the parameters of the input pulse, there is an optimum fiber length at which pulse compression is most effective. From the numerical simulation this length could be estimated to be:

$$L_{opt} \approx 1.4 \sqrt{L_D L_{NL}}. \tag{7.30}$$

7.1. PULSE COMPRESSION

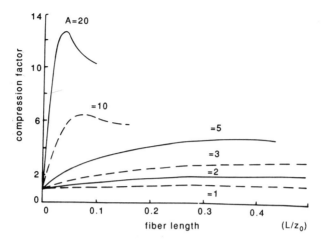

Figure 7.7: Optimum compression factor as a function of the fiber length. Notation: $A \approx 0.6\sqrt{L_D/L_{NL}}$, $z_0 = 0.5 L_D$ (adapted from [356]).

It was found that the corresponding compression factor, using such a fiber length, can be approximated by:

$$K_c \approx 0.37 \sqrt{\frac{L_D}{L_{NL}}}. \qquad (7.31)$$

Figure 7.8(a) shows a typical experimental setup for pulse compression. The input pulse is focused by a microscope objective into a single-mode optical fiber of suitable length. After recollimation, the chirped and temporally broadened pulse is sent through the linear element which, typically, is a grating or prism pair or a sequence of them. To achieve a substantial compression factor input pulses of certain power are necessary, as shown in Figs. 7.7 and 7.6. For compression of fs pulses, peak powers in the kW range are needed. Therefore, to apply the fiber compressor, pulses from most fs oscillators must be amplified first. For ps pulses L_D/L_{NL} takes on large values for peak powers even below 1 kW. It has been therefore possible to compress pulses from a cw Nd:Yag laser ($P_m \approx 100$ W) from 100 ps to 2 ps [360]. The fiber length used in this experiment was 2 km. Starting with 100 kW pulses of 65 fs duration Fujimoto et al. [361] succeeded in generating 16 fs pulses where the fiber length was only 8 mm.

Presently, the shortest light pulses are produced by the compression technique. Fork et al. [100] obtained 6 fs pulses with 65 fs, 300 kW pulses at the input of a 9 mm long fiber. At a wavelength of 620 nm this pulse duration

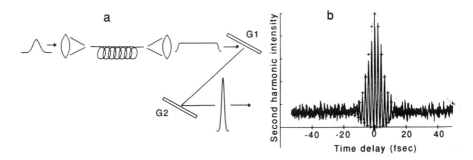

Figure 7.8: Sketch of a fiber-grating compressor (a) and interferometric autocorrelation of a compressed 6 fs pulse (b). For the latter a sequence of prism and gratings was used as linear element (from [100]).

corresponds to only three optical cycles. A corresponding interferometric autocorrelation function is shown in Fig. 7.8(b). As mentioned earlier, the broader the pulse spectrum to be handled, the more important it becomes to match the third-order dispersion of the linear element. Therefore, in this experiment, the authors used a combination of a grating and a prism pair for the independent adjustment of group velocity dispersion and third-order dispersion.

A practical limit for the achievable compression factor and pulse duration is of course given by the maximum power that can be propagated through the fiber without causing damage. However, before this limit is reached, other nonlinear and linear effects have to be considered. These include third-order material dispersion, Raman processes, the $|\tilde{\mathcal{E}}|^4$ dependence of the refractive index, and the effect of a shock term in the wave equation. A detailed discussion can be found, for example, in Refs. [289, 362, 363].

Soliton compression in the anomalous dispersion regime ($k_\ell'' < 0$)

In the spectral region where the nonlinearity (n_2) and dispersion (k_ℓ'') have opposite sign, the pulse propagation may have completely different properties than have been discussed so far. Theoretical studies of the NLSE for this case [212, 364] predicted the existence of pulses either with a constant, or with a periodically reproducing shape. The existence of these solutions, designated as "solitons," can be explained simply as follows. The nonlinearity ($n_2 > 0$) is responsible for spectral broadening and up-chirp. Because of the anomalous dispersion, $k_\ell'' < 0$, which in SQ1 single-mode fibers occurs for $\lambda > 1.31$ μm, the lower frequency components produced in the trailing

7.1. PULSE COMPRESSION

part travel faster than the long wavelength components of the pulse leading edge. Therefore, the tendency of pulse broadening owing to the exclusive action of GVD can be counterbalanced. Of course, the exact balance of the two effects is expected only for certain pulse and fiber parameters. For a rough estimate let us require that the chirp produced in the pulse center by the nonlinearity and the dispersion are of equal magnitude (but of opposite sign). Under this condition the pulse propagates through the fiber without developing a frequency modulation and spectral broadening. Let us use this requirement to estimate the parameters for form-stable pulse propagation. The effect of GVD is to create a pulse broadening and a down-chirp, with the change of the second derivative of the phase versus time equal to:

$$\Delta\left(\frac{\partial^2}{\partial t^2}\varphi(t)\right) = \frac{4k''_\ell}{\tau_{G0}^4}\Delta z \tag{7.32}$$

where we have used the characteristics of Gaussian pulse propagation (cf. Table 1.2). The chirp induced by self-phase modulation is [see Eq. (3.116)]:

$$\Delta\left(\frac{\partial^2}{\partial t^2}\varphi(t)\right) = -\frac{2\pi}{\lambda_\ell}\bar{n}_2\frac{\partial^2 I}{\partial t^2}\Delta z \tag{7.33}$$

Equating both relations for the chirp in the pulse center, we find as condition for the pulse parameters:

$$I_{om}\tau_{G0}^2 = \frac{\lambda_\ell k''_\ell}{\pi \bar{n}_2} \tag{7.34}$$

where I_{om} is the peak intensity.

The exact (analytical) solution of the nonlinear Schrödinger equation (7.11) shows that solitons occur if the pulse amplitude at the fiber input obeys the relation

$$|\tilde{\mathcal{E}}(s)| = \frac{N}{\sqrt{z_c\gamma}}\text{sech}(s), \tag{7.35}$$

where N is an integer and refers to the soliton order, and $z_c \approx \tau_{p0}^2/(1.76k''_\ell)^2 = L_D/1.76^2$. For $N = 1$ the pulse propagates with a constant, stable shape through the fiber. The action of nonlinearity and GVD exactly compensate each other. This shape of the fundamental soliton is:

$$\tilde{\mathcal{E}}(\xi, s) = \frac{1}{\sqrt{z_c\gamma}}\text{sech}(s)e^{-i\xi/2}. \tag{7.36}$$

The solution (7.36) can be verified by substitution of relation (7.36) into Eq. (7.11). Higher-order solitons ($N \geq 2$) periodically reproduce their shape after a distance given by:

$$L_p = \pi z_c/2. \tag{7.37}$$

Optical solitons in fibers were observed by Mollenauer et al. [365, 245]. The potential application in digital pulse-coded communication, has spurred the interest in solitons, which offer the possibility of propagating ultrashort light pulses over thousands of km, while preserving their duration [355].

In relation to pulse compression, it is interesting to note that an arbitrary unchirped input pulse of sufficiently large energy will eventually develop into a soliton. The soliton order N depends on the power of the input pulse. The soliton formation will always lead first to a substantial narrowing of the central pulse peak at a certain propagation length L_S. This narrowing is independent of how complex the following behavior is, provided the amplitude of the input pulse corresponds to $N > 1$. The pulse narrowing was studied experimentally and theoretically in [366]; see Fig. 7.9. It follows that pulses become shorter, the higher their input intensity i.e., the higher the order of the soliton is which they finally develop. Narrowing factors up to 30 were measured. A disadvantage of this method as a compression technique is the relatively poor pulse quality, which manifests itself in broad wings and side lobes.

Soliton narrowing has successfully been applied in connection with a fiber grating setup in two-step compression experiments [360], [288]. By means of this technique, Gouveia-Neto et al. [288] succeeded in compressing 90 ps pulses from a Nd:Yag laser at $\lambda = 1.32$ μm to 33 fs. In the first stage the pulses were compressed to 1.5 ps by using a fiber-grating configuration. Since normal dispersion was required here, a fiber with a zero dispersion wavelength $\lambda_D = 1.5$ μm was chosen. Subsequent propagation of these pulses through 20 m of single-mode fiber with $\lambda_D \approx 1.27$ μm led to a pulse width of 33 fs through soliton narrowing where N was estimated to be 12. One drawback of fiber compressors and, in particular, of multi-stage configurations is the relatively high loss factor. These losses are mainly associated with the coupling of the pulses into the fiber and diffraction at the gratings. The overall transmission in the experiment cited above was about 7%.

7.1.3 Pulse compression using bulk materials

One drawback of fiber compressors is that only pulses of relatively small energy can be handled. With fs input pulses the possible energies do not

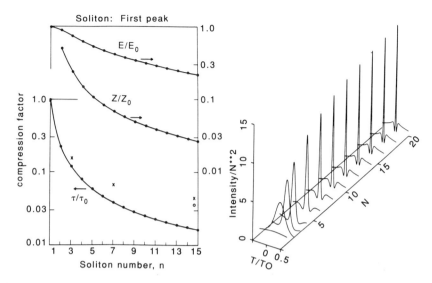

Figure 7.9: Pulse narrowing due to soliton shaping. (a) Calculated parameters of the first optimum narrowing and related experimental data (○, ×). (b) Pulse shape at optimal narrowing (from [366]).

exceed several tens of nJ. Earlier attempts to use bulk materials for the chirping of high energy ps pulses resulted in a relatively poor pulse quality at the compressor output owing to the nonlinear chirp behavior. As discussed before, to obtain an almost linear chirp across the main part of the pulse, a certain ratio of L_D and L_{NL} is necessary. This could be achieved by utilizing pulse propagation in single-mode fibers. The fiber lengths needed become smaller with shorter durations of the input pulse and, for fs pulses, are of the order of several millimeters. From Eq. (7.30) it can easily be seen that $L_{opt} \propto \tau_{p0}$ if the peak intensity is kept constant. Over such propagation distances suitably focused Gaussian beams do not change their beam diameter very much in bulk materials provided self-focusing can be neglected. The latter limits the possible propagation lengths to those shorter than the self-focusing length. The latter in turn can be adjusted by choosing an appropriate spot size of the focused beam, which sets an upper limit for the maximum pulse intensity. The maximum compression factor that can be achieved under such conditions can be estimated to be:

$$K_c \approx 0.3\sqrt{\frac{n_0 \bar{n}_2 P_0}{\lambda_\ell}} \qquad (7.38)$$

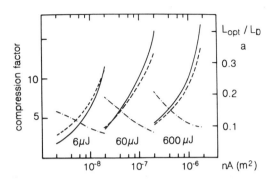

Figure 7.10: Plot of compression parameters after chirping in bulk SQ1 (fused silica) for different pulse energies and $\tau_{P0} = 60$ fs as a function of the beam cross-section at the sample input (location of the beam waist). $-\cdot-\cdot$: Compression factor, —— parameter of the optimum quadratic compressor ($a \approx 3b_2/\tau_{P0}^2$), $---$ normalized optimum sample length ($L_D \approx 3$cm). (From Ref. [367].)

where P_0 is the peak power of the input pulse (cf. [367]). Figure 7.10 shows results of a numerical evaluation of pulse compression using bulk SQ1 fused silica, and 60 fs input pulses of various energies.

Rolland and Corkum [368] demonstrated experimentally the compression of high-power fs light pulses in bulk materials. Starting from 500 μJ, 92 fs pulses from a dye amplifier, they obtained $\approx$ 20 fs compressed pulses at an energy of $\approx$100 μJ. The nonlinear sample was a 1.2 cm piece of quartz and the pulses were focused to a beam waist of 0.7 mm.

At very high intensities a white light continuum pulse can be generated [94]. With fs pulses, self phase modulation is expected to contribute substantially to the continuum generation. There have been successful experiments to compress continuum pulses produced by high-power fs pulses [279].

7.2 Shaping through spectral filtering

On a ps and longer time scale optical pulses can be shaped directly by elements of which the transmission is controlled externally. An example is a Pockels cell placed between crossed polarizers and driven by an electrical pulse. The transients of this pulse determine the time scale on which the optical pulse can be shaped. The advantage of this technique is the possibility of producing a desired optical transmission by synthesizing a certain electrical pulse, as demonstrated in the picosecond scale by Haner and War-

7.2. SHAPING THROUGH SPECTRAL FILTERING

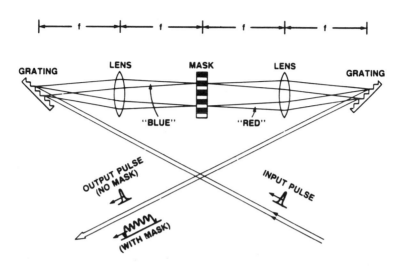

Figure 7.11: Spectral filtering in a dispersion-free grating-lens combination (from [352]).

ren [347]. The speed limitations of electronics have so far prevented the application of this technique to the fs scale.

A technique best suited for the shortest pulses consists of "manipulating" the pulse spectrum in amplitude and phase. This technique was originally introduced for ps light pulses [27, 369, 42] and later successfully applied and improved for fs optical pulses by Thurston et al. [370] and Weiner et al. [352]. The corresponding experimental arrangement is shown in Fig. 7.11.

The pulse to be shaped is spectrally dispersed using a grating or a pair of prisms. The spectrum is propagated through a mask which spectrally filters the pulse. The spectral components are recollimated into a beam by a second grating or pair of prisms. In the arrangement of Fig. 7.11, the two-grating–two-lens combination has zero group velocity dispersion, as can be verified easily by setting $z' = z = 0$ in Eq. (2.90). Each spectral component is focused at the position of the mask (the usual criterion for resolution applies). Since, to a good approximation, the frequency varies linearly in the focal plane of the lens, a variation of the complex transmission across the mask causes a transfer function of the form:

$$\tilde{H}(\Omega) = R(\Omega)e^{-i\Psi(\Omega)} \qquad (7.39)$$

where $R(\Omega)$ represents the amplitude transmission and $\Psi(\Omega)$ the phase change experienced by a spectral component at frequency Ω. These masks can be produced by microlithographic techniques. A pure phase filter, for example, could consist of a transparent material of variable thickness. If we neglect the effects caused by the finite resolution, the field at the device output is:

$$\tilde{E}_{out}(t) = \mathcal{F}^{-1}\{\tilde{E}_{out}(\Omega)\}, \tag{7.40}$$

where

$$\tilde{E}_{out}(\Omega) = \tilde{E}_{in}(\Omega)R(\Omega)e^{-i\Psi(\Omega)}, \tag{7.41}$$

and $\tilde{E}_{in}(\Omega)$ is the field spectrum of the input pulse. In principle, to achieve a certain output $\tilde{E}(t)$, we need to determine $\tilde{E}_{out}$ by a Fourier transform and divide it by the input spectrum $\tilde{E}_{in}$. This ratio gives the transmission function required for the mask. The mask can be a sequence of an amplitude-only and a phase-only filter to generate the desired $R(\Omega)\exp[-i\Psi(\Omega)]$.

A simple slit as mask acts as spectral window. Such spectral windowing can be used to enhance the pulse quality, in particular to reduce small satellites, in fiber grating compressors [371]. One aligns the window so that the wings of the spectrum, which are caused by the nonlinear chirp in the pulse edges, are blocked. The remaining linear chirp can then be compensated by a grating pair.

Many different output fields can be realized by using different mask functions. Of course, the shortest temporal structures that can be obtained are limited by the finite width of the input pulse spectrum. The narrowest spectral features that can be impressed on the spectrum are determined by the grating dispersion, mask structure, and size of the focal spot. Generally, amplitude masks introduce linear losses. For higher overall transmission pure phase masks are advantageous. Figure 7.12 shows some examples of intensity profiles obtained through spectral filtering of 75 fs pulses from a mode-locked dye laser.

Note that the action of any linear element can be interpreted as spectral filtering. If, for instance, the mirror separation d of a Fabry–Perot interferometer is larger than the original pulse width, the output will be a sequence of pulses, separated by $2d$. As explained in the chapter on coherent processes, pulse trains with well-defined phase relations are particularly interesting for coherent excitation in optical spectroscopy.

7.3 Problems

1. A self-phase modulated pulse exhibits a spectrum that shows characteristic modulations, cf. Fig. 7.2. Assume a Gaussian pulse of dura-

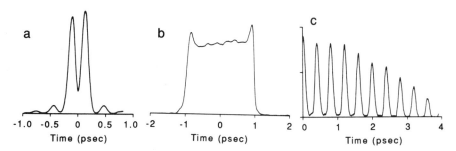

Figure 7.12: Cross correlation measurements of intensity profiles generated by spectral filtering of 75 fs pulses. (a) Half of an originally symmetric spectrum was shifted by π resulting in a zero area pulse. (b) Square pulse produced by an amplitude mask resembling a sinc-function. (c) THz pulse train produced by a pure phase mask. (From [352].)

tion τ_p and peak intensity I_0 which is propagated through a material of length L and nonlinear refractive index coefficient n_2. Explain why for large SPM the number of peaks in the spectrum is roughly given by ϕ_{max}/π. (ϕ_{max} is the maximum induced phase shift which occurs in the pulse center.)

2. Self-phase modulation of Gaussian beams is associated with self-focusing in a bulk material with positive n_2. Estimate the achievable spectral broadening of a Gaussian pulse. Neglect dispersion and require that the beam diameter does not reduce to less than half of its original value.

3. Design a compressor for the pulses of a synchronously pumped and cavity dumped dye laser ($\tau_p = 5$ ps, $W = 5$ nJ, $\lambda_\ell = 600$ nm). A fiber made from fused silica (core diameter 4 μm) and a pair of gratings (grating constant = 2000 mm^{-1}) are to be used. Find a convenient fiber length, the required grating separation, and estimate the achievable compression factor.

4. Find an approximate expression that describes the pulse duration at the output of an n-stage compression unit. Each compression step introduces an energy loss of p. Use the pulse parameters from the previous problem and $p = 0.5$ to argue about the efficiency of such devices.

5. Compare the peak intensity of the fundamental soliton in an optical single-mode fiber as estimated in Eq. (7.34) with the exact value from Eq. (7.35). Explain the difference.

6. It is often desirable to generate a square shaped pulse. One can think of a large number of methods to generate such a signal. The best solution depends often on the initial condition (pulse duration available) and the desired length of the square pulse. A square pulse can be obtained by linear and nonlinear optical processes. The process could be nonlinear absorption (for instance four-photon absorption), self-defocusing through a $n_4 I^4$ process followed by spatial filtering, pulse splitting-sequencing by interferometric delay lines, or spectral filtering as described at the end of this chapter. Compare these methods in terms of energy loss and residual modulation of the "flat" part of the pulse. Assume you have initially a 1 mJ Gaussian pulse of 50 fs duration and want to approximate a square pulse of (a) 40 fs duration; (b) 150 fs duration; and (c) 900 fs duration.

7. How would one generate a train of identical pulses, each being $180°$ out of phase with the previous one?

8. Consider a bandwidth limited pulse of duration τ_p. Is it possible to generate a temporal substructure with transients shorter than τ_p with spectral filtering? Explain your answer.

Chapter 8

Diagnostic Techniques

The femtosecond time scale is beyond the range of standard electronic display instruments. New methods have to be designed to freeze and time resolve events as short as a few optical cycles. We will start by simple, coarse methods that provide some estimate of the pulse duration, and introduce progressively more sophisticated techniques leading to a complete knowledge of the amplitude and phase of fs signals. Any measurement technique introduces some perturbation on the parameter to be measured. This problem is particularly acute in attempting to time resolve fs signals. As we have seen in the chapter on fs optics, reflection and transmission through most optical element will modify the signal to be measured. In addition, most diagnostic schemes involve nonlinear elements which may also have an influence on the amplitude and phase of the pulse to be measured. A careful analysis of the diagnostic instrument is required to find its exact transfer function. The inverse of this instrument transfer function should be applied to the result of the measurement, to obtain the parameters of the signal *prior* to entering the measuring device.

8.1 Intensity correlations

8.1.1 General properties

The temporal profile $I_s(t)$ of an optical signal can be easily determined, if a shorter (reference) pulse of known shape $I_r(t)$ is available. The method is to measure the intensity cross-correlation:

$$A_c(\tau) = \int_{-\infty}^{\infty} I_s(t) I_r(t-\tau) dt \tag{8.1}$$

Let us define the Fourier transforms of the intensity profiles as:

$$\mathcal{I}_j(\Omega) = \int_{-\infty}^{\infty} I_j(t) e^{-i\Omega t} dt, \tag{8.2}$$

where the subscript j indicates either the reference (r) or signal (s) pulse. The Fourier transform of Eq. (8.2) should not be confused with the spectral intensity (proportional to $|\tilde{\mathcal{E}}(\Omega)|^2$). The Fourier transform of the correlation (8.1) is $A_c(\Omega)$, related to the Fourier transforms of the intensities by:

$$A_c(\Omega) = \mathcal{I}_r(\Omega) \mathcal{I}_s^*(\Omega). \tag{8.3}$$

The shape of the signal $I_s(t)$ can be determined by first taking the Fourier transform $A_c(\Omega)$ of the measured cross-correlation, and dividing by the Fourier transform $\mathcal{I}_r(\Omega)$ of the known reference pulse $I_r(t)$. The inverse Fourier transform of the complex conjugate of the ratio $A_c(\Omega)/\mathcal{I}_r(\Omega)$ is the temporal profile $I_s(t)$. In presence of noise, this operation leads to large errors unless the reference function is the (temporally) shortest of the two pulses being correlated (or the function with the broadest spectrum). The ideal limit is of course that of the reference being a delta-function. In the frequency domain, we are dividing by a constant. In the time domain, the shape of the correlation $A_c(\tau)$ is identical to that of the signal $I_s(t)$. Even in that ideal case, the intensity cross-correlation has an important limitation: it does not provide any information on the phase content (frequency or phase modulation) of the pulse being analyzed.

8.1.2 The intensity autocorrelation

A reference pulse much shorter than the signal cannot always be generated. In the ideal cases where such a pulse is available, there is still a need for a technique to determine the shape of the reference signal. It is therefore important to consider the limit where the signal itself has to be used as reference. The expression (8.1) with $I_s(t) = I_r(t) = I(t)$ is called an intensity autocorrelation. An autocorrelation is always a symmetric function – this property can be understood from a comparison of the overlap integral for positive and negative arguments τ. According to Eq. (8.3), the Fourier transform of the autocorrelation is a real function, consistent with a symmetric

8.2. INTERFEROMETRIC CORRELATIONS

function in the time domain. As a result, the autocorrelation provides only very little information on the pulse shape, since an infinity of symmetric and asymmetric pulse shapes can have very similar autocorrelation. Nevertheless, the intensity autocorrelation is the most widely used diagnostic technique, because it can be easily implemented, and is the first tool used to determine whether a laser is producing short pulses rather than intensity fluctuations of a continuous background. Typical examples are given in Section 8.3. The intensity autocorrelation is also used to quote a "pulse duration". The most widely used procedure is to *assume* a pulse shape (generally a sech^2 or a Gaussian shape), and to "determine" the pulse duration from the known ratio between the FWHM of the autocorrelation and that of the pulse. The parameters pertaining to the various shapes are listed in Table 8.1 in Section 8.4.

8.1.3 Intensity correlations of higher order

An intensity correlation of order $(n+1)$ can be defined as:

$$A_n(\tau) = \int_{-\infty}^{\infty} I(t) I^n(t-\tau) dt. \tag{8.4}$$

For $n > 1$, the function defined by Eq. (8.4) has the same symmetry as the pulse. In fact, for a reasonably peaked function $I(t)$, $\lim_{n \to \infty} I^n(t) \propto \delta(t)$, and the shape of the correlation $A_n(\tau)$ becomes a good approximation of the pulse shape I(t). Higher order correlations are very convenient and powerful tools to determine intensity profiles. Because they require nonlinear optical processes of order $(n+1)$, they are limited to high power fs systems.

Correlations with $n = 2$ coupled with spectroscopic measurements have been exploited to provide full characterization of pulses in amplitude and phase. Some examples are analyzed in Section 8.4.5.

8.2 Interferometric correlations

8.2.1 General expression

We have analyzed in Chapter 2 the Michelson interferometer, and defined the field[1] correlation measured by that instrument as:

$$G_1(\tau) = \tilde{A}_{12}^+(\tau) + c.c = \frac{1}{4} \int_{-\infty}^{\infty} \tilde{\mathcal{E}}_1(t) \tilde{\mathcal{E}}_2^*(t-\tau) e^{i\omega_\ell \tau} dt + c.c. \tag{8.5}$$

---
[1] The field correlation is often referred to as a first order correlation.

We have seen also [Eq. (2.4)] that the Fourier transform of $A_{12}^+(\tau)$ is equal to $\tilde{E}_1^*(\Omega)\tilde{E}_2(\Omega)$. Hence, the Fourier transform of the autocorrelation (identical fields) is proportional to the spectral intensity of the pulse. Therefore, a first order field autocorrelation does not carry any other information than that provided by a spectrometer.

In a Michelson interferometer, let us substitute to the detector a second harmonic generating crystal (type I) and a filter to eliminate the fundamental. Instead of the expression (8.5), the detected signal is a second order interferometric correlation, proportional to the function:

$$G_2(\tau) = \int_{-\infty}^{\infty} \{[E_1(t-\tau) + E_2(t)]^2\}^2 dt \tag{8.6}$$

A Mach-Zehnder interferometer can also be substituted for a Michelson interferometer for such a measurement [198]. Substituting for the fields the usual amplitude and phase dependence, leads to the following decomposition for the second order correlation:

$$G_2(\tau) = A(\tau) + \operatorname{Re}\left\{4\tilde{B}(\tau)e^{i\omega_\ell \tau}\right\} + \operatorname{Re}\left\{2\tilde{C}(\tau)e^{2i\omega_\ell \tau}\right\} \tag{8.7}$$

where

$$A(\tau) = \int_{-\infty}^{\infty} dt \left\{\mathcal{E}_1^4(t-\tau) + \mathcal{E}_2^4(t) + 4\mathcal{E}_1^2(t-\tau)\mathcal{E}_2^2(t)\right\} \tag{8.8}$$

$$\tilde{B}(\tau) = \int_{-\infty}^{\infty} dt \left\{\mathcal{E}_1(t-\tau)\mathcal{E}_2(t)\left[\mathcal{E}_1^2(t-\tau) + \mathcal{E}_2^2(t)\right] e^{i[\varphi_1(t-\tau)-\varphi_2(t)]}\right\} \tag{8.9}$$

$$\tilde{C}(\tau) = \int_{-\infty}^{\infty} dt \left\{\mathcal{E}_1^2(t-\tau)\mathcal{E}_2^2(t) e^{2i[\varphi_1(t-\tau)-\varphi_2(t)]}\right\} \tag{8.10}$$

The purpose of the decomposition (8.7) is to show that the correlation has three frequency components[2] centered respectively around zero frequency, around ω_ℓ and $2\omega_\ell$. Most often, the detection system of the correlator will act as a low pass filter, eliminating all but the first term of the expansion. The interferometric correlation reduces then to the sum of a background term and the intensity correlation (labelled $A_c(\tau)$ in Eq. 8.1). The terms A, $\tilde{B}$ and $\tilde{C}$ of the expansion (8.7) can be extracted from a measurement by taking the Fourier transform of the data, identifying the cluster of data near the three characteristic frequencies, and recovering them by successive inverse Fourier transforms.

---

[2]Here "frequency" refers to the variation of the function $G_2(\tau)$ as a function of its argument τ. The latter argument τ is the delay parameter which is continuously tuned in the correlation measurement.

8.2.2 Interferometric autocorrelation

General properties

Let us consider in more detail the particular case of the above expressions for the cross-correlation (8.6) through (8.10) where the two fields $E_1 = E_2 = E$. At $\tau = 0$, the peak value of the function $A(\tau)$ is $6 \int \mathcal{E}^4(t)dt$. For large delays compared to the pulse duration, the cross product term vanishes, leaving a background of $A(\infty) = 2 \int \mathcal{E}^4 dt$. The d.c. term of the interferometric autocorrelation, $A(\tau)$, — which is in fact an intensity autocorrelation — has thus a peak to background of 3 to 1. The measurement leading to $A(\tau)$ [Eq. (8.8)] is generally referred to as the *intensity autocorrelation with background*, as opposed to the *background free autocorrelation* leading to the expression $A_c(\tau)$ (8.1).

When all terms of the autocorrelation are recorded, the constructive interference terms at zero delay add to $16 \int \mathcal{E}^4(t)dt$. The peak to background ratio for the interferometric autocorrelation is thus 8 to 1.

The third term of the expansion of Eq. (8.10) is a correlation of the second harmonic fields. In the absence of phase modulation — i.e., for bandwidth limited pulses — this function is identical to the intensity autocorrelation. This property has been exploited to determine if a pulse is phase modulated or not [372].

As any autocorrelation, the interferometric autocorrelation is a symmetric function. However, as opposed to the intensity autocorrelation, it contains phase information. The shape and phase sensitivity of the interferometric autocorrelation can be exploited to:

1. *qualitatively* test the absence or presence of phase modulation, and eventually determine the type of modulation

2. quantitatively measure a *linear* chirp

3. determine, in combination with the pulse spectrum, the pulse shape and phase by fitting procedures.

Linearly chirped pulses

The sensitivity of the interferometric autocorrelation to chirp is well illustrated by the experimental recordings made on the beam from a Ti:sapphire laser in Ref. [145]. The lower and upper envelopes of the interference pattern split evenly from the background level in Fig. 8.1 pertaining to an unchirped

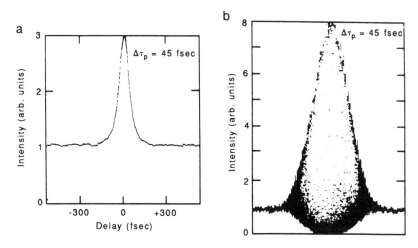

Figure 8.1: Intensity (with background) (a) and interferometric (b) autocorrelation traces for a mode-locked Ti:sapphire laser pulse after extracavity pulse compression. Note the peak to background ratios of 3/1 and 8/1 for the intensity and interferometric autocorrelations, respectively (from [145]).

45 fs pulse. In the case of a phase modulated pulse as in Fig. 8.2, the interference pattern is much narrower than the pulse intensity autocorrelation. The wings of the interferometric autocorrelation are identical to those of the intensity autocorrelation. The level at which the interference pattern starts relative to the peak (2.8/8 in the case of Fig. 8.2) is a measure of the chirp, as explained below.

A simple tabulation of the chirp can be made by considering a linearly chirped Gaussian pulse ($\mathcal{E}(t) = \exp[-(1+ia)(t/\tau_G)^2]$), for which the interferometric autocorrelation can be determined exactly:

$$G_2(\tau) = \left\{ 1 + 2\exp\left[-\left(\frac{\tau}{\tau_G}\right)^2\right] + 4\exp\left[-\frac{a^2+3}{4}\left(\frac{\tau}{\tau_G}\right)^2\right] \cos\frac{a}{2}\left(\frac{\tau}{\tau_G}\right)^2 \cos\omega_\ell\tau \right.$$
$$\left. + 2\exp\left[-(1+a^2)\left(\frac{\tau}{\tau_G}\right)^2\right] \cos 2\omega_\ell\tau \right\} \quad (8.11)$$

A graphical representation of the upper and lower envelopes as a function of the chirp parameter a is shown in Fig. 8.3. Comparison of Figs. 8.2 and 8.3 indicate a chirp parameter of roughly $a = 20$ for the experimental pulse. This is of course only an approximation, but it gives a good estimate of the magnitude of the frequency modulation near the pulse center.

8.2. INTEROMETRIC CORRELATIONS

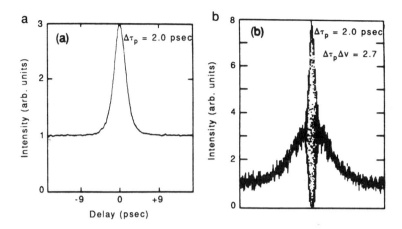

Figure 8.2: Intensity (a) and interferometric (b) autocorrelation traces for a mode-locked phase modulated Ti:sapphire laser pulse (from [145]).

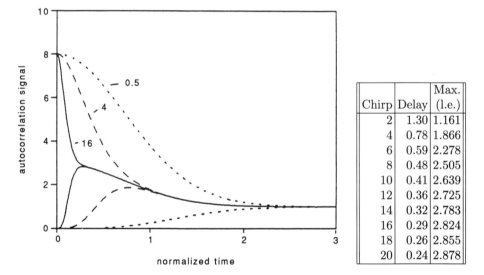

Figure 8.3: Interferometric autocorrelations of Gaussian pulses $\tilde{\mathcal{E}}(t) = \exp\{-(1+ia)(t/\tau_G)^2\}$ for various values of the linear chirp parameter a. The upper and lower envelopes of the autocorrelations are plotted (left) for 3 values of the chirp parameter a. The upper and lower envelopes merge with the intensity autocorrelation. The table on the right shows the position (delay and value) of the maxima of the lower envelope (l.e.) of the interferometric autocorrelation as a function of chirp parameter a.

Averaging over the pulse train

Most often, measurements are an average over a pulse train. It has been demonstrated by Van Stryland [197] that the intensity autocorrelation of pulses with a statistical distribution of pulse durations is shaped like a double sided exponential. This is precisely the shape of the autocorrelation of a single sided exponential pulse. The autocorrelation of the output of synchronously pumped lasers have typically such a double sided exponential shape. It would be incorrect to conclude that the pulse generated by these lasers have a single sided exponential shape. Theoretical simulations [199, 206] have indeed confirmed that fluctuations of the pulse duration along the train are at the origin of the observed autocorrelation.

A similar ambiguity exists in the case of the interferometric autocorrelation of a train of pulses. The measurement can be either interpreted as the interferometric autocorrelation of identical chirped pulses, or as the average of interferometric autocorrelations of bandwidth limited pulses of different frequencies. To illustrate this point, let us consider unchirped Gaussian pulses with a Gaussian distribution of frequencies $F(\Delta\Omega)$ centered around ω_ℓ ($\Omega = \omega_\ell + \Delta\Omega$):

$$F(\Delta\Omega) = \frac{\tau_G}{b\sqrt{\pi}} e^{-(\Delta\Omega\,\tau_G/b)^2}. \tag{8.12}$$

The total autocorrelation is the statistical average of the autocorrelations at each frequency Ω:

$$\begin{aligned}G_2(\tau) &= \int_{-\infty}^{\infty} \left\{ 1 + 2\exp[-(\frac{\tau}{\tau_G})^2] + 4\exp[-\frac{3}{4}(\frac{\tau}{\tau_G})^2]\cos\Omega\tau \right. \\ &\quad + \left. \exp\left[-(\frac{\tau}{\tau_G})^2\right]\cos 2\Omega\tau \right\} F(\Delta\Omega)d\Omega \end{aligned} \tag{8.13}$$

The integration over frequency can easily be performed. For small chirps, we can make the approximation in Eq. (8.11) that $\cos[\frac{a}{2}(\frac{\tau}{\tau_G})^2] \approx 1$. The equation obtained after substitution of (8.12) into Eq. (8.13) and subsequent integration is undistinguishable from Eq. (8.11) for $b = a$. Therefore, it is important to verify that the pulse train is constituted of *identical* pulses, in amplitude (energy), duration and frequency. It is relatively easy to check whether the pulses have constant energy and duration by displaying simultaneously the pulse train in fundamental and second harmonic. If both show no fluctuation, it can be said with reasonable certainty that the intensity autocorrelation represents a single pulse. It can be verified that there are no pulse to pulse variation in frequency by displaying the pulse train on an oscilloscope, after transmission through a spectrometer or reflection off a thin ($\ll 100\mu$m) etalon.

8.3 Measurement techniques

8.3.1 Nonlinear optical processes for measuring fs pulse correlations

In ultrafast optics, second harmonic generation is the most widely used technique for recording second order correlations. Because it is a nonresonant process of electronic origin, the nonlinearity is fast enough to measure pulses down to $10^{-14}s$ duration. While applicable through the IR and visible spectrum, the method is limited at short wavelength ($\lambda < 380nm$) by the UV absorption edge of optical crystals. Techniques that have been used successfully for shorter wavelengths include multi-photon ionization [373] surface second harmonic generation [374], and two-photon luminescence [375]. Third order processes such as the optical Kerr effect have also been applied recently to the diagnostic of fs UV pulses [376, 377, 378]. This third-order correlation as discussed in Section 8.1.3, has the additional advantage of being sensitive to pulse asymmetry.

8.3.2 Recurrent signals

It is assumed here that we dispose of an ensemble of identical fs pulses, so that the correlations can be constructed from a large number of measurements taken for different delay parameters τ.

An example of a simple second order correlator is sketched in Fig. 8.4(a). The beams to be correlated are cross-polarized, and combined with a polarizing beam splitter. An optical delay is used to adjust the delay of the reference signal $I_r(t-\tau)$. The cross-polarized beams are sent orthogonally polarized into a nonlinear crystal phase matched for type two second harmonic generation. If the conditions outlined below are satisfied, the second harmonic signal is proportional to the function $A_c(\tau)$ defined in Eq. (8.1). This measurement — or function — is generally referred to as the "background-free" correlation, as opposed to the correlation with background. The latter is obtained by frequency-doubling the output of a Michelson interferometer (parallel polarization) in a crystal phase matched for type I second harmonic generation. An alternate technique to generate $A_c(\tau)$ is to use beams with parallel polarization intersecting in a nonlinear crystal. The "background

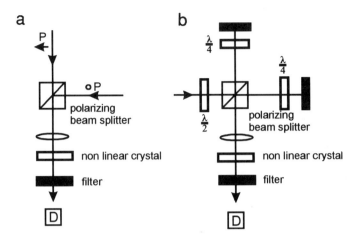

Figure 8.4: Basic intensity cross-correlator, using second harmonic type II detection (a). A polarizing beam-splitter cube combines the beams to be correlated. The second harmonic signal generated by the crystal is proportional to the product of the fundamental intensities along the two orthogonal directions of polarization. The sketch on the right side (b) shows a simple autocorrelator using the same type of detection. The same "recombining" polarizing beamsplitter cube can be used to split the beams into the two arms of the interferometer. Two quarter wave plates are used to rotate the polarization of the beam reflected by the mirrors of the two delay arms by 90°.

free" signal is the second harmonic generated with wave vector $\mathbf{k_2}$ along the bisector of the two wave vectors $\mathbf{k}_s$ and $\mathbf{k}_r$. The crystal orientation has to satisfy the phase matching condition $\mathbf{k_2} = \mathbf{k}_s + \mathbf{k}_r$.

With either background-free techniques, the second harmonic field is proportional to the product of the fundamental fields:

$$\mathcal{E}_{SHG}(\Omega) = \eta(\Omega)\mathcal{E}_s(\Omega)\mathcal{E}_r(\Omega), \tag{8.14}$$

or, for the spectral intensities:

$$I_{SHG}(\Omega) = \frac{2\mu_0 c}{n}|\eta(\Omega)|^2 I_s(\Omega) I_r(\Omega) \tag{8.15}$$

In order for the instrument sketched in Fig. 8.4 to measure the true cross or auto-correlation, it is essential that the second harmonic conversion efficiency $\eta(\Omega)$ be a constant over the frequency range of the combined pulses. Another way to express the same condition, assuming that $\chi^{(2)}$ is frequency independent, is to state that the effective crystal length should be shorter than the coherence length of harmonic generation over the pulse bandwidth (cf. Chapter 3). The "effective length" can be either the physical crystal thickness, the confocal parameter (Rayleigh range ρ_0) of the focused fundamental beams, or the overlapping region of the fundamental pulses. The shorter the crystal, the broader the bandwidth over which phase matched harmonic conversion is obtained, but the lower the conversion efficiency. There is clearly a compromise to be reached between bandwidth and sensitivity. In Eq. (8.15), the bandwidth efficiency factor $\eta(\Omega)$ includes only the frequency dependence of the phase matching condition, and not a finite response time for the harmonic generation process. It is assumed here that the response time of the second harmonic process is much shorter than the pulses to be measured, which is a reasonable assumption since the second order nonlinearity is a nonresonant electronic process.

Provided the pulses can be approximated by a Gaussian, a simple test can be performed to determine whether the proper focusing and crystal thickness has been chosen. In the case of the autocorrelation [Fig. 8.4(b)], a standard spectrometer (a 25 cm spectrometer is generally sufficient) is used to record the spectral intensities of the fundamental and second harmonic [198]. If the conversion efficiency is independent of frequency, the ratio of second harmonic to the *square* of the fundamental spectral intensity will be a constant in the case of Gaussian pulses, according to Eq. (3.91). The spectrum of the second harmonic will be narrower than the squared fundamental spectrum if the effective crystal length is too long. As a consequence of the second harmonic conversion efficiency being frequency dependent, the measured correlation width will be longer than the exact correlation length.

The background-free autocorrelation function A_c [Eq. (8.1)] for a fluctuating cw signal consists of a symmetric "bump" riding on an infinite background (Fig 8.5). The width of the bump is a measure of the temporal width of the fluctuations, and the contrast ratio (peak-to-background ratio of A_c) is a measure of the modulation depth. A 100% modulation depth results in a peak-to-background ratio of 2 to 1 [379]. Any "background free" signal of finite duration results in a function A_c of finite width [Fig 8.5(c)]. If that signal has some fine structure (amplitude modulation), a narrow spike will appear in the middle of the correlation function [Fig 8.5(d)]. This is the coherence spike, typical of a signal consisting of a burst of *amplitude* noise [151]. These considerations do not apply to the phase content or phase coherence of the

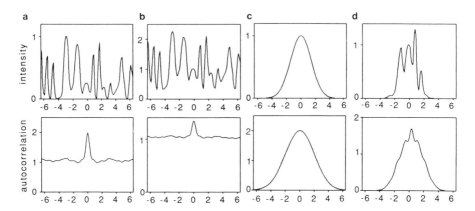

Figure 8.5: Some typical waveforms (intensity versus time) (top) and corresponding intensity autocorrelation $A_c(\tau)$ (bottom). From left to right: (a) continuous signal with 100 % amplitude modulation; (b) noisy cw signal; (c) pulse; and (d) noisy pulse.

pulse: the intensity correlations are the same whether the pulse is at a fixed carrier frequency, or has a random or deterministic frequency modulation.

8.3.3 Single shot measurements

Not all lasers provide a train of identical pulses and/or work at high repetition rate. Pulse to pulse fluctuation can be particularly severe in oscillator-amplifier systems. Single shot autocorrelators are therefore highly desirable.

Intensity autocorrelators

One of the first intensity autocorrelators for mode-locked lasers was a single shot instrument [380]. The beam to be measured is split in two beams, which are thereafter sent with opposite propagation vector into a nonlinear medium. The first autocorrelator was based on two photon excitation rather than second harmonic generation: the medium (for instance a dye solution) was selected for its large two photon absorption and subsequent fluorescence. Because of the larger optical field in the region where the two counterpropagating pulses collide, the observed pattern of two-photon fluorescence essentially displays the intensity autocorrelation (with background). Because of the higher conversion efficiency of SHG, two-photon fluorescence is not widely used in the fs time scale, except in the UV, where no trans-

8.3. MEASUREMENT TECHNIQUES

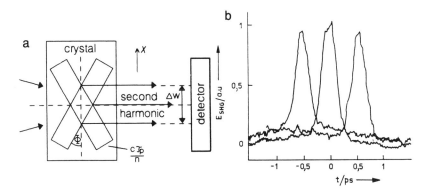

Figure 8.6: (a) Basic principle of the single shot autocorrelator of Jansky *et al* [382]. The nonlinear crystal is oriented for phase matched type I SHG for two beams intersecting at an angle 2Φ. Arrows indicate the propagation direction of the second harmonic (z direction). (b) Typical recording for a 250 fs pulse. Temporal calibration is made by recording the shift of the trace resulting from insertion of a 0.36 mm glass plate in either arm of the autocorrelator (from [383]).

parent nonlinear crystals can be found. To circumvent the difficulty of spatially resolving the μm size of the two-photon fluorescence trace, the beams are made to intersect at a small angle, thereby magnifying the fluorescence trace [381, 375].

Single shot autocorrelators using SHG have also been designed. In an arrangement developed for ps pulses by Jansky *et al.* [382] and Gyuzalian *et al.* [384], the autocorrelation in time is transformed into a spatial intensity distribution. This method has been applied by numerous investigators to the fs scale [385, 386, 383]. The instrument is a typical non-colinear second harmonic autocorrelator. The nonlinear crystal is oriented for phase matched type I SHG for two beams intersecting at an angle 2Φ [Fig. 8.6(a)]. Provided the beam waist in the overlap region w_0 is much larger than the pulse length $v_g\tau_p$, the intensity distribution across the second harmonic beam corresponds to the intensity autocorrelation function. Cylindrical focusing, and a linear array detector can be used to capture the intensity profile of the autocorrelation. A typical measurement is shown in Fig. 8.6(b). The width of the autocorrelation function $\Delta\tau_a$ is proportional to the diameter Δw of the second harmonic beam:

$$\Delta w = \frac{\Delta\tau_a v_g}{\sin\Phi}. \tag{8.16}$$

Let us consider the projection of the two intersecting beams on axis z (along the direction of propagation) and x (the orthogonal direction). The components of the pulse envelope are co-propagating along the z direction, counterpropagating along the x axis. For a pulse with a temporal profile $\tilde{\mathcal{E}}(t)$, the component of the two envelopes propagating along the x direction are $\tilde{\mathcal{E}}_x = \tilde{\mathcal{E}}(t - x/L_x)$ and $\tilde{\mathcal{E}}_{-x} = \tilde{\mathcal{E}}(t + x/L_x)$ with $L_x = v_g/\sin\Phi$. The second harmonic intensity along the x direction is proportional to the product $\tilde{\mathcal{E}}_x \tilde{\mathcal{E}}_{-x}$ [382]. If we consider for instance a linearly chirped Gaussian pulse, the second harmonic just behind the crystal is also a Gaussian:

$$\tilde{\mathcal{E}}_{SHG}(t) \propto \exp\left[-2\left(\frac{t^2}{\tau_G^2} + \frac{x^2 \sin^2\Phi}{v_g^2 \tau_G^2}\right)\left(1 + i\frac{a}{2}\right)\right]. \tag{8.17}$$

The spatial dependence of the SHG intensity is indeed an intensity autocorrelation, expanded transversely by the factor $1/\sin\Phi$. As noted by Jansky et al. [382] and further investigated by Danielius et al. [387], the spatial phase dependence of the second harmonic field indicates that the wavefront is no longer a plane wave, but has a (cylindrical) curvature proportional to the chirp parameter a. Referring to Gaussian beam propagation Eq. (1.127), the curvature is approximately $R = k_\ell v_g^2 \tau_G^2 / 2a \sin^2\Phi$.

An exact analysis of the beam propagation after the crystal can be made by spatial Fourier transforms, as in Section 5.9.2, starting with the wave equation in the retarded frame Eq. (5.71):

$$\frac{\partial}{\partial z}\tilde{\mathcal{E}}_{SHG} = -\frac{i}{2k_\ell}\frac{\partial^2}{\partial x^2}\tilde{\mathcal{E}}_{SHG}. \tag{8.18}$$

Taking the Fourier transform along the transverse spatial coordinate z and integrating along the propagation direction z:

$$\tilde{\mathcal{E}}_{SHG}(k_x, z, t) = \tilde{\mathcal{E}}_{SHG}(k_x, z = 0, t) e^{\frac{i}{2k_\ell} k_x^2 z}. \tag{8.19}$$

Taking the inverse Fourier transform leads to the field distribution after a propagation distance z:

$$\tilde{\mathcal{E}}_{SHG}(x, z, t) = \int_{-\infty}^{\infty} \tilde{\mathcal{E}}_{SHG}(k_x, z = 0, t) e^{-ik_x x} e^{\frac{i}{2k_\ell} k_x^2 z} dk_x. \tag{8.20}$$

Assuming that the overlapping region is short enough to produce an undistorted second harmonic, as discussed in Chapter 3, the second harmonic field at the end of the crystal is:

$$\tilde{\mathcal{E}}_{SHG}(x, z = 0, t) = \eta \tilde{\mathcal{E}}(t - \frac{x}{L_x})\tilde{\mathcal{E}}(t + \frac{x}{L_x}), \tag{8.21}$$

8.3. MEASUREMENT TECHNIQUES

where η is the proportionality factor of Eq. (8.14) assumed to be constant over the frequency range of interest. Taking the Fourier transform gives us the function $\tilde{\mathcal{E}}_{SHG}(k_x, z = 0, t)$ to be inserted in Eq. (8.20):

$$\tilde{\mathcal{E}}_{SHG}(k_x, z = 0, t) = \int \eta e^{ik_x x} \tilde{\mathcal{E}}(t - \frac{x}{L_x}) \tilde{\mathcal{E}}(t + \frac{x}{L_x}) dx. \quad (8.22)$$

The spatial distribution given by Eq. (8.20) focuses after a distance z_F for which the phase factor $(i/2k_\ell)k_x^2 z_F$ compensates the phase factor due to chirp in Eq. (8.17). It is possible to reconstruct the chirp by matching the intensity distributions measured at $z = 0$ and $z = z_F$ to the distributions calculated with help of Eqs. (8.18) through (8.22).

Interferometric autocorrelator

By recording the spatial profile of the second harmonic with interferometric accuracy, Salin et al. [388] showed that it was actually possible to record an interferometric autocorrelation. Because of diffraction effects, as explained earlier, the method is difficult to implement, and does not provide a recording with a 8 to 1 peak to background contrast. Simpler methods are available, making use of the tilt in energy front introduced by any dispersive element. We have seen in the chapter on fs optics that the energy front is tilted with respect to the wavefront by a prism (or any other element introducing angular dispersion). As shown by Szabo et al [36], this property can be exploited to provide a variable delay along a transverse coordinate of the beam. A glass wedge is inserted in one or both arm(s) of a Michelson type autocorrelator (Fig. 8.7). As in the previous method, the spatial (transverse) distribution of second harmonic is proportional to the pulse intensity autocorrelation. Obtaining such an intensity autocorrelation, however, assumes that the beams coming from both arms of the correlator add constructively towards the detector, destructively towards the source. Such a condition is difficult to implement, because it requires sub-wavelength stability and accuracy in controlling either arm. For this particular correlator, it is more convenient to introduce a small tilt of either end mirror of the Michelson. Such a tilt produces a pattern of parallel fringes at the output. Before the frequency doubling crystal, we have thus generated a first order correlation. The second harmonic of such a first order correlation is an interferometric correlation [381]. This arrangement has the advantage that one has complete control over the spacing of the fringes, which can be adjusted to accommodate the spatial resolution of the array detector used in this measurement.

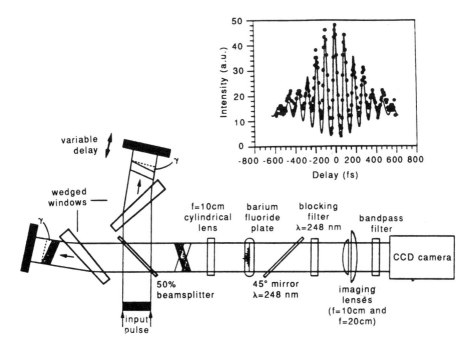

Figure 8.7: Sketch of the single shot interferometric autocorrelator using prisms to transfer the delay to the transverse coordinate of the beam, and typical recording of a 250 fs UV pulse. To record fs pulses, the angle of the prisms should not exceed a few degrees. The fringe spacing is adjusted by the tilt of a mirror in one arm of the autocorrelator (from [381]).

An example of interferometric autocorrelation obtained for a fs UV pulse is shown in Fig. 8.7. For this particular case, the nonlinearity is two-photon fluorescence in BaF_2.

8.4 Pulse amplitude and phase reconstruction

8.4.1 Introduction

Since the second order autocorrelations are symmetric, and do not provide any information about the pulse asymmetry, either an additional measurement or a new technique is required to determine the signal shape. We will start with simple methods that complement the information of the autocorrelations, and proceed with an overview of various methods that have been introduced to provide amplitude and phase information on fs signals. The

8.4. PULSE AMPLITUDE AND PHASE RECONSTRUCTION

ideal diagnostic instrument is obviously one that would give a real time display of all pulse parameters. Because of the ambiguity associated with an average over a large number of pulses, a single shot method is also desirable. Finally, there is a need for a method that does not rely on nonlinear optics, in order to time resolve weak fs transients.

Techniques making direct use of the information available with standard spectrometers and second order autocorrelators are reviewed in the next section. The next approach is to modify the autocorrelator in order to break its symmetry. Femtosecond pulses with a monotonous chirp are the easiest to characterize, by a simple correlation technique using a dispersive interferometric correlator (Section 8.4.3). These are basically correlation techniques in the time domain.

Other techniques measure the amplitude and phase of the pulse in the frequency domain (Section 8.4.6). All the methods cited so far involve some nonlinear detection. As demonstrated by Wong and Walmsley [389], in order to totally characterize an ultrashort pulse, either temporal resolution of the order of the pulse duration, or nonlinear detection is required. The nonlinearity, however, does not need to be optical, but can be incorporated in a fast detector, as will be discussed at the end of this chapter.

8.4.2 Iterative fitting, analytical expressions

The pulse spectrum can complement the information provided by the symmetric second order autocorrelations, in order to determine the signal shape. Simple pulse shapes can be fitted to the autocorrelations and pulse spectra [198], as shown in Fig. 8.8. As a guide to such fitting procedures, analytical expressions [198] for standard pulse shapes and the corresponding spectrum and autocorrelation are given in Table 8.1. For some typical shapes of the temporal intensity profile given in the first column, the spectral intensity (column 2) is used to compute the duration-bandwidth product $\tau_p \Delta \nu$ listed in column 3. The unit for the time t is such that the functional dependence takes the simplest form in column 1. The inverse of that time unit is used as unit of frequency Ω. The most often quoted parameter is the ratio of the FWHM τ_{ac} of the intensity autocorrelation (column 4) to the pulse duration τ_p, and is given in column 5. Finally, the upper and lower envelopes of the interferometric autocorrelation $G_2(\tau)$ can be reconstructed from the expressions given in column 6.

For the cases of complex pulse shapes, a systematic fitting algorithm has been developed by Naganuma *et al.* [372], which converges to the final signal

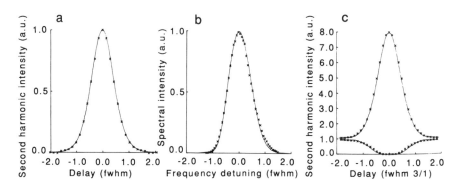

Figure 8.8: Example of pulse shape determination through fitting of the intensity autocorrelation (a), the pulse spectrum (b), and the interferometric autocorrelation (c). The crosses are the experimental data. The solid lines are the functions corresponding to the pulse envelope $\tilde{\mathcal{E}}(t) = \{\exp[-0.15i(t/\tau)^2]\} / \{\exp[-t/0.75\tau] + \exp[t/1.25\tau]\}$.

amplitude and phase function without any *a priori* knowledge of the pulse parameters.

Another method that has been used is to record the spectrum of the second harmonic signal generated in the autocorrelator, for each value of the delay τ. Ishida *et al.* [390, 391] have shown that the amplitude and phase of the pulse can be reconstructed from these sets of measurements. The disadvantage of this method is the large number of data that has to be processed.

8.4.3 Cross-correlation with downchirped pulses

The symmetry of an autocorrelator is broken by insertion of a block of glass in one of its arms (Fig. 8.9). A downchirped pulse[3] can be compressed in the arm of the correlator that contains sufficient positive dispersion. As the delay is being scanned, the measured intensity cross-correlation is an approximation of the pulse intensity profile. From the interferometric cross-correlation, we can easily extract phase information. Indeed, the depth of modulation of the interference pattern is a measure of the frequency offset of the probing (compressed) fs pulse, as compared to the portion of the pulse

---

[3]For the sake of simplicity, we are not considering the insertion of anything more complex than a block of glass — we are therefore limiting ourselves to compressing only downchirped pulses in that arm of the correlator.

8.4. PULSE AMPLITUDE AND PHASE RECONSTRUCTION

$\mathcal{E}^2(t)$	$\|\mathcal{E}(\Omega)\|^2$	$\tau_p \Delta \nu$	$A_c(\tau)$	τ_{ac}/τ_p	$G_2(\tau) - [1 + 3A_c(\tau)]$
e^{-t^2}	$e^{-\Omega^2}$	0.441	$e^{-\tau^2/2}$	1.414	$\pm e^{-(3/8)\tau^2}$
$\mathrm{sech}^2(t)$	$\mathrm{sech}^2(\frac{\pi\Omega}{2})$	0.315	$\frac{3(\tau\mathrm{ch}\tau - \mathrm{sh}\tau)}{\mathrm{sh}^3\tau}$	1.543	$\pm\frac{3(\mathrm{sh}2\tau - 2\tau)}{\mathrm{sh}^3\tau}$
$[e^{t/(t-A)} + e^{-t/(t+A)}] - 1$					
$A = \frac{1}{4}$	$\frac{1+1/\sqrt{2}}{\mathrm{ch}\frac{15\pi}{16}\Omega + 1/\sqrt{2}}$	0.306	$\frac{1}{\mathrm{ch}^3\frac{8}{15}\tau}$	1.544	$\pm 4\left(\frac{\mathrm{ch}\frac{4}{15}\tau}{\mathrm{ch}\frac{8}{15}\tau}\right)^3$
$A = \frac{1}{2}$	$\mathrm{sech}\frac{2\pi}{4}\Omega$	0.278	$\frac{3\mathrm{sh}4x - 8\tau}{4\,\mathrm{sh}^3\frac{4}{3}\tau}$	1.549	$\pm Q$
$A = \frac{3}{4}$	$\frac{1-1/\sqrt{2}}{\mathrm{ch}\frac{7\pi}{16}\Omega - 1/\sqrt{2}}$	0.221	$\frac{2\mathrm{ch}4y + 3}{5\mathrm{ch}^3 2y}$	1.570	$\pm 4\frac{\mathrm{ch}^3 y(6\,\mathrm{ch}2y - 1)}{5\,\mathrm{ch}^3 2y}$

Table 8.1: Typical pulse shapes, spectra, intensity and interferometric autocorrelations. To condense the notation, x has been substituted for $\frac{2}{3}\tau$, y for $\frac{4}{7}\tau$, ch for cosh, sh for sinh. τ_{ac} is the FWHM of the intensity autocorrelation. τ_p is the FWHM of the pulse intensity given in column 1. In the last column, $Q = \pm 4[\tau\mathrm{ch}2\tau - \frac{3}{2}\mathrm{ch}^2 x\mathrm{sh}x(2 - \mathrm{ch}2x)]/[\mathrm{sh}^3 2x]$.

being probed. The phase can easily be extracted from the third term of the expansion (8.7) which is the frequency component of the correlation at $2\omega_\ell$. The Fourier transform of that term $\tilde{C}(\tau)$ given by Eq. (8.10) is:

$$\tilde{C}(\Omega) = \mathcal{F}_s(\Omega)\mathcal{F}_r^*(\Omega) \tag{8.23}$$

where $\mathcal{F}_s$ and $\mathcal{F}_r$ are the Fourier transforms of the square of the pulse amplitude to be measured $[\tilde{\mathcal{E}}_s^2(t)]$ and the compressed pulse $[\tilde{\mathcal{E}}_r^2(t)]$. If the compressed pulse is short enough, the measurement of $\mathrm{Re}\{\tilde{C}(\tau)\}$ gives directly:

$$\mathrm{Re}\{\tilde{C}(\tau)\} \approx \mathcal{E}_s^2(\tau)\cos 2\varphi_s(\tau) \approx A_c(\tau)\cos 2\varphi_s(\tau). \tag{8.24}$$

where $A_c(\tau)$ is the background free portion of the component $A(\tau)$ of the interferometric autocorrelation (8.8).

The pulse is thus completely determined by the measurement of the intensity and interferometric cross-correlation. The approximation can be improved by successive iterations. The complex pulse electric field amplitude $\mathcal{E}_s(t)\exp[i\varphi_s(t)]$ extracted through Eq. (8.24) from the measurement can be

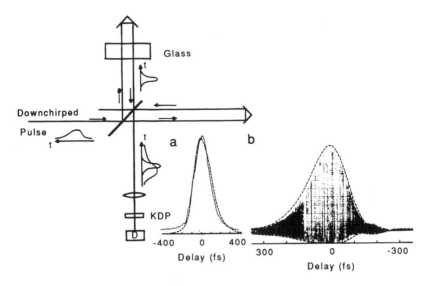

Figure 8.9: Sketch of an asymmetric correlator to determine the amplitude and phase of fs pulses. An example of intensity (a) and interferometric (b) cross-correlation measured with this set-up is also shown. A 5 cm block of BK7 glass was used in one arm of the correlator, resulting in a compression of roughly a factor 2. The asymmetric cross-correlation corresponds to a strongly asymmetric initial pulse shape, which can be approximated by $\mathcal{E}(t) = 1/\{\exp[-(t/40)] + \exp[t/135]\}$ (time in fs). The linear component of the chirp is $\dot{\varphi} \approx -4.5 \cdot 10^{-3} \times t$ (fs^{-1}). Details of the nonlinear phase modulation can be found in [198].

used to compute the compressed pulse (knowing the dispersion and thickness of the glass inserted in the correlator). Having computed the Fourier transform of the compressed pulse, we can directly apply Eqs. (8.3) and (8.23) to obtain a more accurate amplitude and phase profile of the original pulse.

8.4.4 Cross-correlation for arbitrary pulses

It would seem that the instrument described above does not apply to un- or up-chirped pulses, since the pulses will be broadened rather than compressed through positive quadratic dispersion in glass. This broadening can precisely be exploited to measure ultrashort signals. It is a general principle that applies to many different fields of physics, and even in common life situations. Consider the following analogy: trying to observe the structure of a flea with a magnifying glass [Fig. 8.10(a)]. Obviously, this optical in-

8.4. PULSE AMPLITUDE AND PHASE RECONSTRUCTION

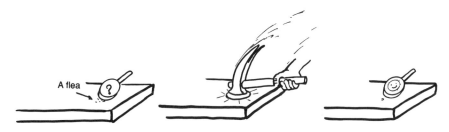

Figure 8.10: Applying a transfer function (b) to magnify the object (a) to be observed. The original object can be reconstructed by applying the inverse transfer function to the observed figure (c).

strument does not have the resolution required to distinguish the details of the object. The solution to this problem is to apply a "reversible" transformation to the object [Fig. 8.10(b)] to create an expanded signal that can be observed [Fig. 8.10(c)]. The final process is to determine the shape of the original object (a), from the knowledge of the expanded object (c) and the inverse transfer function. In the case of fs optics, the choice candidate for the transformation is quadratic dispersion, which can simply be implemented by propagation through glass.

The amount of quadratic dispersion has to be such that the pulse is broadened about 5× in the arm of the correlator containing a glass block. The amplitude reconstruction is difficult with a lesser broadening. On the other hand, a too large amount of glass results in large phase modulation, a low fringe contrast, leading to an inaccurate phase reconstruction.

The data acquisition and reconstruction procedures are similar to the ones of the previous section, with the setup of Fig. 8.9. The Fourier transform of the interferometric correlation is calculated. The d.c. component of this correlation spectrum is isolated, its inverse Fourier transformation performed, yielding the intensity correlation term $A(\tau)$ of Eq. (8.8). The background free portion $A_c(\tau)$ is a first estimate of the intensity profile of the broadened pulse. From the Fourier transform of the interferometric recording, the term near frequency $2\omega_\ell$ is isolated, shifted to frequency $\Omega = 0$, before the inverse Fourier transform is performed, which gives the shape of the envelope of the term $\tilde{C}(\tau)$ in Eq. (8.10). A first approximation of the phase function is:

$$\varphi_2(t) = \frac{1}{2} \arccos \frac{\text{Re}[\tilde{C}(t)]}{A_c(t)} \qquad (8.25)$$

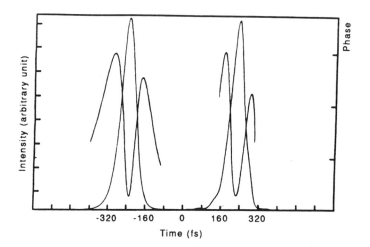

Figure 8.11: Intensity and phase of a 86 fs (FWHM) pulse (left) and its reconstruction (right). The pulse amplitude is the asymmetric function: $\mathcal{E}(t) = 1/[\exp(t/1.25\tau_s) + \exp(-t/0.75\tau_s)]$, with a phase function $\varphi(t) = -t(d|\mathcal{E}|^2/dt)$.

With as first approximation of the broadened pulse $\mathcal{E}_2(t) \approx A_c(t)\exp[\varphi_2(t)]$, the original pulse $\tilde{\mathcal{E}}_1(t)$ is calculated in the frequency domain from the inverse propagation through glass:

$$\tilde{\mathcal{E}}_1(\Omega) = \tilde{\mathcal{E}}_2(\Omega) e^{ik(\Omega)d} \tag{8.26}$$

where d is the thickness of glass. The inverse Fourier transform of the above spectral function is a first approximation of the initial pulse complex amplitude $\tilde{\mathcal{E}}_1(t)$. The accuracy of the result can be improved by successive iterations. Instead of using the background free portion $A_c(\tau)$ as a first estimate of the intensity profile of the broadened pulse, we can now calculate directly the Fourier transform of the broadened pulse intensity by applying Eq. (8.3) through the division $\mathcal{I}_2(\Omega) = A_c(\Omega)/[\mathcal{I}_1(\Omega)]^2$. The amplitude of the broadened pulse is then obtained by taking the square root of the inverse Fourier transform of $\mathcal{I}_2(\Omega)$.

A detailed analysis of this method can be found in Ref. [392]. To demonstrate the fidelity of the method, an example of simulated data and reconstruction is shown in Fig. 8.11. Figure 8.11 demonstrates that a single step is sufficient to obtain an acceptable reconstruction of the original shape and modulation. An arbitrary phase has to be chosen to start the reconstruction. It was taken to be zero to start the single step procedure leading to

8.4. PULSE AMPLITUDE AND PHASE RECONSTRUCTION

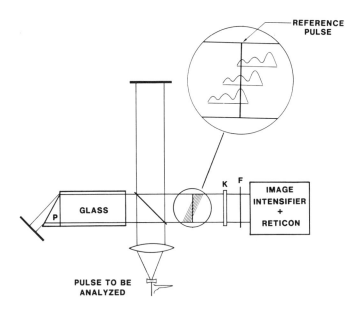

Figure 8.12: Single shot cross-correlator. The prism in one arm of the correlator provides for the variable delay across the x transverse dimension of the beam. Tilt of one mirror of the interferometer provides the interference pattern for the interferometric autocorrelation (from [393]).

the amplitude and phase of the reconstruction shown on the right side of Fig. 8.11.

Single shot implementation

As for the interferometric autocorrelations, the delay variable can be transferred to the transverse coordinate by inserting a glass wedge in one arm of the correlator. Figure 8.12 shows a sketch of a cross-correlator based on this principle [393, 37]. The single shot cross-correlator is more flexible than the scanning version, because the spacing of the fringes is controlled by the tilt of one of the mirrors of this interferometer. It is therefore possible to compromise accuracy and volume of data acquisition.

8.4.5 Method based on a higher order nonlinearity

As mentioned earlier, a measurement based on a higher order nonlinearity can be a close approximation of the pulse intensity profile. Most methods

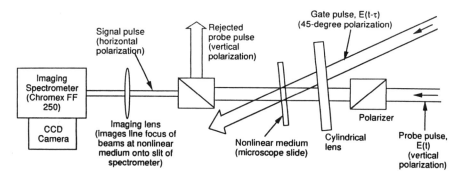

Figure 8.13: Experimental arrangement for frequency resolved optical gating. The pulse to be analyzed $E(t)$ is gated by its delayed replica $E(t - \tau)$ in a Kerr shutter assumed to respond instantaneously (from [377]).

use the nonlinearity to create a shorter pulse, which is then used as a "gate" to probe the pulse under investigation. An example of such a technique is the "frequency resolved optical gating" discussed in the next paragraph. We will conclude this section with a paragraph on another method which has been recently demonstrated. It relies on the determination of the pulse shape from the distortion of the pulse spectrum in a nonlinear material [394].

Frequency resolved optical gating (FROG)

Ishida *et al.* [390] have demonstrated that a tandem combination of autocorrelator and spectrum can be used to extract shape and phase information from ultrashort pulses. In the "frequency resolved optical gating" (FROG), the second order correlator is replaced by a Kerr shutter. This measurement provides directly a representation of the pulse as a two-dimensional function of frequency and time [377, 378]. In the experimental arrangement sketched in Fig. 8.13, the pulse of envelope $\tilde{\mathcal{E}}(t)$ to be measured is gated by its delayed replica in a Kerr medium.

The signal transmitted by the Kerr gate is thus a pulse of electric field (complex) amplitude:

$$\tilde{\mathcal{E}}_s(t, \tau) \propto \tilde{\mathcal{E}}(t) \, g(t - \tau), \qquad (8.27)$$

where the gate function is $g(t - \tau) \approx |\tilde{\mathcal{E}}(t - \tau)|^2$ (approximation valid only for an instantaneous medium response). A recording of the spectrum of the gated pulse versus delay is the two-dimensional function of frequency Ω and

8.4. PULSE AMPLITUDE AND PHASE RECONSTRUCTION

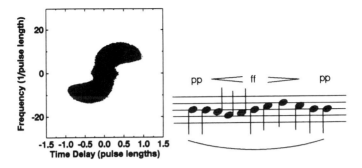

Figure 8.14: (a): Spectrogram of a pulse with strong self phase modulation. The frequency sweeps from -12 to + 12 inverse pulse lengths during the pulse (from [377]). (b): The acoustic analog (see also the reconstruction in Fig. 8.16).

delay τ:

$$S_E(\Omega,\tau) = \left| \int_{-\infty}^{\infty} dt\, \tilde{\mathcal{E}}(t) g(t-\tau) e^{-i\Omega t} \right|^2. \tag{8.28}$$

An optical Kerr gate consists essentially of a Kerr medium between two crossed polarizers. Such a shutter was used by Duguay and Savage [395] to demonstrate picosecond optical sampling. An optical pulse of sufficient intensity and polarized at 45^0 to the input polarizer will make the Kerr medium act as a $\lambda/2$ plate, resulting in total transmission through the second polarizer. To obtain the spectrogram represented by Eq. (8.28), the sequence polarizer — Kerr medium — polarizer that constitute the gate is followed by a spectrometer. The function represented by Eq. (8.28) is well known in acoustics [396] and used to display acoustic waves. Writing of music can be seen as a form of spectrographic representation. The spectrogram of a strongly chirped pulse shown in Fig. 8.14(a) seems identical to the writing of many successive notes [Fig. 8.14(b)]. The difference is only that, in the case of music, the carrier frequency is in the kHz rather than PHz range, and the time delays are seconds rather than fs.

The problem of reconstruction reduces essentially to extracting the function $\tilde{\mathcal{E}}_s(t,\tau)$ from the spectrogram. Indeed, we note that the integral of Eq. (8.27) over the delay τ is simply proportional to the pulse itself [4]

$$\tilde{\mathcal{E}}(t) \propto \int_{-\infty}^{\infty} \tilde{\mathcal{E}}_s(t,\tau) d\tau. \tag{8.29}$$

---

[4]The integration over the variable τ is equivalent to opening the gate function for all times in Eq. (8.27).

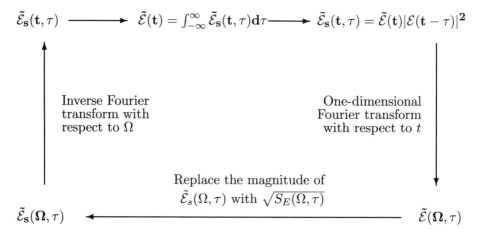

Figure 8.15: Flowchart of a reconstruction algorithm (from [377]).

Let us define the Fourier transform of the function $\tilde{\mathcal{E}}_s(t,\tau)$ with respect to the variable τ as:

$$\tilde{\mathcal{E}}_s(t,\Omega_\tau) = \frac{1}{2\pi}\int_{-\infty}^{\infty}\tilde{\mathcal{E}}_s(t,\tau)e^{-i\Omega_\tau\tau}d\tau. \tag{8.30}$$

The spectrogram [Eq. (8.28)] can be defined in terms of this new quantity:

$$S_E(\Omega,\tau) = \left|\int_{-\infty}^{\infty}dt\int_{-\infty}^{\infty}d\Omega_\tau\tilde{\mathcal{E}}_s(t,\Omega_\tau)e^{-i\Omega t+i\Omega_\tau\tau}\right|^2. \tag{8.31}$$

To extract the unknown signal $\tilde{\mathcal{E}}(t,\Omega_\tau)$ [which leads directly to $\tilde{\mathcal{E}}(t)$] from the spectrogram (8.31) is a two-dimensional (the two dimensions are t and τ) phase-retrieval problem. This (phase retrieval) problem is known to have a unique solution for two and higher dimensions [397, 398]. The flowchart of a reconstruction algorithm is shown in Fig. 8.15 and can be found in Refs. [377, 378]. The reconstruction of the pulse defined by the spectrogram of Fig. 8.14(a) leads to the instantaneous frequency versus time and spectrum of Fig. 8.16.

As mentioned in the introduction of this chapter, any measurement device will affect to a certain extent the parameter to be measured. It is instructive and important to assess the influence of nonlinear interaction on

8.4. PULSE AMPLITUDE AND PHASE RECONSTRUCTION

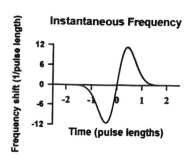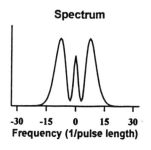

Figure 8.16: Instantaneous frequency versus time and spectrum reconstructed from the spectrogram of Fig. 8.14(a) (from [378]).

the pulse to be measured in the present method. If nonlinear gating via the Kerr effect is used, the probe pulse will be phase modulated by the temporal variation of index induced by the pump pulse ("cross phase modulation"). If d is the thickness of nonlinear medium, $\bar{n}_2$ the nonlinear index in cm^2/W, I_{GATE} the peak intensity of the gating (pump) pulse, the maximum phase shift induced on the probe pulse is:

$$\Delta \Phi \approx \bar{n}_2 I_{GATE} k_\ell d \qquad (8.32)$$

The gating efficiency (ratio of gated probe intensity to incident probe pulse intensity) on the other hand is proportional to the square of the same expression (8.32). If the required accuracy for the phase is 1% (maximum allowable error on $\Delta\Phi$), then the gating intensity and gate thickness d should be chosen such that the gating efficiency does not exceed 10^{-4}.

Spectral measurements on modulated pulses

This method recently demonstrated by Prade et al. [394] is based on the observation of the distortion of the pulse spectrum of an intense pulse acquired after propagation through a transparent nonlinear medium. Even for very simple nonlinearities, such as the Kerr effect, the specrum exhibits a fringe-like structure that is very sensitive to the instantaneous phase and amplitude of the incident pulse, and therefore could be used to provide a "signature" of the incident pulse. The basic measurement is extremely simple: the power spectrum of the pulse is measured before and after propagation through a Kerr nonlinearity. As all higher order methods, this technique is well adapted to high power systems. In the demonstration measurements, pulses from an amplified Ti:sapphire laser (165 fs) of 100 μJ to 1 mJ energy (beam of 13 mm FWHM) were used. The spectrum of the central portion of the beam (defined by a 3 mm aperture) was recorded before and after the sample. The numerical pulse reconstruction algorithm starts by computing an initial pulse envelope from the spectrum of the incident pulse (assuming it to be free of phase modulation). The modulated spectrum is thereafter calculated, compared to the measured spectrum, to make a first determination of the initial phase. Typical reconstruction required about 50 iterations [394].

8.4.6 Detection in the frequency domain

The methods mentioned in the preceding sections are primarily attempts to measure the pulse in the time domain. Another approach is to measure a pulse in the frequency domain, since a signal is completely characterized by its Fourier transform. Because of the broad bandwidth of femtosecond pulses, it would appear that spectral measurements could easily be implemented to determine pulse shapes after inverse Fourier transformation. A simple standard 1/4 meter spectrometer has sufficient resolution to measure the spectral amplitude of a femtosecond pulse. Determination of the phase is however more difficult, and involves either ultrahigh speed or nonlinear detection. Two types of methods are possible to measure the spectral phase: beat note measurements and time domain correlations.

Spectral phase through beat note measurements

The most straightforward technique to determine the phase of a group of oscillators is to measure their beat frequency with a local oscillator. However, the temporal resolution required for the detector is the inverse of the pulse bandwidth, hence a femtosecond resolution. In principle, the speed requirement of the detector can be relaxed if, instead of a fixed frequency local oscillator, one could dispose of an array (in frequency) of oscillators, such that the beat frequency would not exceed the detector bandwidth. Instead of external frequency sources, one could take the reference signal from the pulse itself, i.e., one chooses to beat two adjacent frequency components of the pulse spectrum. Let us consider for instance a simple two prism spectrometer, as sketched in Fig. 8.17. We will assume for simplicity that the prism separation and position of the beam are chosen for zero GVD. After the two prism sequence, in the absence of chirp, the energy front and wavefront are both aligned along the coordinate y perpendicular to the propagation

8.4. PULSE AMPLITUDE AND PHASE RECONSTRUCTION

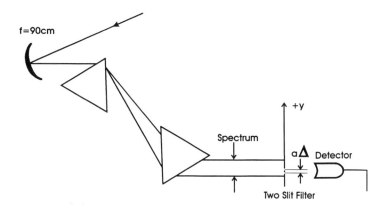

Figure 8.17: Spectroscopic measurement with a pair of prisms. The input beam is focused with a cylindrical mirror to a plane at the output of the prism pair. The prism pair is adjusted for zero group velocity dispersion for the central frequency.

direction of the beam in the plane of the figure. If $\tilde{\mathcal{E}}(\Omega) = \mathcal{E}\exp(i\phi)$ is the Fourier transform of a fs pulse being sent through the spectrometer, the field along the axis y, assumed to be at the focal plane of the spectrometer, is (cf. Chapter 2):

$$\tilde{\mathcal{E}}(\Omega, y) = \mathcal{E}(\Omega)e^{i\phi(\Omega)}e^{-(y+a\Omega)^2/w^2}, \qquad (8.33)$$

where w is the beam waist and a is the dispersive power of the spectrometer having units of *length/frequency*. We have chosen to only keep the spatial dependence of the field in the y-direction for simplicity. Mixing two frequency components Ω_1 and Ω_2 with phase $\phi_1(\Omega_1)$ and $\phi_2(\Omega_2)$ will produce a beat note of frequency $\Omega_2 - \Omega_1$ with phase $\phi_2(\Omega_2) - \phi_1(\Omega_1)$. If the frequencies are sufficiently close so that the measurement can be performed, we will obtain the phase derivative $d\phi/d\Omega$. The phase function $\phi(\Omega)$ can be obtained by integration.

As the position of the double slit is scanned, the detector records a beat note proportional to $\mathcal{E}(y/a)\mathcal{E}(y/a + \Delta)\cos[\Delta t - \Delta\phi'(y/a)]$. The price to pay for the reduced demand on detector speed is that the phase has to be determined by integration of $\phi'(y/a) = d\phi/d\Omega|_{y/a}$.

The method sketched in Fig. 8.17 is still no more than a thought experiment, because of practical limitations on spectrometer size and available detector speeds. The resolution of the spectrometer is determined by the ratio w/a, and the minimum useful separation between a pair of slits is w. With any practical small size spectrometer, the corresponding beat note is still hundreds of gigahertz, a frequency too high for electronic processing.

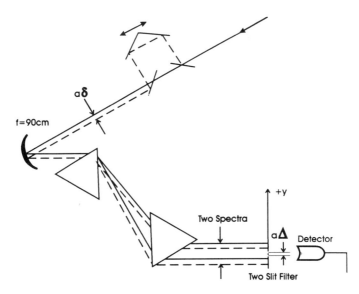

Figure 8.18: A two prism spectrometer disperses the spectrum of an incident pulse along the direction y. The dispersive power of the spectrometer is $a = (y_2 - y_1)/(\Omega_2 - \Omega_1)$ for any two pairs of frequencies Ω_2 and Ω_1. Differential detection is performed between two pairs of frequencies as follows. The input beam is split and spatially shifted by $a\delta$, resulting in the transmission of two frequency components (separated by δ) through each of the two slits located in the focal plane and separated by $a\Delta$. The delay introduced by producing the beam with a transverse shift $a\delta$ has to be compensated by an equal delay (not shown in the figure) in the primary beam. In the actual experiment, the two detectors and the mixer are replaced by an integrated photodiode/mixer.

The detector speed requirements can be reduced further by mixing two adjacent pairs of beat notes [399], and measuring the beat note between the beat signal of each pair. This measurement will lead to the second derivative $d^2\phi/d\Omega^2$. Double integration is required to extract the phase function $\phi(\Omega)$. The implementation of this technique is sketched in Figure 8.18. The input pulse is split into two identical beams, one of them being transversely shifted, for instance by a 90° corner reflector. After passing through the spectrometer, two identical spectra are observed in the focal plane, one being spatially shifted with respect to the other by a small amount $a\delta$. In the approximation of very large dispersive power a, the fields transmitted through the slits can be approximated by sine or cosine functions. With a spacing between the two slits in front of the detector given by $a\Delta$, the four

8.4. PULSE AMPLITUDE AND PHASE RECONSTRUCTION

transmitted field components are:

$$\begin{aligned}
\tilde{E}_1(t,y) &= \mathcal{E}(y/a)e^{i[(y/a)t+\phi(y/a)]} \\
\tilde{E}_2(t,y) &= \mathcal{E}(y/a+\delta)e^{i[(y/a+\delta)t+\phi(y/a+\delta)]} \\
\tilde{E}_3(t,y) &= \mathcal{E}(y/a+\Delta)e^{i[(y/a+\Delta)t+\phi(y/a+\Delta)]} \\
\tilde{E}_4(t,y) &= \mathcal{E}(y/a+\delta+\Delta)e^{i[(y/a+\delta+\Delta)t+\phi(y/a+\delta+\Delta)]}.
\end{aligned} \quad (8.34)$$

The photodiode behind the upper and lower slits record a signal proportional to $S_{1,2}(t,y) = |\tilde{E}_1 + \tilde{E}_2|^2$, and $S_{3,4}(t, y + a\Delta) = |\tilde{E}_3 + \tilde{E}_4|^2$, respectively. The detected signals have high frequency components at frequency δ, with a phase proportional to $\delta\phi'(y/a)$ and $\delta\phi'(y/a + \Delta)$. As noted above, the minimum frequency δ in Eqs. (8.34) is of the order of $w/a \approx 2\pi(100)$ GHz. The relative phase of $S_{1,2}$ and $S_{3,4}$ can be determined by mixing these two signals. An ideal device for this purpose could be the monolithically integrated photodiode/mixer[5] developed by Li et al. [400]. They have shown that this photodiode/mixer combination effectively replaces the non-linear crystal and a photo multiplier tube in standard autocorrelators, making it possible to measure pulses as short as 1.9 ps without the need of complex optical sampling systems. Additional advantages of such a semiconductor device are that it is sensitive to all wavelengths with energy greater than the band-gap energy, and it can detect microwatt optical powers.

Both the photodiode and the mixer being square-law detectors, the output of the circuit is proportional to $||\tilde{E}_1 + \tilde{E}_2 + \tilde{E}_3 + \tilde{E}_4|^2|^2$. Since the frequency of oscillation is the same in both of these expressions, the sum of the two cosines will yield the product of a fast oscillation at frequency δ, and a slow modulation whose argument is $\delta[\phi'(y/a) - \phi'(y/a+\Delta)] \approx -\Delta\delta\phi''(y/a)$. The exact expression is:

$$\begin{aligned}
S &\propto ||\tilde{E}_1 + \tilde{E}_2 + \tilde{E}_3 + \tilde{E}_4|^2|^2 \\
&\propto 54 + 16\cos[\Delta^2 \phi''(y/a)].
\end{aligned} \quad (8.35)$$

In the above equation, we have set $\delta = \Delta$ and have dropped the rapidly oscillating terms which are beyond the speed of any electronic display instrument. When the slits in front of the detector are scanned along the y-direction, the detected signal varies as $\cos[\phi''(y/a)]$. Reconstruction involves twofold integration to obtain $\phi(y/a)$.

---

[5]This device consists of a GaAs Schottky photodiode coupled via a transmission line to a radio-frequency diode with a non-linear characteristic.

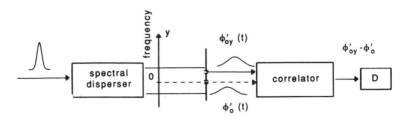

Figure 8.19: Sketch of principle of diagnostic methods in the frequency domain. The pulse to be measured is spectrally dispersed in the y direction. A pair of spectral components is cross-correlated to measure the relative delay between the corresponding wave packets.

Relative phase through time domain correlations

The relative phase of two groups of frequency components corresponds — in the time domain — to a relative delay between these two frequency groups. The basic principle of this method is to measure the relative time of arrival of the various frequency components of the pulse. From this time of arrival one can determine the phase of the corresponding frequency component. The delay is measured by cross-correlating the pulse issued at the slit level y (Fig. 8.17) from the spectrometer, corresponding to the frequency component Ω, with a "reference pulse" issued a slit level $y = 0$, corresponding to the frequency component Ω_0.

Let us refer to the sketch of principle of Fig. 8.19 An "ideal" spectrometer displays a pulse spectrum along a "y" spatial direction (normal to the propagation of the beam). In general, the spectrometer will modify the phase function $\phi(\Omega)$ of the pulse. As in the previous section, we will assume for simplicity an ideal spectrometer with zero group velocity dispersion. Assuming Gaussian optics, the pulse spectrum is displayed as:

$$E_0(x, y, \Omega) = \mathcal{E}(\Omega) e^{-[x^2 + (y - a\Delta\Omega)^2]/w^2 + i\phi(\Omega)} \tag{8.36}$$

where $\mathcal{E}(\Omega) e^{i\phi(\Omega)}$ is the Fourier transform of the complex amplitude function $\mathcal{E}(t) e^{i\varphi(t)}$, and w is the waist of the beam focused after the spectral disperser by the entrance optics. As in the previous section the spectral disperser is characterized by the parameter $a = \Delta y/\Delta \Omega = y/(\Omega - \Omega_0)$ (Ω_0 being a reference frequency corresponding to $y = 0$).

A point detector located at a particular coordinate (x, y) – assumed to be broadband in the spectral region of interest – will measure $S(x,y) = |\int_{-\infty}^{\infty} E_0(x, y, \Omega) d\Omega|^2$ of the expression (8.36). It is only if w is very small

8.4. PULSE AMPLITUDE AND PHASE RECONSTRUCTION

compared to the spectral spread of the beam along the y direction, that the exponential factor $[(y - a\Delta\Omega)/w]^2$ can be assimilated to the "$\delta - function$" $\delta(\Delta\Omega = y/a)$, and $S(x,y) \approx \mathcal{E}(y/a)\exp[-i\phi(y/a)]$. If the pulse bandwidth is approximately $1/\tau_p$, the measurement on an array detector (for instance a reticon) **D** in Fig. 8.19 will be proportional to the pulse spectrum if:

$$w \ll \frac{a}{\tau_p} \tag{8.37}$$

The beam waist w determines also the length of the pulses to be correlated. For each spectral component corresponding to a particular value of y, we find a pulse with temporal characteristic given by the inverse Fourier transform of Eq. (8.36):

$$\tilde{\mathcal{E}}(y,t) = \int \mathcal{E}(\Delta\Omega) e^{(y-a\Delta\Omega)^2/w^2 + i\phi(\Delta\Omega) - i\Delta\Omega t} d\Delta\Omega \tag{8.38}$$

The basic measurement consists in cross correlating the pulse given by Eq. (8.38) for all values of y, with the "reference" pulse given by another value of y - for instance $y = 0$. The only information that needs to be extracted from the correlation is its center of gravity, which, in the frequency domain, corresponds to the first derivative of the spectral phase $\phi(\Delta\Omega)$ versus frequency as shown in Appendix E. For any method trying to retrieve amplitude and phase information, there will be at some point a compromise to reach between the phase and amplitude accuracy. The choice between phase and amplitude accuracy is set by the parameter w. If w is very small, the spectral amplitude is defined with great accuracy. However, for each spectral component, the pulses to be correlated have a very long duration (of the order of a/w). The smaller the value of w, the more difficult it becomes to determine the center of gravity with great accuracy.

Phase reconstruction In the procedure detailed in Appendix E, the spectral field amplitude and phase are replaced in Eq. (8.36) by their first order Taylor expansion around ω_ℓ. After taking the inverse Fourier transform, one finds (Eq. E.4) that any spectral component at y, selected by a slit of width $< w$, is stretched into a Gaussian pulse of duration $\tau_y = 2a/w$. The pulse corresponding to each spectral component at y is centered (in time) at the instant $t_d(y) = \phi'_{\Omega_0 + y/a}$. Knowledge of $t_d(y)$ leads – after integration – to the determination of the spectral phase:

$$\phi(y) = \int_{y=0}^{y} t_d(y') \frac{dy'}{a} \tag{8.39}$$

The y dependence of the time integrated signal obtained by direct detection of the radiation described by Eq. (E.4) (Appendix E) with a photodiode array leads to the pulse spectral amplitude $\mathcal{E}(\Omega)$.

Second order correlation The most direct method to determine the relative maxima of the envelopes is second order cross-correlation. Chilla and Martinez [401] measured the spectral phase of fs pulses through second order (intensity) correlation between each spectral component of the pulse, and a reference spectral component (dashed line in Fig. 8.19). As in the previous methods in the time domain, high peak powers are required since nonlinear optics is involved. The maximum of the cross-correlations for each y corresponds to the time $t_d(y)$. The spectral phase is reconstructed through the integration of Eq. (8.39).

First order correlation Can a direct linear correlation provide the same spectral phase information? Direct linear correlation of the fields of adjacent spectral components will not provide information on the spectral phase, as can be proved by an exact calculation in the frequency domain. The Fourier transform of the reference beam at $y = 0$ is $\tilde{F}_1(\Omega) = \tilde{\mathcal{E}}(\Omega)\exp\{-(a\Omega/w)^2\}$. This field is to be cross-correlated with the field at y, of which the Fourier transform is $\tilde{F}_2(\Omega) = \tilde{\mathcal{E}}(\Omega)\exp\{-[(a\Omega - y)/w]^2\}$. A *linear* correlation in the time domain corresponds to the product of the fields in the frequency domain:

$$\tilde{F}_1(\Omega)\tilde{F}_2^*(\Omega) = |\tilde{\mathcal{E}}(\Omega)|^2 e^{-(\frac{a\Omega}{w})^2} e^{-(\frac{a\Omega-y}{w})^2}. \tag{8.40}$$

Since only the electric field spectral *amplitude* appears in Eq. (8.40), no phase information on the fs pulse to be analyzed can be extracted from this measurement.

Parceval's theorem stating that Eq. (8.40) represents the Fourier transform of the correlation of the two fields applies only if the detector has no temporal resolution. The use of ultrafast detectors (in the 10 ps range) to measure the correlation has been suggested to obtain phase information [389].

Linear correlation with frequency shifted beam A linear field correlation can still provide the phase information, provided the reference pulse is frequency shifted by a small amount Δ. This can be achieved by inserting a fast travelling wave modulator in the reference arm (dashed line in Fig. 8.19).

8.5. PROBLEMS

The Fourier transform of the measurement is now, instead of Eq. (8.40):

$$\tilde{F}_1(\Omega + \Delta)\tilde{F}_2^*(\Omega) = \tilde{\mathcal{E}}(\Omega + \Delta)\tilde{\mathcal{E}}^*(\Omega)e^{-(\frac{a\Omega+a\Delta}{w})^2}e^{-(\frac{a\Omega-y}{w})^2}. \tag{8.41}$$

The fundamental difficulty encountered in the previous paragraph has been eliminated, since the product $\tilde{\mathcal{E}}(\Omega + \Delta)\tilde{\mathcal{E}}^*(\Omega)$ carries spectral phase information. The technical difficulty is to realize a travelling wave modulator capable of shifting the frequency of a pulse of a few ps by a fraction of THz.

The frequency shift is obtained by propagating the reference pulse through an electro-optic material (lithium niobate for instance) to which a linearly time varying electric field is applied. Because of the large shifts required ($\geq$ 500 GHz), special consideration must be given to the device used. Kauffman [402] and Riaziat [403] have constructed modulators which can provide such shifts. However, their devices required sophisticated electronics to generate the microwave field applied to the modulator. A simpler solution is the use of a traveling wave modulator triggered by a photoconductive switch. As demonstrated by Valdmanis *et al* [404], it is possible to create a very fast (10^{-12}s) rise time for the applied electric field by using a silicon or GaAs photoconductive switch mounted adjacent to the electro-optic material.

8.5 Problems

1. Show that the statistical average of the autocorrelation of Gaussian pulses distributed in frequency [distribution given by Eq. (8.12)] is approximately identical to the autocorrelation of a simple Gaussian pulse [perform the integration of Eq. (8.13), and compare to Eq. (8.11)].

2. Consider a Gaussian pulse, of 50 fs duration, with an upchirp corresponding to $a = 1.5$. Analyze analytically the result of a measurement using the cross-correlator in which a bloc of BK-7 glass has been inserted in one arm, as discussed in Section 8.4.4. Calculate the amount of glass required for pulse broadening by a factor 5, after double passage through the glass. Calculate the envelopes $A(\tau_d)$, $B(\tau_d)$, and $C(\tau_d)$ that will be obtained by performing the measurement. Write the expressions corresponding to the various steps of the procedure leading to the reconstruction of the original pulse.

Chapter 9

Measurement Techniques of Femtosecond Spectroscopy

9.1 Introduction

Femtosecond pulses are an ideal tool to investigate ultrafast processes of various origins. There is usually more than one parameter that varies with time in any particular experiment. One of these parameters will often be the position. As an example of the types of time dependence that have to be distinguished, let us consider the example of an apple falling from a tree (Fig. 9.1). A photograph taken with a sufficiently short exposure time can "freeze" the motion of the falling apple as it reaches the position x. We can compare this picture with one of the apple still on the tree, which provides some information about the aging process. To establish either the law of motion $x(t)$ or the temporal behavior of aging, we need to know exactly the time elapsed from the moment the apple was shaken loose from the tree, until the photograph was taken. The standard experimental technique is to *trigger* the fall (for instance, ignite a small explosion) and simultaneously start a clock that synchronizes the shutter of the camera. By triggering the event, we do not have to wait days for the apple to fall down.

A pump-probe femtosecond experiment has analogies as well as fundamental differences with the falling apple measurement. The basic analogy is that a powerful light pulse — usually labeled the "pump pulse" or "excitation pulse" — interacts with the sample and excites it into a non-equilibrium state (Fig. 9.2). The sample thereafter relaxes towards a new equilibrium state. This process can be mapped by sending a second (much weaker) pulse,

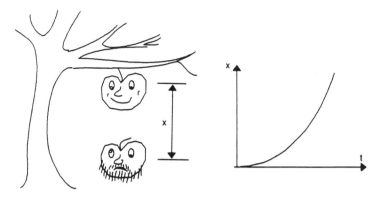

Figure 9.1: The falling apple and the aging apple.

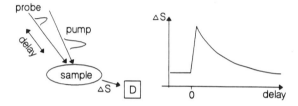

Figure 9.2: Schematic representation of a pump-probe experiment in ultrafast spectroscopy. The pump and probe pulses can be obtained from a single source and delayed with respect to each other in a Michelson or a Mach–Zehnder interferometer for example.

called a probe or test (pulse), onto the sample. The probe is the analogue of the snapshot photograph, aimed at detecting a change of optical properties without disturbing the object under investigation. The difference with the falling apple is that the speed of light is infinite compared to the velocity of the apple, whereas the propagation time of the probe radiation through the sample can be long compared with the event to be observed. Therefore, the geometry of interaction, the angle between probe and pump, the interaction length, and the group velocities in the sample are essential parameters in femtosecond pump-probe experiments.

In a typical pump-probe experiment, the delayed weak pulse probes the change of an optical property ΔS, such as transmission or reflection, induced by the pump. Repeating the experiments for various delays τ_d provides the function $\Delta S(\tau_d)$. Depending on the actual light matter interaction, $\Delta S(\tau_d)$ is related to material parameters such as occupation numbers, carrier density, and molecular orientation. In this chapter we discuss a selection of fs experimental techniques, many of them originally developed for ps and ns spectroscopy.

The obvious temporal limitation of the pump-probe technique is the duration of the pump and probe. The medium preparation should be completed before the material can be probed. If the physics of the interaction is well understood, a theoretical modeling (deconvolution) can provide some interpretation of data corresponding to partial temporal overlap of pump and probe.

A compromise often must be sought between spectral and temporal resolution. Either the probe or the pump pulse has to *select* a specific spectral feature. In order to excite the desired transition, rather than an adjacent one, the excitation spectrum (i.e., the pulse spectral width $\Delta\omega_p$, augmented by the Rabi frequency $\kappa\mathcal{E}$ or power broadening of the transition) should not exceed the separation between lines Δ. The spectral resolution imposes therefore a limit to the temporal resolution, since the pulse duration should not be less than $\tau_p \approx 1/(\Delta + \kappa\mathcal{E})$.

9.2 Data deconvolutions

In most fs time resolved experiments, a signal $S(\tau_d)$ is measured as a function of position or delay τ_d of a reference pulse of intensity $I_{ref}(t)$. We will consider the large class of measurements where the measured quantity $f(t)$ is proportional to the product of a gating function I_g times the physical quantity to be analyzed. The gate $I_g(t)$ is a direct function of the reference intensity $I_{ref}(t)$. Since — as pointed out in the previous chapter — the detection electronics has no fs resolution, the measured signal will be the time integral:

$$S(\tau_d) = \int_{-\infty}^{\infty} I_g(t - \tau_d) f(t) dt. \tag{9.1}$$

Deconvolution procedures should thus be applied to retrieve the physical quantity $f(t)$ from the measurement $S(\tau_d)$. A typical example is a measurement of time resolved fluorescence by upconversion. As detailed in Section 9.7, the detected upconversion radiation results from mixing the signal and the reference in a nonlinear crystal. Therefore, in that particular case, the gate function is the reference pulse itself $I_g(t) = I_{ref}(t)$. It is often assumed that the gating function in the correlation product [Eq. (9.1)] can be approximated by a δ-function. With that simplifying assumption, the signal is directly proportional to the physical parameter to be measured:

$S(\tau_d) \propto f(\tau_d)$. There are, however, very fast events — such as the rise of fluorescence — for which this simplifying assumption is not valid. The exact temporal dependence $f(\tau_d)$ can be extracted from the data if the gating function $I_g(t)$ is known. Indeed, if $\mathcal{I}(\Omega)$ is the Fourier transform of the gate function $I_g(t)$, and $\mathcal{S}(\Omega)$ is the Fourier transform of the measurement, the Fourier transform $f(\Omega)$ of the physical quantity $f(t)$ is just the ratio:

$$f(\Omega) = \frac{\mathcal{S}(\Omega)}{\mathcal{I}(\Omega)}. \tag{9.2}$$

The physical quantity $f(t)$ can be calculated by taking the inverse Fourier transform of Eq. (9.2). This deconvolution technique can be applied in numerous cases where the gate function $I_g(t)$ does not depend on the phase of the interaction.

9.3 Beam geometry and temporal resolution

To obtain a better quantitative understanding of the influence of the beam geometry on the temporal resolution, let us analyze a pump-probe experiment as sketched in Fig. 9.3. The pump pulse creates a small change of the transmission coefficient, $\Delta a(x, y, z, t)$, which is sampled by the time-delayed test pulse. The signal measured by the detector D as function of the delay

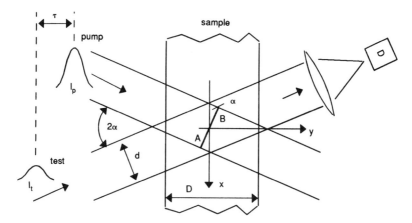

Figure 9.3: Schematic diagram of a pump-probe transmission experiment in non-collinear geometry. The line AB shows the position of the pump pulse maximum at $t = 0$. Refraction at the sample interfaces has been neglected.

9.3. BEAM GEOMETRY AND TEMPORAL RESOLUTION

τ_d can be written as

$$\begin{aligned}\mathcal{W}_t(\tau_d) &= \mathcal{W}_{t0} + \Delta\mathcal{W}_t(\tau_d) \\ &\propto \int_{-\infty}^{\infty} dt \iiint dx\,dy\,dz\,[1 + a_o + \Delta a(x,y,z,t)]\,I_t(x,y,z,t-\tau_d)\end{aligned} \quad (9.3)$$

where a_o is the transmission coefficient in the absence of the pump and $|a_0| \ll 1$ has been assumed. In the overlapping volume of the two beams, a complex mixing of spatial and temporal effects occurs. We want to derive conditions under which the excitation geometry does not affect substantially the outcome of the experiment. For simplicity, the beam profiles of pump and test pulse are assumed to be uniform and of rectangular shape, the temporal profiles are Gaussian of equal FWHM τ_p, and the overlapping region is symmetric with respect to the sample center. The time axis is chosen such that the pump pulse maximum reaches the origin of the coordinate system at $t = 0$, and the test pulse reaches the origin at $t = \tau_d$. The sample response is assumed to follow the pump pulse instantaneously, and we expect a signal $\Delta\mathcal{W}_t(\tau_d)$ resembling the pulse autocorrelation in the absence of geometrical effects. An increase of the correlation FWHM is then a measure of the loss in temporal resolution of any pump-probe experiment due to geometrical effects.

In the following considerations we will omit constants for the sake of brevity. The delay-dependent part of the measured signal is

$$\Delta\mathcal{W}_t(\tau_d) \propto \int_{-\infty}^{\infty} dt \int dx \int dy\, I_t(x,y,t-\tau_d) I_p(x,y,t) \quad (9.4)$$

where we have already carried out the z-integration yielding a constant. The pulses propagate through the sample with the group velocity v_g. Lines of equal intensity (parallel to AB in Fig. 9.3) obey the equation

$$(x - v_g t \sin\alpha) = -(y - v_g t \cos\alpha)\frac{\cos\alpha}{\sin\alpha} \quad (9.5)$$

for the pump pulse and

$$[x + v_g(t - \tau_d)\sin\alpha] = [y - v_g(t - \tau_d)\cos\alpha]\frac{\cos\alpha}{\sin\alpha} \quad (9.6)$$

for the test pulse. Hence, the corresponding pulse intensities which are needed in the integral (9.4) can be written as

$$I_p = I_{p0} \exp\left\{-\frac{4\ln 2}{(v_g \tau_p)^2}[y\cos\alpha + x\sin\alpha - v_g t]^2\right\} \quad (9.7)$$

$$I_t = I_{t0} \exp\left\{-\frac{4\ln 2}{(v_g \tau_p)^2}[y\cos\alpha - x\sin\alpha - v_g(t - \tau_d)]^2\right\}. \quad (9.8)$$

Inserting these expressions into Eq. (9.4) and carrying out the time integration yields after some algebra

$$\Delta W_t(\tau_d) \propto e^{-2\ln 2(\tau_d/\tau_p)^2} \int_{-y_m}^{y_m} dy \int_{l(y)}^{u(y)} dx \exp\left\{-\frac{8\ln 2}{(\tau_p v_g)^2}\left[x^2 \sin^2\alpha - v_g x \tau_d \sin\alpha\right]\right\} \quad (9.9)$$

where

$$y_m = \min\left(\frac{d}{2\sin\alpha}, \frac{D}{2}\right) \quad (9.10)$$

$$l(y) = -\frac{d}{2\cos\alpha} + y\tan\alpha \quad (9.11)$$

$$u(y) = \frac{d}{2\cos\alpha} - y\tan\alpha. \quad (9.12)$$

The value of y_m depends on whether or not the overlapping area of the beams is completely inside the sample. The upper and lower limit of the x-integration form the diamond-shaped boundary of the overlapping region as sketched in Fig. 9.3.

The exponential function in front of the integrals is the autocorrelation function of a Gaussian pulse and represents the result of an ideal measurement where the geometrical effects do not play a part, i.e., the spatial integration yields a constant which does not depend on τ_d. This is obviously the case for a collinear beam geometry ($\alpha = 0°$). For all other cases one can evaluate Eq. (9.9) numerically. Figure 9.4 shows the FWHM of $\Delta W_t(\tau_d)$ normalized to its value at $\alpha = 0°$ as a function of α and for different values of the parameter $K = v_g \tau_p / d$. The latter describes the ratio of the geometrical pulse length and the beam width. As can be seen, the shorter the pulses at a given beam width, the more critical becomes the beam geometry in a noncollinear experiment. The temporal broadening of ΔW_t can be substantial, causing a loss in time resolution of a pump-probe experiment.

9.4. TRANSIENT ABSORPTION SPECTROSCOPY

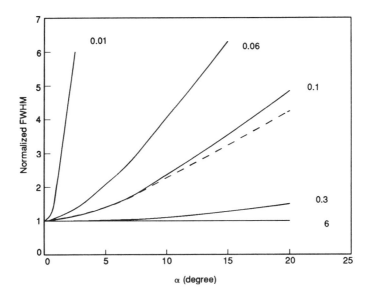

Figure 9.4: FWHM of $\Delta W_t(\tau_d)$ according to Eq. (9.9), normalized to its value at $\alpha = 0°$ and shown as a function of the half-angle α between pump and probe pulse. The curves are depicted for different values of the ratio of the geometrical pulse length and beam width, $K = v_g \tau_p / d$. The sample thickness was chosen to be $D = 3 v_g \tau_p$. For $K = 0.1$, a second curve for $D = 10 v_g \tau_p$ is also shown for comparison (from [405]).

The effect of the crystal thickness, on the other hand, is small at moderate values of the angle α. In the following sections we will always assume an experimental geometry which justifies neglecting these geometric effects.

9.4 Transient absorption spectroscopy

Transient absorption spectroscopy is a widely used form of a pump-probe technique. As a simple example to illustrate the method, we consider an ensemble of two-level systems at resonance with a fs pulse source. With all the particles in the ground state at thermal equilibrium, the sample acts as a saturable absorber. The physical quantity to be determined is the energy relaxation time T_1 of the excited state. This parameter is to be extracted from the measurement of attenuation of the probe versus delay.

A typical experimental arrangement is sketched in Fig. 9.5. In order for the probe to be much weaker than the pump, the reflectivity of the beam

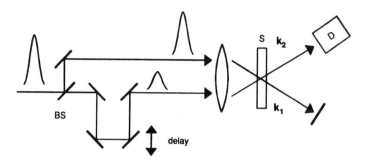

Figure 9.5: Typical geometry for measuring transient absorption. The two relatively delayed pulses are spatially separated before being sent onto the focusing optics, to provide for **k** vector separation in the sample S.

splitter (called BS in Fig. 9.5) should be larger than 0.5. Since only the transmission of the *probe* is measured, there is a need to devise a means to shield the pump pulse from the detector. Since pump and probe have the same wavelength, one is left with the following choices:

- separation by polarization
- separation by wave vector
- temporal separation in combination with a gated detector.

In the example sketched in Fig. 9.5, it is the wave vector separation that is used.

In a typical experiment, the first (pump) pulse will saturate the absorber, and the delayed probe pulse will sample the absorption coefficient. The interpretation of the data is straightforward if the transition can be considered as homogeneously broadened. For delays longer than the phase relaxation time T_2 of the transition, the probe pulse samples the absorption coefficient α:

$$\alpha = \sigma \Delta \gamma \tag{9.13}$$

where σ is the absorption cross-section and $\Delta \gamma$ is the population difference (density) between the upper and lower level of the transition. According to the rate equation (3.50), after excitation, the absorption coefficient relaxes exponentially with time:

$$\alpha(\tau_d) = \alpha_0 + \Delta\alpha \, e^{-\tau_d/T_1}, \tag{9.14}$$

9.4. TRANSIENT ABSORPTION SPECTROSCOPY

where $\Delta\alpha$ is the change in absorption due to the pump pulse, τ_d the delay of the probe relative to the pump, and T_1 is the energy relaxation time of the absorbing transition.

Extraction of T_1 from the measurement can be made under a variety of experimental conditions. In the considerations that follow, we will not attempt to select the experimental conditions for best signal-to-noise ratio, but the ones that lead to the simplest analytical expression relating the measurement to T_1, without the need for numerical modeling. We assume a uniform beam profile. In addition to being "optically thin," the sample thickness d is assumed to be negligible compared with the overlap length of pump and probe beams. Finally, the pump and probe pulse duration is assumed to be much shorter than the relaxation time to be measured ($\tau_p \ll T_1$), to avoid the need of decorrelation procedures. The completion of the pumping process is taken as time origin. The measured signal $S(\tau_d)$ is the energy of the transmitted probe versus delay. For a probe signal of energy density $W = \int I dt$ sent through a sample of thickness d:

$$S(\tau_d) = A \int_{-\infty}^{\infty} I(t - \tau_d) e^{\alpha(t)d} dt$$
$$\approx AW e^{\alpha(\tau_d)d} \approx AW[1 + \alpha(\tau_d)d], \qquad (9.15)$$

where A is the beam cross-section. Inserting Eq. (9.14) into Eq. (9.15) yields:

$$S(\tau_d) = AW\left[1 + \alpha_0 d + \Delta\alpha d\, e^{-\tau_d/T_1}\right] \qquad (9.16)$$
$$\approx S_{-\infty} + AW\Delta\alpha d e^{-\tau_d/T_1}, \qquad (9.17)$$

where $S_{-\infty}$ is the probe transmission in the absence of the pump pulse. The energy relaxation time T_1 can be obtained directly from a logarithmic plot of $S(\tau_d) - S_{-\infty}$ versus τ_d. The key approximation in Eq. (9.15) is that the temporal variations of the absorption coefficient be slow compared to the duration of the probe pulse. A numerical deconvolution of the data can be made if this latter condition is not satisfied.

The expression (9.14) is only valid for delays sufficiently large such that the excitation of the pump has dephased before the arrival of the leading edge of the probe ($\tau_d \gg T_2 + \tau_p$). For short delays that do not satisfy the latter condition, the probe coherently interacts with the polarization created by the pump. For pump and probe collinear and having the same polarization, the induced dipoles created by the pump will be in or out of phase with the probe field, depending on whether the delay is an even or odd

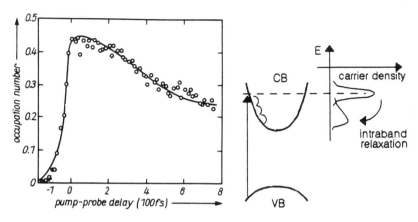

Figure 9.6: Transient transmission in CdSe. The population numbers in the states excited by the pump can be determined from the measured changes in transmission of the 100 fs probe pulses at 618 nm. The change in occupation numbers is a measure of intraband relaxation (from [406]).

number of half wavelengths. The transmitted energy versus delay will have an interference-like pattern similar to that observed in a "zero-area pulse" experiment (see Chapter 4 and the end of this chapter). This pattern is often referred to as the "coherent spike" of a pump-probe experiment. In the case of non-collinear pump-probe experiments, a "transient grating" is created by the spatial-temporal superposition of the probe and the polarization created by the pump. In the latter geometry, the coherent spike can be explained as a result from partial diffraction of the pump pulse into the direction of the probe.

As a typical example of transient absorption, Fig. 9.6 shows the absorption recovery of a mixed crystal, CdS_xSe_{1-x}, after excitation with a 618 nm pulse. The pump pulse creates free carriers, i.e., electrons in the conduction band and holes in the valence band, which occupy states and subsequently increase the transmission of a test pulse at the corresponding wavelength. As the carriers relax towards the bottom of the band, the transmission decreases. The decay time is a measure of the intraband relaxation.

Much more information can be gained by using a fs white light continuum instead of a probe at the excitation frequency. In the previous example, the test pulse monitored the change in carrier density only for specific states above the band gap. A continuum fs pulse can probe simultaneously all states in a broad energy range, providing detailed information on the time dependent carrier density distribution. An example of a pump-probe trans-

9.5 Transient polarization rotation

A linearly polarized pump pulse can induce anisotropy in a sample, which can be probed subsequently with a delayed pulse. Anisotropy means here that the transmission depends on the polarization of the probing radiation. The decay in anisotropy can often be related to orientational relaxation of the dipoles excited by the pump. Such measurements have been applied successfully in the fs scale to the determination of momentum (**k**-space) relaxation of photo-excited electrons in condensed matter (for instance, GaAs [407]).

A polarization rotation can also be induced in transparent media. The pump pulse acts through the optical Kerr effect causing birefringence. This polarization anisotropy can be seen as a polarization-direction dependence of the refractive index experienced by the probe.

A standard experimental arrangement to measure the temporal change in pump induced polarization anisotropy in shown in Fig. 9.7(a). Subsequent to the excitation by the pump pulse, a probe pulse — of the same wavelength — is sent with its electric field vector oriented at an angle χ (typically 15°) with respect to that of the pump. Let $\mathcal{E}_p$ be the input electric field amplitude of the probe. As in Ref. [408], we define the polarization angle as the arctangent of the ratio of the components of the *electric field vector* axes parallel ($\mathcal{E}_{p\parallel} = \mathcal{E}_p \cos \chi$) and perpendicular ($\mathcal{E}_{p\perp} = \mathcal{E}_p \sin \chi$) to the pump polarization.

As a result of the induced anisotropy, the polarization vector of the probe rotates. In a complex representation, the probe electric field vector changes from $\mathcal{E}_p e^{i\chi}$ to $\mathcal{E}'_p e^{i(\chi+\epsilon)}$. If the change in probe amplitude is small, the rotation angle ϵ of the probe polarization can easily be approximated by projecting the changes in parallel ($\Delta \mathcal{E}_{p\parallel}$) and perpendicular ($\Delta \mathcal{E}_{p\perp}$) components of the probe onto a direction orthogonal to the pump, as sketched in Fig. 9.7(b):

$$\epsilon = \frac{\Delta \mathcal{E}_{p\parallel} \sin \chi}{\mathcal{E}_p} + \frac{\Delta \mathcal{E}_{p\perp} \cos \chi}{\mathcal{E}_p}. \tag{9.18}$$

If $a_\parallel$ and $a_\perp$ are the absorption coefficients respectively parallel and perpendicular to the direction of pump polarization, $\Delta \mathcal{E}_{p\parallel} = \frac{1}{2} a_\parallel \mathcal{E}_{p\parallel} \Delta z$ and $\Delta \mathcal{E}_{p\perp} = -\frac{1}{2} a_\perp \mathcal{E}_{p\perp} \Delta z$ are the changes of the parallel and perpendicular components of the field. Substituting in Eq. (9.18), we find for the rotation of

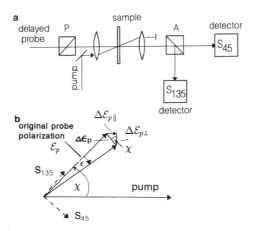

Figure 9.7: (a) Experimental setup to measure pump-induced birefringence. The Glan polarizer P sets the polarization of the probe at an angle χ with respect to that of the pump. The analyzer A extracts the components of the transmitted probe at 45° and 135° of the pump polarization. (b) Sketch showing the relation between the rotation angle ϵ of the probe $\mathcal{E}_p$ and the induced anisotropy in absorption.

the probe polarization ϵ:

$$\epsilon = \frac{\sin 2\chi}{4} \int_0^d (\alpha_\| - \alpha_\perp) dz. \tag{9.19}$$

The experimental technique to measure the polarization rotation [Fig. 9.7(a)] consists of measuring the polarization component at 135° (S_{135}) and at 45° (S_{45}) with respect to the pump polarization. The ratio R of the two intensities is the square of the tangent of the angle $45° + \chi + \epsilon$ of the probe in these new coordinates:

$$R = \tan^2\left(\chi + \frac{\pi}{4} + \epsilon\right) \approx R_0 \left(1 + \frac{4\epsilon}{\cos 2\chi}\right), \tag{9.20}$$

where R_0 is the ratio of intensities S_{135}/S_{45} between the two orthogonal directions in the absence of pump field. The polarization rotation ϵ is thus:

$$\epsilon = \frac{\cos 2\chi}{4}\left(\frac{R}{R_0} - 1\right). \tag{9.21}$$

The scattered pump intensity adds a noise component to the signals S_{135} and S_{45}. The angle χ has to be sufficiently large, such that the scattered noise

9.6. TRANSIENT GRATING TECHNIQUES

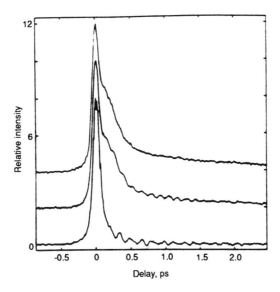

Figure 9.8: Optical Kerr signals for the liquids CH_2Cl_2, $CHCl_3$ and CCl_4 (top to bottom). Different time constants can be identified for each sample. They represent a complex interplay of intramolecular processes as well as interaction with the local environment (from [409]).

from the pump be negligible compared with the measured probe component $\mathcal{E}_p^2 \sin^2 \chi$. An angle of $\chi = 15^o$ is chosen in most applications [408, 407]. The measured anisotropy (rotation in probe beam polarization of the order of one degree) decays with probe delay and is a measure of relaxation processes following the excitation. An example of a polarization rotation measurement is shown in Fig. 9.8.

9.6 Transient grating techniques

9.6.1 General technique

There are numerous variations of transient grating techniques, providing a wide array of information on sample properties. A general review of these techniques is given in a book by Eichler *et al.* [410]. The basic experimental setup is sketched in Fig. 9.9. Two pump pulses of different propagation direction overlap in the sample. If their relative delay (τ_1) is less than the phase relaxation time of the interaction, they produce an interference pat-

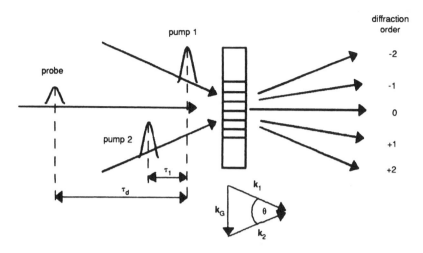

Figure 9.9: Schematic representation of a transient grating experiment.

tern of the sample excitation. The modulation of the sample excitation can manifest itself in a periodically changing transmission (amplitude grating) and/or a refractive index (phase grating). The grating vector is:

$$\mathbf{k}_G = \mathbf{k}_1 - \mathbf{k}_2, \tag{9.22}$$

where $\mathbf{k}_{1,2}$ are the propagation vectors of the pump pulses in the sample. The existence and dynamics of the grating can be probed by the diffraction that a delayed (weak) probe pulse experiences. A series of detectors can probe simultaneously the behavior in several diffraction orders. If the excitation is weak, the absorption and refractive index modulation are small $[(\Delta\alpha/\alpha_0) \ll 1, (\Delta n/n) \ll 1]$, and the diffracted probe intensity (first order) is given by [410]:

$$\frac{\Delta I_{diff}}{I_p} \propto (\Delta n)^2 + \left(\frac{\lambda_p}{2\pi}\right)^2 (\Delta\alpha)^2. \tag{9.23}$$

One has generally to distinguish between two mechanisms for the decay of diffraction efficiency with delay:

1. The pump-induced changes in the sample relax locally. For example, if $\Delta\alpha$ and Δn are the result of free carrier generation in a semiconductor, carrier relaxation towards the original equilibrium state will lower the modulation depth. The diffraction as a function of delay provides information on the carrier relaxation time.

9.6. TRANSIENT GRATING TECHNIQUES

2. The sample excitation diffuses spatially (non-local mechanism). In the example of free carrier generation in a semiconductor, the pump modulates the carrier density, and thus triggers diffusion of carriers into the low excitation regions (minima of the induced grating). The result is a gradual "wash-out" of the modulation, and decline of the diffraction signal. In many cases, the diffusion process can be described by a diffusion equation with a characteristic diffusion constant. From the characteristic decay time of the diffraction efficiency and the grating period one can determine the diffusion constant.

Both processes (1) and (2) have to be taken into account in the data evaluation. To distinguish between the local and non-local relaxation mechanisms (in particular when they occur on a comparable time scale), a series of measurements can be performed at various angles θ between the two pump beams inducing the grating. Since the grating constant is modified by changing the angle θ [Eq. (9.22)], the decay component due to diffusion will also be modified. The local relaxation component to the decay should not depend on the angle θ. Another possibility to distinguish between local and non-local contributions to the decay is to compare transmitted (zero diffraction order) and diffracted signals.

Grating techniques also provide the possibility of measuring coherent effects by varying the delay τ_1 between the two pump pulses, at constant probe delay τ_d. The first arriving pump pulse generates a polarization oscillation in the sample which decays with the characteristic transverse relaxation time T_2. The second pump interferes with this polarization, which results in a modulation of the excitation (e.g., occupation numbers). The modulation depth and thus the diffraction efficiency experienced by the probe pulse and measured as function of τ_1 contain information on T_2.

The actual data evaluation in a transient grating experiment can be complex and requires a detailed model of the processes involved. An example of determination of phase relaxation times using collinear counter-propagating pump pulses is detailed in the next subsection.

9.6.2 Degenerate four-wave mixing (DFWM)

In this particular variation of transient grating experiment, the two pump pulses are two strong counter-propagating waves $\tilde{\mathcal{E}}_{p1}(t)\exp[i(\omega_\ell t - k_p z)]$ and $\tilde{\mathcal{E}}_{p2}(t)\exp[i(\omega_\ell t + k_p z)]$. The probe wave is sent along an intersecting direction x and has as electric field $\tilde{\mathcal{E}}_3(t - \tau_d)\exp[i(\omega_\ell t - k_x x)]$. The nonlinear interaction results in the generation of a signal $\tilde{\mathcal{E}}_4(t)\exp[i(\omega_\ell t + k_x x)]$, which, for momentum conservation, is counter-propagating to the probe direction (Fig. 9.10). In the case of continuous waves, and, for instance, a quadratic nonlinearity, it can be shown that the wavefront of the generated signal wave $\tilde{\mathcal{E}}_4$ is the reversed of the wavefront of the probe $\tilde{\mathcal{E}}_3$ [10]. This property of spatial phase conjugation does not transpose directly in the time domain. Temporal phase conjugation is chirp reversal, which can be shown to occur only when the following conditions are met [411]:

- instantaneous nonlinearity,
- medium thickness $\ll$ than the pulse length,
- weak interaction ($|\tilde{\mathcal{E}}_4| \ll |\tilde{\mathcal{E}}_3|$).

It can easily be seen that if all but the second condition are met, each depth of the medium will generate a DFWM signal, resulting in a square pulse $\tilde{\mathcal{E}}_4$ with a length equal to twice the sample thickness [411].

We have so far assumed that all three waves meet simultaneously in the nonlinear medium. Interesting information on the dynamics of the interaction can be gathered from the study of the DFWM signal when all three waves are applied in a particular time sequence.

We assume in the following discussion that the nonlinear medium is shorter than the optical pulses and is either at single- or at two-photon resonance with the radiation. Let us first consider the case of a single-photon resonant absorber being excited first by a weak probe, followed by

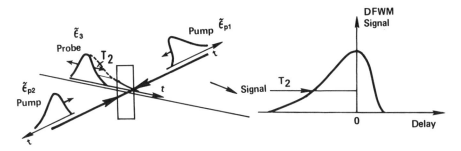

Figure 9.10: Coherent single-photon resonant DFWM. The probe pulse is trailed by a polarization wave, that forms a population grating with one of the pump pulses that follows. The other pump pulse scatters off that grating into the direction from which the probe originates. The *rise* of the signal energy versus delay is thus a measure of the phase relaxation time of the single-photon resonance.

9.6. TRANSIENT GRATING TECHNIQUES

two simultaneous strong counter-propagating pump pulses (Fig. 9.10). As we have seen in Chapter 4, the short pulse creates a pseudo-polarization $\tilde{Q}_3 = w_0 \sin\theta_0 \exp[-ik_x x]$ that decays with a characteristic time T_2. If a strong pump pulse enters the interaction region within that characteristic time, it will form a population grating [as seen from the third Bloch equation (4.7)] corresponding to the interferences between waves of vector $\mathbf{k}_x$ and $\mathbf{k}_z$. If the second pump pulse impinges on this grating, it will be diffracted along the opposite direction as the signal (wave vector $-\mathbf{k}_x$) according to the second Bloch equation (4.6). The longer T_2 is, the more the probe can be launched in advance of the two pump pulses, and still produce a signal. As illustrated in Fig. 9.10, the rise-time of the signal versus delay is a measure of the phase relaxation time of a single photon transition.

The same experiment performed on a two-photon resonant transition, as sketched in Fig. 9.11(a), leads to different results and interpretation. Since the interaction is a two-photon process, the weak probe alone can-

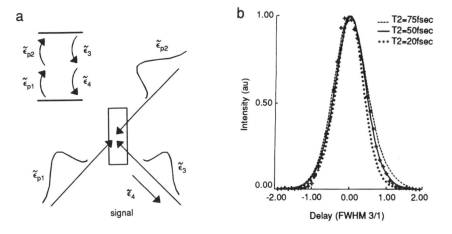

Figure 9.11: Coherent two-photon resonant DFWM. (a) The counter-propagating pump pulses create a two-photon excitation of frequency $2\omega_\ell$, which decays with the two-photon phase relaxation time $T_{2(2ph)}$. The signal is two-photon stimulated emission induced by subsequent passage of the probe pulse. It is thus the *fall* of the signal energy versus delay that is a measure of the phase relaxation time of the two-photon resonance. (b) Intensity of the DFWM signal versus probe delay. The sample is the saturable absorber jet of a mode-locked dye laser. The crosses indicate the experimental data points. Theoretical curves for the two-photon resonant interaction are plotted for three values of the phase relaxation time: 20 fs (dotted), 50 fs (solid), and 75 fs (dashed).

not have any significant effect on the system, and there will be no signal if the probe is ahead of the pump pulses. We have seen in Chapter 4 that for a two-photon transition, Bloch's equations (4.6), (4.7) apply, except that the driving term is proportional to the square of the field. The two counter-propagating pump pulses can produce a two-photon excitation oscillating at $2\omega_\ell$ [see Eqs. (4.95)], which will decay with the phase relaxation time $T_{2(2ph)}$ of the two-photon transition. One component of this two-photon excitation, ϱ_{12}, with no spatial modulation (zero spatial frequency), will interact with a probe to generate a counter-propagating signal by two-photon stimulated emission. A probe pulse sent through the interaction region with a subsequent delay τ_d will induce a signal by two-photon stimulated emission, $\tilde{\mathcal{E}}_4 \propto \varrho_{12}(\tau_d)\tilde{\mathcal{E}}_3^*$. Since the probe field corresponds to a phase factor $\omega_\ell + k_x x$ and the two-photon excitation to a phase factor $2\omega_\ell$, the signal E_4 has a phase factor $2\omega_\ell - \omega_\ell - k_x x = \omega_\ell - k_x x$, which describes a wave propagating in the direction opposite to the probe. Since the two-photon excitation ϱ_{12} is the amplitude of an off-diagonal matrix element decaying with a two-photon phase relaxation time $T_{2(2ph)}$, the two photon stimulated emission being proportional to ϱ_{12} will only exist within $T_{2(2ph)}$ of the pump excitation. In the case of two-photon resonance, it is thus the trailing edge of the signal energy versus delay that is a measure of the phase relaxation time $T_{2(2ph)}$.

An example of determination of phase relaxation times through DFWM is given in Fig. 9.11(b). In this particular case, the sample is the saturable absorber jet (dye DODCI) of a mode-locked dye laser [170]. The two pump pulses are the counter-propagating pulses circulating inside the dye laser cavity. The probe is taken from one of the outputs of the dye laser, and focused with a 25 mm focal distance lens into the interaction region. Figure 9.11(b) shows the average intensity of the signal retro-reflected into the probe direction, as a function of the delay of the probe. The leading edge of the signal matches exactly the instantaneous response, given the pulse shape $\tilde{\mathcal{E}}(t) = \exp[-0.15ix^2]/\{\exp[-1.33x] + \exp[0.8x]\}$ (where $x = t/\tau_s$, and the pulse FWHM is $1.72\tau_s = 76$ fs). The instantaneous response, for the single-photon transition model, is calculated by taking the steady state solution of Bloch's equations (4.10) and (4.11) for the field consisting of the sum of the probe and pump fields. The trailing edge of the DFWM signal versus delay shows clearly the effect of a two-photon coherence. Following the procedure outlined above, the DFWM signal can be calculated as a function of delay for the two-photon excitation [170]. The result of the calculation (for the particular pulse shape mentioned above) is plotted in Fig. 9.11(b) for three values of the two-photon phase relaxation time $T_{2(2ph)} = 20$ fs (dotted line), $T_{2(2ph)} = 50$ fs (solid line), and $T_{2(2ph)} = 75$ fs (dashed line).

The experiment thus indicates a two-photon resonant DFWM and a phase relaxation time of 50 fs (decay of the DFWM signal versus delay) for the two-photon transition. There is no resolvable effect of a single-photon resonant DFWM (rise-time of DFWM signal versus delay). The dominance of the two-photon enhancement of DFWM in DODCI at 620 nm is confirmed by theory. Simple numerical estimates [170] indicate that indeed, the contribution of the two-photon resonance dominates the DFWM signal.

9.7 Femtosecond resolved fluorescence

If an excitation is followed by fluorescence (luminescence), the time resolved measurement of the transients of this radiation provides useful information on the evolution of occupation numbers and relaxation channels. Streak cameras are often used to measure fluorescence decay. The temporal resolution of this instrument is limited to approximately one half of a picosecond. As noted earlier, all-optical techniques are needed to obtain even better time resolution. The general method of correlation introduced in Chapter 8 applies also to fluorescence measurements.

A pump pulse provides the time dependent excitation to be analyzed. The radiation to be measured as a function of time is correlated with a delayed replica of the pump (reference pulse). This cross-correlation is achieved by up-converting (sum frequency generation) the fluorescence with the fs reference pulse. This technique, pioneered by Mahr and Hirsch [412] with ps pulses, was first applied to the fs range to measure the risetime of fluorescence in organic dyes [413].

The basic experimental setup as sketched in Fig. 9.12 includes a polarizing beam splitter, two quarter wave plates, and a nonlinear crystal for type II sum frequency generation. Type II sum frequency generation is essential to provide an optimum signal-to-background ratio. In the first experiment, an unamplified dye laser at 620 nm was used. After the calcite polarizing beam splitter, one of the polarization components of the main pulse is focused into the sample, e.g., a concentrated solution of oxazine dye in ethylene glycol. The backscattered fluorescence radiation (at ω_f) is collected by the focusing lens and recollimated towards the calcite prism and the nonlinear detection. In this reflective geometry, the temporal resolution is limited by the optical depth of the sample or the confocal parameter of the focused

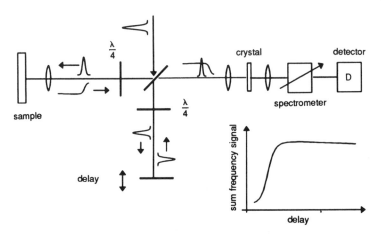

Figure 9.12: Setup for femtosecond resolved detection of fluorescence

beam, whichever is shorter. With the concentrated solution of oxazine dyes used, the optical depth of the sample was approximately 2 μm, limiting the temporal resolution to 6 fs. A half wave plate can be introduced before the calcite polarizer to control the fraction of radiation sent to the sample.

The nonlinear crystal generates the sum frequency (of intensity I_{sum}) of the radiation components polarized along two orthogonal axes. If $I_{ref}(t-\tau_d)$ is the reference signal delayed by an amount τ_d and polarized along $\hat{x}$, and $I_s(t)$ is the fluorescence signal polarized along the orthogonal direction $\hat{y}$:

$$I_{sum}(\tau_d) \propto \int_{-\infty}^{\infty} I_{ref}(t-\tau_d) I_s(t) dt. \tag{9.24}$$

To reach the ultimate resolution, the bandwidth of the conversion process should be larger than the bandwidth of the reference pulse. Both the imperfection of the crystal and the nonperfect rejection factor of the polarizing beam splitter contribute to a fraction ϵ_y of the reference beam polarized along the axis y, and hence a (τ_d independent) background signal:

$$I_b \propto \int_{-\infty}^{\infty} \epsilon_y I_{ref}^2(t-\tau_d) dt. \tag{9.25}$$

The optical quality of the nonlinear crystal is essential in this experiment, since it helps discriminate between the signal and a background caused by second harmonic generation of the more intense reference beam. Additional background rejection can be obtained by spectrally separating the gated fluorescence (at $\omega_\ell + \omega_f$) from the second harmonic of the reference signal (at $2\omega_\ell$). The bandwidth of this filter should correspond to the pulse spectral bandwidth $\Delta\omega_p$ in order to ensure a temporal resolution given by the pulse duration. The focusing lens can be replaced by a parabolic mirror which collects fluorescence from a larger solid angle [414].

The number of upconverted photons per excitation pulse can formally be written as

$$N_{up} \approx V_1 V_2 Q \left(\frac{\nu_F \mathcal{W}_0}{\hbar \omega_F}\right) \frac{\Delta\omega_p}{\Delta\omega_F} \frac{\tau_p}{T_F}, \qquad (9.26)$$

where V_1 is the linear loss of the experimental setup, V_2 is the fractional solid angle (i.e., solid angle divided by 4π) from which the focusing optics gathers the fluorescence, the term in parentheses describes the total number of fluorescence photons excited, $\Delta\omega_p/\Delta\omega_F$ is the fraction of the fluorescence spectrum which is upconverted and reaches the detector, τ_p/T_F with T_F as fluorescence lifetime is the fraction of fluorescence within the time window set by the pulse, and Q is the conversion efficiency of the sum frequency generation. For an upconversion experiment using a passively mode-locked dye laser to resolve the fluorescence dynamics of an organic dye and urea as nonlinear crystal, the following parameters are typical: $V_1 = 10^{-1}$, $V_2 = 2\ 10^{-3}$, $\mathcal{W}_0 = 150$ pJ, $\tau_p = 100$ fs, $T_F = 1$ ns, $Q = 5\ 10^{-4}$, $\nu_F = 1$. This yields $N_1 = 1.5\ 10^{-4}$ upconverted photons per pump pulse photon. This weak signal is detectable because the repetition rate of the source is $\approx 10^8$ Hz, resulting in a photon flux of $1.5\ 10^4$ s^{-1}. Figure 9.13 shows as an example the onset of fluorescence for the dye oxazine 720.

9.8 Photon echoes

Photon echo is the standard method [415] — directly derived from spin echoes [416] — to determine the phase relaxation time T_2 of a transition. In the basic photon echo experiment, a sequence of two pulses is sent through the sample. Ideally, the first pulse will be a "$\pi/2$ pulse," and the second a "π pulse."

In an inhomogeneously broadened medium, the $\pi/2$ pulse excites the electric dipoles to oscillate with their characteristic frequency ω_0. Immediately after excitation all dipoles are in phase and the macroscopic polarization is maximum. As time progresses, because of their different eigenfrequencies, the dipoles dephase relative to each other. The macroscopic polarization is damped with a time constant given by the inverse of the width of the

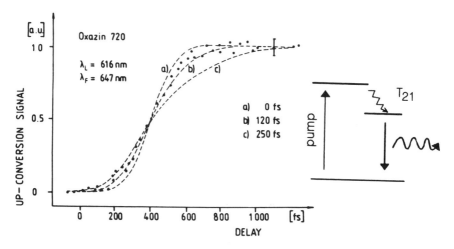

Figure 9.13: Fluorescence versus reference pulse delay. The number of photon counts is plotted versus delay. Theoretical curves corresponding to different relaxation times T_{21} of a simple three-level model system are shown.

inhomogeneous line profile $g_{inh}(\omega_0 - \omega_{ih})$. The individual dipole groups still oscillate, with an amplitude damped with the phase relaxation time T_2 corresponding to the homogeneous line profile. The π pulse at delay τ_d adds a phase of π to each individual oscillator, which causes them to add again in phase after a time τ_d ($2\tau_d$ after the $\pi/2$ pulse). The associated macroscopic polarization results in a collective radiation effect called an "echo." The explanation of the echo in the Bloch vector model is as follows.

After the first pulse ($\pi/2$ pulse), the pseudo-polarization vector is aligned along the v axis, as shown in Fig. 9.14(a). Each component of the line $g_{inh}(\omega_0 - \omega_{ih})$ will precess around the "w axis" — thus in the u-v plane — at an angular velocity $(\omega_0 - \omega_\ell)$, for a time equal to the delay τ_d between the $\pi/2$ and π pulses. The effect of the π pulse, however, is to create the mirror image of the component of the pseudo-polarization vector, with respect to the v axis, as shown in Fig. 9.14(b). Each component of the line $g_{inh}(\omega_0 - \omega_{ih})$ is subsequently precessing at the velocity $(\omega_\ell - \omega_0)$, hence reversing the course of the previous "spreading." After a time τ_d following the π pulse, the components of the pseudo-polarization vector will be lined up again, resulting in a macroscopic echo signal [Fig. 9.14(c)]. The only decay mechanism for the echo is the nonreversible homogeneous decay. The pseudo-polarization vector of initial amplitude $\mathcal{P}_0$, after the delay $2\tau_d$, has been reduced exponentially to $\mathcal{P}_0 \exp(-2\tau_d/T_2)$; hence the echo intensity decays

9.8. PHOTON ECHOES

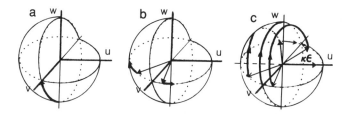

Figure 9.14: Photon echoes: A $\pi/2$ pulse (a) creates a macroscopic polarization (pseudo-polarization vector aligned along the v axis). In the interval between the two pulses, the components of the pseudo-polarization vector precess (b). After a π pulse is applied, the spreading process is reversed (c).

as $\exp(-4\tau_d/T_2)$. It should be noted that the $\pi/2$ and π areas need not be reached in order to observe an echo. The amplitude however, is maximum for this particular choice.

The dephasing in condensed matter at room temperature is extremely fast. The challenge in photon echo measurement is to resolve the fast decaying echo from scattering from the tail of the preceding π pulse. The various possibilities to separate the signal are:

1. temporal gating of the echo

2. **k** vector separation

3. separation by polarization

4. separation by focalization

The first of these techniques — to our knowledge — has not yet been applied to fs photon echo techniques. The necessary time resolution could be achieved by up-converting the echo using type II second harmonic generation with a delayed excitation pulse (of polarization orthogonal to that of the echo). The other three techniques are commonly used. If $\mathbf{k_1}$ is the wave vector of the first ($\pi/2$) pulse, and $\mathbf{k_2}$ the wave vector of the second one, it can be shown that the angle of emission of the third pulse (the echo) is twice the angle between the direction of the two first pulses — hence $\mathbf{k_e} = 2\mathbf{k_2} - \mathbf{k_1}$ (cf. Fig. 9.15). This property can be easily understood from momentum conservation considerations. Indeed, a photon echo is a particular case of degenerate four-wave mixing experiment, in which the first two waves form

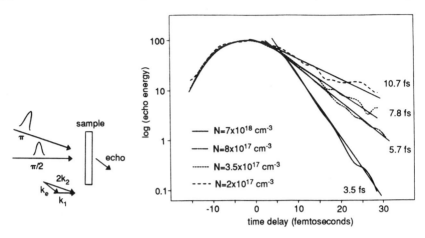

Figure 9.15: Photon echo applied to GaAs. Left: wave vector diagram. Right: echo amplitude versus delay for different carrier densities (adapted from [419]).

a grating. The latter waves do not need to be simultaneous; their interval only needs to be shorter than the phase relaxation time, to form a grating. The period of that grating is $\mathbf{k_2} - \mathbf{k_1}$. The second pulse $\mathbf{k_2}$ scatters off that grating, to generate a first-order diffracted wave in the direction $\mathbf{k_e} = \mathbf{k_2} - (\mathbf{k_1} - \mathbf{k_2})$. The latter property is directly related to the focusing properties of the echo. If the radius of curvature of the first $\pi/2$ pulse is R_1, and that of the π pulse R_2, the echo has the wavefront curvature given by [417]:

$$\frac{1}{R_e} = \frac{2}{R_2} - \frac{1}{R_1} \tag{9.27}$$

Polarization can also be used to distinguish the echo from the intense excitation pulses. The polarization dependence of the echo has been investigated by Alekseef and Evseev [418] and shown to depend on the total angular momentum number J of each of the two levels involved in the transition. For a $J=1/2 \to J=1/2$ transition, the polarization of the echo makes an angle 2ψ with that of the first (linearly polarized) pulse (ψ being the angle between the polarization of the first and second pulse). In the latter case also, a linearly polarized pulse following a circularly polarized pulse, produces a photon echo with circular polarization. For transitions $J=0 \leftrightarrow J=1$ and $J=1 \to J=1$, the echo has the polarization of the second pulse, with an amplitude proportional to $\cos\psi$.

9.9. ZERO-AREA PULSE PROPAGATION

None of the echo separation techniques totally eliminates the background due to the first pulses. The duration of the exciting pulses should be shorter than the phase relaxation T_2 to be measured.

The fs photon echo technique has been applied to the study of dephasing of band-to-band transitions in the direct gap semiconductor GaAs [420]. Dephasing in this system is due to momentum relaxation of the carriers, as verified by an independent method that specifically probes momentum relaxation (see previous sections). The data (Fig. 9.15) indicate a carrier concentration dependent phase relaxation time ranging from 14 to 44 fs [419, 420], fitting the power law $T_2 = 6.2 \; 10^6 \times N^{-0.3}$ (T_2 in fs, concentration of excited carriers N in cm^{-3}). This power law is characteristic of a three dimensional screening. A similar experiment performed on two-dimensional multiple quantum wells gave a density law $T_2 = 6.8 \; 10^7 \times N^{-0.55}$, reflecting a two-dimensional screening of carriers [421].

In summary, the photon echo method is quite powerful and useful for the determination of relaxation times longer than the pulse duration. It has been one of the most commonly used. The interpretation of the measurement becomes difficult when the pulse spectrum covers several transitions with different dipole moment. The method described below does not have these limitations.

9.9 Zero-area pulse propagation

The photon echo experiment is based on a sequence of two non-overlapping pulses with no phase relationship. An essential feature of coherent excitation is that the excitation depends on the phase of the applied signal. We have seen in Chapter 4 that a sequence of two pulses 180° out of phase applied at resonance to a two-level system, will return that system to the ground state. There will be no energy loss for this particular pulse sequence, while there will be maximum absorption if the pulses are in phase. The contrast in absorption for the "in phase" pulse sequence — as opposed to the sequence of pulses out of phase — can be used as a measure of coherent interaction, and to determine T_2. The experimental setup consists essentially of a Michelson or a Mach–Zehnder interferometer (Fig. 9.16).

The measurement is particularly simple and clear in the case of a single homogeneously broadened line. A linear (i.e., with a small area pulse) measurement provides all the information needed in that case. The zero-area pulse sequence has zero spectral Fourier component at the average pulse fre-

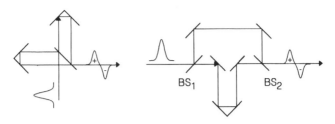

Figure 9.16: Michelson or Mach–Zehnder interferometers for the generation of zero-area pulses. The beam splitters BS_1 and BS_2 should be identical, in order to produce a zero-area pulse. The field envelopes of the pulses are shown.

quency. The *linear* absorption for that pulse sequence — when applied at resonance with the line — is proportional to the spectral overlap of the line and the pulse spectrum. For $T_2 = \infty$, the infinitely narrow line coincides with the "node" of the spectrum of the zero-area pulse, and there is no absorption. The smaller T_2, the broader the line and its overlap with the pulse spectrum. With decreasing T_2, the ratio of absorption for an "out of phase" (zero-area) pulse sequence to the absorption for an "in phase" pulse sequence will also decrease. An illustration of such a measurement in Li-vapor is shown in Fig. 9.17. The energy of the second harmonic of the transmitted pulse sequence is plotted as function of the delay between the two components of the pulse. In the time domain the experiment can be explained as follows. The first signal emerging out of the interferometer of Fig. 9.16 excites the resonant transition in lithium vapor. The induced dipoles reradiate a field which opposes the applied field and therefore cause absorption. The energy stored in the medium will be restituted to the second signal emerging out of the interferometer if the latter is 180° out of phase with the first pulse (the reradiated field adds in phase with the applied electromagnetic signal). Maximum absorption occurs for in-phase pulse sequences. The signal versus delay should therefore show an interference pattern with a periodicity in delay equal to the light period.

The constructive/destructive interferences that extend beyond the region of pulse overlap decay with the collision time of the resonant sodium atoms with a buffer gas (Ar, 1000 torr pressure). Second harmonic detection was used in that particular example [422]. By using second harmonic detection, the transition between the region corresponding to pulse interferences, and coherent interaction effects, can easily be identified. For delays smaller than the pulse duration, the pulse interference pattern is an interferometric autocorrelation (see Chapter 8).

9.9. ZERO-AREA PULSE PROPAGATION

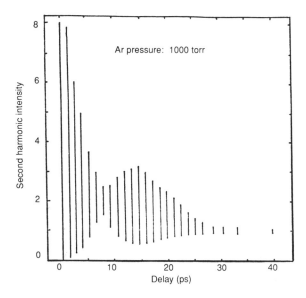

Figure 9.17: Second harmonic detection of the transmission of a zero-area pulse sequence as a function of delay through lithium vapor in the presence of argon as buffer gas. The vertical lines indicate the contrast between in- and out-of-phase transmission. The second harmonic of the transmitted zero-area pulse versus delay is recorded. The advantage of the second harmonic detection is that the first portion of the curve is approximately the interferometric autocorrelation of the pulse. The transmission corresponding to out-of-phase pulse sequence is the lower envelope near zero delay (weaker pulse because of destructive interference) and becomes the upper envelope for larger delays (larger transmission on resonance for out-of-phase pulse sequences).

In the case of an inhomogeneously broadened line, the phase dependence of the interaction disappears in the weak pulse limit. We have seen in Chapter 4 that the weak pulse absorption is proportional to the spectral overlap of the line and pulse. As shown by Eq. (4.35), in the case of infinite inhomogeneous broadening and no saturation:

$$\frac{dW}{dz} \propto \int_{-\infty}^{\infty} g_{inh}|\mathcal{E}(\Omega)|^2 d\Omega = -\alpha_0 W, \tag{9.28}$$

and the absorbed energy is independent of the phase content of the pulse. There is, however, a difference in *nonlinear* transmission [for which the approximation of Eq. (9.28) does not apply] of "in phase" and "out of phase" pulse sequences, even in the case of infinite inhomogeneous broadening. If each half of the pulse sequence has an area between 0 and π, the zero-area pulse sequence will be absorbed more strongly than the "in phase" pulse sequence. The physical reason can be explained simply by considering the originally uniform absorption spectrum [Fig. 9.18(a)] in which the first pulse burns a hole, which is seen by the second pulse as an inverted homogeneously broadened line [Fig. 9.18(b)]. For the "in phase" pulse sequence, there is less absorption because of the reduced absorbing spectral components at the pulse average frequency [Fig. 9.18(c)]. In contrast, the zero-area pulse sequence does not have spectral components overlapping with the center of the hole [Fig. 9.18(d)]. However, if each half of the pulse sequence is a π pulse, the system will be returned to ground state independently of the relative phase of the pulses.

The phase relaxation time T_2 can be extracted [423] by measuring the ratio of the energy transmission factor $\Delta W/W$ for a sequence of pulses $180°$ out of phase to the same transmission factor for the "in phase" pulse

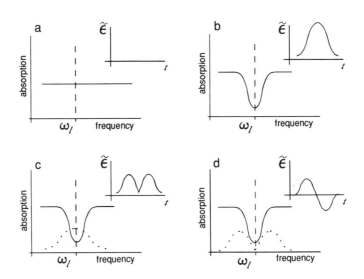

Figure 9.18: Initially (a), the line profile is uniform. At $t = t_1$, the first of a two-pulse sequence burns a hole in the uniformly inhomogeneously broadened absorption line (b). The second pulse no longer sees a uniform line, but an inverted homogeneously broadened line. The absorption will be smaller for the in-phase pulse sequence (c) which has more spectral components (dotted line) overlapping with the center of the hole, than with the out-of-phase pulse sequence (d).

9.9. ZERO-AREA PULSE PROPAGATION

sequence, as a function of total energy W in the pulse sequence. The corresponding values of α_π/α_0 are plotted in Fig. 9.19 for three values of the phase relaxation time T_2.

The main advantages of zero-area pulse excitation as applied to the determination of phase relaxation times are:

- the pulses of the sequence can overlap

- phase relaxation times shorter than the pulse duration can be measured

- the experimental technique is particularly simple

- a 180° pulse sequence has zero area for transitions of different degeneracy and dipole moment

The last property results in an easier interpretation of the data when the measurement covers more than one type of transition. The extension of this method to molecular multiphoton transitions has been discussed in Refs. [132, 424]. In addition to the measurement of a dephasing time for a multiphoton transition, the pattern of absorption versus relative phase of the pulse sequence can be used to identify the type of resonance [424].

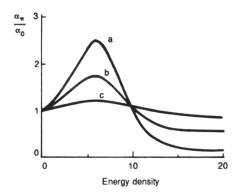

Figure 9.19: Ratio of the relative energy transmission for the 180° out-of-phase α_π to the in-phase α_0 pulse sequence, as a function of pulse energy W, for various values of the phase relaxation time T_2, for an absorption line with infinite inhomogeneous broadening. The pulses are Gaussian, with a temporal separation equal to twice their duration (FWHM). The phase relaxation times are 100× (a), 6× (b), and 1× (c) the pulse duration.

430   CHAPTER 9. MEASUREMENT TECHNIQUES

9.10 Impulsive stimulated Raman scattering

9.10.1 General description

Some molecular vibrations — for instance the stretching mode of a symmetric diatomic molecule such as N_2 — cannot be directly excited by a resonant electromagnetic field. However, such dipole forbidden transitions between states of equal parity and angular momentum can be accessed by a transition involving two photons. In resonant Raman scattering, the difference between the optical frequencies of the two photons involved in the transition is equal to the frequency of the mechanical vibration being excited. Let us consider, for instance, a molecular vibration of frequency ω_{21} between two states $|2\rangle$ and $|1\rangle$ of identical parity [Fig. 9.20(a)]. The molecule can be brought in the vibrationally excited state $|2\rangle$ by a succession of optical excitations via the dipole allowed transitions $|1\rangle \to \langle \ell|$ and $\langle \ell| \leftarrow |2\rangle$, involving photons of frequencies $\omega_{\ell 1}$ and $\omega_{\ell 2}$.

Because of the broad bandwidth of the fs pulse, stimulated Raman scattering can occur through the mixing of various spectral components of the ultrashort optical pulse [Fig. 9.20(b)]. For example, a molecular vibration is initiated by the sudden impulse exerted by the electric field of the pulse. The sample selects a pair of frequency components whose difference is in resonance with the eigenfrequencies of a Raman transition. This type of Raman scattering is called "impulsive stimulated Raman scattering" [425]. The fs

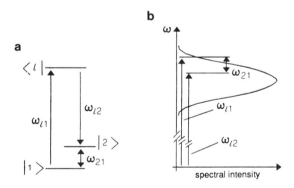

Figure 9.20: (a) Excitation of a Raman transition $|1\rangle \to |2\rangle$ via an electronically excited state $\langle \ell|$. (b) For a fs pulse of average carrier frequency ω_ℓ, the two frequency components of the Raman transition are contained within the pulse spectrum. The medium itself selects the frequency pairs suitable to drive the Raman transition.

9.10. IMPULSIVE STIMULATED RAMAN SCATTERING

excitation makes it possible to excite in phase a macroscopic ensemble of vibrating molecules. In solids, it is a coherent excitation of lattice vibrations that is achieved.

It is generally not possible to achieve a complete excitation of the Raman transition with a single fs pulse. Many Raman-active modes can sometimes be accessed by the same fs pulse. However, if the process of impulsive stimulated Raman scattering is repeated at each cycle $2\pi/\omega_v$ of a Raman transition, the motion will be enhanced. The selectivity of the process is also increased by the periodic excitation [426].

Impulsive stimulated Raman scattering can be used to analyze vibrational motions — for instance to determine their decay through pump-probe techniques. Synchronous excitation by a train of pulses can lead to substantially larger amplitudes of motion. This excitation process can generate high-frequency vibration. A train of pulses spaced by a picosecond can generate THz LO phonons in semiconductors, which have a wavelength in the 100 Å range, and can therefore be used for high-resolution imaging in solids.

9.10.2 Detection

The change in matter properties associated with the Raman excitation can be probed in a variety of ways. One can, for instance, probe an induced birefringence, in which case the rotation of the probe polarization will be measured, as detailed in Section 9.5. In parallel polarization (probe polarization parallel to that of the pump), the attenuation of the probe will be modulated with delay, because the probe pulse can also induce Raman transitions. The phase of the oscillations of attenuation versus delay of the probe depends on the particular spectral component that is being probed. In the example reproduced in Fig. 9.21, the pump and probe have the same polarization and are sent nearly collinearly through a sample of liquid CH_2Br_2 [425]. The transmitted probe is dispersed by a monochromator. Two frequency components (609 nm and 620 nm) are displayed as a function of delay in Fig. 9.21.

Both spectral components are seen to oscillate with delay at the molecular vibration frequency, but with opposite phase. A simple explanation is that at a particular delay, the position of the vibrating coordinates is such that the 609 nm radiation is absorbed, and the 620 nm reinforced by the Raman transition. For a delay corresponding to half a vibration cycle later, the 609 nm transition will be reinforced, and the 620 nm attenuated.

So far we have assumed a single pump pulse to induce the Raman signal. A standing wave pattern can also be generated for the impulsive stimulated

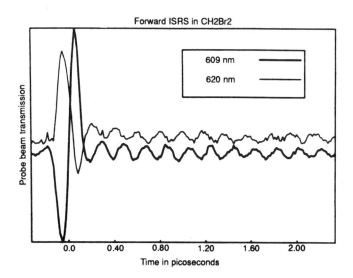

Figure 9.21: Transmission versus delay for two spectral components of the probe signal, for a sample of CH_2Br_2 pumped by a 65 fs pulse of a few μJ energy at 615 nm. The excitation and probe pulses are focused to a 200 μm spot size, at an angle of $5°$, into a 2 mm sample cuvette (from [425]).

Raman signal, either through a periodic configuration of the sample, or through the use of two intersecting pump pulses.

An example of sample periodicity is a multiple quantum well structure, of which the spacing between wells is made to match the wavelength of the phonon to be generated. The phonon can be generated by a train of fs pulses spaced by the phonon period, tuned to the intraband absorption in the quantum wells. The periodic spatial structure that is excited is responsible for the spatial coherence of the phonon [427]. The excitation by a periodic pulse sequence ensures temporal coherence of the created phonons.

It is also possible to create a standing wave Raman excitation with two intersecting pump pulses of the same frequency [425]. The temporal evolution of the vibration is easily analyzed through diffraction of a probe pulse by the standing wave pattern.

9.10. IMPULSIVE STIMULATED RAMAN SCATTERING

9.10.3 Theoretical framework

The same density matrix formalism as in Chapters 3 and 4 can be used to describe impulsive stimulated Raman scattering. As in Fig. 9.20, we will consider Raman transitions between a ground state $|1\rangle$ and a first excited state $|2\rangle$ of a vibrational mode with frequency ω_{21}. The states $|1\rangle$ and $|2\rangle$ are infrared inactive, i.e., there is no dipole allowed transition $|1\rangle \to |2\rangle$. Coupling between these two states can occur via any electronic state $\langle \ell |$. All states $|\ell\rangle$ connected to $|1\rangle$ and $|2\rangle$ via a dipole transition will contribute to the Raman transition. We assume the optical field E to be off-resonant with all single-photon transitions. For this assumption to hold, the detuning of the intermediate levels $\langle \ell |$ has to exceed several pulse bandwidths.

The evolution of the system is described by the density matrix equations 4.1. For the particular level system being considered:

$$\frac{\partial \rho_{12}}{\partial t} - i\omega_{21}\rho_{12} = -\frac{iE}{\hbar}\sum_{\ell}(\rho_{1\ell}p_{\ell 2} - p_{1\ell}\rho_{\ell 2})$$

$$\frac{\partial \rho_{22}}{\partial t} = -\frac{iE}{\hbar}\sum_{\ell}(\rho_{2\ell}p_{\ell 2} - p_{2\ell}\rho_{\ell 2})$$

$$\frac{\partial \rho_{1\ell}}{\partial t} - i\omega_{\ell 1}\rho_{1\ell} = -\frac{iE}{\hbar}\sum_{j}(\rho_{1j}p_{j\ell} - p_{1j}\rho_{j\ell}), \qquad (9.29)$$

where the sum over j applies to any level connected to $\langle \ell |$ by a dipole transition, including levels 1 and 2. A similar equation applies for $\rho_{2\ell}$.

As we have seen in Chapter 4, it is more convenient to decompose the off-diagonal matrix elements $\rho_{1\ell}$ and $\rho_{2\ell}$ into an envelope and fast varying phase term. For instance:

$$\rho_{1\ell} = \varrho_{1\ell}\, e^{i\omega_\ell t} \qquad (9.30)$$

and a similar equation for $\rho_{2\ell}$. Substituting Eq. (9.30) into the third equation (9.29), and keeping only the levels 1 and 2 as dipole connected levels to levels ℓ:

$$\frac{\partial \varrho_{1\ell}}{\partial t} + i(\omega_\ell - \omega_{\ell 1})\varrho_{1\ell} = -\frac{i\tilde{\mathcal{E}}}{2\hbar}(\rho_{11}p_{1\ell} + \rho_{12}p_{2\ell}). \qquad (9.31)$$

The assumption of the intermediate levels $\langle \ell |$ being off-resonance enables us to use the adiabatic approximation. This is a standard approximation used routinely in the context of deriving interaction equations in condition of two (and more) photon resonance [428]. A detailed analysis of the use of the adiabatic approximation in the context of two-photon transitions can be found in [428]). Essentially, the second term in the left hand-side of Eq. (9.31) dominates, and we can approximate $\varrho_{1\ell}$ by its steady state value:

$$\varrho_{1\ell} = \frac{-\tilde{\mathcal{E}}(\rho_{11}p_{1\ell} + \rho_{12}p_{2\ell})}{2\hbar(\omega_\ell - \omega_{\ell 1})}, \qquad (9.32)$$

and a similar equation for $\varrho_{2\ell}$. Substituting into the first equation (9.29), we find the evolution equation for the coherent Raman excitation:

$$\frac{\partial \rho_{12}}{\partial t} - i\omega_{21}\rho_{12} = -\frac{i}{4\hbar^2}\tilde{\mathcal{E}}\tilde{\mathcal{E}}^* \sum_\ell \left(\rho_{11}\frac{p_{1\ell}p_{\ell 2}}{\omega_\ell - \omega_{\ell 1}} - \frac{p_{1\ell}p_{\ell 2}}{\omega_\ell - \omega_{\ell 2}}\rho_{22} \right) \qquad (9.33)$$

Of particular interest is the amplitude of the off-diagonal element ρ_{12}. Let us define a (complex) amplitude ϱ_{12} similarly as in Eq. (9.30): $\rho_{12} = \varrho_{12} \exp(i\omega_{12}t)$. In addition, to simplify the discussion, let us assume, that there is only one level $\langle\ell|$ that dominates the interaction. We note that $\omega_\ell - \omega_{\ell 2} = (\omega_\ell - \omega_{\ell 1})[1 + \omega_{21}/(\omega_\ell - \omega_{\ell 1})]$. Substituting in Eq. (9.33) yields:

$$\frac{\partial \varrho_{12}}{\partial t} e^{i\omega_{21}t} = \frac{ip_{1\ell}p_{\ell 2}}{4\hbar^2(\omega_\ell - \omega_{\ell 1})} \tilde{\mathcal{E}}\tilde{\mathcal{E}}^* (\varrho_{22} - \rho_{11}) \qquad (9.34)$$

where $\varrho_{22} = \rho_{22}/[1 + \omega_{12}(\omega_\ell - \omega_{\ell 1})]$. We recognize in Eq. (9.34) a Rabi frequency similar to the two-photon Rabi frequency discussed in Chapter 4:

$$\frac{p_{1\ell}p_{\ell 2}}{4\hbar^2(\omega_\ell - \omega_{\ell 1})}\tilde{\mathcal{E}}(t)\tilde{\mathcal{E}}^*(t) = \frac{r_{12}}{\hbar^2}\tilde{\mathcal{E}}(t)\tilde{\mathcal{E}}^*(t). \qquad (9.35)$$

The evolution equations for the density matrix components can be rewritten:

$$\begin{aligned}\frac{\partial \varrho_{12}}{\partial t} &= i\frac{r_{12}}{\hbar^2}\tilde{\mathcal{E}}\tilde{\mathcal{E}}^* e^{-i\omega_{21}t}[\varrho_{22} - \rho_{11}] \\ \frac{\partial \rho_{22}}{\partial t} &= -2\mathrm{Im}\left[\frac{r_{12}}{\hbar^2}\tilde{\mathcal{E}}\tilde{\mathcal{E}}^* \varrho_{12}e^{i\omega_{21}t}\right].\end{aligned} \qquad (9.36)$$

The form of the set of equations (9.36) is similar to Bloch's equations (4.6) and (4.7). In the weak pulse approximation ($\rho_{11} \approx 1$), after passage of the fs excitation, the off-diagonal matrix element oscillates at the Raman frequency:

$$\begin{aligned}\rho_{12} &\approx -e^{i\omega_{21}t}\left[\int_{-\infty}^\infty \frac{r_{12}}{\hbar^2}\tilde{\mathcal{E}}(t')\tilde{\mathcal{E}}^*(t')e^{-i\omega_{21}t'}dt'\right] \\ &= -e^{i\omega_{21}t}\left[\int_{-\infty}^\infty \frac{r_{12}}{\hbar^2}\tilde{\mathcal{E}}(\Omega)\tilde{\mathcal{E}}^*(\Omega - \omega_{21})d\Omega\right].\end{aligned} \qquad (9.37)$$

9.10. IMPULSIVE STIMULATED RAMAN SCATTERING

Equation (9.37) is obtained by integrating the first differential equation (9.36) with $\rho_{11} \approx 1$ and $\rho_{22} \approx 0$. It can be seen immediately from the convolution product in Eq. (9.37) that, for efficient Raman excitation, the pulse spectrum should be broad compared with the Raman frequency ω_{21}. Indeed, for $\omega_{21} \gg \tau_p^{-1}$, there is no overlap between $\tilde{\mathcal{E}}(\Omega)$ and $\tilde{\mathcal{E}}(\Omega - \omega_{21})$. The dimensionless quantity

$$\theta_R = \int_{-\infty}^{\infty} \frac{r_{12}}{\hbar^2} \tilde{\mathcal{E}}(\Omega) \tilde{\mathcal{E}}^*(\Omega - \omega_{21}) d\Omega \tag{9.38}$$

is the analogue of the tipping angle of the polarization in the Bloch vector model (cf. Chapter 4). We recognize from the analogy between Eqs. (9.36) and Bloch's equations (4.6) and (4.7), and the description of the vector model in Chapter 4, that an angle θ_R on the order of unity will be required to bring the ground state population to a vibrational excited state of energy $\hbar\omega_{21}$. It is left as a problem at the end of this chapter to demonstrate that the convolution product in θ_R can be maximized by using, instead of a single pulse, a train of pulses spaced in time by $\tau_d = 2n\pi/\omega_{21}$ (n integer). The increase in selectivity can be inferred from the form of θ_R in the frequency domain [Eq. (9.38)]: θ_R vanishes for a pulse spacing $\Delta t \neq 2\pi/\omega_{21}$, in the case of a large number of pulses and undamped oscillations. The technique of using a synchronized pulse train can also lead to much larger amplitudes of motion than a single pulse [426]. Methods of generating such pulse trains have been presented in Chapter 7.

9.10.4 Single pulse shaping versus mode-locked train

The expression (9.38) for θ_R can be maximized by a pulse train at any submultiple of the frequency $\omega_{21}/2\pi$. With ω_{21} in the THz range, one technique is to "shape" a fs pulse into a sequence of pulses. Another possibility is to tune the mode-locked period of the laser to $T = 2n\pi/\omega_{21}$ (n integer). Such a technique is reminiscent of high-resolution coherence spectroscopy, where the repetition rate of mode-locked trains is tuned to a submultiple of an atomic resonance, leading, for instance, to enhanced quantum beats [429]. With advances in semiconductor lasers, repetition rates in the GHz to THz range are accessible with fs pulses. The repetition rate of passively mode-locked lasers can also be tuned continuously by adjusting the cavity length [168].

A question that arises is: what is the lowest repetition rate that can be used to excite a particular resonance ω_{21}? That question can be simply answered by modeling the resonant system by a classical oscillator, driven by an infinite series of δ functions separated by a time T. Each successive pulse excites the particular oscillation corresponding to the resonance. This oscillation is represented in the classical model by the displacement x of an oscillator of mass m, restoring force Kx, and damping constant b. The oscillation is not completely damped before the time of arrival of the next pulse, which, if T is a multiple of $2\pi/\omega_{21}$, will reinforce the motion. After an infinite number of driving pulses, the damped oscillation between two successive driving pulses will be stationary (see Fig. 9.22). The periodic driving force is represented by a series of δ functions: $F = F_0 \sum_{j=-N}^{N} \delta(t - jT)$. The equation of motion for the classical oscillator is:

$$m\ddot{x} + b\dot{x} + Kx = F_0 \sum_{j=-N}^{N} \delta(t - jT). \tag{9.39}$$

Simple Fourier transformation of this equation leads to a solution for the amplitude $x(t)$:

$$x(t) = \frac{F_0}{2\pi} \sum_j I_j \tag{9.40}$$

with

$$I_j = \int_{-\infty}^{\infty} \frac{e^{i\omega(t-jT)}}{(K - m\omega^2) + ib\omega} d\omega. \tag{9.41}$$

The integrand I_j in Eq. (9.41) has two poles at $\omega = i\Gamma \pm \omega_{21}$, and $\omega_{21}^2 = K/m - \Gamma^2$ and $\Gamma = b/2m$. The stationary solution for the oscillator is found

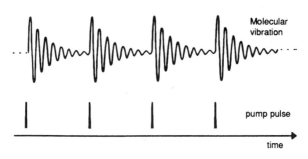

Figure 9.22: Damped molecular vibration, following impulsive stimulated Raman excitation by a train of ultrashort pulses. The amplitude of the oscillation will be maximum for a pulse separation equal to a multiple of the period of the molecular vibration.

by contour integration and summation of the geometric series of index j:

$$x(t) = \frac{1}{2}A(T)e^{-\Gamma t + i\omega_{21} t} + c.c. \tag{9.42}$$

with

$$\begin{aligned} A(T) &= \frac{iF_0}{\omega_{21}} \frac{1}{1 - e^{\Gamma T + i\omega_{21} T}} \\ &= \frac{iF_0}{\omega_{21}} \frac{1 - e^{\Gamma T - i\omega_{21} T}}{1 + e^{2\Gamma T} - 2e^{\Gamma T} \cos \omega_{21} T}. \end{aligned} \tag{9.43}$$

$A(T)$ is essentially the amplitude of the first cycle of oscillation. Its value is maximum and equal to $iF_0/[\omega_{21}(1-e^{\Gamma T})]$ when $\omega_{21}T = 2n\pi$, and minimum, equal to $iF_0/[\omega_{21}(1+e^{\Gamma T})]$ for $\omega_{21}T = 2(n+1)\pi$. The modulation depth $(1 - e^{\Gamma T})/(1 + e^{\Gamma T})$ is thus determined solely by the damping rate and the period of the driving force T. When driving a system at a subharmonic of the resonant frequency, the term $\omega_{21}T$ in Eq. (9.43) can be very large ($\omega_{21}T = 2n\pi$, with n a large integer). The "resonances" (values of the periodicity T that satisfy the resonance condition) are closely spaced. The damping factor Γ determines which subharmonic N can still be used to drive effectively the resonance ω_{21}. Each δ-function sets off an oscillation, which should not be completely damped before being reinforced by the next δ function.

9.11 Self-action experiments

Pump-probe experiments are intended to provide information on linear and nonlinear properties of matter. As noted earlier, there is a fundamental temporal limitation: The pump or excitation process has to be completed before the medium can be probed. One could try to obtain similar information on the properties of matter by measuring the time resolved fields of a pulse incident, reflected and/or transmitted by a thin sample (Fig. 9.23), using some of the techniques outlined in Chapter 8.

In the case of a linear interaction with the medium, the problem is analogous to the analysis of a linear circuit. For instance, referring to Chapter 1 [Eqs. (1.46) through (1.51)], the complex dielectric constant $\tilde{\epsilon}(\Omega) = [1 + \tilde{\chi}(\Omega)]$ can be extracted by taking the Fourier transform $\tilde{E}(\Omega)$ of the incident (i) and transmitted (t) fields:

$$1 + \tilde{\chi}(\Omega) = -\frac{c^2}{z^2 \Omega^2} \ln^2 \left[\frac{\tilde{E}_t(\Omega)}{\tilde{E}_i(\Omega)} \right] \tag{9.44}$$

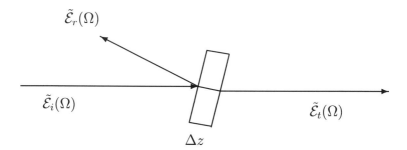

Figure 9.23: For linear systems, and some simple nonlinear systems, the complex susceptibility can be completely determined from single pulse transmission (reflection) measurements, provided the amplitude and phase of the incident, transmitted, and reflected signals are completely determined.

where z is the sample thickness.

There is no simple algorithm that can solve the general problem of retrieval of a nonlinear susceptibility $\chi^{(n)}(\Omega)$ from a series of measurements of incident, transmitted, and reflected fields. Some assumptions have to be made — for instance, that all nonlinear susceptibilities except the third order, $\chi^{(3)}$, can be neglected. Measurement of the third harmonic transmitted field $\tilde{\mathcal{E}}_{3\omega}$ leads to a determination of the third-order susceptibility:

$$\chi^{(3)}(\Omega) = \frac{2c\Delta z}{\omega_\ell} \frac{\tilde{\mathcal{E}}_{3\omega}(\Omega)}{\tilde{\mathcal{E}}^2(\Omega)}. \qquad (9.45)$$

The transmission measurements provide information on the bulk properties of the sample. Temporal properties at the surface can be analyzed by measuring the reflected field. For instance, at normal incidence, the reflection coefficient is approximately $\chi_s(\Omega)/[4 + \chi_s(\Omega)]$, where $\chi_s(\Omega)$ is the complex susceptibility at the surface (assumed to be $\ll 1$). In the presence of resonances, these complex susceptibilities may have a complicated functional dependence on the optical field.

9.12 Problems

1. Referring to Section 9.3, derive in detail Eqs. (9.5) through (9.9). Find the effect of the beam geometry on the temporal resolution for a square temporal profile, and square spatial profile in x and y.

9.12. PROBLEMS

2. A transient grating experiment is performed with a semiconductor. Let us assume that we have an amplitude grating only and that the carrier density $n(x,t)$ obeys the equation for ambipolar diffusion (one-dimensional model):

$$\frac{\partial n}{\partial t} - D\frac{\partial^2 n}{\partial x^2} = 0. \tag{9.46}$$

Derive a formula that relates the diffraction signal versus τ_d to the diffusion parameter D to be determined. From the diffusion parameter one can then obtain the carrier mobility $\mu = \frac{e}{k_B T}D$, where e is the electron charge, k_B Boltzmann's constant, and T the temperature.

3. Prove that in a transient grating experiment the diffraction of a probe pulse measured as function of the delay between the two pump pulses contains the information on the transverse relaxation time T_2. Assume an ensemble of homogeneously broadened two-level systems, weak excitation, thin samples.

4. The purpose of this problem is to compare impulse stimulated Raman scattering by either a single pulse or a train of identical pulses. The period of the Raman oscillation to excite is 1 ps, and its damping time is 500 ps. The molecular system has a resonant absorption at 750 nm. The laser system delivers a Gaussian pulse of 50 fs duration, 1 nJ energy, at 770 nm, focused into the sample with a beam waist of $w_0 = 200\mu m$. The dipole moment of the transitions $p_{1\ell} = p_{\ell 2} = 6\ 10^{-29}$ Cm. Calculate the off-diagonal matrix element ρ_{12} resulting from the excitation by the Gaussian pulses. Assume next each fs pulse is replaced by a train of 10 Gaussian pulses of 50 fs duration, but of 0.1 nJ energy each. Calculate the off-diagonal matrix element as a function of the period of this pulse train (in the range 1–10 ps).

Chapter 10

Examples of Ultrafast Processes in Matter

10.1 Introduction

A microscopic analysis of many fundamental processes in matter starts at the ps or fs time scale. Primary events associated with macroscopic transformations that appear relatively slow, such as chemical reactions, photosynthesis, phase changes, and human vision, evolve on a fs time scale. A mere listing of all processes in biology, chemistry, and physics that are being actively investigated is already beyond the scope of this book. A periodic update of these topics can be found in the proceedings of the international conferences "Ultrafast Phenomena" held every even year in the U.S. and "Ultrafast Phenomena in Spectroscopy" held every odd year in Europe. A review of several topics can be found in [5].

Rather than to attempt an extensive review, this chapter will focus on a few examples of ultrafast events in matter, and their measurement. We shall proceed by order of material systems of increasing complexity. The simplest system is the single atom, in which wave packets representing the motion of the electron in a Rydberg orbit can be analyzed with ultrafast techniques. Next, we proceed from the single atom to simple molecules to dissociating molecules — a step towards chemical reactions. The next form of arrangement of atoms is condensed matter, in which fs techniques are particularly powerful in analyzing changes of phase. Finally, biological systems offer the ultimate in molecular complexity. Femtosecond techniques are an essential tool in unraveling, for example, the primary processes of vision and photosynthesis.

10.2 Ultrafast transients in atoms

10.2.1 The classical limit of the quantum mechanical atom

Bohr's model of the hydrogen atom was based on the concept of the electron describing a classical trajectory in the attractive potential of the nucleus. The quantization relation introduced empirically by Bohr (see, for instance Ref. [125]) states that the angular momentum of the orbits is quantized:

$$\frac{m_e m_p}{m_e + m_p} vr = pr = \hbar n \qquad (10.1)$$

where v is the radial velocity of the electron (p its linear momentum) along the orbit of radius r, m_e and m_p are the masses of the electron and proton, and n is the quantum number. The classical picture of the orbiting electron violates the uncertainty principle for small quantum numbers n. In order to be able to describe the electron motion by a classical trajectory, the uncertainty in position (Δr) and momentum (Δp) should be smaller than r and p, respectively, or:

$$\frac{\Delta r}{r} \frac{\Delta p}{p} \ll 1 \qquad (10.2)$$

$$\gg \frac{\hbar}{n\hbar}. \qquad (10.3)$$

The last inequality (10.3) is simply obtained by substituting the uncertainty principle and the quantization condition (10.1) into the classical representation condition (10.2). The two conditions (10.2) and (10.3) are only compatible for large values of the quantum number n, or large orbits. States characterized by a high principal quantum number are called Rydberg states. The classical orbit becomes a reasonable approximation for these states with large quantum number n.

10.2.2 The radial wave packet

A fs pulse cannot be used to excite an atom from its ground state to a single Rydberg state, because Rydberg states are closely spaced as compared to the bandwidth of ultrashort pulses. Instead, a fs pulse will excite a superposition

10.2. ULTRAFAST TRANSIENTS IN ATOMS

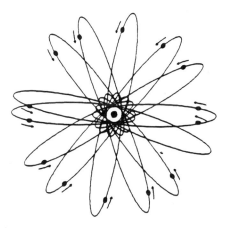

Figure 10.1: An ensemble of classical Kepler orbits make up the radially localized wave packet created by fs excitation of the ground state atom. The major axes of the ellipses are randomly distributed over all directions with a $\sin^2\theta$ distribution, but each electron has the same angular coordinate in various ellipses (from [430]).

of many Rydberg states. This superposition is a wave packet localized in the *radial* coordinate. The period of oscillation corresponds to the period of the Kepler orbit of a classical particle with the energy corresponding to that of the average Rydberg state excited.

The experimental technique to observe the radial motion of the electrons along these Kepler orbits is a pump-probe experiment. The pump pulse excites the atoms to a superposition of Rydberg states. The number of ions (or free electrons) produced by a subsequent probe is recorded as a function of delay. As explained below, the number of ions can be related to the position (velocity) of the electron along its Kepler orbit [430, 431, 432].

There is no localization in the angular coordinates. If the ground state is an S state, the states in the wave packet superposition are P-states with various principal quantum numbers. Each of these states has as an angular dependence proportional to the square of a single spherical harmonic, a dependence in $\sin^2\theta$ in the case of the $l = m = 1$ state (where l and m are the usual eigenvalues of the orbital angular momentum and its projection along a z axis). The classical description of a Kepler orbit applies: Rather than a single orbiting electron, we should visualize an ensemble of non-interacting particles orbiting the nucleus, with their principal axes distributed according to the $\sin^2\theta$ distribution (Fig. 10.1). This "radial Rydberg wave packet" will move in the effective atomic potential between the two classical turning points. It is a radial wave packet, because only a few angular momentum eigenstates can be excited (selection rule $l \to l \pm 1$), and the angular coordinates of the Rydberg electron are delocalized in a quantum mechanical sense [433]. Each of these orbits correspond to approximately the same energy, hence the same classical period. Therefore, with all particles moving in phase along the various elliptical orbits, they arrive at the same time close to the nucleus, as illustrated in Fig. 10.1. To the motion of the charged particle is associated an electric current $\mathbf{J} = e\mathbf{v}$ proportional to its velocity $\mathbf{v}$. The Rydberg wave packet is excited by a pump pulse. The energy absorbed by a delayed probe pulse of electric field $\mathbf{E}_p$ is proportional to $\mathbf{J} \cdot \mathbf{E}_p$. The absorbed energy is large if the delay is such that the pulse reaches the atoms with the electron near the nucleus (maximum velocity), and substantial ionization will result. At the other turning point far away from the core, the electron is nearly a free particle (which will not absorb radiation).

The photoionization versus probe delay is shown in Fig. 10.2. The signal oscillates at the classical orbital frequency. However, because the Rydberg states are not equally spaced in frequency, the states get out of phase, and the wave-packet decays. The observed decay of the wave-packet shown in Fig. 10.2 is not an irreversible process: After a large number of cycles, the components of the wave packet come back in phase [434], a process called "revival" of the wave packet. One can also observe "fractional revivals" [432]. For instance, during the one-half fractional revival, every other state in the superposition comes into phase, leading to the formation of two wave packets. Experimental evidence of the formation of two wave packets is the change in oscillation frequency to twice the orbital frequency in Fig. 10.2.

10.2.3 The angularly localized wave packet

Radial localization was obtained by creating a superposition of states corresponding to a large radial quantum number n, spanning a group of values Δn. Similarly, angular localization will require the superposition of excitations to a large angular momentum l, spanning a group of values Δl. Since a single photon carries only one unit of angular momentum, many photons are required to reach the high angular momentum states from the ground state. The technique devised by Yeazell and Stroud [435] is to excite sodium atoms from the ground state to the $n \approx 50$ manifold of states via a two-photon transition, using circularly polarized light at 483.7 nm from an excimer pumped dye laser. A radio-frequency (rf) field is used to create the high angular momentum wave packet through 30-photon excitation from the $50d$ state to

10.2. ULTRAFAST TRANSIENTS IN ATOMS

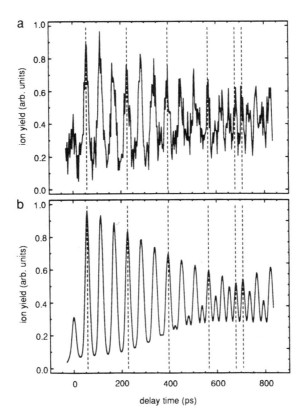

Figure 10.2: Photoionization signal as a function of probe delay. (a) Experimental recording. (b) Theoretical simulation. The Rydberg wave packet spans $\Delta n = 5$ and is centered around the Rydberg state $n = 72$ of atomic potassium (wavelength of pump pulse 285.6 nm). The probe is at 571.2 nm. (From Ref. [432].)

states grouped around $l = 32$ ($n = 50$, $29 < l < 37$, $m = l$). The orientation of the wave packet lies along the direction of the rf-field vector at the time of the optical excitation. Since the precession and rate of dispersion of the wave packet are very slow (order of ms), detection can be made through ionization with a pulsed electric field [435]. The wave packet is localized in the angular direction, but not in the radial direction. The classical description is that of an ensemble of elliptical orbits, all with their axes aligned along the direction of the rf field. However, the phases of the motion along the ellipses are not determined, resulting in an elliptical distribution in space that is approximately stationary in time.

Localization in the radial and angular coordinates, which leads to the observation of an atomic electron on a classical orbit, has not been achieved as of today. However, techniques involving ultrashort optical and electrical pulses have been proposed by Gaeta et al. [436].

10.3 Ultrafast processes in molecules

10.3.1 Observation of molecular vibrations

Simple molecules

When single atoms combine to form molecules, additional internal degrees of freedom, such as rotations and vibrations, arise, with transients in the picosecond and femtosecond range.

Instead of an electron moving in the field of an atom, we shall now consider the case of an atom in a molecule. In the example of Fig. 10.4, a molecule is irradiated by a fs pulse of wavelength λ_1 corresponding to a single photon excitation to a non-dissociative state **B**. For the purpose of illustration let us consider as specific system the I_2 molecule for which the potential curves are reproduced in Fig. 10.3. A fs pulse is used to excite the $\mathbf{X}(v" = 0) \rightarrow \mathbf{B}(v' = n)$ transition in the 500–600 nm wavelength range (where v characterizes the vibronic excitation). Owing to the broad excitation spectrum, the fs pulse creates a coherent superposition of vibronic states of mean quantum number n. Note that during the short fs interaction the nuclear motion can be neglected, which corresponds to a vertical transition in Figs. 10.4 or 10.3. The time evolution of this system can be viewed as the motion of this wave packet in the molecular potential. The classical limit is the mechanical (harmonic) oscillation with a characteristic vibration frequency ω_{vib}. As in the case of the electron in a Rydberg atom, the periodic motion of the wave packet can be observed with fs techniques.

There are several techniques available to monitor the quantum state of excited molecules. They are based on the fact that the interaction strength with a second light pulse depends on the instantaneous location and shape of the wave packet. If the experiment is carried out in a molecular beam, a delayed fs pulse can be used to excite from state **B** to a dissociative state (Fig. 10.3(a) shows the potential surfaces for the I_2 molecule). The fragments can be monitored with a mass spectrometer. If the measurement is carried out in a cell, the population of the **B** state can be observed simply through the fluorescence from the **B** state to the ground state. To probe the dynamics of the vibration, the molecule can be irradiated by probe pulses identical to

10.3. ULTRAFAST PROCESSES IN MOLECULES

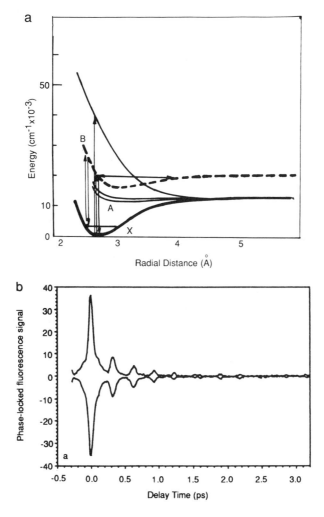

Figure 10.3: Study of the potential surface of I_2, using a pair of identical pulses with adjustable relative phase and delay. (a) Sketch of the potential surfaces. (b) Envelope of the fluorescence signal (only the contribution due to the two-pulse excitation) from the **B** state, corresponding to an in-phase sequence (upper envelope) and an out-of-phase sequence (lower envelope) (adapted from [437]).

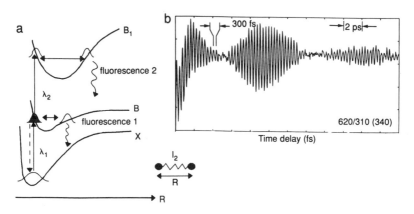

Figure 10.4: (a) Sketch of the bound potential energy surfaces relevant to the study of iodine through excited state fluorescence. (b) Fluorescence from the excited state B_1 of I_2 as function of the delay between the two excitation pulses of wavelength λ_1 and λ_2 (adapted from [438]).

those that created the excitation (except for the timing and phase). Since the excited state still represents a stable molecule, return to ground state will be possible at periodic intervals corresponding to the vibrational period of the electronically excited I_2 molecule (Fig. 10.3).

The experimental technique is essentially that of the zero-area pulse experiment described in the previous chapter. In the case of I_2, the excitation wavelength should be in the range of 608 to 613 nm. The zero-area pulses are generated in a Mach–Zehnder interferometer and sent through a 5 cm long room temperature I_2 cell at 0.25 torr [437]. The fluorescence is detected at a right angle. The return to ground state will occur if the delayed probe of the same wavelength as the pump is π out of phase with the exciting pulse. If instead the probe is in phase with the first pulse, the excitation will be reinforced. As in the case of the radial Rydberg wave packet, the classical picture for an oscillating particle fully applies. The envelope of the fluorescence pattern for in-phase and out-of-phase pulse sequences is shown in Fig. 10.3(b) (from [437]). The successive peaks are separated by 278 fs. This spacing corresponds to the superposition of the vibrational levels of the **B** state pumped by excitation at 611.2 nm from the thermally populated levels of the ground state. The classical approximation is that we are seeing the period of the oscillation of the excited molecule, corresponding to a vibration frequency of 3.6 THz.

Another possibility to measure the dynamics of the wave packet is to probe the excitation of state **B** into a bound state B_1 [Fig. 10.4(a)] with

10.3. ULTRAFAST PROCESSES IN MOLECULES

a time delayed pulse of different frequency ($\lambda_2 = 310$ nm) [438]. This was done by measuring the fluorescence from state $\mathbf{B}_1$ as a function of the delay between the excitation pulses, as shown in Fig. 10.4(b). The short time oscillation has the period of vibration of the molecule, or period of wave packet motion in the **B** state. The periodical behavior of the oscillation period is due to a revival of the wave packet (see, for instance, Ref. [434]). The wave packet consists of a finite number of *nearly* equally spaced energy states (anharmonic potential). This causes the wave packet to spread as time progresses so that it is no longer localized. As a result, the periodic behavior of the fluorescence disappears. However, since only a finite number of states is excited by the fs pulse and forms the wave packet, a rephasing of the states occurs after a certain time period. As in the case of the Rydberg states, the wave packet again becomes localized, which manifests itself in an increased modulation amplitude of the fluorescence.

Complex molecules

Vibrations and other motions The interpretation of the vibrational studies is particularly simple for isolated diatomic molecules. Molecular vibration and vibrational relaxation, of course, occur in more complex systems, too. As an example, let us consider organic dye molecules in solution. As outlined in previous chapters these systems have gained importance as laser dyes and saturable absorbers, and have therefore been extensively studied. Because of the large number of internal degrees of freedom and the strong interaction with the solvent, the damping of coherently excited wave packets and the vibrational relaxation often proceed on a sub-picosecond time scale. Wise *et al.* [439] observed a damped sinusoidal decay in a pump-probe absorption experiment. For the dye Nile Blue, for example, they could identify eight different oscillation frequencies, documenting the large manifold of molecular eigenmodes of this complex system. Femtosecond techniques have also been successfully applied to the spectroscopic characterization of clusters, see, for example, [440].

An absorption spectrum of a dye solution taken with an ordinary spectrophotometer typically exhibits a resonance corresponding to the $S_0 \rightarrow S_1$ transition with a spectral width of several tens of nanometers. This broad absorption profile results from a very large number of rotational and vibrational states within one electronic state. An interesting question is whether the transition is homogeneously or inhomogeneously broadened. As explained in Chapter 3 (Fig. 3.2), the answer depends on the time scale on which

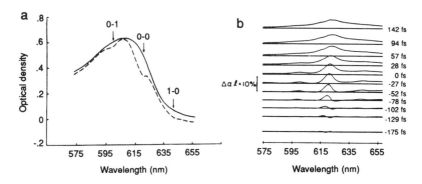

Figure 10.5: (a) Absorption spectrum of cresyl violet near the region of pumping. The dashed line illustrated qualitatively the spectral modification immediately after the 60 fs pump pulse. (b) Differential absorption spectra for successive delay increments after excitation of cresyl violet with a 60 fs pump pulse at 618 nm (from [441]).

the experiment is performed. A convenient experimental technique is time resolved hole burning.

Hole burning Hole burning or saturation spectroscopy is the standard technique to determine the homogeneous linewidth (T_2^{-1}) in gases and vapors. In the case of condensed matter, a fs variation of that technique can be used. First, an intense fs pump pulse is applied to saturate a particular transition. Let us consider as a specific example a hole burning experiment performed on cresyl violet [441]. The 60 fs wide pump pulse (centered at 618 nm) excites the $S_0(v=0) \to S_1(v=0)$ transition of the molecule (Fig. 10.5). Since the occupation numbers of the $v=0$ transition in the S_0 and S_1 electronic state are modified, an absorption change in the $0 \to 1$ and $1 \to 0$ transition is also observed immediately after excitation. A fraction of the pump pulse is chirped and compressed — and hence spectrally broadened — to probe the modified absorption spectrum of the sample. Twelve millimeters of fiber and a pair of gratings compress that fraction of the pump down to 10 fs, using the technique outlined in Chapter 6. For every delay increment, the difference spectrum (with and without pump) is recorded. The resulting plot reproduced in Fig. 10.5 shows clearly that three successive holes are burnt in the absorption profile. The inverse of the linewidth of the hole indicates a phase relaxation time of $T_2 = 75$ fs. A plot of the differential absorption versus time also shows a fast transient. The decay of

10.3. ULTRAFAST PROCESSES IN MOLECULES

the hole structure with time is a measure of the cross relaxation. The red shift of the peak in the differential absorption indicates vibrational relaxation. The spectral feature gives in this case a more positive identification of the homogeneous broadening than the more complex temporal transient.

10.3.2 Chemical reactions

One of the great frontiers in chemistry is detailed experimental investigations of chemical reactions in progress from reactants through a transition state to products. Until recently, understanding of the evolution of the transition state relied almost exclusively on theoretical treatments. For a three-atom system with a small number of electrons, calculations may provide potential energy surfaces on which to compute classical trajectories to simulate chemical reactivity. To adequately reflect observable chemical phenomena, the accuracy of these energy potential surfaces needs to be on the order of 1 kcal/mole, an appalling figure for spectroscopists, since it corresponds to a spectral uncertainty of 350 cm^{-1} (or 10^{13} s^{-1})! The uncertainty is even worse for more complex molecular systems with many more internal degrees of freedom, hence the need for an experimental technique that will directly measure the potential energy surfaces of the transition states.

Femtosecond pulses offer the possibility of separating the electronic and nuclear parts of the wave function, and therefore work directly within the framework of the Born–Oppenheimer approximation (see, for instance, [125]). One of the new methods discussed in the beginning of this section is to coherently excite or de-excite a transition from a ground to a higher energy potential surface. The advantage of the fs pulse excitation is that no substantial change in nuclear coordinates can take place during the interaction with light.

For the study of chemical reactions, the experimental difficulty is that measurements cannot be made on a single isolated molecule, and it seems difficult at best to synchronize (for instance) pairs of molecules involved in bimolecular reactions. In the case of unimolecular reactions, however, it is possible to use a femtosecond pulse to initiate synchronously the dissociation of a group of molecules (excitation to a repulsive potential surface V_1), and to monitor subsequently their evolution to products with delayed probe pulses [442] as sketched in Fig. 10.6.

Potential curves have been extracted from such measurements [444]. The probe pulse induces transitions from the repulsive potential surface V_1 (along which the dissociating molecule is moving, following excitation by the pump),

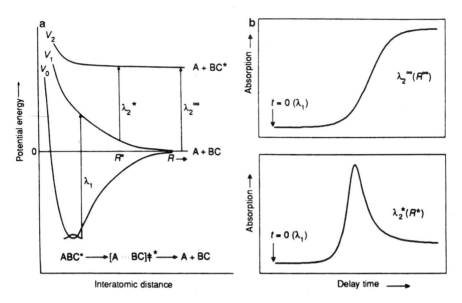

Figure 10.6: Pump-probe experiment to observe the transition region of a reaction. (a) The molecule is first excited by a pump pulse at λ_1 from the potential energy curves for the bound molecule V_0, to the dissociative state V_1. After a delay τ_d during which the fragment evolves along the repulsive potential V_1, a probe pulse at λ_2^* excites the complex to the dissociative (eventually ionizing) potential surface V_2. As the fragments recoil, the pulse at λ_2^* (or λ_2^∞) probes the transition region (or the free fragments). (b) The expected fs transients, signal versus delay τ_d at λ_2^* and λ_2^∞ (adapted from [443]).

to another potential surface V_2. For a given delay τ_d, the probe absorption versus probe wavelength will be an approximation of a step function, where the wavelength at the step is a measure of the energy difference between potential energy surfaces at the particular delay τ_d. It is usually not practical to measure the absorption of the probe with low-density molecules. Instead, one can determine the number of molecules excited to the V_2 potential surface by laser induced fluorescence. Another possibility is to use a photoionizing probe (in which case the number of transitions is directly measured by an ion count). This technique has been first applied to a detailed study of the unimolecular dissociation [442]:

$$\text{ICN}^* \rightarrow \text{I} + \text{CN}. \tag{10.4}$$

10.3. ULTRAFAST PROCESSES IN MOLECULES

The value of the resonance energy versus delay is a measure of the *difference* between the potential energy surfaces $V_2 - V_1$. In order to obtain an absolute measurement of an energy potential curve, it is necessary to know the shape of the upper curve V_2, or to make the assumption that this upper curve is flat. For the particular experiment reported in [442], the variation of the potential surface V_2 should not exceed 100 cm^{-1} [442]. Another approach is to use a theoretical model to calculate the upper surface. However, since the potential variation is on the order of 6000 cm^{-1} over the range of interest [442], the required accuracy if on the order of a few percent. Procedures to invert the data to obtain the energy surfaces for the ICN reaction have been developed by Bernstein and Zewail [444]. Because of the large energy changes along the potential energy surface in a short delay, the probe pulse duration has to be selected to obtain the optimal combination of temporal and spectral resolution. We refer the reader interested in a general overview of fs probing of dissociative reaction to [443, 445, 446].

The technique of probing chemical reactions has been successfully applied to unimolecular dissociations. The possibility of using a femtosecond technique to study bimolecular reactions at the individual collision level is complicated by the difficulties of spatial and temporal synchronization. One way to overcome this problem is through the use of van der Waals complexes of weakly bound molecular clusters. In these complexes the moieties are held in a reasonably well-defined geometry, so that the prospective bimolecular reagents or their precursors may be frozen into a convenient geometry in preparation for reaction initiation. There are well established techniques to produce clusters of heterodimers [447, 448]. Once frozen collision complexes have been prepared by supersonic nozzle expansion of appropriate compounds, a bimolecular chemical reaction can be initiated by a fs photodissociation pulse producing a pair of reagents. Such a technique has been applied for the first time [449] to the reaction

$$H + OCO \rightarrow [HOCO]^{\ddagger} \rightarrow OH + CO. \tag{10.5}$$

The van der Waals "precursor molecule" was [IH$\cdots$OCO] formed in a free-jet expansion of a mixture of HI and CO_2 in an excess of helium carrier gas. To clock the reaction, an ultrashort laser pulse photodissociates HI, ejecting an H atom towards the O atom of the CO_2. The delayed probe detects the formation of OH. Such an experiment establishes clearly that the reaction proceeds via an intermediate state, as shown in Eq. (10.5), and gives values for the lifetime of the intermediate complex [HOCO]‡.

Femtosecond techniques are not limited to the *observation* of chemical reactions, but can even be exploited to influence the course of the reaction (see, for instance, Ref. [450]). This can open new relaxation channels or increase the yield of certain reaction products.

10.3.3 Molecules in solution

Considerable progress has been made towards the microscopic understanding of molecular vibration and chemical reactions in solution. For instance, we have shown at the beginning of this section techniques to study the wavepacket dynamics of the nuclear motions of iodine in the **B** state, in the collision-free limit. These techniques can be applied to solutions of different densities, and liquids. The primary effects of the solvent on the fs wave packet are dephasing, energy relaxation, caging and recombination [451]. Except for collision induced rapid nonradiative transition in the liquid state, which cause the main fluorescence emission to originate from a lower transition, the experimental techniques are similar to the one used in the gaseous phase.

Femtosecond techniques have also been applied to more complex chemical problems, such as the study of photodissociation. The influence of the solvent on the dynamics of photodissociation of ICN can be dramatic [452]. The knowledge gained of how the solvent influences the decay of photofragment translation and rotation is useful in understanding the dynamics of thermally activated chemical reactions [452]. Theoretical simulations have indeed shown that the fluctuations in reactant and product translational and rotational motions of thermally activated reactions proceed on the fs scale [453]. A description of this work is clearly beyond the scope of this book.

10.4 Ultrafast processes in solid state materials

10.4.1 Excitation across the band gap

Femtosecond techniques made it possible to resolve fundamental interaction mechanisms and times in solids at room temperature. These processes are of tremendous importance. For instance, they determine physical limits for speed and miniaturization in semiconductor devices. Figure 10.7 illustrates essential processes in semiconductors, following optical excitation above the band gap.

10.4. ULTRAFAST PROCESSES IN SOLID STATE MATERIALS

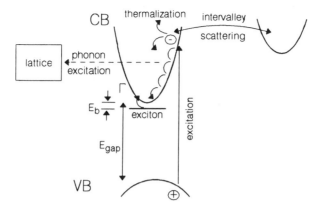

Figure 10.7: Simplified diagram of ultrafast processes occurring after above–band gap excitation in semiconductors.

An ultrashort light pulse of frequency ω_ℓ creates electron–hole pairs in states above the band gap. Their mean excess energy is $\Delta E = \hbar(\omega_\ell - \omega_{gap})$, and their initial energy distribution resembles the excitation spectrum. With large excited carrier densities, mainly carrier–carrier scattering leads to a thermalization within the Γ-valley without changing the mean carrier energy. This means that some carriers scatter out of their initial states, so that the distribution of occupied states becomes broader. Such processes are generally associated with momentum relaxation and are responsible for the dephasing of the polarization. Corresponding T_2 times can be measured by means of photon-echo experiments as described in the previous chapter. The temperature which can be attributed to the thermalized electronic system can exceed the lattice temperature by far. Depending on the band structure, and photon energy, inter-valley scattering can occur. Energy due to inelastic electron–phonon collisions is given to the lattice (heating), and the carriers relax into states at the bottom of the band. The Fermi distribution which is finally reached can be characterized by a temperature which is equivalent to the lattice temperature. If the excitation density is sufficiently high, a local change in the lattice temperature can readily be observed. Extremely high excitation can even result in melting. While the initial carrier scattering proceeds on a time scale of tens of fs or less, the intraband energy relaxation times can amount to a few ps.

10.4.2 Excitons

Another interesting feature of the excitation spectrum of solids is the exciton resonance. Excitons can be viewed as an electron–hole pair bound together through the Coulomb attraction, with properties similar to a hydrogen atom. Because of the positive Coulomb interaction, the corresponding energy levels are below the band gap (cf. Fig. 10.7). If the energy of the exciton is raised by an amount larger than the binding energy (E_b), the bound systems decays into a free electron and hole (exciton ionization). Such a process can be induced, for example, by longitudinal optical (LO) phonon scattering and typically proceeds on a time scale of about 100 fs in bulk materials at room temperature.

Owing to the strong excitonic oscillator strength and nonlinear susceptibilities, transient properties of excitons have attracted much attention over the last few years. In particular in multiple quantum well (MQW) structures, the exciton resonances can be clearly distinguished from the bulk absorption at room temperature. Figure 10.8 displays the absorption spectrum of a CdZnTe–ZnTe MQW and the results of a pump-probe experiment [454]. The pump spectrum was chosen to excite predominantly excitons. The differential transmission at the exciton resonance shows a fast increase and a partial recovery. Its dynamics can be explained by exciton excitation, exciton ionization due to LO-phonon scattering, and the presence of a coherent artifact. The increase of the transmission at photon energies which probe the occupation of states at the bottom of the bands ($\lambda = 610$ nm) is a direct indication of the exciton ionization into free carriers. The characteristic ionization time was determined to be about 110 fs [454].

10.4.3 Intraband relaxation

Intraband relaxation processes can conveniently be observed using pump-probe absorption techniques. A pump pulse of certain energy creates carriers at corresponding states above the band gap. Temporally delayed probe pulses of various frequencies test the occupation of states at different energies above the gap. The results of such an experiment for $Al_{0.2}Ga_{0.3}As$ [455] are shown in Fig. 10.9. A quantitative evaluation of the data is rather complicated, in view of the complexity of the processes involved in highly excited semiconductors. The interested reader is referred to the book by Haug and Koch [456]. Qualitatively, however, the time resolved transmission data follow a pattern consistent with the basic properties of the band model.

10.4. ULTRAFAST PROCESSES IN SOLID STATE MATERIALS

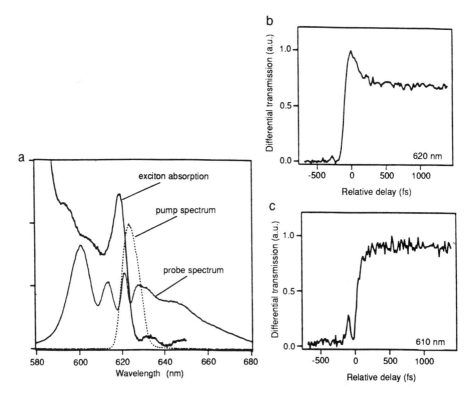

Figure 10.8: (a) Room temperature absorption spectrum of a CdZnTe–ZnTe MQW and the spectra of the 80 fs pump pulse and 14 fs probe pulse. The latter is a self-phase modulated and compressed part of the pump pulse. (b) Differential transmission at 620 nm and (c) 610 nm for a pump excitation level of 2×10^{11} carriers/cm^2. The wavelength filtering was done after the sample with a filter with a bandwidth of ≈ 8 nm. (From [454].)

A rapid transmission change occurs not only at the excitation energy, but over a broader spectral range, indicating a thermalization within a time range significantly shorter than 100 fs. At 1.88 eV and 1.94 eV, a reduced change in transmission can be attributed to the cooling of the electronic system through energy transfer to the phonon system (lattice). This cooling results in a relaxation of carriers towards the bottom of the band, thus emptying higher energy states. The increase of transmission at 1.78 eV accounts for the increase of occupied states at the band edge, with a characteristic time constant of 1 to 2 ps. The transmission features observed at 2.07 eV are

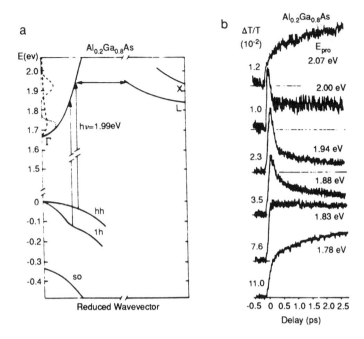

Figure 10.9: (a) Band structure for $Al_{0.2}Ga_{0.8}As$. The excitation photon energy being 1.99 eV, absorption from the light hole (ℓh) and heavy hole (hh) subbands into the Γ band are allowed. The carrier distribution is sketched for different instants (upper left corner). (b) Relative transmission change as seen by the test pulse for different photon energies and as function of the delay with respect to the pump pulse (adapted from [455]).

explained by intervalley scattering ($\Gamma \leftrightarrow L$) and confirmed by additional probing of the split-off transition [455].

10.4.4 Phonon dynamics

Phonons represent lattice vibrations. Just as in the case of molecular vibrations, they can be probed either by Raman techniques (frequency domain spectroscopy) or by ultrafast probing (time domain spectroscopy). The latter has the additional advantage of being able to retrieve not only the amplitude, but also the phase of the vibration. The theoretical basis for coherent excitation of phonons was established in Section 9.10. The phonon vibration of frequency ω_{phonon} is excited by pairs of spectral components of the pulse spectrum ω_1 and ω_2 such that $\omega_{phonon} = \omega_2 - \omega_1$. The measurement of such

10.4. ULTRAFAST PROCESSES IN SOLID STATE MATERIALS 459

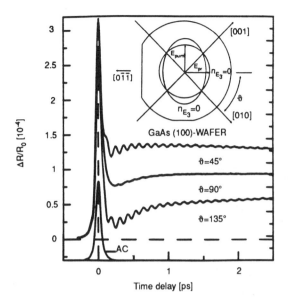

Figure 10.10: Relative reflection change of a [100] GaAs crystal excited by a 50 fs pulse at 2 eV. The generated carrier density is 10^{18} cm^{-3}. ϑ is the angle between the probe polarization and the [010] crystal axis. AC denotes the autocorrelation of the pump (probe) pulse (from [457]).

collective atomic motion in crystals can be performed in reflection as well as in transmission (see, for example, Kütt et al. [457] and references therein). The vibrations are observable through optical probing because the atomic displacements directly affect the band structure, and consequently the dielectric function through the deformation potential and electro-optic coupling. In addition, in polar crystals, direct excitation of phonons is possible by an electric field containing suitable frequency components.

Transient reflectivity measurements performed on GaAs are presented in Fig. 10.10. A 50 fs pump pulse at 2 eV [457] is followed by an orthogonally polarized probe. The [010] crystal axis is oriented at an angle $\vartheta = 45°$ and $135°$ with respect to the probe polarization. After an initial peak, the reflectivity versus delay shows an oscillatory behavior, with a characteristic frequency of 8.8 THz that matches the frequency of the longitudinal (LO) phonons in GaAs. The dependence of the modulation amplitude on ϑ results from the electrooptic effect, which is here responsible for the phonon-induced reflectivity change.

10.4.5 Laser-induced surface disordering

In a strongly absorbing material, the energy deposited in a small surface layer by a short light pulse can locally raise the temperature beyond the melting point. What is the response of matter to a δ-function impulse of energy, sufficient to cause melting? "How fast does melting occur?" is a question of fundamental interest, which involves changes in order and structure. In addition to the possibility of observing melting through "femtosecond photography" [458], nonlinear techniques sensitive to the material symmetry can be applied [459]. Results of an experiment to monitor changes in symmetry in GaAs during melting are shown in Fig. 10.11. In GaAs, melting can be considered as a transition from a non-centrosymmetric material to an isotropic liquid. The second-order nonlinear susceptibility $\chi^{(2)}$ is therefore expected to change from a relatively large value (for the crystal) to (almost) zero (for the liquid) during the phase transition. This second-order susceptibility can be monitored by measuring the second harmonic in reflection generated by a delayed probe [461, 462, 463]. The reflectivity for the fundamental increases from the solid reflectance value to that of liquid GaAs with a characteristic time of about 1 ps. On the other hand, the second harmonic signal drops substantially on a time scale of about 100 fs. These data suggest an intermediate state between the non-centrosymmetric crystal structure and the molten material. Note that a transition to a centrosymmetric crystal would only require a small displacement of the atoms and

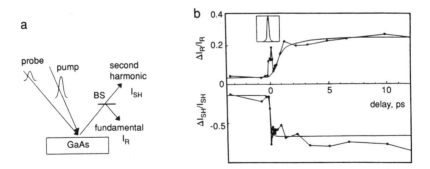

Figure 10.11: (a) Experimental setup to monitor ultrafast phase changes on a GaAs surface. A strong pump pulse induces the melting. (b) The upper curve shows the reflectivity of a delayed probe. The lower curve is a plot of the second harmonic (in reflection) of the probe signal, which is a measure of the change in symmetry associated with the phase transition (adapted from [460]).

substantially less energy than required for the actual bond breaking that occurs with melting. The availability of fs x-ray pulses will make it possible to explore ultrafast symmetry changes directly via x-ray diffraction.

10.5 Primary steps in photo-biological reactions

In the progression of increasingly complex systems, we have come to the role of fs tools in analyzing the most complex biological systems. The two most important biological problems presently connected to fs spectroscopy are photosynthesis and vision. In both cases, light energy is converted to biochemical energy, either for the purpose of energy storage/transfer, or for the purpose of detection. The primary processes in the complex chain of reactions following light absorption, in vision or photosynthesis, takes place on a fs time scale. The quantum yield of these ultrafast transformations is remarkably high — typically between 50 and 100%.

10.5.1 Femtosecond isomerization of rhodopsin

The primary process of vision takes place in rhodopsin, a pigment embedded in the membranes of specialized photoreceptor cells, the rod and cone cells of the retina. The role of the pigment is light absorption followed by a molecular conformational change, which leads eventually to a change in membrane potential. This change in electrical potential across the photoreceptor cells is eventually transmitted to the nervous system [464]. We are interested here in the primary process of vision, which is the isomerization of the pigment following absorption of a photon.

The pigment is a complex molecule called rhodopsin consisting of an "opsin" protein bound to the 11-cis form of retinal chromophore. The absorption band of rhodopsin peaks at 500 nm which corresponds to the peak sensitivity of vision. This main absorption corresponds to a transition from the S_0 ground state to a S_1 excited state in the potential energy surface representation of Fig. 10.12(a). The potential energy is plotted as a function of an angular torsional coordinate of the molecule. Absorption of a photon at 500 nm is followed by isomerization to the red-absorbing trans-isomer bathorhodopsin. The classical representation of the transformation is a "twist" of the chain [Fig. 10.12(b)]. The potential surfaces as a function of the corresponding coordinate angle have a minimum corresponding to the cis- and trans- configurations.

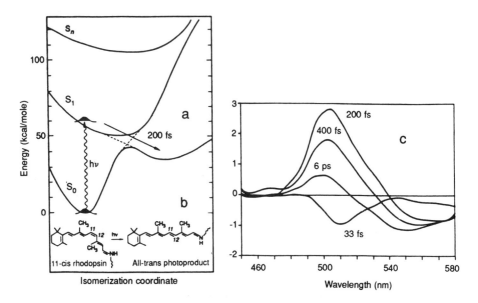

Figure 10.12: Schematic representation of the photo-isomerization reaction of rhodopsin. (a) Ground and excited state potential energy curves as a function of the torsional coordinate. The spectrum is red-shifted after absorption of a photon at 500 nm. The classical sketch of the cis-trans transformation is shown in (b). (c) Difference spectra measurements of 11–cis-rhodopsin at various delays following a 35 fs pump pulse at 500 nm ($\approx$10 fs probe). (From [465].)

The quantum efficiency of this reaction is exceptionally high (0.67). The radiation lifetime of the excited state of rhodopsin is 10^{-8} s. The extinction coefficient is $6.4\ 10^4\ \mathcal{M}^{-1}\mathrm{cm}^{-1}$, a typical value for a strongly absorbing dye.

The reaction of photo-isomerization was studied through transient transmission spectroscopy through a jet of rhodopsin [466]. Adequate spectral selectivity was achieved with a pump pulse of 35 fs at 500 nm. A 10 fs probe pulse in the range of 450 nm to 570 nm was used. The differential spectrum versus delay shown in Fig. 10.12 indicate disappearance of the 500 nm peak, and increased absorption at 530 nm, in the first 150 fs following excitation. The speed of that isomerization calls for a better classical representation of the cis versus trans configuration than Fig. 10.12(b). It is doubtful that the large motion of nuclei implied by the sketch could take place in a time as short as 100 fs.

10.5.2 Photosynthesis

Photosynthesis is the process by which plants convert solar energy into chemical energy. Its importance is obvious, since it is at the origin of life on our planet. This topic is too vast to be adequately covered in a section of this book. A general overview of the topic can be found in a review article by Fleming and Grondelle [467], and in topical books [464, 5].

There are pigment–protein complexes called reaction centers, where a directional electron transfer takes place across a biological membrane. Light harvesting molecules ("antenna" chlorophylls) transfer electronic excitation energy to a special pair (P in the sketch of Fig. 10.13) of chlorophyll molecules, which acts as the primary electron donor. The latter transfers an electron to a pheophytin (H_A) within 3 ps, and from it to a quinone (Q_A) in 200 ps, thence to the other quinone Q_B, hence establishing a potential difference across a biological membrane. Biochemical reactions that store the energy subsequently occur with these separated charges.

The energy dissipation in the first processes should be small (about 0.25 eV) as compared to the excitation energy (1.38 eV), in order to minimize the waste of excitation energy. The electron transfer should be fast in order to compete with fluorescence and radiationless decay.

The complexity of the problem can be appreciated by looking at the representation of the molecular structure of a bacterium's photosynthetic reaction center, which was determined to atomic resolution by Deisenhover

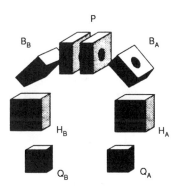

Figure 10.13: Sketch of the molecular arrangement of the four bacteriochlorophylls (P, B_A, B_B), the two bacteriopheophytins (H_A, H_B), and the two quinones (Q_A, Q_B) in reaction centers (from Ref. [468]).

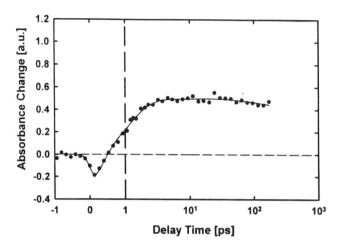

Figure 10.14: Transient absorption changes for native reaction centers at 1.02 μm. The large initial absorption increase has a time constant of 0.9 ps. It is related to the absorption change due to the formation of a bacteriochlorophyll anion (from Ref. [470]).

and Michel [469, 467]. A block diagram of the electron carrying pigments in the reaction center is shown in Fig. 10.13.

Recent transient absorption experiments [470] have concentrated on the fast initial electron transfers. In a model proposed by Zinth et al. [468, 470], the bacteriochlorophyll anion B_A^- is created in the first 3 ps reaction. The subsequent electron transfer to the bacteriopheophytin H_A is faster, taking only 0.9 ps.

After the main absorption band of the pigment is pumped, a probe is sent in the near IR, where the bacteriochlorophyll anions ($P^+B_A^-$) have a strong absorption. The transient absorption change at 1.02 μm is shown in Fig. 10.14. The 0.9 ps time constant would correspond to the electron transfer from the bacteriochlorophyll to the bacteriopheophytin ($P^+H_A^-$).

Chapter 11

Generation of Extreme Wavelengths

Presently, the generation of fs light pulses in lasers covers a spectral range from the UV to the NIR. Figure 11.1 gives an overview on laser materials matched to specific wavelength ranges. Nonlinear optical processes such as sum and difference frequency generation and parametric generation are typically used to extend the spectral range covered by laser materials.

Beyond the traditional nonlinear optics, new techniques have emerged leading to electromagnetic pulses of extremely short (x-ray) and extremely long wavelengths (far infrared or FIR). These novel sources in turn opened completely new application fields, for example, time resolved x-ray spectroscopy, short pulse (single cycle) radar, and FIR coherent spectroscopy. Femtosecond light pulses can generate the shortest electrical pulses, which in turn serves to characterize the fastest electronic components where purely electronic means (must) fail. Moreover, with fs optical pulses very short acoustic pulses can be produced and launched into materials for diagnostic purposes. The geometrical length of these pulses is only a few nm. In this chapter we shall discuss the basic principles leading to the formation of pulses at these extreme carrier frequencies. The relevant processes are a fascinating example of the complexity of the interaction of ultrashort light pulses with matter. From a naive view point all four types of pulses arise simply when a fs pulse is focused onto a solid surface. Another area where fs light pulses help to push the limits of our present knowledge is the physics of extremely intense electromagnetic fields (larger than the atomic field strengths) which can be reached at the focus of high-power pulses.

466 CHAPTER 11. EXTREME WAVELENGTHS

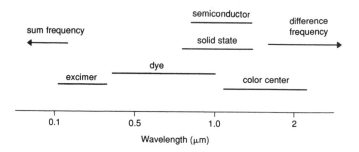

Figure 11.1: Femtosecond pulse generation in different wavelength regions.

11.1 Generation of terahertz radiation

Two mechanisms can lead to terahertz radiation through focalization of an ultrashort light pulse onto a sample: (i) optical rectification and (ii) a radiative current transient. In the original experiment on optical rectification using femtosecond light pulses Auston et al. [471] demonstrated the generation of THz waves through Cherenkov radiation. A fs pulse with an energy of about 100 pJ was focused into a LiTaO$_3$ crystal (Fig. 11.2). The optical pulse propagating through the crystal produces a polarization pulse via optical rectification. The latter is a second-order nonlinear optical process which occurs simultaneously with second harmonic generation. Indeed, an instantaneous polarization quadratic in the electric field is:

$$P_{2\omega_\ell} = \epsilon_0 \chi_{(2)} \mathcal{E}^2(t) \cos^2[\omega_\ell t + \varphi(t)]$$
$$= \frac{1}{2} \epsilon_0 \chi_{(2)} \mathcal{E}^2(t) \{\cos 2[\omega_\ell t + \varphi(t)] + 1\} \quad (11.1)$$

which includes a second harmonic term centered at $2\omega_\ell$ and a dc field of the same amplitude centered at zero frequency. Optical rectification can also be understood as difference frequency generation. Provided that the response is instantaneous, the frequency domain form of Eq. (11.1) is

$$P(\omega_d = \omega_1 - \omega_2) = \epsilon_0 \chi^{(2)} E(\omega_1) E(\omega_2) \quad (11.2)$$

where ω_1 and ω_2 can be any two frequencies from the pulse spectrum. Thus, the difference frequency ω_d can cover a spectral range from zero to several THz, which corresponds to far-infrared (FIR) radiation. For a Gaussian pulse with a temporal intensity profile $I(t) \propto \exp\left[-4\ln 2(t/\tau_p)^2\right]$, the THz radiation can be described as a single cycle infrared radiation of frequency $\sqrt{\ln 2}(2\tau_p)^{-1}$ [472].

11.1. GENERATION OF TERAHERTZ RADIATION

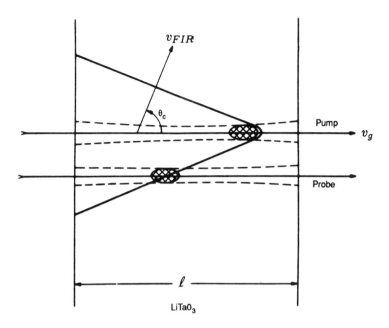

Figure 11.2: Schematic diagram for the generation of terahertz radiation through optical rectification (adapted from [471]). A time delayed optical pulse sampled the FIR field by probing the induced birefringence.

The optical rectification pulse is basically an electric dipole field that moves with the group velocity of the optical pulse. At a given position along its path in the crystal, this dipole generates a field that travels with the group velocity v_{FIR} associated with its low carrier frequency (of the order of the inverse pulse duration τ_p^{-1}). For LiTaO$_3$, $v_{FIR} \approx 0.153c$, which is rather low because (quasi-resonant) lattice vibrations contribute substantially to the dispersion behavior in the FIR spectral range. Since the group velocity of the optical pulse is $v_g \approx 0.433c$, we have an interesting situation where the source velocity is larger than the velocity of the emitted wave. This condition leads to the formation of an electromagnetic shock wave (Cherenkov radiation) propagating on a conical surface in the crystal. The characteristic angle between the surface normal of the cone and its symmetry axis (propagation direction of the optical pulse) is

$$\cos\theta_c = \frac{v_{FIR}}{v_g} \qquad (11.3)$$

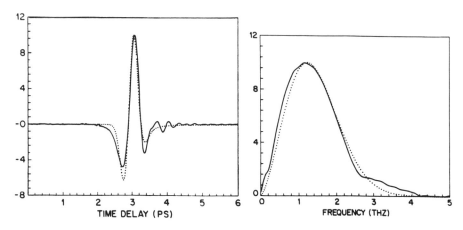

Figure 11.3: Temporal behavior of the THz field and the corresponding Fourier spectrum obtained from optical rectification (from [471]).

which is 69° for LiTaO$_3$ [471]. The electric field of the far-infrared pulse transient was measured using a second fs optical pulse which probed the FIR-induced birefringence. By varying the time delay between pump and probe pulse the FIR field can be sampled. This detection gives a complete recording of the FIR waveform — hence, the FIR radiation is completely determined in amplitude and phase. Figure 11.3 displays the temporal behavior of the FIR field as well as its Fourier spectrum. If the (external) angle of incidence of the optical pulse is chosen to be $\alpha = 51°$, a portion of the Cherenkov cone propagates normally to the crystal surface [473] and thus can propagate into free space (air). In this manner the crystal acts like an emitter for THz radiation.

Instead of a moving dipole in a dielectric medium, an ultrashort electrical pulse on a coplanar transmission line can also serve as source for THz radiation, as reported by Fattinger and Grischkowsky [474]. The physical situation is sketched in Fig. 11.4. The transmission line consisted of two 5 μm wide, 5 μm thick aluminum lines separated by 10 μm. The substrate was heavily implanted silicon on sapphire to ensure a short carrier life time of about 600 fs [475]. The latter is crucial for generating the short electrical transients and for a short response time of the detection switch.

More recently even simpler techniques turned out to be effective means to generate THz radiation. These include biased metal–semiconductor interfaces [475] and semiconductor surfaces between biased metal electrodes [476]

11.1. GENERATION OF TERAHERTZ RADIATION

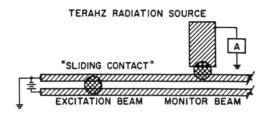

Figure 11.4: THz radiation through ultrafast switching of a charged coplanar transmission line. A fs pulse shortens the transmission line creating a subps electrical pulse. Its transients after propagation over a certain length L can be measured by a time delayed second pulse that excites a fast photoconductive switch (from [474]).

which is similar to a technique already used by Mourou et al. [477] to produce ps microwave pulses. Their common operational principle rests on the production of a fast carrier transient in an external bias field, following the optical excitation. The rise time is roughly given by the carrier generation process, i.e., by the optical pulse duration. The fall time is limited by the finite transit time of the carriers across the region of the electric field and by the carrier lifetime, whichever is shorter.

Surprisingly, semiconductor surfaces excited above the band gap and without external bias voltage were found to be emitters of THz radiation, too [478]. A static build-in field normal to the semiconductor air surface acts as bias and may drive the carrier current. The origin of this field is a downward bending of the conduction and valence band creating a charge depletion region near the semiconductor surface. The generation of photo-carriers in this layer then initiates an electron and hole current in opposite directions.

A coherent contribution to this FIR generation process on a fs time scale was recently suggested and analyzed [479]. The effect relies on the formation of an instantaneous nonlinear polarization due to excited electron–hole pairs and the coherent evolution of this system. These coherent effects — as is the case for any type of coherent light-matter interactions, cf. Chapter 4 — play an important role if the Rabi frequency becomes comparable or larger than the dephasing rate T_2^{-1}. Therefore, the coherent processes are expected to produce a significant contribution to the THz emission process when very short and/or powerful optical pulses are used [479]. The FIR radiation originating from optical rectification, as mentioned above, is another example of a coherent generation process.

The sources of FIR radiation described so far are mainly point sources. An extended source (area of linear dimensions that are large compared with the THz wavelength) has been demonstrated by sending fs pulses of very large energy onto a large area biased semiconductor [480]. Such an extended source has interesting far- and near-field properties. The radiation is coherently emitted from an area of cm dimensions and has diffraction properties similar to a Gaussian beam of cm beam waist at a wavelength of approximately 0.3 mm (Raleigh range of the order of one meter). The inconvenience of this source is the large optical power (distributed over the large area of the emitter) required. However, because of the large excitation energy and area, the emitted FIR power is considerably larger than with the point source devices [479], [480]. The THz pulses can therefore be detected directly with bolometric structures. Pyroelectric detectors can even be used for nJ and larger energies [479]. The bandwidth limitation associated with the photoconducting receiver structures is eliminated by recording THz pulse interferograms directly. To this aim, the FIR radiation is sent into a Michelson interferometer and the output signal is measured as a function of length detuning. A corresponding interferogram of a THz pulse with a field FWHM as short as 85 fs is shown in Fig. 11.5.

The large majority of the THz emitters can be considered as point sources (dimensions $\ll$ wavelength). For most applications, it is desirable to collimate the emitted waves with a system of lenses or mirrors, propagate the beam over long distances, and finally detect it. A simple means of collimating the beam is to attach a lens, for example a silicon lens, directly to the radiation source [481, 482, 473]. Collimation of the point source radiation was achieved by a combination of silicon lenses and parabolic mirrors [481], [483] as sketched in Fig. 11.6. The transmitted radiation is focused by a combination of parabolic mirror and lens onto a photoconducting antenna which produces a transient bias voltage across the 5 μm gap [Fig. 11.6(b)]. The transient voltage is gated by a photoconductive switch driven by a time delayed optical pulse. The recorded average current versus delay is a convolution among the FIR field, the optical pulse, and the response of the receiver. It is the last that limits the temporal resolution of the detection system.

An interesting application of the THz source is coherent rotational spectroscopy [484]. The THz radiation was propagated through an 88 cm long sample cell. The time domain spectrometer consists of a transmitter and a receiver part which are replicas of each other. The rotational spectrum of N_2O between 0 and 1 THz consists essentially of regularly spaced narrow

11.1. GENERATION OF TERAHERTZ RADIATION

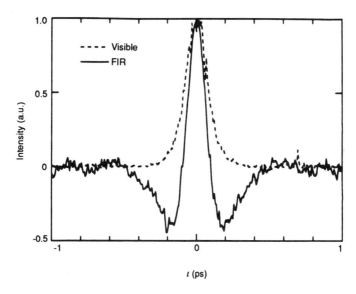

Figure 11.5: Second-order autocorrelation of the visible excitation pulse (broken line) and interferogram of the FIR pulse (solid line). The THz pulse was produced from an unbiased, n-type InP (111) wafer doped at 10^{17} cm^{-3} (from [479]).

lines [484]. The rotational frequencies are given by:

$$\nu_R = 2B(J+1) - 4D(J+1)^2 \tag{11.4}$$

where J is the rotational quantum number, B the rotational constant and D the centrifugal stretching constant. As many as 70 transitions can be excited simultaneously by the THz pulse. Bloch's equations (cf. Chapter 4) apply to this system. They can be simplified assuming small pulse area and long T_2 time. Indeed, at a pressure of 600 torr (which corresponds to an optical thickness of the sample $\alpha d = 1.2$), the dephasing time due to collisions is 65 ps, thus much longer than the exciting pulse. Each of the excited molecules, for a particular J value, acts as a vibrating dipole. These dipoles — nearly equally spaced in frequency — radiate in phase following the pulsed excitation (free induction decay), resulting in a train of ultrashort pulses (Fig. 11.7). The directly transmitted pulse is followed by a series of THz pulses at a repetition rate of 25.1 GHz, equal to the frequency separation between adjacent lines [$2B$ in Eq. (11.4) if the anharmonicity is negligibly small], as shown in Fig. 11.7(a). Since the incident and transmitted fields are measured with high precision (the signal-to-noise ratio in this experiment

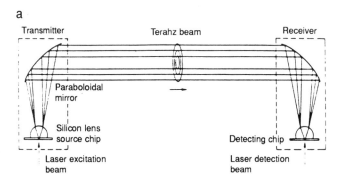

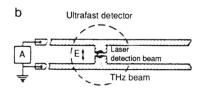

Figure 11.6: (a) Setup for generating and transmitting ultrashort THz radiation. (b) Detector for the time resolved measurement of the THz field. (From [481].)

is better than 20,000), in amplitude and phase (the detection measures the field magnitude rather than the radiation intensity), accurate fit with the theory is possible [Fig. 11.7(b) as compared to Fig. 11.7(a)]. In addition to a verification of the rotational constant B, the data lead to an improved determination of the centrifugal stretching constant $D = 5.28$ kHz, and the exact interatomic distances in the molecule (N $\leftrightarrow$ N $= 1.125$ Å; and N $\leftrightarrow$ O $= 1.191$ Å) [484]. An example of measurement and fitting of successive pulses in the train is shown in Fig. 11.7(c) and (d).

11.2 Generation of ultrafast x-ray pulses

When a powerful light pulse is focused onto a solid, a high-temperature plasma is produced, which can emit x-rays. Stimulated by laser fusion research, the physics of laser induced x-ray generation has been extensively investigated. The availability of high-energy (millijoule) tabletop lasers has made laser produced plasmas a unique and practical source of ultrashort

11.2. GENERATION OF ULTRAFAST X-RAY PULSES

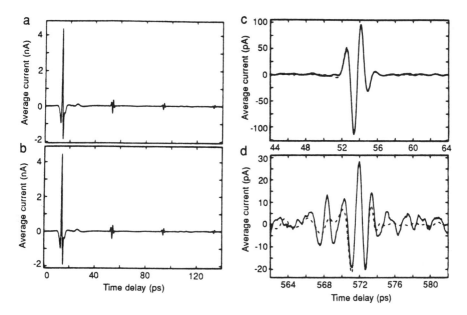

Figure 11.7: Coherent THz rotational spectroscopy of N_2O. THz radiation transmitted (a) and calculated (b) through N_2O. Measurement (solid line) and calculation (dashed lines) of (c) the first, and (d) the 14th radiated coherent pulse (from [484]).

x-ray pulses [485, 486, 487, 488, 489, 490, 491]. A diagram of optical pulse induced x-ray production is sketched in Fig. 11.8(a). In the absorption process, which takes place in a thin surface layer, the optical energy is initially transferred to the electronic system. Depending on the absorption characteristics of the target material and the incident laser intensity, one- and multi-photon processes excite electrons and heat the electronic system rapidly to high temperatures. These electrons are able to ionize the atoms creating a hot plasma. The cooling of this laser produced plasma is accompanied by the emission of a burst of x-rays. Figure 11.8(b) shows the time integrated spectral distribution from a Si target as was obtained by Murnane et al. [492].

Several processes are believed to contribute to the x-ray emission spectrum:

- the emission of distinct lines resulting from transitions between inner subshells of the ions

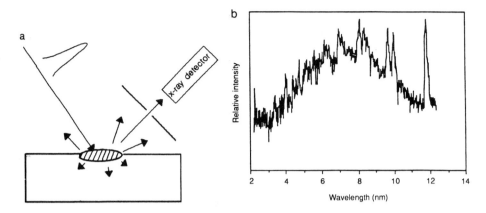

Figure 11.8: (a) Sketch of laser generated x-ray plasmas. (b) Time integrated soft x-ray emission. Pulses of 160 fs duration and 5 mJ of energy were focused onto a Si target at an intensity of 10^{16} W/cm^2 (from [492]).

- broadband emission due to recombination of unbound (free) electrons and ions
- broadband emission consisting of bremsstrahlung radiation.

There are substantial differences between the plasma generation with pulses on the order of ps or longer and fs pulses. The physical difference is related to the time scale of the plasma expansion for a distance equal to the absorption length, into the surrounding vacuum. Expansion velocities are comparable with the sound velocity, typically about 0.1 nm/fs, while the absorption length of metals is of the order of 10 nm. The energy will thus be deposited in a time shorter than the expansion time with a 100 fs pulse. A fs pulse thus interacts with a solid or near-solid density plasma. With ps and longer pulses, the leading part generates an expanding plasma that creates a density gradient above the target surface. The later parts of the pulse are absorbed in this low density plasma up to a density at which the electron plasma frequency equals the laser frequency.

Roughly, the duration of the emitted x-ray burst follows the duration of the optical excitation pulse if it is of ps or longer duration. To obtain sub-ps x-ray pulses, fs excitation pulses are needed to produce a fast enough rise of the electron temperature, and the plasma must cool rapidly to terminate the emission. While the former is responsible for the rise time of the x-ray pulse, it is the latter that will determine its decay characteristics. There are a number of processes which allow a rapid cooling of the plasma if excited to high densities [492]. Among them are high thermal and pressure gradients which drive the hot electrons into the bulk and expand the plasma into the vacuum. Moreover, since the electron temperature exceeds the temperature of the atoms and ions by far, inelastic collisions between them and the electrons are an efficient cooling mechanism. Using 160 fs excitation pulses, x-ray pulses as short as ~ 1.1 ps could be measured [492], where this value represents the detector limit of the x-ray streak camera. While the corresponding spectrum was in the soft x-ray range, a two-order of magnitude larger excitation density ($\sim 10^{18}$ W/cm^2) and a heavy metal target (Ta) yielded hard x-rays in a range between 20 keV and 2 MeV [488]. Conversion efficiencies of about 0.3% were observed.

11.3 Generation of ultrashort acoustic pulses

An acoustic wave is a strain or shear wave which can be produced by piezoelectric transducers in mechanical contact with the material. The transducers are usually driven by rf voltages. Such techniques have gained importance for the design of acousto-optical modulators for actively mode-locked lasers, for example, cf. Chapter 5. Moreover, acoustic pulse propagation, scattering, and reflection can conveniently be used for material characterization.

The generation of acoustic waves with (fs) optical pulses is another example that illustrates the complexity of processes associated with the interaction of light pulses with matter. Let us assume that a fs pulse is incident on a solid surface of a highly absorbing material and that its energy density is below the threshold for plasma generation and other irreversible processes. Two effects are responsible for launching an ultrasonic wave (pulse) into the material. Their relative importance depends on the material properties and the parameters of the (optical) excitation pulse (see, for example, Grahn et al. [493] and references therein).

(i) If the pulse is absorbed in a semiconductor, a high-density distribution of electron–hole pairs is created in a thin layer at the material surface. This electron–hole plasma changes locally the effective potential which determines the arrangement of the atoms in the lattice. The resulting stress is given by $\sigma = n(dV_g/d\eta)$ where n is the excited carrier density and $dV_g/d\eta$ is the deformation potential.

(ii) The electrons excited above the band gap relax mainly because of interaction with the lattice through electron–phonon collisions. Consequently, the temperature of the excited volume rises, producing an elastic stress. The stress amplitude is directly related to the thermal expansion coefficient and the absorbed energy density. For a more detailed analysis one has to account for a possible diffusion of the excited carriers during the electron–phonon interaction. This may increase the effective excited region.

Figure 11.9 illustrates a simple model that explains the acoustic pulse formation. At $t = 0$ elastic stress is generated with a depth profile that follows approximately the absorbed energy density. This stress distribution acts as source for a strain wave (pulse) propagating into the material and towards the interface. Upon reflection at the interface, the latter experiences a phase change of π. This gives rise to the final shape of the strain pulse as shown in Fig. 11.9 for $t = d/v_{ac}$. According to our model the width of the stress pulse is determined by the absorption length z_a for the optical pulse if the propagation length of the strain wave during the optical excitation is smaller than z_a. If v_{ac} is the sound velocity this condition can be expressed as

$$\tau_p < \frac{z_a}{v_{ac}}. \qquad (11.5)$$

For an order of magnitude estimate let us assume $z_a \approx 10^{-8}$ m and $v_{ac} \approx 10^4$ m/s, which yields $\tau_p < 1$ ps. For optical pulses shorter than 1 ps, the duration of the acoustic pulse depends mainly on material parameters rather than on the particular shape of the exciting pulse. In reality, of course, the

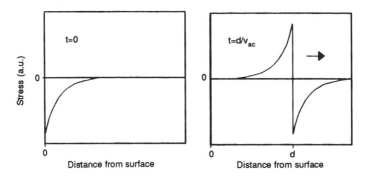

Figure 11.9: Propagation of a stress pulse generated at $t = 0$ through absorption of an optical pulse. (Adapted from [493].)

11.3. GENERATION OF ULTRASHORT ACOUSTIC PULSES

formation of the stress is not instantaneous. This concerns, for example, the finite thermalization time of excited carriers (∼0.1...2 ps) and the rise of the temperature, respectively. Such effects, in addition, limit the achievable duration of the acoustic pulse.

It is worth mentioning that the geometrical length of the acoustic pulse amounts only to few nm, i.e., a fraction of an optical wavelength, which enables a variety of attractive applications. One example is shown in Fig. 11.10. A pump pulse creates an acoustic pulse that subsequently bounces back and forth in a thin material layer while being damped. This could be monitored by testing the reflectivity of the surface with a time-delayed probe pulse, see Fig. 11.10(b). The temporal separation between the peaks in the reflectivity change corresponds to the transit time of the acoustic pulse in the film. From the decrease in the peak height, information on the damping of acoustic waves can be obtained.

Periodic structures [494] such as multiple quantum wells can be used to generate acoustic waves with some degree of spatial coherence. For instance, the energy can be deposited in an array of multiple quantum wells (MQW) spaced by the wavelength of a longitudinal optical (LO) phonon generated in the process of thermalization of the excited carriers. Such a structure is the acoustical analogue of the distributed feedback laser (cf. Chapter 5): the LO phonons add coherently along the normal to the MQW planes. The frequency of the phonon can be in the hundreds of GHz, with a corresponding wavelength of the order of 100 Å. The phonon generation can be understood

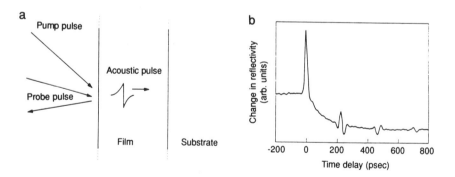

Figure 11.10: (a) Diagram for the generation and detection of acoustic pulses. (b) Time dependence of the reflectivity change of a 220 nm As_2Te_3 film as seen by the probe pulse. The "echoes" corresponding to subsequent reflections of the acoustic pulse at the film–air interface appear clearly as reflection changes (from [493]).

as a particular case of impulsive Raman scattering. As mentioned in Chapter 9, coherent addition of the phonons at the frequency ν_{ph} will be achieved if a train of pulses separated by the period $1/\nu_{ph}$ is used instead of a single pulse. The pulse shaping techniques mentioned in Chapter 7 can generate such pulse trains. Femtosecond technology applied to MQW structures can thus lead to the generation of temporally and spatially coherent acoustic waves of very short ($\approx 100 \text{Å}$) wavelength.

11.4 Generation of ultrafast electric pulses

The rapid progress in microelectronics not only makes circuits smaller and more powerful but also faster. The fastest all-electronically produced transients (≈ 1 ps) are still a few orders of magnitude slower than what can be obtained by all-optical techniques. Of course, optics and electronics cover different applications and compete directly only in a limited application field. In many cases a combination of optical and electronic means can be regarded as the optimum. In light of this, to generate ultrashort electrical transients, one can advantageously use fs light pulses to trigger photoconductive switches. Such switches made from semiconductors and driven by ps optical pulses were first demonstrated by Auston [495] and Lawton and Scavannec [496]. The development of fs pulse sources and progress in material fabrication have since made possible the production of sub-ps electrical pulses [497]. These pulses are employed to test ultrafast electronic circuits and components and measure their temporal response; see [498] for example.

The basic operational principle of a photoconductive switch can be explained by means of Fig. 11.11. A metallic microstrip line on top of a semiconductor is interrupted by a narrow gap. The implementation of the switch in a high-speed transmission line such as a strip line is necessary to propagate ultrafast electric pulses while limiting the broadening effect of dispersion. The photoconductor can also be a thin film deposited on an insulator. The bottom of this substrate is metal-coated and grounded. The strip line is connected to a bias voltage V_b. Because of the low (ideally zero) dark conductivity of the gap, the output voltage is zero. When a light pulse of suitable frequency is focused into the gap free carriers are generated which increase the conductivity. Subsequently, a current develops which can be measured as a certain output voltage V_{out}. In the first demonstration of a ps pulse triggered switch [495], the voltage was again set to zero by a second light pulse of certain delay but of longer wavelength (1.06 μm) which

11.4. GENERATION OF ULTRAFAST ELECTRIC PULSES

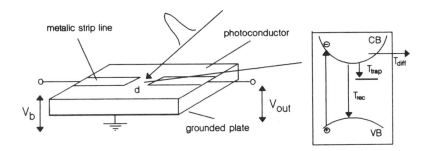

Figure 11.11: Sketch of a photoconductive switch. The inset shows the excitation and relaxation processes of the semiconductor.

produced a short circuit between the excited surface layer and the grounded plate. This was possible since the absorption length of the 0.53 μm excitation pulse in Si was much smaller than that of the "turn-off" pulse. Without the second pulse, the drop of the output voltage is given by the decrease of free carriers in the gap. This in turn is determined by local processes such as carrier recombination and carrier trapping, and by nonlocal processes such as carrier diffusion.

An exact analysis of the switching behavior in the fs regime is a rather difficult task. It requires not only the solution of Maxwell's equations with corresponding initial and boundary conditions, but also an accurate modeling of the matter response to the fs excitation pulse. In order to explain the basic operational principle, however, a strongly simplified approach is possible and provides satisfactory results. We will briefly explain this model for a photoconductive switch in a transmission line and follow the discussion of Auston [499, 500]. The main idea is to model the switch by an equivalent circuit consisting of a capacitor of capacitance C and a parallel (time varying) resistor of conductance $G(t)$; see Fig. 11.12. The conductance can be written as

$$G(t) = G_0 + g(t) \tag{11.6}$$

where G_0 is the dark conductivity and $g(t)$ is the pulse induced conductance. The latter can be expressed in terms of the free carrier concentration n and the electron (hole) mobility μ_n (μ_p), as

$$g(t) = \frac{1}{V_g^2} \int d^3x (ne\mu_n + ne\mu_p)|\tilde{E}|^2 \tag{11.7}$$

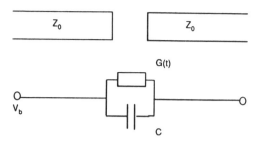

Figure 11.12: Equivalent circuit for a photoconductive switch in a transmission line. (Adapted from [499].)

where V_g is the voltage across the gap, E is the local field strength in the gap and d^3x denotes integration over the excited volume layer. The local field is a complicated function of the gap geometry and the carrier concentration and distribution. In order to get an explicit expression for the photoconductance let us assume that the perturbation by the pulse is small so that $g \ll G$ holds.[1] The electric field in the gap can then be approximated by $E = V_g/d$ where d is the gap length. Assuming homogeneous gap illumination and complete absorption in a thin surface layer, the total number of excited carriers is given by $(1 - r^2)\mathcal{W}/(\hbar\omega_\ell)$ where r^2 is the reflection coefficient and $\mathcal{W}/(\hbar\omega_\ell)$ is the number of incident photons. Inserting the values for the number of carriers and the electric field thus obtained in Eq. (11.7) yields for the photoconductance after pulse absorption at $t = 0$

$$g(t > 0) = \frac{1 - r^2}{d^2} e(\mu_n + \mu_p) \frac{\mathcal{W}}{\hbar\omega_\ell}. \qquad (11.8)$$

This expression shows the importance of high carrier mobilities for the switch sensitivity.

Let us next investigate the fundamental switching properties by idealizing the process of carrier generation and using the equivalent circuit model. For a step function conductance change

$$G(t) = \begin{cases} 0 & t < 0 \\ G_1 & t \leq 0 \end{cases} \qquad (11.9)$$

---

[1] This simplifying assumption is for the purpose of analytical evaluation only: It is neither desired for high switching efficiencies nor typical for ultrashort pulse excitation.

11.4. GENERATION OF ULTRAFAST ELECTRIC PULSES

and a dc bias voltage V_b, the transmitted voltage signal for a photoconductor in a transmission line of impedance Z_0 is [499]

$$V(t) = \frac{V_b}{2} \frac{2Z_0 G_1}{1 + 2Z_0 G_1} \left\{ 1 - \exp\left[-(1 + Z_0 G_1) \frac{t}{CZ_0} \right] \right\} \quad (11.10)$$

where Z_0 is the impedance of the strip line. Obviously the transmitted signal increases with G_1 and saturates at $V_b/2$ for $Z_0 G_1 \gg 1$. For small excitation, i.e., $Z_0 G_1 \ll 1$, the signal's rise time is limited by the capacitance and is expected to decrease with increasing G_1.

As mentioned previously, the drop in the transmitted signal can be controlled with such material parameters as carrier recombination, trapping, and diffusion. These parameters depend sensitively on the material and the fabrication process. Diffusion and recombination times typically range well above several 10 ps and thus would not lead to a sub-ps signal switch-off. Additional relaxation channels can be opened by the introduction of local defects which act as trapping centers. These defects are implanted by doping with impurities or through radiation damage. Effective carrier lifetimes as short as 600 fs were measured for radiation damaged silicon on sapphire [475]. Other examples are CdTe [501] and GaAs [502] where recombination centers were introduced during the material growth process. With the implementation of such techniques there is usually a trade-off to be made between the decrease of the free carrier lifetime and a decrease of the mobility. These materials, however, allowed the generation of sub-ps electrical pulses [497, 501]. An example is shown in Fig. 11.13.

An important question is how to transmit and measure sub-ps electrical transients with THz bandwidths. Strip lines like those depicted in Fig. 11.11 have too large a dispersion to be ideal transmitters for sub-ps electrical pulses. Mode dispersion is the most severe problem. For a single mode to exist, the distance between the metal strip and the grounded plate has to be very small (a few microns) which is difficult to achieve (mechanical tolerances). Electrical pulses of less than 10 ps double their length after propagation distances shorter than 1 mm [503].

Better results have been obtained with coplanar strip lines deposited on top of the photoconductor as depicted in Fig. 11.13. Their geometrical separation can easily be controlled during the fabrication process, which is essential for single mode propagation. Using "sliding contact" excitation the effective capacitance was found to be zero to first order [504] which is most desirable for short electric transients, cf. Eq. (11.10). But even with

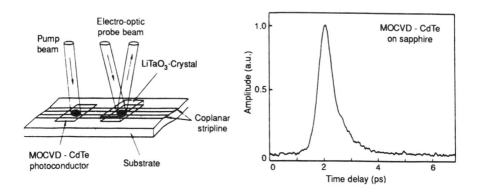

Figure 11.13: Sub-ps electrical pulse generated with a CdTe coplanar strip line and 100 fs excitation pulses. Left: experimental setup. Right: correlation signal obtained through electro-optic sampling. (From [501].)

coplanar strip lines, sub-ps electrical pulses (or slopes) cannot be propagated over more than several mm [505]. As mentioned at the beginning of this chapter, they can be sent through considerable distances in air or a dielectric bulk medium if suitable antenna–receiver structures are used. There are speculations about utilizing soliton mechanisms to transmit sub-ps electrical pulses over long distances in strip lines [506]. In analogy to solitons in optical fibers, this requires an interplay between nonlinear and linear processes.

Because of the dispersion problem, the measurement of sub-ps electrical pulses has to be performed close to the location of their generation. This can be done by implementing a second photoelectric switch in the transmission line. The voltage pulse generated in the first gap, propagates to the second gap which is illuminated by a time delayed second pulse [495]. The voltage signal at the output of such a device measured as a function of the delay time is thus an autocorrelation of the electrical pulse. More recently, electro-optic techniques have been applied to measure cross-correlations between the electrical and optical pulses. If the substrate has a large enough electro-optic coefficient, the electrical pulse induced birefringence can be probed by a time delayed optical pulse of suitable wavelength [404]. What is measured then is the polarization rotation experienced by a time delayed test pulse. Another approach is to bring an electro-optic material in the vicinity of the location to be probed [404]. The experiment depicted in

11.4. GENERATION OF ULTRAFAST ELECTRIC PULSES

Fig. 11.13 utilizes this method. For good spatial resolution, the probe can be a needle produced from a suitable material such as $LiTaO_3$. The advantage of this technique is that the behavior of the electrical pulse can be sampled along the transmission line and possibly in following electronic components.

Chapter 12

Miscellaneous Applications

In previous chapters we have seen the role of femtosecond pulses in basic research. Ultrashort pulses are not limited to esoteric research on ultrafast events. We want to emphasize here more down-to-the-earth applications, for which the femtosecond source has practical advantages.

The topics covered are short pulse imaging, fs laser gyro, and solitons. The short physical length of fs pulses makes them ideal candidates for three-dimensional imaging, either to acquire depth resolution through range gating, or to discriminate against scattering. Femtosecond ring lasers have proven to be unique laser gyros. This property extends to a new method of phase sensitive spectroscopy. Very high peak intensities can be reached at moderate pulse energies with ultrashort pulses. The "physics of high fields" has become experimentally accessible through intense fs pulses. We refer to [507] for a detailed review of this new area of physics. At more moderate pulse energies (in the mJ range), ultraviolet fs pulses appear to be ideal candidates for the triggering of spark gaps and lightning discharges. A review of this area is found in [508]. Finally, the ability to make "solitons" led to a new pulsed code communication system with optical fibers.

12.1 Imaging

12.1.1 Introduction

It does not come as a surprise that ultrashort pulses contribute to the most fundamental function of light: imaging. The intensity information of light is sufficient to record two-dimensional images. Additional information provided by the phase of the optical field makes it possible to record an image along all three space coordinates. This technique, combining phase and amplitude retrieval of the light scattered by objects, is called holography, and is the most accurate of all macroscopic imaging methods. It can easily measure deformations much smaller than one wavelength. The price to pay for the high accuracy of holography is that an excessive amount of data has to be recorded. Since all the three-dimensional information of the object has to be stored in a single recording, high optical energy densities are used, with the possibility of laser damage if the material absorbs light.

"Range gating" is another method to obtain depth information, by measuring the transit time of the radiation from the source to the object and thereafter to the detector. If — as is most often the case — the source and detector are co-located, the distance z from source to object is simply $c\times$ (round trip time)$/2$. This technique has been used since World War II for localizing and tracking moving objects. The resolution has shifted from meter (radar) to centimeter (lidar). Femtosecond pulses offer the possibility of a depth resolution of a few microns. Another function of ultrashort range gating is to discriminate against scattering, as will be shown later in this chapter.

12.1.2 Range gating with ultrashort pulses

A basic sketch of principle for range gating 3D images with fs pulses is shown in Fig. 12.1. The source fs beam is split into a reference and an object beam. The reference beam, after an appropriate delay line, triggers the optical gate at a time t_d. The light backscattered from various depths z of the object reaches the optical gate at time intervals spaced out by $2z/c$. A particular depth is selected by the opening instant t_d. The signal S received on the detector, as a function of delay τ, for a particular position (x, y) in the transverse plane of the beam, is the correlation of the gating function $g(t)$ and the intensity from the object $I_s(t)$:

$$S(\tau) = \int_{-\infty}^{\infty} I_s(t) g(t-\tau) dt. \tag{12.1}$$

If the gating function can be assimilated to a δ-function with respect to the variations of the signal $I_s(t)$, the measured transmitted signal is simply $S(\tau) \approx I_s(\tau) = I_s(z/c)$, where the depth of observation z is determined by the position of the reference mirror (delay). In general, the higher the order of the gating process is, the better is the depth resolution. For instance, if three-photon interaction ($\omega_d = 2\omega_r + \omega_s$, where ω_d, ω_r and ω_s are respectively

12.1. IMAGING

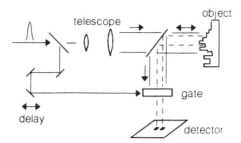

Figure 12.1: Recording of a three-dimensional object through range gating with ultrashort pulses. The gate is "opened" by a fs reference pulse derived from the illuminating source and appropriately delayed. An ultrafast gating function can be achieved, for example, with a Kerr gate and a nonlinear crystal for sum frequency generation. In the latter case, transverse resolution is generally obtained by scanning a narrow beam across the object.

the frequencies of the detected, reference, and signal photons) is used to gate the signal light, since $g(t) \propto I_r(t)^2$, the gating signal is approximately $\sqrt{2}\times$ shorter than the signal pulse. A compromise has to be reached: The higher the order of the gating process, the greater the required reference intensity. Given the power limitation of the source, the intensity requirement for the reference pulse limits the beam cross-section that can be utilized. Lateral scanning is therefore often used to obtain the transverse information on the object.

Bruckner [509] proposed using a fast Kerr shutter to gate the radiation reflected by index discontinuities in eyes. A Kerr shutter consists of a Kerr liquid between crossed polarizers. The shutter is "opened" by an intense ultrashort pulse inducing birefringence in the liquid. The gating time is either the pulse duration or the response time of the liquid, whichever is longer. Kerr gates have been applied to picosecond gating [510, 511]. Femtosecond temporal resolution can be achieved, for example, by gating through second harmonic or sum frequency generation [512, 393]. The technique consists of generating a second harmonic signal $I_2(t)$ proportional to the product of a reference pulse $I_r(t)$, derived directly from the source, and the signal $I_s(t)$. The second harmonic energy $S_{2\omega}(\tau)$ recorded as a function of reference delay τ is simply the correlation function defined in Chapter 8 (Section 8.1). This technique has been applied first in one dimension to fibers, to locate defects in fibers and connectors with a resolution of the order of a few microns [513]. The method has been extended to three dimensions (three-dimensional imaging of the eyes) by scanning the beam transversely [514]. The transverse resolution, limited by the size of the beam, could be improved by illuminating each point of the 3D object and using tomographic reconstruction algorithms [512].

Linear correlation techniques, such as heterodyning, can also be applied. Here the reference pulse is frequency-shifted. Gating is achieved by interfering reference and object pulse and detection at the heterodyne frequency. Because no nonlinear optical processes are involved, these linear techniques are very sensitive even at low illumination power. A particular example is Doppler interferometry. The backscattered signal is mixed with the reference signal, which is continuously scanned. Because of the scanning, the reference is Doppler-shifted, and the mixing produces a beat note observable only in the regions where the reference and signal are coherent with each other. In this particular application, either ultrashort pulses, or light with a short coherence length are used. The technique was first developed for fibers and integrated optics structures [515] and latter extended to eyes [516].

Speed is an essential element in 3D imaging of in vivo biological objects. There is a compromise between speed and sensitivity: Very sensitive detectors require a longer integration time. One possibility to reduce the time needed for data acquisition is to reduce the multidimensional scanning to only one dimension (the depth). It is possible in the case of nonlinear gating (second harmonic or parametric generation) to record a single-shot transverse picture for each depth increment. Direct recording of 2D images in "depth slices" has been demonstrated with high-contrast objects [393, 392]. The optical arrangement is sketched in Fig. 12.2. The laser beam is expanded to the size of the object after being split by a calcite prism (polarizing beam-splitter) between a reference and probing beam. The amount of beam splitting is controlled by a half-wave plate, in order to have the maximum probing intensity that the sample can accommodate. A quarter-wave plate in both the reference and object arms ensures that the returning beams are re-directed toward the detection. The backscattered signal beam and the orthogonally polarized retro-reflected reference are sent into a nonlinear crystal cut for type-II phase matched second harmonic generation. Assuming the reference has a uniform transverse intensity profile, the second harmonic contains the image information contained in the fundamental beam. The time of arrival of the reference ultrashort pulse determines the depth d at which a cross-section through the object is imaged into the CCD, as illustrated in Fig. 12.2.

12.1. IMAGING

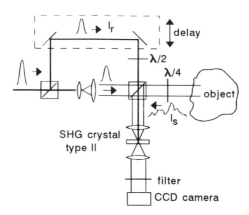

Figure 12.2: Typical setup for recording successive two-dimensional "slices" of a transparent 3D object. The backscattered radiation I_s from the object is stretched out in time, corresponding to the time of arrival from various depths. In the nonlinear crystal, the second harmonic, being proportional to the product of the reference I_r and signal I_s, selects a portion of the signal corresponding to a certain depth (set by the reference delay).

The reason for expanding the object beam *after* the polarizing beam splitter is to have uniform illumination of the object beam by the reference beam in the crystal. Indeed, the narrower reference beam has a long Rayleigh range ρ_0 and large waist w_0 as compared to the volume of focused object beams (from the various sections of the object). The crystal is assumed to be sufficiently long.

The ultrashort pulse source used for the preliminary tests was a fs dye laser operating at 620 nm [393]. Urea crystals were chosen for these tests as being the only phase-matchable type II crystals at that wavelength. Unfortunately, good-quality urea crystals are not readily available. Despite these limitations, a spatial transverse resolution of the order of 100 μm has been achieved in experiments with a configuration in which the reference and object beams have the same size [393, 392]. The titanium sapphire femtosecond lasers appear promising for this particular application, because of the better transmission of biological tissues in its wavelength range of 750 nm to 850 nm, and the possibility of using KDP crystals for the gating.

There is a subtle interplay among sensitivity, depth, and transverse resolution in the setup shown in Fig. 12.2. One cannot have the three parameters simultaneously optimized. For instance, an optimum conversion efficiency of one single second-harmonic photon for one signal photon can be achieved, provided the crystal length and the reference power density are sufficient. A simple estimate given as a problem at the end of this chapter illustrates this problem. A minimum crystal length is required to achieve single photon upconversion (i.e., one second harmonic photon for each signal photon). But to a minimum crystal length, there corresponds a phase matching bandwidth, hence a limitation to the temporal resolution of the up-conversion. In addition, the waist of the reference beam acts as a spatial filter, limiting the transverse resolution of the imaging system.

12.1.3 Imaging through scatterers

One of the main medical motivations for this type of research is early detection of breast cancer. The photon energy of visible and infrared light is too small for direct ionization of most tissues. Hence, an optical method seems to be an attractive and safe alternative to x-rays. Large differences in absorption have been reported between in vivo normal tissue and some types of tumor [517].

In medical and biological imaging, the biggest challenge is generally posed by scattering. To illustrate the effect of scattering on short light pulses, let us consider a femtosecond pulse being incident on a slab of thickness L made of isotropic scatterers as shown in Fig. 12.3. If we time resolve the transmission, we observe a peak on the leading edge of the transmitted pulse, which corresponds to the unscattered light ("ballistic" component) followed by a broad distribution of scattered light. Diffraction-limited resolution in an imaging process can only be obtained with the ballistic light. The latter can be separated from the diffuse light by appropriate time gating. The ballistic component of the transmitted light is attenuated exponentially:

$$I_{\text{ball}} = I_0 e^{-\mu_s L}, \tag{12.2}$$

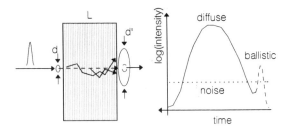

Figure 12.3: Sketch of pulse propagation through a scattering medium.

12.1. IMAGING

where $\mu_s = l_t^{-1}$ is the scattering coefficient, and l_t is the scattering mean free pathlength. There have been numerous attempts to visualize objects embedded in dense scatterers by range gating the backscattered or transmitted ballistic radiation [511, 518, 392]. Very high sensitivity is required in order to compensate for the large attenuation of the return signal. In the case of nonlinear detection, the second harmonic that is recorded is proportional to the product of a reference intensity by the weak backscattered radiation. Therefore, high peak powers (of the reference signal) is required to obtain a good conversion efficiency at the detection.

Problems arise if the scatterer is dense and I_{ball} approaches the noise level of the detection system. Here techniques are needed which provide not only the time gating but also an optical amplification. Nonlinear techniques, such as Raman amplification, and linear methods, such as heterodyning, have been applied successfully [519, 520, 521, 522] leading to micrometer resolution through dense scatterers. The ultimate limit to the resolution is the quantum noise (photon shot noise) which essentially implies that at least one photon should be detected per element or pixel of the image. In the case of biological and medical samples that can only withstand average powers of irradiation of a few mW, these noise considerations limit the applicability of diffraction limited imaging to scattering densities $\mu_s L \leq 35$.

The multiply scattered photon path can be described by a diffusion model [523]. For $L \gg l_t$, it can be shown that a collimated input beam of diameter $d < l_t$ at the sample input broadens to a diameter d', which is approximately given by [524]:

$$d' = 0.2L \qquad (12.3)$$

if we refer to the early-arriving scattered light that exceeds the detection noise and to illumination intensities below the critical values for biomedical samples. The quantity d' gives a reasonable measure of the resolution which can be achieved in imaging an object buried in a dense scatterer. With a sample thickness of several mm to several cm, the achievable resolution cannot be better than a few mm.

Several approaches are being attempted to utilize the large diffuse light component for imaging through very dense scatterers, such as several cm of tissue. One direction that promises to improve the resolution is to use the earlier portion of the scattered light, which may or may not follow a diffusion-like path [525, 520] for imaging.

An overview of current activities in the field of imaging with short light pulses can be found in Ref. [526].

12.1.4 Prospects for four-dimensional imaging

Ultrashort pulses can be used as a substitute for holographic techniques to record three-dimensional images, provided the object does not move on the time scale of the ultrashort pulses. Holography with ultrashort pulses should be used to record the temporal evolution of ultrafast 3D events. A method called "light-in-flight holography" (LIFH) has been proposed by N. Abramson [527, 528] to convert and store the rapid time information obtained with ultrashort illumination holography into space information. The basic principle is that the holographic fringes can only be recorded if the object and reference beams arrive simultaneously at the recording medium. An ultrashort reference beam sent at oblique incidence on the recording medium sets a time axis (Fig. 12.4). Point A is illuminated first and point B last after a time interval $\Delta t = D \sin(\theta/c)$. At any point on the line AB, a hologram of the object beam corresponding to a particular instant within the time interval Δt is recorded.

This technique has been used to detect the first arriving light through scattering media [529, 530] — the "ballistic" component cited earlier. The holographic data can be recorded on a CCD camera (provided the reference and object beam make a sufficiently small angle for the fringes to be resolved by the camera), and reconstructed through numerical fast Fourier transform [530, 531]. The advantage of the electronic recording and numerical processing is the possibility to integrate a large number of successive (reconstructed) images. The speckle pattern is averaged out if the time interval between exposures is greater than the correlation time of the speckle.

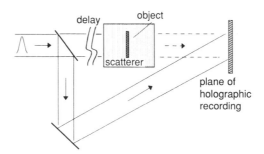

Figure 12.4: Light-in-flight holography illustrated for a simple one-dimensional transparent object (uniform in the transverse dimension of the object).

12.1. IMAGING

12.1.5 Microscopy

Laser-scanning microscopy (see, for example, [532]) is ideally suited to combine microscopic imaging with fs pulse illumination. There are several attractive application fields of femtosecond microscopy — (i) nonlinear microscopy, (ii) microscopy with simultaneous space and time resolution, and (iii) microscopy of structures immersed in a scattering environment; for a detailed review, see [522]. In nonlinear microscopy the image signal is generated by a nonlinear optical process, such as surface second-harmonic generation and two-photon excited fluorescence. An image is a map of the distribution of the corresponding nonlinear susceptibility. The two-photon fluorescence is particularly attractive for microscopy of biological cells [533] because of the depth selectivity of the excitation process. Femtosecond pulses are needed because of their great peak power at comparatively small pulse energy (i.e., small heat consumption in the specimen).

Simultaneous μm spatial and temporal resolution is of great desire for the inspection of ultrafast opto-electronic circuits. Another direction is to combine techniques of ultrafast spectroscopy with spatial resolution — to monitor, for example, diffusion and relaxation of excited carriers in semiconductors. In fluorescence microscopy additional information can be gained by measuring the lifetime of the fluorescence. Since this relaxation depends sensitively on the interaction of the fluorescing dye with the environment, the "lifetime" image can describe local field and ion concentrations in cells [534].

Another example where scanning can be ideally complemented by correlation is confocal imaging of objects buried under scattering layers. Confocal microscopy distinguishes itself by its depth selectivity [see Fig. 12.5].

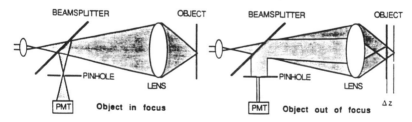

Figure 12.5: Schematic diagram of (reflection) confocal microscopy. Only light from layers that are in-focus can pass through the pinhole and reach the detector. The beam (or object) is scanned in transverse direction to obtain a two-dimensional image of the layer which is displayed on a computer.

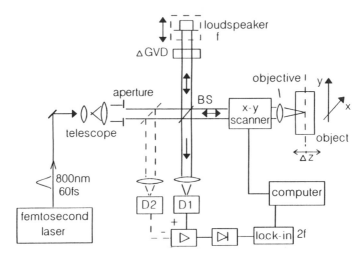

Figure 12.6: Schematic diagram of a correlation microscope based on heterodyne detection (from [536]).

Enhanced depth selectivity and optical amplification of the image signal are desirable for imaging through scattering layers which strongly attenuate the ballistic light. Both aspects can be addressed with scanning microscopy based on a sensitive correlation technique, such as heterodyning [535]. A realization of such a microscope is shown in Fig. 12.6. It consists of a Michelson interferometer which contains a scanning microscope in one arm and a piezo-electric transducer for Doppler-shifting the reference pulse in the other arm. The role of the pinhole is played by the coherent overlap of the plane reference wave and the image light. A maximum heterodyne signal is obtained if the wave front from the object is plane (parallel to the reference wave front), that is, if the object is in focus.

Assuming a layer of thickness L with scattering coefficient μ_s on top of an object with reflectivity R, the image signal of the correlation microscope can be written as:

$$S_d \propto \left(\mathcal{E}_{r0}\mathcal{E}_{s0}e^{-2\mu_s L}\right)\left(R \otimes \tilde{h}^2\right)\left[\frac{\langle \mathcal{E}(t)\mathcal{E}(t-\tau)\rangle}{\mathcal{E}_{r0}\mathcal{E}_{s0}}\right]. \quad (12.4)$$

$\mathcal{E}_{r0}$, $\mathcal{E}_{s0}$ are the amplitudes of the reference and the object pulse, respectively, $\tilde{h}$ is the amplitude point spread function (APF) of the objective, $\otimes$ describes

12.1. IMAGING

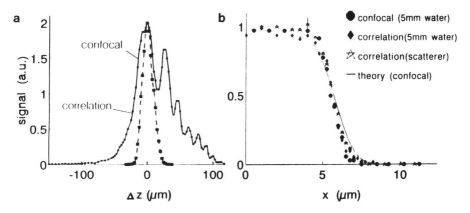

Figure 12.7: (a) Depth scan through a 10 mm glass layer that introduces spherical aberration and decreases the depth resolution of the confocal microscope (from [536]). (b) Scan over a straight edge. The scatter density was $2\mu_s L \approx 20$.

convolution and $\langle \rangle$ denotes correlation. As can be seen from the first term in Eq. (12.4), the attenuation of the ballistic light can be compensated by a large enough amplitude of the reference wave (optical amplification). The second term is the convolution of the object response with the APF. This term is essentially the square root of the response of a confocal microscope and describes the transverse and depth resolution. Additional depth selectivity and discrimination of scattered light from layers close to the object is possible due to the correlation (third) term in Eq.(12.4). It is nonzero only if the length mismatch of reference and image arm is smaller than the pulse duration (coherence length).

The depth resolution of a microscope is usually measured by scanning a reflecting object through the focus. The improved depth resolution of the correlation as compared to the confocal microscope is shown in Fig. 12.7(a). Figure 12.7(b) illustrates the transverse resolution as measured by scanning the beam focus over a straight edge buried under 5 mm of scattering material (96 nm latex spheres dissolved in water). If the system is able to detect ballistic light, there is no loss in resolution. Microscopic techniques through scatterers have great potentials for noninvasive imaging of biological and medical samples, see, for example, [537]. Figure 12.8 shows the images from a confocal and a correlation microscope of a cell layer of a leaf. The depth position of the layer was 80 μm from the lower epidermis.

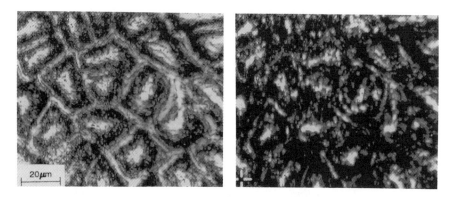

Figure 12.8: Correlation (left) and confocal (right) image of a cell layer 80 μm buried under the lower epidermis (from [538]).

12.2 Femtosecond laser gyroscopes

12.2.1 Introduction

The mechanical gyroscope is an instrument based on the conservation of angular momentum of a spinning wheel. The fixed orientation of the angular momentum provides information on the motion of a moving frame of reference. In a ring laser, the two counterrotating beams of the same frequency form a ring standing-wave pattern, fixed in an absolute frame of reference. A detector can sense the motion of a rotating laboratory frame through the motion of these interference fringes. Another point of view, from within the laboratory frame, is that the two counterrotating beams are resonating in a cavity that is lengthened in the sense of rotation, shortened in the other direction. Hence, the corresponding modes will be shifted in frequency, a shift that can be observed as a beat note between the two laser outputs.

Compact single-mode He–Ne ring lasers are currently used as navigation gyroscopes in commercial aircrafts. It may seem unlikely that fs lasers can play a role in the measurement of ultraslow motion, a field traditionally reserved for lasers with ultranarrow bandwidths. However, continuous-wave laser gyros are plagued by a phenomenon called lock-in: the response of continuous-wave laser gyros is zero for a range of small rotation rates. This dead band is due to the scattering of one circulating beam of the ring laser into the other direction. This weak coupling may "injection lock" the

12.2. FEMTOSECOND LASER GYROSCOPES

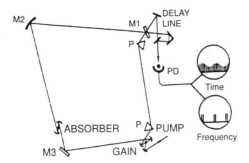

Figure 12.9: A mode-locked ring dye laser used as a laser gyro. The laser is mounted on a rotating table (not shown). The delay line on one of the output beams ensures that the two counterpropagating pulses overlap on the detector PD. The two insets illustrate the detection in the time domain (a train of pulses with a modulated envelope) and in the frequency domain (splitting of the cavity modes).

counterpropagating modes, i.e., force each of them to operate at the same frequency as radiation injected from the other mode.[1]

The "lock-in" problem can be avoided with ultrashort pulse lasers, where the two counterpropagating pulses meet in only two places. If the phase coupling at these meeting points can be avoided, the dead band can be eliminated. For instance, there is no coupling if the two pulses meet in vacuum. There is no phase coupling for a phase conjugated interaction through degenerate four-wave mixing, such as occurs in a saturable absorber dye jet [170]. A fs dye laser gyro is shown in Fig. 12.9 and will be explained in the next subsection in more detail.

The "gyro response" is merely the manifestation of a difference in frequency between the modes of the clockwise and counterclockwise cavities. Usually, the frequencies of the two sets of modes are exactly equal, unless a "nonreciprocal" element is present in the cavity. The cavity imbalance due to the nonreciprocity is measured as a beat frequency between the two waves. Rotation is only one example of nonreciprocal effects, establishing a difference in optical pathlength for the cavities of the clockwise and counterclockwise waves. Since fs pulses do not share the same portion of the cavity at the same time, it is relatively easy to modify the optical pathlength of only one of the two circulating pulses. The fs ring laser can be used as a

---

[1]Hence the label "mode-locking" which is sometimes given to this effect, since the scattering of one circulating mode locks the frequency of the other. This is not to be confused with the mode-locking creating ultrashort pulse trains.

particularly sensitive probe of changes in induced phase (pathlength, phase change upon reflection, index of refraction). Nonreciprocal differences in phase of 10^{-6} have been measured, as will be shown later. This corresponds to rotation rates less than that of the earth, or changes of index of refraction of 10^{-10} over 1 cm, or linear motions of 0.01 Å.

12.2.2 Rotation sensing

Let us consider for the purpose of illustration a circular cavity of perimeter P and radius a. If the cavity is rotated at an angular velocity Ω (rad/s), it will appear extended by $a\Omega P/c$ to the wave traveling in the sense of rotation, and shrunk by the same amount to the oppositely traveling wave. If the two counterpropagating waves are interfered on a photodetector, a beat note equal to the frequency difference of the longitudinal modes is detected. The measured beat frequency is proportional to the rotation rate and is given by:

$$\Delta\nu = \frac{2a\Omega}{\lambda} = \frac{4A}{\lambda P}\Omega = \mathcal{R}\Omega, \tag{12.5}$$

where A is the area enclosed by the ring and λ is the operating wavelength of the gyroscope. The definition $\mathcal{R} = 4A/(\lambda P)$ is valid for cavities of arbitrary shape [539, 540].

In general, there is some coupling between the two counterpropagating laser beams which may inhibit the measurement of nonreciprocal effects. Optical elements or the gain medium may scatter some of the clockwise beam into the counterclockwise direction. When the effect of the scattering dominates that of the nonreciprocity, the injected scattering may force the counterpropagating waves to operate at the same frequency. If r is the scattering amplitude for one field into the counterpropagating field, an upper limit for the minimum detectable rotation rate Ω_c is given by [541]:

$$\Omega_c = \frac{rc\lambda}{2A}. \tag{12.6}$$

A typical example of the lock-in effect is shown in Fig. 12.10(a) for a 4 m perimeter cw CO_2 laser cavity shaped as a diamond [542]. For this particular laser using standard optics, the scattering from the mirrors and the gain medium locks the frequencies of the counterpropagating modes for rotation rates between $-6°/s$ and $+6°/s$. Departure from linearity can be observed for rotation rates between $-10°/s$ and $+10°/s$.

12.2. FEMTOSECOND LASER GYROSCOPES

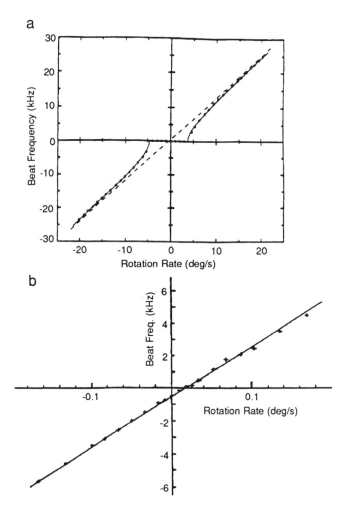

Figure 12.10: Gyroscopic response of a conventional (cw) CO_2 laser (a) (from [542]) compared to that of the fs ring dye laser (b) (from [171]). The beat frequency (in kHz) is plotted as a function of the rotation rate (in °/s). In (b), the solid line corresponds to the calculated scale factor. Note the difference in scales for the fs and CO_2 lasers, and the absence of a dead band for the fs laser.

As mentioned earlier, the use of fs pulses can reduce or eliminate the dead band by reducing the region of mutual coupling from the whole cavity to a submillimeter length of pulse overlap. Dead band elimination [173, 171] is achieved with the passively mode-locked dye laser shown in Fig. 12.9. The two crossing points of the pair of counterpropagating pulses are at the absorber jet (DODCI in ethylene glycol) and between a prism and an output coupler. When the two trains of output pulses (approximately 100 fs duration) are made to interfere on a detector PD, a beat note appears as a modulation of the pulse train envelope, as shown in the insert of Fig. 12.9. Even in the absence of rotation, the difference in index of refraction in the gain jet for the two counterpropagating pulse trains can cause a beat note of several tens of kHz [543].

The sensitivity of the mode-locked laser for making gyroscopic measurements is demonstrated in Fig. 12.10(b). In contrast to the CO_2 laser of similar shape and construction, there is no measurable dead band. All data for the frequency beat note fall along a straight line with a slope equal to the expected scale factor $\mathcal{R} = 31.40$ kHz/(°/s). The bandwidth of slightly less than 100 Hz for the beat note sets a detection limit for the system shown in Fig. 12.9. This limit can be attributed to the mechanical vibrations of the components of this laser, which has no active mode stabilization. With modest cavity stabilization, one can reduce the minimum detectable beat frequency by 2–3 orders of magnitude, making it possible to detect rotations on the order of 10^{-5} °/s.

12.2.3 Measurement of changes in index

The exceptional phase and optical length sensitivity of the same femtosecond ring laser can be exploited to investigate the response of electro-optic elements and ultrafast photodetectors. To demonstrate the concept, a Pockels cell oriented as a phase modulator is inserted in the cavity as shown in Fig. 12.11. An avalanche photodiode detects the pulse train from the laser and applies an electrical pulse to the Pockels cell. Appropriate optical delay ensures temporal coincidence of the electrical pulses and *one* of the cavity pulses at the electro-optic crystal. Because the other cavity pulse always reaches the cell between electrical pulses, the two counterpropagating pulses experience different indices in the cell. The optical length of the cavity is therefore different for the two senses of circulation of the intracavity pulses. Therefore, interfering the two outputs on a detector will result in a beat

12.3. SOLITONS

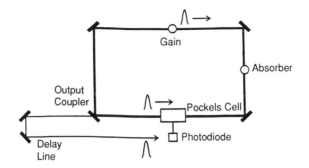

Figure 12.11: The output of the clockwise pulse of the ring cavity reaches the avalanche photodiode shortly before the time of arrival of the counterclockwise pulse in the Pockels cell. The electrical pulse from the photodiode is applied to the Pockels cell, resulting in a change in index, and therefore also in cavity length for the counterclockwise pulse.

frequency between the two outputs that is equal to the difference in the longitudinal mode frequencies.

Figure 12.12(a) shows a plot of the beat frequency versus the amplitude of the electric pulse applied to the modulator. By varying the optical delay, one can record the temporal response of the detector/electro-optic crystal combination as shown in Fig. 12.12(b). The temporal resolution is about 300 ps, limited by the detector, cable, and the pulse transit time through the electro-optic crystal. The intrinsic resolution of the method, however, is in the fs regime, limited only by the pulse duration. This arrangement is a fast and sensitive tool for studying the intrinsic response of photodetectors (by measuring directly the change of index due to the generated carriers) or photodetector/modulator combinations.

12.3 Solitons

The concept of solitons has already appeared in various chapters of this book. For instance, we have seen in Chapter 5 how an elementary model for a laser cavity, including only self-phase modulation and dispersion, leads to the nonlinear Schrödinger equation, which has some simple steady-state "soliton" solutions. The same model applied to fibers finds stable pulse shapes propagating without distortion over long distances. We have seen in Chapter 7 how the solitons could be used for the shaping of fs pulses. This application transcends the fs time domain: In communication, pulse

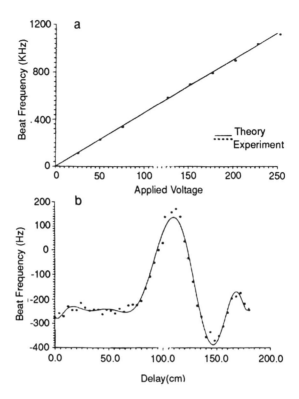

Figure 12.12: Beat note versus amplitude of the signal from the avalanche photodiode (a), all other parameters being kept constant. (b) Beat note versus the optical delay of the pulse impinging on the photodiode.

durations of the range of 20 to 80 ps are propagated without distortion through tens of km of fibers, or over 10^6 pulse lengths. The solitons used for pulse-coded communication which are *solitons in the time domain* will be briefly reviewed later. The nonlinear Schrödinger equation was first derived and solved by Zacharov and Shabat [212] in the context of self-focusing and self-filamentation. The solutions of this equation describing stable filaments are *solitons in the space domain*. The high intensities of fs pulses also lead to spatial solitons, such as the recently observed filamentation in air of fs pulses [544, 545, 546]. A more complex problem is that of solitons both in the temporal and spatial domain, which we will discuss at the end of this section.

12.3. SOLITONS

12.3.1 Temporal solitons

We have seen in Chapter 7 that ultrashort pulses of sufficient intensity launched in a single-mode fiber above the zero-dispersion wavelength evolve into a soliton. As mentioned earlier, since the soliton maintains its characteristics over long distances, it is an ideal signal for pulse-coded long-distance communication. Since any wavelength above the zero-dispersion point can be used, wavelength multiplexing is possible. The following problems must be overcome for-long distance propagation:

1. Decrease of the soliton pulse energy over long distances due to linear fiber losses.

2. Variations of the soliton propagation velocity resulting in timing jitter, hence loss of information.

3. Sliding of the soliton frequency, resulting in a mixture of adjacent frequency channels.

The first problem is solved with erbium-doped fibers (cf. Chapter 5) used as optical amplifiers. In a recent test of a trans-Pacific soliton link [547], an Er-doped amplifier was located every 26 km to restore the original soliton pulse energy, as sketched in Fig. 12.13.

Gordon and Haus [548] showed that the timing jitter is related to a jitter in pulse central frequency. They found that a certain component of

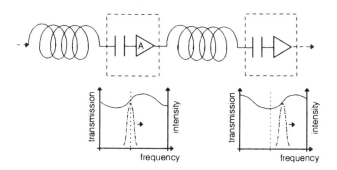

Figure 12.13: Optical soliton transmission line, showing schematically two units of fiber-repeater. Each unit consists of a single-mode, low-loss communication fiber, a Fabry–Perot filter, and an erbium-doped amplifier. The soliton frequency makes its way through the successive transmission peaks, while the noise sees the overall attenuation of the overlapping filters.

noise added to the soliton will instantly shift its optical frequency, and consequently its velocity, hence its time of arrival. A consequence of the Gordon–Haus theory is that a soliton has the property to shift its average frequency toward the frequency of maximum transmission of a filter. Let us consider, for instance, a soliton whose frequency is slightly shifted from the filter peak transmission. The differential loss across the soliton spectrum, *in conjunction* with the ability of the nonlinear effect to generate new frequency components, provides a force to push the soliton back towards the filter peak. A frequency filter can therefore solve the time jitter problem by preventing the soliton frequency from drifting around.

In an actual system, to accommodate frequency multiplexing, a Fabry–Perot filter is used at the same location as the amplifier. Each transmission peak of the Fabry–Perot interferometer defines a particular communication channel. One disadvantage of the optical amplifiers is that the noise is also amplified. Because of the property of the soliton to shift its frequency in the presence of differential noise, filters of slightly different frequency can be put at the successive amplifier location. The soliton makes its way through the different filters located at each successive amplifier, while the noise, being linear, sees the attenuation provided by the overlapping filters centered at different frequencies. Typical etalons [547] used as filters are Fabry–Perots of 1.5 mm spacing and 9% reflectivity. The fiber can support communication channels at frequency intervals of 100 GHz, which is the free spectral range of this etalon.

In a recent example of trans-Pacific communication, the frequency of the etalons at successive amplifiers was shifted by 0.18 GHz at each successive amplifier spaced 26 km apart. The total shift over the 9,000 km trans-Pacific distance is still smaller than the 100 GHz frequency spacing between channels [547]. The soliton pulse duration was 16 ps at 1557 nm. The fiber had a average dispersion $D = 0.5$ ps/(nm km). We recall (cf. Fig. 7.4) that the parameter $D = dk'_\ell/d\lambda = -(2\pi c_0/\lambda^2)k''_\ell$ is generally used to characterize fibers, because it relates directly to the group delay (in ps) per nm of bandwidth, and per km propagation length.

While this particular example is not specifically in the femtosecond time scale, the concept and implementation originate directly from the theories on fs pulse propagation of Chapters 1 and 2, and the pulse compression techniques of Chapter 7. Stable soliton propagation requires a balance of positive (negative) self-phase modulation and negative (positive) group velocity dispersion, and negligible losses. The diffraction losses are eliminated in fibers by confining the high-intensity pulse in a waveguide. In a bulk material, a

12.3. SOLITONS

mechanism for transverse confinement of the beam is required to compensate for diffraction losses. Such a mechanism is provided by self-focusing and self-filamentation, a problem addressed in the next subsection.

12.3.2 Spatial solitons

Spatial solitons relate to the confinement of pulses in a self-guided waveguide. Femtosecond pulses are a primary source leading to self-trapped filaments because of their high peak power. Stable filaments have even been observed in air [549]. The potential applications for soliton trapping in air involve propagation and partial ionization over long distances. Filaments have been shown to trigger and guide high-voltage discharges [544]. A subset of that application may be the guidance of particle beams. Another potential application relates to atmospheric probing for pollution monitoring.

Chiao, Garmire, and Townes [550] showed that the propagation equations for a time-independent field, in the presence of a self-focusing nonlinearity, reduce to the nonlinear Schrödinger equation. As we have seen in Chapters 5 and 7, the nonlinear Schrödinger equation has stationary solutions. These solution were precisely investigated in the context of self-focusing [212]. Steady-state solutions, however, are not a proof of the existence of stable filaments, in particular for pulsed radiation. Akhmanov et al. [551] showed that a negative nonlinearity of order larger than n_2 resulted in the formation of stable filaments. As the beam collapses because of self-focusing, the intensity on axis increases until the self-defocusing (for instance, from a negative term in $\bar{n}_4 I^2$) dominates the self-focusing due to the positive $\bar{n}_2 I$ term. The filament stabilizes at a diameter w such that the defocusing and focusing are in equilibrium. Using for an estimate the critical power $P_{cr} = 3.77\lambda_\ell^2/(8\pi n_0 \bar{n}_2)$ [see Eq. (3.133)] as power trapped in the filament:

$$I = \frac{\bar{n}_2}{\bar{n}_4} = \frac{2}{\pi w^2} P_{cr};$$

$$w = \sqrt{\frac{3.77\bar{n}_4}{n_0}} \frac{\lambda_\ell}{2\pi \bar{n}_2} \qquad (12.7)$$

Such filaments have been observed with continuous radiation in materials of large, slow nonlinearities, such as suspensions of latex spheres, aerosols, and microemulsions [552, 553]. In the case of pulsed radiation, the existence of filaments has been questioned. Instead, it has been postulated that only a "moving focus," which in solids leaves a filament-like trace, can be generated with pulses [554, 555]. The nonlinearity of air has long been known to be sufficient to reach the critical power with intense laser pulses [556, 557]. It is only recently that femtosecond pulses have been trapped in a self-focused filament [544, 545, 546].

In the case of air, a higher-order negative index contribution results from field ionization by the intense laser pulse. Ionization contributes to the defocusing $\Delta n_{df}(I)$ through the creation of an electron plasma (negative index) and the shift in absorption edge towards shorter wavelength (because of the replacement of neutral molecules by ions). Depending on the wavelength, the mechanism of ionization can be n-photon absorption (in which case the negative index contribution Δn_{df} is of order n), or field ionization (in which case the negative index contribution Δn_{df} is a steep threshold function). The stability of the filament can be attributed to a balance between self-focusing due to the Kerr lensing effect and self-defocusing from the induced plasma frequency. These filaments were observed over a distance of approximately 30 m, with 10 mJ, 200 fs pulses from a Ti:sapphire laser–amplifier chain [544, 545, 549]. At that wavelength (790 nm), the dispersion of air is $k_\ell'' \approx 0.002$ fs^2/cm, corresponding to a dispersion length $L_D = \tau_p^2/k_\ell'' = 200$ km. All dispersion effects of the atmosphere are thus negligible at that wavelength.

12.3.3 Spatial and temporal solitons

As shown in Table 6.2, the dispersion k_ℓ'' of air is approximately 1 fs^2/cm at 248 nm. For a 30 fs pulse at that wavelength, $L_D \approx 9$ m, and significant pulse broadening should occur over distances of the order of 10 m with fs pulses. Therefore, the existence of filaments over tens of meters requires that the pulse be trapped in space *and* in time.

Self-focusing with ultrashort pulses in media with group velocity dispersion brings another element of complexity: It is essential that the energy remain confined in time during the beam collapse process [102, 105, 558, 106]. The possibility of having a space-time filament or a "light bullet" has recently been suggested [107, 559, 560, 561]. Such light-bullet filaments were recently observed in air at 248 nm [562]. During femtosecond pulse propagation, the spectral broadening due to the self-phase modulation (SPM) *accelerates* the pulse duration broadening with distance due to group velocity dispersion (GVD). For the spatial beam characteristics, the positive n_2 of air tends to focus the beam, opposing diffraction.

Observations of self-focusing and filamentation have been made with fs pulses at 248 nm, with pulse energies in excess of 0.5 mJ [562]. Since the beam had a non-uniform transverse profile, the laser beam breaks into several filaments. The filament diameter remains practically constant over several meters. The average diameter of the filament is about 220 μm, and with less than 0.5 mJ energy trapped in it. The size is about three times larger than the 80 μm filaments observed at 800 nm [549]. The peak power corresponding to a 0.1 mJ, 50 fs pulse is much larger than the critical power for self-focusing $[P_{cr} \approx (1.22\lambda)^2 \pi/(32 n_0 \bar{n}_2)$, see Chapter 3, with $\bar{n}_2$ of air being 8×10^{-19} cm^2/W). The observation of filamentation with fs pulses at 248 nm is an indication that simultaneous trapping in time and space is possible.

There are two mechanisms that would tend to prevent the existence of a stable filament. First, diffraction should broaden the filament by a factor of $\sqrt{2}$ in a distance $\rho_0 \approx 60$ cm. Second, the pulse intensity on axis — hence the self-focusing effect — should also decrease because of pulse broadening by dispersion as mentioned earlier. A simple physical argument can be given to explain the stabilization mechanism of filaments in space and time. As in the case of spatial solitons, the self-defocusing effect due to the negative contribution of the free electrons (created by multiphoton ionization on the beam axis) compensates the self-focusing effect. The light pulses creates in air a "hollow waveguide" [563]. If we consider the temporal index of refraction variation on axis, an initial increase (due to $\bar{n}_2 I$) is followed by a steep decrease in index of refraction as the negative contribution of the photoionized electrons dominate. A *negative* self-phase modulation is thus induced on the pulses on axis, which, as we have seen in Chapter 7, can result in the formation of stable temporal solitons in the presence of the positive dispersion of air.

12.4 Problem

Consider a range gating setup like that in Fig. 12.1, with second-harmonic generation as gating mechanism. Both the reference pulse and the backscattered radiation are focused, with crossed polarization, into a second-harmonic crystal phase matched for second-harmonic generation type II at 620 nm. The crystal is urea, with a nonlinear coefficient of $d = 1.04\,10^{-23}$ C/V^2, and an index of refraction of 1.48. Given that both the reference beam and the signal beam are collimated with a diameter of 0.7 mm, and focused into the crystal with a 2.5 cm focal distance lens, find the peak power of the reference beam required to achieve single photon up-conversion. What should the crystal length be? Given that crystal length, is there a limitation to the depth resolution of this 3D imaging system? Is there a limitation to the transverse resolution? Calculate these limitations.

Appendix A

Phase Shifts upon Transmission − Reflection

A.1 The symmetrical interface

Let us consider first the very simple situation sketched in Fig. A.1. The interface can be a mirror with a reflecting coating on the front face and an antireflecting coating on the back face. We are only interested in fields propagating *outside* the mirror. The energy conservation relation between the reflected (field reflection coefficient $\tilde{r}$) and transmitted (field transmission coefficient $\tilde{t}$) waves implies:

$$|\tilde{r}|^2 + |\tilde{t}|^2 = 1 \tag{A.1}$$

Another relation can be found by adding another incident field of amplitude 1 (beam 2 in the figure), and taking advantage of the symmetry. Summing the intensities:

$$|\tilde{r} + \tilde{t}|^2 + |\tilde{r} + \tilde{t}|^2 = 2 \tag{A.2}$$

Combination of Eqs. (A.1) and (A.2) leads to

$$2[\tilde{r}\tilde{t}^* + \tilde{r}^*\tilde{t}] = 0, \tag{A.3}$$

which implies that the phase shifts upon transmission and reflection are complementary:

$$\varphi_r - \varphi_t = \frac{\pi}{2} \tag{A.4}$$

APPENDIX A. PHASE SHIFTS UPON TRANSMISSION

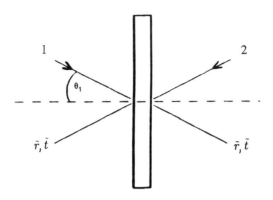

Figure A.1: Reflection and transmission by a plane mirror between between two identical media

It is because of the latter phase relation that the antiresonant ring reflects back all the incident radiation, and has zero losses if $|\tilde{r}|^2 = |\tilde{t}|^2 = 0.5$.

A.2 Coated interface between two different dielectrics

Let us consider – as in Fig. A.2 – a partially reflecting coating at an interface between air (index 1) and a medium of index n. A light beam of amplitude $\mathcal{E}_1 = 1/\sqrt{\cos\theta_1}$ is incident from the air, at an angle of incidence θ_1. The transmitted beam is refracted at the angle θ_2, and has an amplitude $\tilde{t}_1/\sqrt{\cos\theta_1}$. The reflected beam has amplitude $\tilde{r}_1/\sqrt{\cos\theta_1}$. The energy conservation (energy density × beam cross section) leads to the relation:

$$|\tilde{r}_1|^2 + |\tilde{t}_1|^2 \frac{n \cos\theta_2}{\cos\theta_1} = 1 \tag{A.5}$$

We have a similar energy conservation equation for a beam of amplitude $\mathcal{E}_2 = 1/\sqrt{n\cos\theta_2}$ incident at an angle θ_2 on the dielectric/air interface:

$$|\tilde{r}_2|^2 + |\tilde{t}_2|^2 \frac{\cos\theta_1}{n \cos\theta_2} = 1 \tag{A.6}$$

The amplitude of the reflection coefficient is equal on both sides of the interface. For the phase, the only sign relation consistent with energy conservation in a Gires-Tournois interferometer, and with the known phase shift

A.2. COATED INTERFACE BETWEEN DIELECTRICS

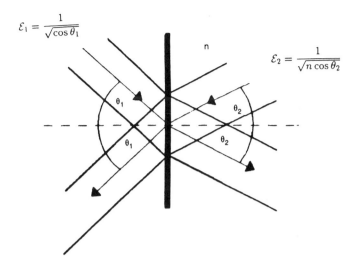

Figure A.2: Reflection and transmission at an interface

on pure dielectric interfaces, is:

$$\tilde{r}_1 = -\tilde{r}_2^*, \tag{A.7}$$

or, $r_1 = r_2$, with the relation between phase angles:

$$\varphi_{r,1} = -\varphi_{r,2} - \pi \tag{A.8}$$

The amplitudes of the transmission coefficients are not equal, but in the ratio $t_2/t_1 = n \cos\theta_2 / \cos\theta_1$.

In order to find a relation between the phase shift upon transmission and reflection, we consider the energy conservation for light incident from the upper half of the symmetric figure (the axis of symmetry being the dashed normal to the interface):

$$1 + 1 = \cos\theta_1 \left| \frac{\tilde{r}_1}{\sqrt{\cos\theta_1}} + \frac{\tilde{t}_2}{\sqrt{n\cos\theta_2}} \right|^2$$
$$+ n\cos\theta_2 \left| \frac{\tilde{r}_2}{\sqrt{n\cos\theta_2}} + \frac{\tilde{t}_1}{\sqrt{\cos\theta_1}} \right|^2 \tag{A.9}$$

Taking into account the energy conservation relations (A.5) and (A.6), leads to the following trigonometric relations between phase shifts upon transmis-

sion and reflection:
$$\frac{\cos(\varphi_{r,1} - \varphi_{t,2})}{\cos(\varphi_{r,2} - \varphi_{t,1})} = -1, \qquad (A.10)$$

which leads to the relation between phase angles:

$$\varphi_{t,2} - \varphi_{r,2} = \varphi_{r,2} - \varphi_{t,1} + (2n+1)\pi \qquad (A.11)$$

A direct consequence of this phase relation is:

$$\tilde{t}_1 \tilde{t}_2 - \tilde{r}_1 \tilde{r}_2 = 1 \qquad (A.12)$$

Appendix B

Fs Dye Laser Configurations

Linear cavities

To illustrate some alignment problems common to cavities with folding curved mirrors, we will take as an example the dye laser cavity sketched in Fig. B.1. If d is the spacing between mirrors M_1 and M_2, the distance between the point source in the jet and mirror M_2 is taken to be $(d - R_1)$ (this choice limits to two the number of images of the fluorescence, since the source is then imaged upon itself through mirror M_1). Geometrical optics is used to calculate the sagittal and tangential images of the fluorescence spot through M_2. For the minimum spot size condition of Fig. 5.23, the correct inter-mirror spacing d is obtained if the vertical image of the fluorescence is at a point A (Fig. B.1) 344 mm away from M_2. The pump geometry is thereafter optimized by minimizing the threshold of a simple 3 mirror cavity obtained by positioning a flat (parallel faces) mirror at position A. The laser beam from this three mirror cavity is used to align the antiresonant ring. The spacing between the mirrors of the antiresonant ring is adjusted for maximum feedback into the three mirror laser.

The astigmatism of successive folding angles is cumulative. Therefore, one would expect that the addition of an additional fold (Fig. B.2) would further degrade the quality of the beam emitted by the laser. As in the previous example, simple intuitive rules fail in a cavity of that complexity. A complete mapping of the cavity parameters for different cavity geometries has shown [168] that it is possible to have simultaneously a round output beam, and a very small (round) waist in the absorber jet. Table B.1 (from [149]) gives some geometrical parameters of a 2 m cavity,

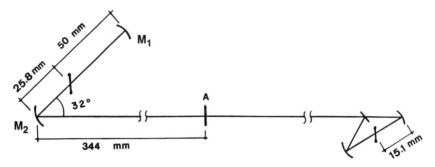

Figure B.1: Alignment procedure for the simple linear antiresonant ring laser

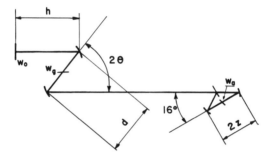

Figure B.2: Antiresonant ring laser cavity for hybrid mode-locking

with folding mirrors radii of 75 mm at the amplifier section, and 50 mm at the antiresonant ring section, as a function of the (equal) folding angles at the amplifier section, the distance between folding mirrors d, the distance h from the output mirror to the first folding mirror, and the spacing 2z between mirrors at the antiresonant ring. s_a is the ratio of the beam cross section in the amplifier jet, to that in the absorber. As in the previous example, it can be seen that, despite the large angle of the antiresonant ring triangle (16^0), very tight focusing is possible in the absorber jet, without the insertion of any astigmatism compensating element in the cavity.

For complex ring or linear cavities such as used for mode-locked dye lasers, algebraic manipulation languages are a very useful tools. However, they are not a sufficient substitute for human manipulation. The accuracy of numerical calculations of the elements of the final ABCD matrix is strongly dependent on the particular form of the analytical expressions, and on the order of calculations. A cavity will be stable if the beam waist calculated

θ	s_a	Waist area in ABS (μm)^2	Spot size in the absorber w_H	w_V	Output eccentricity	d (cm)	z (cm)	h (cm)	Stability range (mm)	Remarks
5.5	540	8.25	2.30	2.28	−0.886	7.801	2.542	42.5	0.002	large eccentricity small stability
8	38.6	30.65	4.04	4.83	0.724	7.643	2.559	40	0.564	
13	95	15.38	3.135	3.124	0.741	7.520	2.558	45	0.052	
14	400	6.26	2.00	1.99	−0.828	7.488	2.557	40	0.006	round spot in both jets
14	150	4.35	1.66	1.66	−0.626	7.706	2.533	40	0.001	tight focusing in absorber
15	8.3	51.50	4.65	7.00	−0.02	7.817	2.551	40	0.27	round output beam
15.42	6.75	40.69	5.09	5.09	0.14	7.752	2.545	45.62	0.22	round colimated beam between amplifier and absorber
15.42	21.09	12.12	2.72	2.83	0.46	7.760	2.530	43.75	0.27	
16.5	2,000	3.03	1.39	1.38	−0.78	7.385	2.562	42.5	0.003	largest s in that range of θ
16.5	15.6	107.	25.41	2.68	0.003	7.618	2.556	42.5	0.09	round output beam

Table B.1: Geometric parameters of the antiresonant ring laser (from ref. [149]).

from Eq. (5.77) is real, or:

$$\left(\frac{A+D}{2}\right)^2 \leq 1 \tag{B.1}$$

The stable region may not be found if A and D are not calculated with an accuracy better than 0.2 %. If no care is taken to reorganize the raw expression from the algebraic manipulation language, the expression for A and D may contain differences of large terms — such as $(Lf - Ld_1)$, where L is the cavity length, and $2d_1$ is the distance between 2 mirrors of radius of curvature $2f$. Because of the large number of such expressions, even double precision may not be sufficient to calculate the matrix elements A, B, C, D with an accuracy of only 0.2%. All such differences have to be eliminated, factoring out small terms such as $(d_1 - f)$.

Ring cavities

As shown at the beginning of this section, the focusing optics for two dye jets will have an overwhelming contribution to the astigmatism of the cavity, as compared to that of the thin dye jets and the pairs of prisms. However, as in the case of the linear laser, a complete cavity calculation may show stable regions where contributions to the astigmatism originating from different parts of the cavity cancel each other, resulting in very sharp focusing in one of the dye jets. As in the case of the linear cavity, the origin is chosen at mid-point between the absorber mirrors (distance between mirrors: $2z$, the radius of curvature of mirrors M_1 and M_2 is R_1). The angles of incidence on these mirrors are both equal to θ_1. The angles of incidence on the two mirrors M_4 and M_5 of curvature R_3 and spacing $2d$ of the amplifier section are both equal to θ_3 . The two sections are connected in one direction by a straight propagation matrix of distance e, in the other direction by a propagation matrix of distance v, a reflection on a mirror M_3 of curvature R_2, at an incidence θ_2 , and a propagation for a distance w. Since the two arms of the cavity should be in a length ratio of 1 to 3, the distance w can be expressed as a function of the other parameters ($w = 3e + R_1 + R_3 - v$). The resulting ABCD matrix can be used to represent various ring configurations as in Fig. B.3(a) (where the mirror of curvature R_2 could be in position M_3 or M_6, and the only acceptable solutions are those for which there are three beams waists in the plane of the figure) or in Fig. B.3(b) (where the radius of

APPENDIX B. FS DYE LASER CONFIGURATIONS

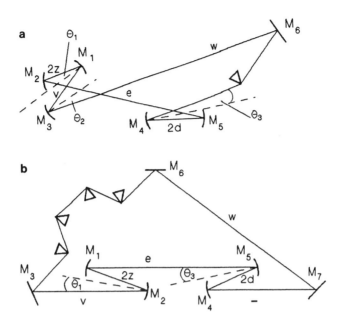

Figure B.3: Two common ring cavity configurations (from ref. [149]).

curvature R_2 is chosen to be infinite). The same procedure as in the former subsection can be used, namely to optimize the inter absorber mirror spacing $2z$ for best round spot (smallest absolute value of the difference between sagittal and tangential beams waists), for every value of the other parameters (distance between amplifier mirrors d, distance v, angles θ_1 and θ_3). Some typical results are summarized in Table B.2. The cavity of Fig. B.3(b) gives a minimum spot of equal dimension in the absorber and amplifier sections. A tighter focusing can be obtained in the case of the cavity of Fig. B.3(a) in the absorber jet ($60\mu m^2$). Even smaller spot sizes ($34\mu m^2$) are possible for smaller angles θ_1, as indicated in Table B.2. However, the minimum angle of 5^0 may not be sufficient for beam clearance (at the jet and at the mirrors) Much tighter focusing, and largest values of the "s parameter" are obtained if the curved mirror of focal distance R_2 is located in position M_6 [Fig. B.3(b)]. As in the case of the linear cavity, the smallest spot sizes are not necessarily found for small angles of incidence on the curved mirrors. The very small round focal spot in the absorber obtained for $\theta_1 = 13^0, \theta_3 = 20^0$, (Table B.2) is an example of geometrical astigmatism compensation.

Typical Geometric Parameters of the Ring Laser

θ_1	θ_3	R_2 (m)	v (m)	s_a	Waist area in ABS $(\mu m)^2$	Spot size in the absorber w_H	w_V	d (cm)	z (cm)	Third beam waist H at (cm)	w (mm)	Third beam waist V at (cm)	w (mm)	Stability range (mm)	Remarks
9	10	1	0.5	1.3	520	10.0	33	2.506	1.429	105	0.054	N	N	0.2	
13	10	1	0.5	1	940	24	25	2.548	1.438	263	0.425	N	N	0.055	
9	13	1	2	104	7.2	3.8	1.2	2.588	1.528	46	0.085	N	N	0.01	
9	15	1	2	54	7.8	4.5	1.1	2.495	1.537	38	0.089	132	0.055	0.017	
13	20	1	1.75	26	14.3	1.1	8.3	2.741	1.487	86	0.045	N	N	0.023	
13	20	1	2.08	620	8	2.3	2.2	2.789	1.486	84	0.112	N	N	0.005	
17	10	1	2	12	44.5	3.5	8.1	2.633	1.467	61	0.14	N	N	0.036	
5	10	∞	—	0.15	264	58	2.9	2.524	1.518	N		N	N	0.011	Smallest gain spot (40 μm²)
5	10	∞	—	7	34	12	1.8	2.534	1.517	N		N	N	0.011	
9	10	∞	—	2	63.3	12.6	3.2	2.519	1.531	11	0.3	N	N	0.078	Smallest absorber spot

$\theta_2 = 8.3°$
"N" indicates no third beam waist

Table B.2: Geometrical parameters of the ring laser. The unit of R_2 is [m].

Appendix C

Four Photon Coherent Interaction

The procedure to calculate the coherent four-photon resonant coherent interaction is detailed in Refs. [136, 564]. The density matrix of the multilevel atomic system is reduced to a two by two matrix associated with the resonant levels. The temporal evolution of the density matrix elements associated with all the off-resonant levels is calculated by an adiabatic expansion to fourth order [136, 564]. We refer to the literature for the "straightforward but tedious" derivation of the time evolution of the diagonal (ρ_{44}, ρ_{00}) and off-diagonal [$\rho_{04} = \varrho_{04} \exp(4i\omega_\ell t)$] elements. We will instead concentrate on the physical meaning of the various terms entering these equations. In analogy with the two-photon coherent interaction equations, we introduce an amplitude function for the four-photon excitation in the atom $\tilde{Q}_4 = 2i\varrho_{04}$. In addition to the fundamental frequency ω_ℓ, fields at the third harmonic ($\omega_3 = 3\omega_\ell$) and the fifth harmonic ($\omega_5 = 5\omega_\ell$) frequency are also generated.

One of the main driving term of the electronic excitation ρ_{04} oscillating at $4\omega_\ell$ is a term proportional to the fourth power of the applied field. The time evolution of the amplitude $\tilde{Q}_4$ of the harmonic electronic excitation of the atom is given by the system of interaction equations:

$$\left\{\frac{\partial}{\partial t} + i(\omega_{04} - 4\omega_\ell + \delta\omega_2) + \frac{\gamma_0 + \gamma_4}{2} + \frac{1}{T_2}\right\} \tilde{Q}_4$$
$$= (\rho_{44} - \rho_{00}) \left\{\tilde{V}_1^4 + (\xi_{40} + a_3|\tilde{V}_1|^2)\tilde{V}_1\tilde{V}_3 + (\zeta_{40}^* + c_5^*|\tilde{V}_1|^2)\tilde{V}_1^*\tilde{V}_5 + d_5^*\tilde{V}_1^2\tilde{V}_5\tilde{V}_3^*\right\}$$
$$\text{(C.1)}$$

$$\frac{\partial \rho_{44}}{\partial t} + \gamma_4 \rho_{44} + \frac{\rho_{44}}{T_1}$$
$$= \mathrm{Re}\left\{Q_4^*[\tilde{V}_1^4 + (\xi_{40} + a_3|\tilde{V}_1|^2)\tilde{V}_1\tilde{V}_3 + (\zeta_{40}^* + c_5^*|\tilde{V}_1|^2)\tilde{V}_1^*\tilde{V}_5 \right.$$
$$\left. + d_5^*\tilde{V}_1^2\tilde{V}_5\tilde{V}_3^*]\right\} \tag{C.2}$$

$$\frac{\partial(\rho_{44} + \rho_{00})}{\partial t} = -\gamma_4\rho_{44} - \gamma_0\rho_{00}, \tag{C.3}$$

where ξ_{40} is the ratio of the sum frequency ($\omega_\ell + \omega_3$) to four photon excitation coefficients, a_3 being the intensity dependent part of this ratio; ζ_{40} and b_3 are the ratio — and the intensity dependent part thereoff — of the difference frequency ($\omega_5 - \omega_\ell$) to four photon excitation coefficients; d_5 is the ratio of the excitation process ($2\omega_\ell + \omega_5 - \omega_3$) to the process ($4\omega_\ell$).

In the above equations (C.1) through (C.3), the fields $\tilde{V}_i$ are in units of the Rabi cycle for the four photon transition:

$$\tilde{V}_i^4 = \frac{2r_{04}}{(2\hbar)^4}\tilde{\mathcal{E}}_i^4 = \frac{p_{0f}p_{fk}p_{kj}p_{j4}}{(\omega_{j1} - 3\omega_\ell)(\omega_{k0} - 2\omega_\ell)(\omega_{f0} - \omega_\ell)}\frac{\tilde{\mathcal{E}}_i^4}{8\hbar^4}, \tag{C.4}$$

where $\tilde{\mathcal{E}}_i$ is the amplitude of the field at $\omega_{i\ell} = i\omega_\ell$, and the summation convention over the repeated indices is assumed. Equation C.4 defines a complex parameter r_{04}. All the fields are expressed in frequency units through the conversion factor $(r_{04})^{1/4}/\hbar$.

As in the case of the two-photon resonant two level system, the detuning is time dependent through the Stark shift $\delta\omega_2$:

$$\delta\omega_2 = [\alpha_0' - \alpha_4']|\tilde{V}_1|^2 \tag{C.5}$$

where the "prime" indicate the real part of the *normalized* susceptibility α_i given by:

$$\alpha_i = \frac{1}{\sqrt{r_{04}}}\sum_k\left[\frac{|p_{ki}|^2}{(\omega_{ki} - \omega_\ell)} + \frac{|p_{ki}|^2}{(\omega_{ki} + \omega_\ell)}\right]. \tag{C.6}$$

The coherence is expressed by the function $\tilde{Q}_4 = 2i\rho_{04}e^{-4\omega_\ell t}$, which will be the main driving term of Maxwell's equations. For instance, for the fundamental:

$$\frac{\partial \tilde{V}_1}{\partial z} = \frac{\mu\omega_\ell cN}{2}\frac{2\sqrt{2r_{04}}}{\hbar}[\tilde{V}_1^{*3}\tilde{Q}_4$$
$$+(\xi_{40} + a_3|\tilde{V}_1|^2)\tilde{Q}_4\tilde{V}_3^* + (\zeta_{40}^* + c_5^*|\tilde{V}_1|^2)\tilde{Q}_4^*\tilde{V}_5 + 2d_5\tilde{V}_1^*\tilde{V}_3\tilde{V}_5^*\tilde{Q}_4$$
$$+\frac{3}{4}\frac{a_3}{r_{04}}\tilde{V}_3\tilde{V}_1^{*2}\rho_{11}] + \frac{\mu\omega_\ell cN}{2}[\alpha_0(\omega_\ell)\rho_{00} + \alpha_4(\omega_\ell)\rho_{44}]\tilde{V}_1, \tag{C.7}$$

where α_0 and α_4 are the linear susceptibilities of levels 0 and 4, respectively.

APPENDIX C: FOUR PHOTON COHERENT INTERACTION

In Eq. (C.7), α_3 is the (complex) polarizability, responsible for *nonresonant* third harmonic generation:

$$\alpha_3 = \frac{p_{0f} p_{fk} p_{kj} p_{j0}}{(\omega_{f0} - \omega_\ell)(\omega_{k0} - 2\omega_\ell)(\omega_{j0} - 3\omega_\ell)}. \tag{C.8}$$

As source terms opposing or enhancing the fundamental field we recognize in Eq. (C.7) all the combination of amplitude terms for which the corresponding phase factor is at the frequency ω_ℓ: $\tilde{Q}_4 \tilde{V}_1^{*3}$ ($4\omega_\ell - 3\omega_\ell$), $\tilde{Q}_4 \tilde{V}_3^*$ ($4\omega_\ell - \omega_{3\ell}$), $\tilde{Q}_4 \tilde{V}_1^* \tilde{V}_3 \tilde{V}_5^*$ ($4\omega_\ell + \omega_{3\ell} - \omega_{5\ell} - \omega_\ell$), and $\tilde{V}_3 \tilde{V}_1^{*2}$ ($\omega_{3\ell} - 2\omega_\ell$). The latter term, in contrast to the formers, does not involve the resonant 4-photon process, but is a second order Raman term associated with non-resonant third harmonic generation. The non-resonant third harmonic generation term can interfere constructively or destructively with the resonant generation process, as can be seen in Maxwell's equations for the field at $\omega_{3\ell}$:

$$\frac{\partial \tilde{V}_3}{\partial z} = \frac{\mu \omega_\ell c N}{2} \frac{2\sqrt{2 r_{04}}}{\hbar} [(\xi_{40} + \alpha_3 |\tilde{V}_1|^2) \tilde{Q}_4 \tilde{V}_1$$
$$+ 2 d_5^* \tilde{V}_1^2 \tilde{V}_5 \tilde{Q}_4$$
$$+ \frac{3}{4} \frac{\alpha_3}{r_{04}} \tilde{V}_1^3 \rho_{11}] + \frac{\mu \omega_\ell c N}{2} [\alpha_0(\omega_{3\ell}) \rho_{00} + \alpha_4(\omega_{3\ell}) \rho_{44}] \tilde{V}_3. \tag{C.9}$$

A detailed and complete set of equations can be found in [136, 564]. The object of this display of equations is not to confuse the reader, but point out to interesting transient phenomena and intriguing possibilities of new applications for femtosecond pulses.

Appendix D

Dyes and Solvents

Abbreviations for the dyes used in Chapter 5

Abbreviation	Full name
RhB	Rhodamine B or Rhodamine 610
Rh 110	Rhodamine 110 or Rhodamine 560
Rh6G	Rhodamine 6G or Rhodamine 590
DODCI	3,3'-diethyloxadicarbocyanine iodide
DQOCI	1,3'-diethyl-4,2'-quinolyoxacarbocyanine iodide
SRh101	Sulforhodamine 101 or sulforhodamine 640
DOTCI	3,3'-diethyloxatricarbocyanine iodide
DCCI	or DCI, 1,1'-diethyl-2,4'-carbocyanine iodide
DQTCI	1,3'-diethyl-4,2'-quinolthiacarbocyanine iodide
DDI	= DDCI 1,1'-diethyl-2,2'-dicarbocyanine iodide
HITCI	1,1',3,3,3',3'-hexamethylindotricarbocyanine iodide
DaQTeC	2-(p-dimethylaminostyryl)-benzothiazolylethyl iodide

Appendix E

Reconstruction of the Spectral Phase

This appendix contains some of the calculation steps involved in the detection in frequency domain presented in Section 8.4.6.

With the assumption that the spectral amplitude and phase $\mathcal{E}(\Omega)$ and $\phi(\Omega)$ do not change significantly over the spectral width w/a, we can substitute in Eq.(8.36) the first order approximation:

$$\mathcal{E}(\Omega) = \mathcal{E}|_{(\Omega_0+y/a)} + \frac{\partial \mathcal{E}(\Omega)}{\partial \Omega}\bigg|_{(\Omega_0+y/a)}[\Omega - (\Omega_0 + y/a)]$$
$$= \mathcal{E}_{0y} + \mathcal{E}'_{0y}[\Omega - (\Omega_0 + y/a)]$$

$$\phi(\Omega) = \phi|_{(\Omega_0+y/a)} + \frac{\partial \phi(\Omega)}{\partial \Omega}\bigg|_{(\Omega_0+y/a)}[\Omega - (\Omega_0 + y/a)]$$
$$= \phi_{0y} + \phi'_{0y}[\Omega - (\Omega_0 + y/a)], \qquad (E.1)$$

where Ω_0 is a reference frequency corresponding to $y = 0$. Substituting the expansion into Eq. (8.36), we obtain:

$$\tilde{\mathcal{E}}_{0y}(x, y, \Omega) = \mathcal{E}_{0y} \exp\left\{-\frac{x^2}{w^2} - \frac{(y - a\Delta\Omega)^2}{w^2} + i\phi_{0y} + i\phi'_{0y}(\Delta\Omega - y/a)\right\}$$
$$(E.2)$$

Taking the inverse Fourier transform:

$$\tilde{\mathcal{E}}_{0y}(x,y,t) = \mathcal{E}_{0y} e^{-\frac{x^2}{w^2} + i\phi_{0y} - i\phi'_{0y} y/a} \int e^{-\frac{(y-a\Delta\Omega)^2}{w^2} + i\phi'_{0y}\Delta\Omega} e^{-i\Omega t} d\Omega \quad (E.3)$$

$$= \mathcal{E}_{0y} e^{-\frac{x^2}{w^2} + i\phi_{0y} + i\Omega_0 t - i\frac{y}{a} t} \frac{w}{a\sqrt{2}} e^{-(t-\phi'_{0y})^2/4(a/w)^2}$$

$$= E_{0y} e^{-\frac{x^2}{w^2} - \frac{(t-\phi'_{0y})^2}{\tau_y^2} - i\frac{y}{a} t + i\phi_{0y} + i\Omega_0 t}, \quad (E.4)$$

where $E_{0y} = \frac{\sqrt{2}}{\tau_y} \mathcal{E}_{0y}$.

Equation (E.4) shows that any component y of the pulse is stretched into a Gaussian pulse of duration $\tau_y = 2a/w$. The pulse corresponding to each spectral component y is centered (in time) at the instant $t_d(y) = \phi'_{\Omega_0 + y/a}$. Knowledge of $t_d(y)$ leads – after integration – to the determination of the spectral phase.

Bibliography

[1] J. Orear. *Physik*. Carl Hansen Verlag, Munchen, Vienna, 1982.

[2] R. A. Serway, C. J. Moses, and C. A. Moyer. *Modern Physics*. Saunders College Publishing, Philadelphia, New York, 1989.

[3] S. L. Shapiro, editor. *Ultrashort Light Pulses*. Springer Verlag, Berlin, Heidelberg, New York, 1977.

[4] J. Herrmann and B. Wilhelmi. *Lasers for Ultrashort Light Pulses*. Akademie-Verlag, Berlin, 1987.

[5] W. Kaiser and D. H. Auston, editors. *Ultrashort Laser Pulses*, volume 60 of *Topics in Applied Phys.* Springer-Verlag, Berlin, 1988.

[6] Eugene Hecht and Alfred Zajac. *Optics*. Addison-Wesley, Menlo Park, California, 1979.

[7] Willian H. Press, Brian P. Flannery, Saul E. Teukolsky, and William P. Vetterling. *Numerical Recipes*. Cambridge University Press, New York, 1986.

[8] J. D. Jackson. *Classical Electrodynamics*. John Wiley & Sons, New York, 1975.

[9] M. Born and E. Wolf. *Principles of Optics – Electromagnetic theory of propagation, interference and diffraction of light*. Pergamon Press, Oxford, New York, 1980.

[10] Y. R. Shen. *The Principles of Nonlinear Optics*. John Wiley & Sons, New York, 1984.

[11] M. Schubert and B. Wilhelmi. *Nonlinear Optics and Quantum Electronics*. John Wiley & Sons, New York, 1978.

[12] Kamke. *Differentialgleichungen*. Geest and Portig, Leipzig, 1969.

[13] Govind P. Agrawal. *Nonlinear Fiber Optics*. Academic Press Inc., Boston, San Diego, 1989.

[14] A.E.Siegman. *Lasers*. University Press, Oxford, 1986.

[15] M.V.Klein and T.E.Furtak. *Optics*. John Wiley & Sons, New York, Chichester, Brisbane, Toronto, Singapur, 1986.

[16] B. H. Kolner and M. Nazarathy. Temporal imaging with a time lens. *Optics Lett.*, 14:630–632, 1989.

[17] I. P. Christov. Propagation of femtosecond light pulses. *Optics Commun.*, 53:364–367, 1985.

[18] F. J. Duarte. Dye Lasers. In F. J. Duarte, editor, *Tunable Lasers Handbook*. Academic Press, Boston, 1994.

[19] S. DeSilvestri, P. Laporta, and O. Svelto. The role of cavity dispersion in cw mode-locked lasers. *IEEE J. Quantum Electron.*, QE-20:533–539, 1984.

[20] W. Dietel, E. Dopel, K. Hehl, W. Rudolph, and E. Schmidt. Multilayer dielectric mirrors generated chirp in femtosecond dye ring lasers. *Optics Comm.*, 50:179–182, 1984.

[21] W. H. Knox, N. M. Pearson, K. D. Li, and C. A. Hirlimann. Interferometric measurements of femtosecond group delay in optical components. *Optics Lett.*, 13:574–576, 1988.

[22] K. Naganuma, K. Mogi, and H. Yamada. Group delay measurement using the fourier transform of an interferometric cross-correlation generated by white light. *Optics Lett.*, 17:393–395, 1990.

[23] K. Naganuma and H. Yasaka. Group delay and alpha-parameter measurement of 1.3 micron semiconductor traveling wave optical amplifier using the interferometric method. *IEEE J. Quantum Electron.*, 27:1280–1287, 1991.

[24] A. M. Weiner, J. G. Fujimoto, and E. P. Ippen. Femtosecond time-resolved reflectometry measurements of multiple-layer dielectric mirrors. *Opt. Lett.*, 10:71–73, 1985.

[25] A. M. Weiner, J. G. Fujimoto, and E. P. Ippen. Compression and shaping of femtosecond pulses. In D. H. Auston and K. B. Eisenthal, editors, *Ultrafast Phenomena IV*, pages 11–15, New York, 1984. Springer-Verlag.

[26] J.-C. Diels, W. Dietel, E. Dopel, J. J. Fontaine, I. C. McMichael, W. Rudolph, F. Simoni, R. Torti, H. Vanherzeele, and B. Wilhelmi. Colliding pulse femtosecond lasers and applications to the measurement of optical parameters. In D.H. Auston and K. B. Eisenthal, editors, *Ultrafast Phenomena IV*, pages 30–34, New York, 1984. Springer-Verlag.

[27] J. Desbois, F. Gires, and P. Tournois. A new approach to picosecond laser pulse analysis and shaping. *IEEE J. Quantum Electron.*, QE-9:213–218, 1973.

[28] J. Kuhl and J. Heppner. Compression of fs optical pulses with dielectric multi-layer interferometers. *IEEE J. Quantum Electron.*, QE-22:182–185, 1986.

[29] M. A. Duguay and J. W. Hansen. Compression of pulses from a mode-locked He-Ne laser. *Appl.Phys.Lett.*, 14:14–15, 1969.

[30] J. Heppner and J. Kuhl. Intracavity chirp compensation in a colliding pulse mode-locked laser using thin-film interferometers. *Appl.Phys.Lett.*, 47:453–455, 1985.

[31] Z. Bor. Distortion of femtosecond laser pulses in lenses and lens systems. *Journal of Modern Optics*, 35:1907–1918, 1988.

[32] Z. Bor. Distorsion of femtosecond laser pulses in lenses. *Optics Letters*, 14:119–121, 1989.

[33] M. Kempe, W. Rudolph, U. Stamm, and B. Wilhelmi. Spatial and temporal transformation of femtosecond laser pulses by lenses and lens systems. *J. Opt. Soc. Am.*, B9:1158–1165, 1992.

[34] M. Kempe and W. Rudolph. The impact of chromatic and spherical aberration on the focusing of ultrashort light pulses by lenses. *Opt. Lett.*, 18:137–139, 1993.

[35] M. Kempe and W. Rudolph. Femtosecond pulses in the focal region of lenses. *Phys. Rev.*, A48:4721–4729, 1993.

[36] G. Szabo, Z. Bor, and A. Mueller. Phase-sensitive single pulse autocorrelator for ultrashort laser pulses. *Opt. Lett.*, 13:746–748, 1988.

[37] J.-C. Diels, N. Jamasbi, C. Yan, and M. Lai. Ultrafast detection of weak signals. In *Proceedings of the SPIE meeting on Semiconductor Diagnostics*, volume 943, pages 190–194, Newport Beach, CA, April 1988. SPIE.

[38] Zs. Bor, S. Szatmari, and A. Muller. Picosecond pulse shortening by travelling wave amplified spontaneous emission. *Applied Physics B*, 32:101–104, 1983.

[39] E. B. Treacy. Optical pulse compression with diffraction gratings. *IEEE J. Quantum Electron.*, QE-5:454–460, 1969.

[40] F. J. Duarte and J. A. Piper. Dispersion theory of multiple prisms beam expanders for pulsed dye lasers. *Optics Comm.*, 43:303–307, 1982.

[41] R. L. Fork, O. E. Martinez, and J. P. Jordan. Negative dispersion using pairs of prism. *Opt. Lett.*, 9:150–152, 1984.

[42] C. Froehly, B. Colombeau, and M. Vampouille. Shaping and analysis of picosecond light pulses. *in Progress of Modern Optics*, Vol. XX:115–125, 1981.

[43] O. E. Martinez. Grating and prism compressor in the case of finite beam size. *J. Opt. Soc. Am. B*, 3:929–934, 1986.

[44] O. E. Martinez, J. P. Gordon, and R. L. Fork. Negative group-velocity dispersion using diffraction. *J. Opt. Soc. Am.*, A1:1003–1006, 1984.

[45] F. J. Duarte. Generalized multiple-prism dispersion theory for pulse compression in ultrafast dye lasers. *Opt. and Quant.Electr.*, 19:223–229, 1987.

[46] V. Petrov, F. Noack, W. Rudolph, and C. Rempel. Intracavity dispersion compensation and extracavity pulse compression using pairs of prisms. *Exp. Technik der Physik*, 36:167–173, 1988.

[47] C. P. J. Barty, C. L. Gordon III, and B. E. Lemoff. Multiterawatt 30-fs Ti:sapphire laser system. *Optics Lett.*, 19:1442–1444, 1994.

[48] P. F. Curley, Ch. Spielmann, T. Brabec, F. Krausz, E. Wintner, and A. J. Schmidt. Operation of a fs Ti:sapphire solitary laser in the vicinity of zero group-delay dispersion. *Opt. Lett.*, 18:54–57, 1993.

[49] M. T. Asaki, C. P. Huang, D. Garvey, J. Zhou, H. Kapteyn, and M. M. Murnane. Generation of 11 fs pulses from a self-mode-locked Ti:sapphire laser. *Opt. Lett.*, 18:977–979, 1993.

[50] O. E. Martinez. 3000 times grating compressor with positive group velocity dispersion: application to fiber compensation in the 1.3-1.6 μm region. *IEEE J. of Quantum Electron.*, 23:59–64, 1987.

[51] M. Pessot, P. Maine, and G. Mourou. 1000 times expansion compression of optical pulses for chirped pulse amplification. *Optics Comm.*, 62:419, 1987.

[52] S. A. Akhmanov, A. P. Sukhorov, and A. S. Chirkin. Stationary phenomena and space-time analogy in nonlinear optics. *Sov. Phys. JETP*, 28:748–757, 1969.

[53] E. Doepel. Matrix formalism for the calculation of group delay and group-velocity dispersion in linear optical elements. *J. Mod. Opt.*, 37:237–242, 1990.

[54] O. E. Martinez. Matrix formalism for pulse compressors. *IEEE J.Quantum Electron.*, QE-24:2530–2536, 1988.

[55] O. E. Martinez. Matrix formalism for dispersive cavities. *IEEE J.Quantum Electron.*, QE-25:296–300, 1989.

[56] S. P. Dijaili, A. Dienes, and J. S. Smith. ABCD matrices for dispersive pulse propagation. *IEEE J. Quantum Electron.*, QE-26:1158–1164, 1990.

[57] A. G. Kostenbauder. Ray-pulse matrices: A rational treatment for dispersive optical systems. *IEEE J. Quantum Electron.*, QE-26:1148–1157, 1990.

[58] F. J. Duarte. Multiple-prism dispersion and 4×4 ray transfer matrices. *Opt. Quant. Electr.*, 24:49–53, 1991.

[59] O. Hittmair. *Quantum Mechanics*. Thiemig, Munich, 1972.

[60] L. M. Frantz and J. S. Nodvik. Theory of pulse propagation in a laser amplifier. *J. of Appl. Physics*, 34:2346–2349, 1963.

[61] D. Kuehlke, W. Rudolph, and B. Wilhelmi. Calculation of the colliding pulse mode locking in cw dye ring lasers. *IEEE J. Quant. Electr.*, QE-19:526–533, 1983.

[62] V. Petrov, W. Rudolph, and B. Wilhelmi. Chirping of femtosecond light pulses passing through a four-level absorber. *Opt. Commun.*, 64:398–402, 1987.

[63] V. Petrov, W. Rudolph, and B. Wilhelmi. Chirping of ultrashort light pulses by coherent interaction with samples with an inhomogeneously broadened line. *Sov. J. Quant. Electr.*, 19:1095–1099, 1988.

[64] W. Rudolph. Calculation of pulse shaping in saturable media with consideration of phase memory. *Opt. Quant. Electr.*, 16:541–550, 1984.

[65] W. Boyd. *Nonlinear Optics*. Academic Press, New York, 1991.

[66] R. C. Miller. Second harmonic generation with a broadband optical maser. *Phys. Lett.*, A26:177–178, 1968.

[67] J. Comly and E. Garmire. Second harmonic generation from short light pulses. *Appl. Phys. Lett.*, 12:7–9, 1968.

[68] S. A. Akhmanov, A. S. Chirkin, and A. P. Sukhorukov. Second harmonic generation of broadband light pulses. *Sov. J. Quantum Electron.*, 28:748–759, 1968.

[69] W. H. Glenn. Second harmonic generation by ps optical pulses. *IEEE J. Quantum Electron.*, QE-5:281–290, 1969.

[70] Y. Wang and R. Dragila. Efficient conversion of picosecond laser pulses into second-harmonic frequency using group-velocity-dispersion. *Phys. Rev. A*, 41:5645–5649, 1990.

[71] F. Zernicke. Refractive indices of ADP and KDP between 200nm and 1500nm. *J. Opt. Soc. Am.*, 54:1215–1220, 1964.

[72] M. Choy and R. L. Byer. Accurate second-order susceptibility measurements of visible and infrared optical pulses. *Phys. Rev.*, B14:1693–1706, 1976.

BIBLIOGRAPHY

[73] K. Kato. Second harmonic generation to 2058 Å in β-bariumborate. *IEEE J. Quantum Electron.*, QE-22:1013–1014, 1986.

[74] Y. N. Karamzin and A. P. Sukhorukov. Limitation on the efficiency of frequency doublers of picosecond light pulses. *Sov. J. Quantum Electron.*, 5:496–500, 1975.

[75] R. C. Eckardt and J. Reintjes. Phase matching limitations of high efficiency second harmonic generation. *IEEE J. Quantum Electron.*, QE-20:1178–1187, 1984.

[76] N. C. Kothari and X. Carlotti. Transient second harmonic generation: influence of effective group velocity dispersion. *J. Opt. Soc. Am. B*, 5:756–764, 1988.

[77] D. Kuehlke and U. Herpers. Limitation of the second harmonic conversion of intense femtosecond pulses. *Opt. Commun.*, 69:75–80, 1988.

[78] G. Szabo and Z. Bor. Broadband frequency doubler for femtosecond light pulses. *Appl. Phys.*, B50:51–54, 1990.

[79] O. E. Martinez. Achromatic phase matching for second harmonic generation of femtosecond light pulses. *IEEE J. Quantum Electron.*, QE-25:2464–2468, 1989.

[80] Th. Hofmann, K. Mossavi, F. K. Tittel, and G. Szabo. Spectrally compensated sum-frequency mixing scheme for generation of broadband radiation at 193 nm. *Optics Lett.*, 17:1691–1693, 1992.

[81] A. Stabinis, G. Valiulis, and E. A. Ibragimov. Effective sum frequency pulse compression in nonlinear crystals. *Optics Comm.*, 86:301–306, 1991.

[82] A. Umbrasas, J. C. Diels, G. Valiulis, J. Jacob, and A. Piskarskas. Generation of femtosecond pulses through second harmonic compression of the output of a Nd:YAG laser. *Opt. Lett.*, to be published, 1995.

[83] A. Mokhtari, L. Fini, and J. Chesnoy. Efficient frequency mixing by a cw femtosecond laser synchronously pumped by a frequency doubled Nd:Yag laser. *Opt. Commun.*, 61:421–424, 1987.

[84] T. Elsaesser and M. C. Nuss. Femtosecond pulses in the mid-infrared generated by down conversion of a traveling wave dye laser. *Opt. Lett.*, 16:411–413, 1991.

[85] T. M. Jedju and L. Rothberg. Tunable femtosecond radiation in the mid-infrared for time resolved absorption in semiconductors. *Appl. Opt.*, 27:615–618, 1988.

[86] R. Laenen, H. Graener, and A. Laubereau. Broadly tunable femtosecond pulses generated by optical parametric oscillation. *Opt. Lett.*, 15:971–973, 1990.

[87] D. C. Edelstein, E. S. Wachmann, and C. L. Tang. Broadly tunable high repetition rate femtosecond optical parametric oscillator. *Appl. Phys. Lett.*, 54:1728–1730, 1988.

[88] A. Jankauskas, A. Piskarskas, D. Podenas, A. Stabinis, and A. Umbrasas. New nonlinear optical methods of powerful femtosecond pulse generation and dynamic interferometry.

In Z. Rudzikas, A. Piskarskas, and R. Baltramiejunas, editors, *Proceedings of the V international symposium on Ultrafast Phenomena in Spectroscopy*, pages 22–32, Vilnius, 1987. World Scientific.

[89] R. R. Alfano, Q. X. Li, T. Jimbo, J. T. Manassah, and P. P. Ho. Induced spectral broadening of a weak ps pulse in glass produced by an intense ps pulse. *Opt. Lett.*, 14:626–628, 1986.

[90] W. Rudolph, J. Krueger, P. Heist, and W. Wilhelmi. Ultrafast wavelength shift of light induced by light. In *ICO-15*, page 1319. SPIE, 1990.

[91] D. Anderson and M. Lisak. Nonlinear asymmetric self phase modulation and self steepening of pulses in long waveguides. *Phys. Rev. A*, 27:1393–1398, 1983.

[92] W. H. Knox, R. L. Fork, M. C. Downer, R. H. Stolen, and C. V. Shank. Optical pulse compression to 8 fs at a 5 kHz repetition rate. *Appl. Phys. Lett.*, 46:1120–1121, 1985.

[93] J. E. Rothenberg and D. Grischkowsky. Observation of the formation of an optical intensity shock and wave breaking in the nonlinear

propagation of pulses in optical fibers. *Phys. Rev. Lett.*, 62:531–533, 1989.

[94] R. R. Alfano, editor. *The Supercontinuum Laser Source.* Springer, New York, Berlin, 1989.

[95] R. R. Alfano and S. L. Shapiro. Observation of self-phase modulation and small-scale filaments in crystals and glasses. *Phys. Rev. Lett.*, 24:592–594, 1970.

[96] P. Heist, D. Lap, F. Noack, W. Rudolph, and T. Schroeder. Generation of tunable fs pulses of high energy. In E. Klose and B. Wilhelmi, editors, *Ultrafast Phenomena in Spectroscopy*, pages 19–22. Springer, 1990.

[97] J. H. Glownia, J. Misewich, and P. P. Sorokin. Ultrafast ultraviolet pump probe apparatus. *J.Opt.Soc.Am.*, B3:1573–1579, 1986.

[98] E. Tokunaga, Q. Terasaki, and T. Kobayashi. Induced phase modulation of chirped continuum pulses studied with a femtosecond frequency-domain interferometer. *Optics Lett.*, 18:370–372, 1992.

[99] W. H. Knox, M. C. Downer, R. L. Fork, and C. V. Shank. Amplifier femtosecond optical pulses and continuum generation at 5 kHz repetition rate. *Opt. Lett.*, 9:552–554, 1984.

[100] R. L. Fork, C. H. Cruz, P. C. Becker, and C. V. Shank. Compression of optical pulses to six femtoseconds by using cubic phase compensation. *Opt. Lett.*, 12:483–485, 1987.

[101] J. H. Marburger. In J. H. Sanders and S. Stendholm, editors, *Progr. Quantum Electron.*, volume 4, page 35. 1977.

[102] J. E. Rothenberg. Space-time focusing: breakdown of the slowly varying envelope approximation in the self-focusing of femtosecond pulses. *Optics Lett.*, 17:1340–1342, 1992.

[103] I. P. Christov. Space-time focusing of femtosecond pulses in a Ti:sapphire laser. *Optics Lett.*, 20:309–311, 1995.

[104] J. E. Rothenberg. Pulse splitting during self-focusing in normally dispersive media. *Optics Lett.*, 17:583–586, 1992.

[105] P. Chernev and V. Petrov. Self focusing of light pulses in the presence of normal group-velocity dispersion. *Optics Letter*, 17:789–791, 1992.

[106] G. G. Luther, J. V. Moloney, A. C. Newell, and E. M. Wright. Self-focusing threshold in normally dispersive media. *Opt. Lett.*, 19:862–864, 1994.

[107] Y. Silberberg. Collapse of optical pulses. *Optics Letters*, 15:1282–1284, 1990.

[108] D. E. Edmundson and R. H. Enns. Robust bistable light bullets. *Opt. Lett.*, 17:586–588, 1992.

[109] C. Lecompte, G. Mainfray, C. Manus, and F. Sanchez. Laser temporal coherence effects on multiphoton ionization processes. *Phys. Rev. A*, A-11:1009, 1975.

[110] J.-L. De Bethune. Quantum correlation functions for fields with stationary mode. *Nuovo Cimento*, B 12:101–117, 1972.

[111] J.-C. Diels and J. Stone. Multiphoton ionization under sequential excitation by coherent pulses. *Phys. Rev. A*, A31:2397–2402, 1984.

[112] M. Lai and J.-C. Diels. Interference between spontaneous emission in different directions. *Am. Journal of Physics*, 58:928–930, 1990.

[113] M. Lai and J.-C. Diels. Ring-laser configuration with spontaneous noise reduced by destructive interference of the laser outputs. *Phys. Rev. A*, A42:536–542, 1990.

[114] L. Allen and J. H. Eberly. *Optical Resonances and Two-level Atoms*. John Wiley & Sons, 1975.

[115] F. Bloch. Magnetic resonances. *Phys. Rev.*, 70:460, 1946.

[116] R. P. Feynman, F. L. Vernon, and R. W. Hellwarth. Geometrical representation of the Schroedinger equation for solving maser problems. *J. Appl. Phys.*, 28:49–52, 1957.

[117] S. McCall and E. L. Hahn. Self-induced transparency. *Phys. Rev.*, 183:457, 1969.

[118] J.-C. Diels and E. L. Hahn. Carrier-frequency distance dependence of a pulse propagating in a two-level system. *Phys. Rev. A*, A-8:1084–1110, 1973.

[119] J.-C. Diels and E. L. Hahn. Phase modulation propagation effects in ruby. *Phys. Rev. A*, A-10:2501–2510, 1974.

[120] J.-C. Diels and E. L. Hahn. Pulse propagation stability in absorbing and amplifying media. *IEEE J. of Quantum Electronics*, QE-12:411–416, 1976.

[121] F. P. Mattar and M. C. Newstein. Transverse effects associated with the propagation of coherent optical pulses in resonant media. *IEEE J. of Quantum ELectronics*, QE-13:507, 1977.

[122] V. Petrov and W. Rudolph. The chirped steady state pulse in an amplifier with loss. *Physics Letters A*, 145:192–194, 1990.

[123] F. T. Arecchi and R. Bonifacio. Theory of optical maser amplifiers. *IEEE J. of Quantum ELectronics*, QE–1:169–178, 1965.

[124] J. A. Armstrong and E. Courtens. π pulses in homogeneously broadened amplifiers. *IEEE J. of Quantum ELectronics*, QE-4:411–416, 1968.

[125] C. Cohen-Tannoudji, B. Diu, and F. Laloe. *Quantum Mechanics*. John Wiley & Sons, New York, 1977.

[126] J-C. Diels. Efficient selective optical excitation for isotope separation using short laser pulses. *Phys. Rev. A*, 13:1520–1526, 1976.

[127] J.-C. Diels and S. Besnainou. Multiphoton coherent excitation of molecules. *J. Chem. Phys.*, 85:6347–6355, 1986.

[128] G. Herzberg. *Molecular Spectra and Molecular Structure: I. Spectra of Diatomic Molecules*. Van Nostrand, New York, 1950.

[129] C. H. Townes and A. L. Schawlow. *Microwave Spectroscopy*. McGraw-Hill, New York, 1955.

[130] Walter Gordy and Robert L. Cook. *Microwave Molecular Spectra*. John Wiley & Sons, New York, 1970.

[131] J. N. Franklin. *Matrix Theory*. Prentice Hall, Englewood Cliff, N. J., 1968.

[132] J.-C. Diels, J. Stone, S. Besnainou, M. Goodman, and E. Thiele. Probing the phase coherence time of multiphoton excited molecules. *Optics Communications*, 37:11–14, 1981.

[133] J.-C. Diels and A. T. Georges. Coherent two-photon resonant third and fifth harmonic VUV generation in metal vapors. *Phys. Rev. A*, 19:1589–1591, 1979.

[134] J.-C. Diels, N. Nandini, and A. Mukherjee. Multiphoton coherences. *Physica Acta*, T23:206–210, 1988.

[135] A. Mukherjee, N. Mukherjee, J.-C. Diels, and G. Arzumanyan. Coherent multiphoton resonant interaction and harmonic generation. In G. R. Fleming and A. E. Siegman, editors, *Picosecond Phenomena V*, pages 166–168, Berlin, 1986. Springer-Verlag.

[136] N. Mukherjee, A. Mukherjee, and J.-C. Diels. Four-photon coherent resonant propagation and transient wave mixing: Application to the mercury atom. *Phys. Rev. A*, A38:1990–2004, 1988.

[137] J.-C. Diels. Two-photon coherent propagation, transmission of 90^0 phase shifted pulses, and application to isotope separation. *Optics and Quantum Electronics*, 8:513, 1976.

[138] J.-C. Diels. Application of coherent interactions to isotope separation. In Plenum Publishing Co., editor, *Proceedings of the 4th Rochester conference on coherence and quantum optics*, page 707, 1978.

[139] J. E. Rothenberg. Self-induced heterodyne: the interaction of a frequency-swept pulse with a resonant system. *IEEE J. of Quantum Electron.*, QE-22:174–179, 1986.

[140] J. E. Rothenberg and D. Grischkowsky. Subpicosecond transient excitation of atomic vapor and the measurement of optical phase. *J. of Optical Soc. B*, 4:174–179, 1987.

[141] R. L. Fork and C. V. Shank. Generation of optical pulses shorter than 0.1 ps by colliding pulse mode-locking. *Appl. Phys. Lett.*, 38:671, 1981.

[142] A. Finch, G. Chen, W. Steat, and W. Sibbett. Pulse asymmetry in the colliding-pulse mode-locked dye laser. *J. Mod. Opt.*, 35:345–349, 1988.

[143] H. Kubota, K. Kurokawa, and M. Nakazawa. 29 fs pulse generation from a linear cavity synchronously pumped dye laser. *Opt.L ett.*, 13:749–751, 1988.

[144] Ch. Spielmann, F.Krausz, T. Brabec, E. Wintner, and A. J. Schmidt. Femtosecond pulse generation from a synchronously pumped ti;sapphire laser. *Opt. Lett.*, 16:1180, 1991.

[145] D. E. Spence, P. N. Kean, and W. Sibbett. 60-fs pulse generation from a self-mode-locked Ti:sapphire laser. *Optics Lett.*, 16:42–44, 1991.

[146] A. Stingl, Ch. Spielmann, F. Krausz, and R. Szipocs. Generation of 11 fs pulses from a Ti:sapphire laser without the use of prisms. *Optics Lett.*, 19:204–406, 1994.

[147] A. Stingl, M. Lenzner, Ch. Spielmann, F. Krausz, and R. Szipoecs. Sub-10-fs mirror-dispersion-controlled Ti:sapphire laser. *Optics Lett.*, 20:602–604, 1995.

[148] S. V. Chekalin, P. G. Kryukov, Yu. A. Matveetz, and O. B. Shatherashvili. The processes of formation of ultrashort laser pulses. *Opto-Electronics*, 6:249–261, 1974.

[149] J.-C. Diels. Femtosecond dye lasers. In F. Duarte and L. Hillman, editors, *Dye Laser Principles: With Applications*, chapter 3, pages 41–132. Academic Press, Boston, 1990.

[150] G. H. C. New. Mode-locking of quasi-continuous lasers. *Optics Comm.*, 6:188–193, 1972.

[151] G. H. C. New. Pulse evolution in mode-locked quasi-continuous lasers. *IEEE J. of Quantum Electron.*, QE-10:115–124, 1974.

[152] H. A. Haus. Theory of mode-locking with a slow saturable absorber. *IEEE J. of Quantum Electron.*, QE-11:736–746, 1975.

[153] J.-C. Diels, J. J. Fontaine, I. C. McMichael, W. Rudolph, and B. Wilhelmi. Experimental and theoretical study of a femtosecond laser. *Sov. J. Quantum electron.*, 13:1562–1569, 1983.

[154] W. Rudolph and B. Wilhelmi. Calculation of light pulses with chirp in passively mode-locked lasers taking into account the phase memory of absorber and amplifier. *Appl. Phys.*, B-35:37–44, 1984.

[155] J.-C. Diels, W. Dietel, J. J. Fontaine, W. Rudolph, and B. Wilhelmi. Analysis of a mode-locked ring laser: chirped-solitary-pulse solutions. *J. Opt. Soc. Am. B*, 2:680, 1985.

[156] H. A. Haus and Y. Silberberg. Laser mode-locking with addition of nonlinear index. *IEEE J. of Quantum Electron.*, QE-22:325–331, 1986.

[157] Xin Miao Zhao and J.-C. Diels. Stability study of fs dye laser with self-lensing effects. *J. Optical Society America B*, 10(7):1159–1165, 1993.

[158] J. L. A. Chilla and O. E. Martinez. Spatial-temporal analysis of the self-mode-locked Ti:sapphire laser. *Journal of the Optical Society B*, 10:638–643, 1993.

[159] U. Keller, G. W. 'tHooft, W. H. Knox, and J. E. Cuninham. Femtosecond pulses from a continously self-starting passively mode-locked Ti:sapphire laser. *Opt. Lett.*, 16:1022–1024, 1991.

[160] P. W. Smith, Y. Silberberg, and D. Q. B. Miller. Mode-locking of semiconductor diode lasers using saturable excitonic nonlinearities. *J. Opt. Soc. Am.*, B-2:1228–1235, 1985.

[161] U. Keller. Ultrafast all-solid-state technology. *Appl. Phys. B*, 58:347–363, 1994.

[162] L. R. Brovelli, U. Keller, and T. H. Chiu. Design and operation of antiresonant Fabry–Perot saturable semiconductor absorbers for mode-locked solid-state lasers. *J. Opt. Soc. Am. B*, 12:311–322, 1995.

[163] K. A. Stankov. A new mode-locking technique using a nonlinear mirror. *Opt. Commun.*, 66:41, 1988.

[164] K. A. Stankov. Pulse shortening by a nonlinear mirror mode locker. *Appl. Opt.*, 28:942–945, 1989.

[165] J.-C. Diels, P. Dorn, M. Lai, W. Rudolph, and X. M. Zhao. Femtosecond sagnac interferometry. In J.-L. Martin, A. Migus, G. A. Mourou, and A. H. Zewail, editors, *Ultrafast Phenomena VIII*, pages 120–123, Antibes, France, 1992. Springer Verlag, Berlin.

[166] I. S. Ruddock and D. J. Bradley. Bandwidth-limited subpicosecond pulse generation in mode-locked cw dye lasers. *Appl. Phys. Lett.*, 29:296, 1976.

[167] A. E. Siegman. An antiresonant ring interferometer for coupled laser cavities, laser output coupling, mode-locking, and cavity dumping. *IEEE J. Quantum Electron.*, QE-9:247–250, 1973.

[168] N. Jamasbi, J.-C. Diels, and L. Sarger. Study of a linear femtosecond laser in passive and hybrid operation. *J. of Modern Optics*, 35:1891–1906, 1988.

[169] J. A. Valdmanis, R. L. Fork, and J. P. Gordon. Generation of optical pulses as short as 27 fs directly from a laser balancing self-phase modulation, group velocity dispersion, saturable absorption, and saturable gain. *Opt. Lett.*, 10:131–133, 1985.

[170] J.-C. Diels and I. C. McMichael. Degenerate four-wave mixing of femtosecond pulses in an absorbing dye jet. *Journal of the Opt. Soc. of Am.*, B3:535–543, 1986.

[171] Ming Lai, Jean-Claude Diels, and Michael Dennis. Nonreciprocal measurements in fs ring lasers. *Optics Letters*, 17:1535–1537, 1992.

[172] J.-C. Diels, P. Dorn, M. Lai, W. Rudolph, and X. M. Zhao. Femtosecond Sagnac interferometry. In J.-L. Martin, A. Migus, G. A. Mourou, and A. H. Zewail, editors, *Ultrafast Phenomena VIII*, pages 120–123, Antibes, France, 1992. Springer Verlag, Berlin.

[173] M. L. Dennis, J.-C. Diels, and M. Lai. The femtosecond ring dye laser: a potential new laser gyro. *Optics Letters*, 16:529–531, 1991.

[174] W. H. Knox. Femtosecond intracavity dispersion measurements. In J. L. Martin, A. Migus, G. A. Mourou, and A. H. Zewail, editors, *Ultrafast Phenomena VIII*, pages 192–193, Berlin, 1992. Springer.

[175] J. P. Gordon and R. L. Fork. Optical resonator with negative dispersion. *Opt. Lett.*, 9:153–156, 1984.

[176] M. Yamashita, K. Torizuka, and T. Sato. Intracavity femtosecond pulse compression with the addition of highly nonlinear organic material. *Optics Lett.*, pages 24–26, 1988.

[177] J. Kafka and T. Baer. Multimode mode-locking of solid state lasers. In *1991 Technical Digest Series: CLEO'91*, volume 10, page 2, Baltimore, 1990. Optical Society of America.

[178] H. W. Kogelnik and T. Li. Laser beams and resonators. *Appl. Opt.*, 5:1550–1567, 1966.

[179] A. Agnesi and G. C. Reali. Analysis of unidirectional operation of Kerr lens mode-locked ring oscillators. *Optics Comm.*, 110:109–114, 1994.

[180] E. P. Ippen, L. Y. Liu, and H. A. Haus. Self-starting condition for additive-pulse mode-locked lasers. *Opt. Lett.*, 15:183–185, 1990.

[181] J. Goodberlet, J. Wang, J. G. Fujimoto, and P. A. Schulz. Starting dynamics of additive-pulse mode locking in the Ti:Al_2O_3 laser. *Opt. Lett.*, 15:1300–1302, 1990.

[182] Jyhpying Wang. Analysis of passive additive pulse mode locking with eigenmode theory. *IEEE J. of Quantum Electron.*, 28:562–568, 1992.

[183] F. Krausz, T. Brabec, and Ch. Spielmann. Self-starting passive modelocking. *Optics Letters*, 16:235–238, 1991.

[184] K. Tamura, J. Jacobson, E. P. Ippen, H. A. Haus, and J. G. Fujimoto. Unidirectional ring resonators for self-starting passively mode-locked lasers. *Optics Lett.*, 18:220–222, 1993.

[185] V. Petrov, W. Rudolph, U. Stamm, and B. Wilhelmi. Limits of ultrashort pulse generation in cw mode-locked dye lasers. *Physical Review A*, 40:1474–1483, 1989.

[186] H. Avramopoulos, P. M. W. French, J. A. R. Williams, G. H. C. New, and J. R. Taylor. Experimental and theoretica studies of complex pulse evolution in a passively mode locked ring dye laser. *IEEE J. of Quantum Electron.*, 24:1884–1890, 1988.

[187] D. Kuehlke, W. Rudolph, and B. Wilhelmi. Influence of transient absorber grating on the pulse parameters of passively mode-locked ring lasers. *Appl. Phys. Lett.*, 42:325–327, 1983.

[188] D. Kuehlke and W. Rudolph. On downchirped pulses from a cw ring dye laser in the CPM regime. *Optical and Quantum Electron.*, 16:57–64, 1984.

[189] O. E. Martinez, R. L. Fork, and J. P. Gordon. Theory of passively mode locked lasers for the case of a nonlinear complex propagation coefficient. *J.Opt.Soc.Am.*, B2:753–760, 1985.

[190] T. R. Zhang, G. Focht, P. E. Williams, and M. C. Downer. Theory of intracavity frequency doubling in passively modelocked femtosecond lasers. *IEEE J.Quant.Electron.*, QE-24:1877–1883, 1988.

[191] H. A. Haus and E. P. Ippen. Self-starting of passively modelocked lasers. *Optics Lett.*, 16:1331–1333, 1991.

[192] V. Petrov, W. Rudolph, and B. Wilhelmi. Computer simulation of passive mode locking of dye lasers with consideration of coherent light-matter interaction. *Optical and Quantum Electron.*, 19:377–384, 1987.

[193] M. C. Downer, R. L. Fork, and M. Islam. 3 MHz amplifier for femtosecond optical pulses. In D. H. Auston and K. B. Eisenthal, editors, *Ultrafast Phenomena IV*, pages 27–29, Berlin, Heidelberg, 1984. Springer Verlag.

[194] V. Petrov, W. Rudolph, and B. Wilhelmi. Evolution of chirped light pulses and the steady-state regime in passively mode-locked femtosecond dye lasers. *Rev. Phys. Appl.*, 22:1639–1650, 1987.

[195] M. Beck and I. A. Walmsley. The role of amplitude and phase shaping in the dispersive pulse regime of passively mode locked dye lasers. *IEEE J. Quantum Electron.*, QE-28:2274–2284, 1992.

[196] H. Avramopoulos and R. L. Fork. Bandwidth limitations and distinct operating regimes of passively mode locked dye lasers using strong phase shaping. *J. Opt. Soc. Am.*, B8:119–122, 1991.

[197] E. W. Van Stryland. The effect of pulse to pulse variation on ultrashort pulsewidth measurements. *Opt. Comm.*, 31:93–94, 1979.

[198] J.-C. Diels, J. J. Fontaine, I. C. McMichael, and F. Simoni. Control and measurement of ultrashort pulse shapes (in amplitude and phase) with femtosecond accuracy. *Applied Optics*, 24:1270–1282, 1985.

[199] G. H. C. New and J. M. Catherall. Advances in the theory of mode-locking by synchronous pumping. In *Ultrafast Phenomena V*, pages 24–26, Berlin, 1986. Springer Verlag.

[200] U. Stamm. Numerical analysis of pulse generation in synchronously mode locked cw dye lasers. *Appl. Phys. B*, B45:101–108, 1988.

[201] V. A. Nekhaenko, S. M. Pershin, and A. A. Podshivalov. Synchronously pumped tunable picosecond lasers (review). *Sov. J. Quantum Electron.*, 16:299–315, 1986.

[202] N. J. Frigo, H. Mahr, and T. Daly. A study of forced mode locked cw dye lasers. *IEEE J. of Quantum Electron.*, QE-17:1134, 1977.

[203] A. M. Johnson and W. M. Simpson. Tunable femtosecond dye laser synchronously pumped by the compressed second harmonic of Nd:YAG. *J. Opt. Soc. Am. B*, B4:619–125, 1985.

[204] G. Angel, R. Gagel, and A. Laubereau. Generation of femtosecond pulses by a pulsed laser system. *Optics Comm.*, 63:259–263, 1987.

[205] L. Turi and F. Krausz. Amplitude modulation mode locking of lasers with regenerative feedback. *Appl. Phys. Lett.*, 58:810, 1991.

[206] J. M. Catherall and G. H. C. New. Role of spontaneous emission in the dynamics of mode-locking by synchronous pumping. *IEEE Journal of Quantum Electronics*, QE-22:1593, 1986.

[207] D. S. Peter, P. Beaud, W. Hodel, and H. P. Weber. Passive stabilazation of a synchronously pumped mode locked dye laser with the use of a modified outcoupling mirror. *Optics Lett.*, 16:405, 1991.

[208] J. P. Ryan, L. S. Goldberg, and D. J. Bradley. Comparison of synchronous pumping and passive mode-locking cw dye lasers for the generation of picosecond and subpicosecond light pulses. *Opt. Commun.*, 27:127–132, 1978.

[209] B. Couillaud, V. Fossati-Bellani, and G. Mitchel. Ultrashort pulse spectroscopy and applications. In *SPIE Proceedings*, volume 533, page 46. Springer-Verlag, 1985.

[210] M. C. Nuss, R. Leonhart, and W. Zinth. Stable operation of a synchronously pumped colliding-pulse mode-locked ring dye laser. *Opt. Lett.*, 10:16–18, 1985.

[211] Z. Bor, K. Osvay, H. A. Hazim, A. Kovacs, G. Szabo, B. Racz, and O. E. Martinez. Adjustable prism compressor with constant transit

time for synchronously pumped mode locked laser. *Optics Comm.*, 90:70–79, 1992.

[212] V. E. Zakharov and A. B. Shabat. Exact theory of two-dimensional self-focusing and one-dimensional self-modulation of waves in nonlinear media. *Sov. Phys. JETP*, 34:62–69, 1972.

[213] C. S. Gardner, G. Green, M. Druskal, and R. Miura. Method for solving the Korteweg-deVries equation. *Phys. Rev. Lett.*, 19:1095, 1967.

[214] J.-C. Diels and H. Sallaba. Black magic with red dyes. *J. Opt. Soc. Am.*, 70:629, 1980.

[215] J.-C. Diels, J. Menders, and H. Sallaba. Generation of coherent pulses of 60 optical cycles through synchronization of the relaxation oscillations of a mode-locked dye laser. In R. M. Hochstrasser, W. Kaiser, and C. V. Shank, editors, *Picosecond Phenomena II*, page 41, Berlin, 1980. Springer-Verlag.

[216] F. Salin, P. Grangier, G. Roger, and A. Brun. Observation of high-order solitons directly produced by a femtosecond ring laser. *Phys. Rev. Lett.*, 56:11321135, 1986.

[217] T. Tsang. Observation of higher order solitons from a mode-locked Ti:sapphire laser. *Optics Lett.*, 18:293–295, 1993.

[218] Y. Ishida, N. Sarukura, and H. Nakano. Soliton like pulse shaping in a passively mode locked Ti:sapphire laser. In C. B. Harris, E. P. Ippen, G. A. Mourou, and A. H. Zewail, editors, *Ultrafast Phenomena VII*, pages 75–77. Springer-Verlag, 1990.

[219] A. V. Babushkin, N. S. Vorob'ev, A. M. Prokhorov, and M. Ya. Shchelev. Stable picosecond $YAlO_3$:Nd crystal laser with hybrid mode-locking and passive intracavity feedback utilizing a GaAs crystal. *Sov. J. Quantum Electron.*, 19:1310–1311, 1989.

[220] O. E. Martinez and L. A. Spinelli. Deterministic passive mode locking of solid-state lasers. *Appl. Phys. Lett.*, 39:875–877, 1981.

[221] K. Burneika, R. Grigonis, A. Piskarskas, G. Sinkyavichyus, and V. Sirutkaĭtis. Highly stable subpicosecond neodymium (Nd^{3+}) glass laser with passive mode-locking and negative feedback. *Kvantovaya Elektron.*, 15:1658–1659, 1988.

[222] P. Heinz, W. Kriegleder, and A. Laubereau. Feedback control of an actively-passively mode-locked Nd:glass laser. *Applied Physics A*, 43:209–212, 1987.

[223] J. Schwartz, W. Weiler, and R. K. Chang. Laser-pulse shaping using inducible absorption with controlled q-switch time behavior. *IEEE J. of Quantum Electronics*, QE-6:442–450, 1970.

[224] A. Hordvik. Pulse stretching utilizing two-photon-induced light absorption. *IEEE J. of Quantum Electronics*, QE-6:199–203, 1970.

[225] A. Agnesi, A. Del Corno, J.-C. Diels, P. Di Trapani, M. Fogliani, V. Kubecek, G. C. Reali, C.-Y. Yeh, and Xin Miao Zhao. Generation of extended pulse trains of minimum duration by passive negative feedback applied to solid state q-switched lasers. *IEEE J. of Quantum Electron.*, 28:710–719, 1992.

[226] D. H. Auston, S. McAfee, C. V. Shank, E. P. Ippen, and O. Teschke. Picosecond spectroscopy of semiconductors. *Solid State Electronics*, 21:147, 1978.

[227] M. Sheik-Bahae, D. J. Hagan, and E. W. Van Stryland. Dispersion and band-gap scaling of the electronic kerr effect in solids associated with two-photon absorption. *Phys. Rev. Lett*, 65:96–99, 1990.

[228] W. A. Schroeder, T. S. Stark, M. D. Dawson, T. F. Boggess, A. L. Smirl, and G. C. Valley. Picosecond separation and measurement of coexisting photo-refractive, bound-electronic, and free-carrier grating dynamics in GaAs. *Optics Lett.*, 16:159–161, 1991.

[229] L. C. Foster, M. D. Ewy, and C. B. Crumly. Laser mode locking by an external Doppler cell. *Appl. Phys. Lett.*, 6, 1965.

[230] R. W. Dunn, A. T. Hendow, J. G. Small, and E. Stijns. Gas laser mode-locking using an external acoustooptic modulator with a potential application to passive ring gyroscopes. *Applied Optics*, 21:3984–3986, 1982.

[231] K. J. Blow and D. Wood. mode-locked lasers with nonlinear external cavities. *J. Opt. Soc. Am. B*, 5:629–632, 1988.

[232] K. J. Blow and B. P. Nelson. Improved mode locking of an f-center laser with a nonlinear nonsoliton external cavity. *Opt. Lett.*, 13:1026–1028, 1988.

[233] P. N. Kean, X. Zhu, D. W. Crust, R. S. Grant, N. Langford, and W. Sibbett. Enhanced mode locking of color-center lasers by coupled-cavity feedback control. *Opt. Lett.*, 14:39, 1989.

[234] J. Mark, L. Y.Liu, K. L. Hall, H. A. Haus, and E. P. Ippen. Femtosecond laser pumped by a frequency-doubled diode-pumped Nd:YLF laser. *Opt. Lett.*, 14:48–50, 1989.

[235] C. P. Yakymyshyn, J. F. Pinto, and C. R. Pollock. Additive-pulse mode-locked NaCl:OH$^-$ laser. *Opt. Lett.*, 14:621–623, 1989.

[236] P. M. French, J. A. R. Williams, and R. Taylor. Femtosecond pulse generation from a titanium-doped sapphire laser using nonlinear external cavity feedback. *Opt. Lett.*, 14:686, 1989.

[237] J. Goodberlet, J. Wang, J. G. Fujimoto, and P. A. Shulz. Femtosecond passively mode-locked Ti:sapphire laser with a nonlinear external cavity. *Optics Lett.*, 14:1125–1127, 1989.

[238] J. Goodberlet, J. Jacobson, J. G. Fujimoto, P. A. Schulz, and T. Y. Fan. Self-starting additive-pulse mode-locked diode-pumped Nd:YAG laser. *Opt. Lett.*, 15:504–507, 1990.

[239] L. Y. Liu, J. M. Huxley, E. P. Ippen, and H. A. Haus. Self-starting additive-pulse mode-locking of a Nd:YAG laser. *Optics Lett.*, 15, 1990.

[240] J. M. Liu and J. K. Chee. Passive mode locking of a cw Nd:YLF laser with a nonlinear external coupled cavity. *Opt. Lett.*, 15:685, 1990.

[241] J. Goodberlet, J. Jackson, J. G. Fujimoto, and P. A. Schultz. Ultrashort pulse generation with additive pulse modelocking in solid state lasers: Ti:Al$_2$O$_3$, diode pumped Nd:YAG and Nd:YLF. In C. B. Harris, E. P. Ippen, G. A. Mourou, and A. H. Zewail, editors, *Ultrafast Phenomena VII*, page 11, Berlin, 1990. Springer-Verlag.

[242] F. Krausz, Ch. Spielmann, T. Brabec, E. Wintner, and A. J. Schmidt. Subpicosecond pulse generation from a Nd:glass laser using a nonlinear external cavity. *Opt. Lett.*, 15:737–739, 1990.

[243] W. Sibbett. Hybrid and passive mode-locking in coupled-cavity lasers. In C. B. Harris, E. P. Ippen, G. A. Mourou, and A. H. Zewail, editors, *Ultrafast Phenomena VII*, pages 2–7, Berlin, 1990. Springer-Verlag.

[244] E. P. Ippen, H. A. Haus, and L. Y. Liu. Additive pulse mode locking. *J. Opt. Soc. Am. B*, 6:1736, 1989.

[245] L. F. Mollenauer and R. H. Stolen. The soliton laser. *Opt. Lett.*, 9:13, 1984.

[246] D. E. Spence, P. N. Kean, and W. Sibbett. Sub-100fs pulse generation from a self-modelocked titanium-sapphire laser. In *Postdeadline papers of the Conference on Lasers and Electro-Optics CLEO'90*, pages 619–620, Baltimore, VA, 1990.

[247] H. W. Kogelnik, E. P. Ippen, A. Dienes, and C. V. Shank. Astigmatically compensated cavities for cw dye lasers. *IEEE Journal of Quantum Electron.*, QE-8:373–379, 1972.

[248] R. Danielius, A. Piskarskas, A. Stabinis, G. Banfi, P. di Trapani, and R. Righini. Travelling wave parametric generation of widely tunable, highly coherent femtosecond light pulses. *Journal of the Opt. Soc. Amer. B*, 10:2222–2232, 1993.

[249] T. Nishikawa, N. Uesugi, and J. Yumoto. Parametric superfluorescence in $KTiOPO_4$ pumped by 1 ps pulses. *Appl Phys. Lett.*, 58:1943–1945, 1991.

[250] K. Kato and M. Masutani. Widely tunable 90^0 phase-matched KTP parametric oscillator. *Optics Letters*, 17:178–179, 1992.

[251] P. E. Powers, R. J. Ellingson, W. S. Pelouch, and C. L. Tang. Recent advances of the Ti:sapphire-pumped high repetition rate femtosecond optical parametric oscillator. *J. of the Opt. Soc. Am. B*, 10:2162–2167, 1993.

[252] P. E. Powers, R. J. Ellingson, W. S. Pelouch, and C. L. Tang. Recent advances of the ti:sapphire-pumped high-repetition-rate femtosecond optical parametric oscillator. *J. Opt.Soc. Am. B*, 10:2162–2167, 1993.

[253] E. S. Wachmann, D. C. Edelstein, and C. L. Tang. Continuous- wave mode-locked and dispersion compensated fs optical parametric oscillator. *Optics Lett.*, 15:136–139, 1990.

[254] W. S. Pelouch, P. E. Powers, and C. L. Tang. Ti:sapphire pumped, high repetition rate, femtosecond optical parametric oscillator. *Optics Lett.*, 17:1070–1072, 1992.

[255] P. Powers, S. Ramakrishna, C. L. Tang, and K. L. Cheng. Optical parametric oscillation with $KTiOAsO_4$. *Optics Lett.*, 18:1171–1173, 1993.

[256] A. Nebel, C. Fallnich, and R. Beigang. Efficient noncritically phase-matched mode-locked ps and fs optical parametric oscillator in ktp. In *CLEO, JWB5*, volume 11, page 282, Baltimore, MD, 1993. Optical Society of America.

[257] Q. Fu, G. Mak, and H. M. van Driel. High power, 62 fs infrared optical parametric oscillator synchronously pumped by a 76-MHz Ti:sapphire laser. *Opt. Lett.*, 17:1006, 1992.

[258] E. Gaizauskas, A. Piskarskas, A. Stabinis, and K. Staliunas. Theoretical analysis of the generation of femtosecond pulses in synchronously pumped parametric oscillators for the optical range. *Lit. Fiz. Sb. (Soviet Physics – Collection)*, 28:516–519, 1988.

[259] E. Gaizauskas, A. Piskarskas, and K. Staliunas. Generation of femtosecond pulses in mode-locked optical parametric oscillators. *Lit. Fiz. Sb. (Soviet Physics – Collection)*, 28:75–79, 1988.

[260] J. D. V. Khaydarov, J. H. Andrews, and K. D. Singer. Pulse compression in a synchronously pumped optical parametric oscillator from group-velocity mismatch. *Opt. Lett.*, 19:831–833, 1994.

[261] A. Umbrasas, J. C. Diels, J. Jacob, and A. Piskarskas. Parametric oscillation and compression in KTP crystals. *Opt. Lett.*, 19:1753–1755, 1994.

[262] Y. Wang and B. Luther Davies. Frequency-doubling pulse compressor for picosecond high power neodymium laser pulses. *Opt. Lett.*, 17:1459–1461, 1992.

[263] P. Heinz, A. Laubereau, A. Dubietis, and A. Piskarskas. Fiberless two-step parametric compression of sub-picosecond laser pulses. *Lithuanian Phys. Rev.*, 33:314–317, 1993.

[264] J. P. Heritage and R. K. Jain. Subpicosecond pulses from a cw. mode locked dye laser. *Appl. Phys. Lett.*, 32:101, 1978.

[265] W. H. Knox and F. A. Beisser. Two-wavelength synchronous generation of femtosecond pulses with less than 100 fs jitter. *Optics Lett.*, 17:1012–1014, 1992.

[266] D. R. Dykaar and S. B. Darack. Cross-mode locking: two-color femtosecond operation of one Ti:sapphire laser. *Opt. Lett.*, 18:634–636, 1993.

[267] M. R. X. de Barros and P. C. Becker. Two-color synchronously mode-locked femtosecond Ti:sapphire laser. *Opt. Lett.*, 18:631–633, 1993.

[268] J. M. Evans, D. E. Spence, D. Burns, and W. Sibbett. Dual-wavelength self-mode-locked Ti:sapphire laser. *Opt. Lett.*, 18:1074–1076, 1993.

[269] J. D. Kafka, J. W. Pieterse, and M. L. Watts. Two-color subpicosecond optical sampling technique. *Optics Letters*, 17:1286–1288, 1992.

[270] N. Luo, F. Ratsavong, and G. W. Robinson. Dual Ti:sapphire laser system for ultrafast pump/probe studies. OSA Annual Meeting, paper *ThY5*, 1992.

[271] Zs. Bor. A novel pumping arrangement for tunable single picosecond pulse generation with a N_2 laser pumped distributed feedback dye laser. *Optics Comm*, 29:103–108, 1979.

[272] F. P. Schaefer. New methods for the generation of ultrashort laser pulses. In H. Walther and K. W. Rothe, editors, *Laser Spectroscopy IV, Proceedings of the fourth Int. Conf. Rottach-Egern*, pages 590–596, Berlin, 1979. Springer-Verlag.

[273] Zs. Bor and A. Muller. Picosecond distributed feedback dye lasers. *IEEE Journal of Quantum Electronics*, QE-22:1524–1533, 1986.

[274] Zs. Bor and F. P. Schaefer. New single-pulse generation technique for distributed feedback dye lasers. *Applied Physics B*, B31:209–213, 1983.

[275] S. Szatmari and F. P. Schaefer. Subpicosecond, widely tunable distributed feedback dye laser. *Appl. Phys.*, B46:305–311, 1988.

[276] P. H. Chin, P. Pex, L. Marshall, F. Wilson, and R. Aubert. A high energy, electromagnetically tuned single mode, short cavity picosecond dye laser system: design and performances. *Optics and Lasers in Engineering*, 10:55–68, 1989.

[277] H. P. Kurz, A. J. Cox, G. W. Scott, D. M. Guthals, H. Nathel, S. W. Yeh, S. P. Webb, and J. H. Clark. Amplification of tunable, picosecond pulses from a single mode, short cavity dye laser. *IEEE J. of Quantum Electron.*, QE-21:1795–1798, 1985.

[278] N. D. Hung and Y. H. Meyer. Tunable sub-100 fs dye laser pulses generated with a nanosecond pulse pumping. *Appl. Phys.*, B55:409–412, 1992.

[279] P. Simon, S. Szatmari, and F. P. Schafer. Generation of 30 fs pulses tunable over the visible spectrum. *Opt. Lett.*, 16:1569–1571, 1991.

[280] P. P. Vasil'ev. Ultrashort pulse generation in diode lasers. *Optical and Quantum ELectron.*, 24:801–824, 1992.

[281] G. A. Alphonse, D. B. Gilbert, M. G. Harvey, and M. Ettenberg. High power superluminescent diodes. *IEEE J. of Quantum Electron.*, 24:2454–2458, 1988.

[282] P. J. Delfyett, L. Flores, N. Stoffel, T. Gmitter, N. Andreadakis, G. Alphonse, and W. Ceislik. 200 femtosecond optical pulse generation and intracavity pulse evolution in a hybrid modelocked semiconductor diode laser- amplifier system. *Optics Lett.*, 17:670–672, 1992.

[283] P. J. Delfyett, C. H. Lee, L. Flores, N. Stoffel, T. Gmitter, N. Andreadakis, G. Alphonse, and J. C. Connolly. Generation of subpicosecond high-power optical pulses from a hybrid mode-locked semiconductor laser. *Optics Lett.*, 15:1371–1373, 1990.

[284] D. S. Chemla and D. A. B. Miller. Room-temperature excitonic nonlinear-optical effect in semiconductor quantum-well structures. *J. Opt. Soc. Am.*, B-2:1155–1173, 1985.

[285] S. Gupta, J. F. Whitaker, and G. A. Mourou. Ultrafast carrier dynamics in III-V semiconductors grown by molecular-beam epitaxy at very low substrate temperatures. *IEEE J. of Quantum Electron.*, 28:2464–2472, 1992.

[286] Y. K. Chen, M. C. Wu, T. Tanbun-Ek, R. A. Logan, and M. A. Chin. Subpicosecond monolithic colliding-pulse mode-locked muliple quantum well lasers. *Appl. Phys. Lett.*, 58:1253–1255, 1991.

[287] C. F. Lin and C. L. Tang. Colliding pulse mode-locking of a semiconductor laser in an external ring cavity. *Appl. Phys. Lett.*, 62:1053–1055, 1993.

[288] A. S. Gouveia-Neto, A. S. L. Gomes, and J. B. Taylor. Generation of 33 fs pulses at 1.32 μm through a high order soliton effect in a single mode optical fiber. *Opt. Lett.*, 12:395–397, 1987.

[289] W. Rudolph and B. Wilhelmi. *Light Pulse Compression*. Harwood Academic, Chur, London, Paris, New York, 1989.

[290] B. Zysset, P. Beaud, and W. Hodel. Generation of optical solitons in the wavelength region 1.37—1.43 μm. *Appl. Phys. Lett.*, 50:1027–1029, 1987.

[291] J. D. Kafka and T. Baer. Fiber raman soliton laser pumped by a Nd:YAG laser. *Optics Lett.*, 12:181–183, 1987.

[292] I. N. Duling III. Subpicosecond all-fiber erbium laser. *Electron. Lett.*, 27:544–545, 1991.

[293] I. N. Duling. All fiber ring soliton laser mode locked with a nonlinear mirror. *Optics Lett.*, 16:539–541, 1991.

[294] S. P. Craig-Ryan and B. J. Ainslie. Glass structure and fabrication techniques. In P. W. France, editor, *Optical Fiber Lasers and Amplifiers*, pages 50–78. Blackies & Son, Ltd., Glasgow, 1991.

[295] I. N. Duling, R. P. Moeller, W. K. Burns, C. A. Villaruel, L. Goldberg, E. Snitzer, and H. Po. Output characteristics of diode pumped fiber ase sources. *IEEE J. Quantum Electron.*, 27:995–1003, 1991.

[296] N. J. Doran and D. Wood. Nonlinear-optical loop mirror. *Opt. Lett.*, 13:56–58, 1988.

[297] M. Hofer, M. E. Fermann, F. Haberl, M. H. Ober, and A. J. Schmidt. Mode locking with cross-phase and self-phase modulation. *Opt. Lett.*, 16:502–504, 1991.

[298] M. E. Fermann, F. Haberl, M. Hofer, and H. Hochreiter. Nonlinear amplifying loop mirror. *Opt. Lett.*, 15:752–754, 1990.

[299] M. H. Ober, F. Haberl, and M. E. Fermann. 100 fs pulse generation from an all-solid-state Nd:glass fiber laser oscillator. *Appl. Phys. Lett.*, 60:2177–2179, 1992.

[300] M. L. Dennis and I. N. Duling. Role of dispersion in limiting pulse width in fiber lasers. *Appl. Phys. Lett.*, 62:2911–2913, 1993.

[301] M. L. Dennis and I. N. Duling. Experimental study of sideband generation in femtosecond fiber lasers. *IEEE J. Quantum Electron.*, 30:1469–1477, 1993.

[302] M. Nakazawa, E. Yoshida, and Y. Kimura. Generation of 98 fs optical pulses directly from an erbium-doped fibre ring laser at 1.57 μm. *Electron. Lett.*, 29:63–65, 1993.

[303] M. L. Dennis and I. N. Duling. Third order dispersion effects in femtosecond fiber lasers. *Opt. Lett.*, 19:1750–1752, 1993.

[304] P. Heist, W. Rudolph, and B. Wilhelmi. Amplification of femtosecond light pulses. In W. Koehler, editor, *Laser Handbook vol. 2*, pages 62–73. Vulkan Verlag, 1991.

[305] J. D. Simon. Ultrashort light pulses. *Rev. Sci. Instrum.*, 60:3597–3620, 1989.

[306] W. H. Knox. Femtosecond optical pulse amplification. *IEEE J. of Quantum Electron.*, 24:388–397, 1988.

[307] S. Szatmari, F. P. Schaefer, E. Mueller-Horsche, and W. Mueckenheim. Hybrid dye-excimer laser system for the generation of 80 fs pulses at 248 nm. *Opt. Commun.*, 63:5–10, 1987.

[308] P. Maine, D. Strickland, P. Bado, M. Pessot, and G. Mourou. Generation of ultrahigh peak power pulses by chirped pulse amplification. *IEEE J. Quantum Electron.*, QE-24:398–403, 1988.

[309] J. H. Glownia, J. Misewich, and P. P. Sorokin. 160 fs XeCl excimer amplifier system. *J. Opt. Soc. Amer. B*, 4:1061, 1987.

[310] S. Szatmari, B. Racz, and F. P. Schaefer. Bandwidth limited amplification of 220 fs pulses in XeCl. *Opt. Commun.*, 62:271–276, 1987.

[311] A. Penzkofer and W. Falkenstein. Theoretical investigation of amplified spontaneous emission with picosecond light pulses in dye solutions. *Opt. Quant. Electr.*, 10:399–405, 1978.

[312] A. Hnilo and O. E. Martinez. On the design of pulsed dye laser amplifiers. *IEEE J.Quantum Electron.*, QE-23:593–599, 1987.

[313] G. P. Agrawal and N. A. Olsson. Self-phase modulation and spectral broadening of optical pulses in semiconductor laser amplifiers. *IEEE J. Quant. Electr.*, QE-25:2297–2306, 1989.

[314] W. Dietel, E. Doepel, V. Petrov, C. Rempel, W. Rudolph, B. Wilhelmi, G. Marowsky, and F. P. Schaefer. Self phase modulation in a femtosecond dye amplifier with subsequent compression. *Appl. Phys.B*, 46:183–185, 1988.

[315] P. Heist, W. Rudolph, and V. Petrov. Combined self-phase modulation and amplification of femtosecond light pulses. *Appl. Phys. B*, 49:113–119, 1989.

[316] Y. Shimoji, A. T. Fay, R. S. F. Chang, and N. Djeu. Direct measurement of the nonlinear refractive index of air. *J. Opt. Soc. Am. B*, 6(11):1994–1998, 1989.

[317] D. C. Hutchings, M. Sheik-Bahae, D. J. Hagan, and E. W. Van Stryland. Kramers-kronig relations in nonlinear optics. *Opt. and Quant. Electr.*, 24:1–30, 1992.

[318] M. Sheik-Bahae, A. A. Said, and E. W. Van Stryland. High sensitivity single beam n_2 measurement. *Opt. Lett.*, 14:955, 1989.

[319] Ming Lai, J.-C. Diels, and Chi Yan. A transversally pumped 11-pass amplifier for femtosecond optical pulses. *Applied Optics*, 30:4365–4367, 1991.

[320] T. Damm, M. Kaschke, F. Noack, and B. Wilhelmi. Compression of ps pulses from a solid-state laser using self-phase modulation in graded index fibers. *Opt. Lett.*, 10:176–178, 1985.

[321] G. Boyer, M. Franco, J. P. Chambaret, A. Migus, A. Antonetti, P. Georges, F. Salin, and A. Brun. Generation of 0.6mj pulses of 16fs duration through high-repetition rate amplification of self-phase modulated pulses. *Appl.Phys.Lett.*, 53:823–825, 1988.

[322] S. Watanabe, A. Endoh, M. Watanabe, and N. Sakura. Terawatt XeCl discharge system. *Opt. Lett.*, 13:580–582, 1988.

[323] A. J. Taylor, T. R. Gosnell, and J. P. Roberts. Ultrashort pulse energy extraction measurements in XeCl amplifiers. *Opt. Lett.*, 15:39–41, 1990.

[324] P. Heist, W. Rudolph, and B. Wilhelmi. Amplification of femtosecond light pulses. *Exper. Tech. Phys.*, 38:163–188, 1990.

[325] K. Mossavi, Th. Hoffmann, and F. K. Tittel. Ultrahigh-brightness, femtosecond ArF excimer laser system. *Appl. Phys. Lett.*, 62:1203–1205, 1993.

[326] R. L. Fork, C. V. Shank, and R. Yen. Amplification of 70 fs optical pulse to gigawatt powers. *Appl. Phys. Lett.*, 41:223–225, 1982.

[327] C. Rolland and P. B. Corkum. Amplification of 70 fs pulses in a high repetition rate XeCl pumped dye laser amplifier. *Opt. Commun.*, 59:64–67, 1986.

[328] T. Turner, M. Chatalet, D. S. Moore, and S. C. Schmidt. Large gain amplifier for subpicosecond optical pulses. *Optics Lett.*, 11:357–359, 1986.

[329] D. B. McDonald and C. D. Jonah. Amplification of ps pulses using a copper vapor laser. *Rev.Sci.Instrum.*, 55:1166, 1984.

[330] D. Nickel, D. Kuehlke, and D. von der Linde. Multipass dye cell amplifier for high repetition rate optical pulses. *Opt. Lett.*, 14:36–39, 1989.

[331] E. V. Koroshilov, I. V. Kryukov, P. G. Kryukov, and A. V. Sharkov. Amplification of fs pulses in a multi-pass dye amplifier. In T. Yajima, K. Yoshihara, C. B. Harris, and S. Shionoya, editors, *Ultrafast Phenomena VI*, pages 22–24, Berlin, 1984. Springer.

[332] D. S. Bethune. Dye cell design for high-power low divergence excimer-pumped dye lasers. *Appl. Opt.*, 20:1897–1899, 1981.

[333] M. A. Kahlow, W. Jarzeba, and T. B. DuBruil. Ultrafast emission spectroscopy by time gated fluorescence. *Rev. Sci. Instrum.*, 59:1098–1109, 1988.

[334] P. Bado, M. Boovier, and J. S. Coe. Nd:YLF mode locked oscillator and regenerative amplifier. *Opt. Lett.*, 12:319–321, 1987.

[335] J. E. Murray and D. J. Kuizenga. Regenerative compression of laser pulses. *Appl. Phys. Lett.*, 12:27–30, 1987.

[336] T. Sizer II, J. D. Kafka, A. Krisiloff, and G. Mourou. Generation and amplification of subps pulses using a frequency doubled Nd:Yag pumping source. *Opt. Commun.*, 39:259–262, 1981.

[337] I. N. Duling II, T. Norris, T. Sizer, P. Bado, and G. Mourou. Kilohertz synchronous amplification of 85 fs optical pulses. *J. Opt. Soc. Am. B*, 2:616–621, 1985.

[338] M. Pessot, J. Squier, G. Mourou, and G. Vaillancourt. Chirped-pulse amplification of 100-fs pulses. *Optics Lett.*, 14:797–799, 1989.

[339] J. Squier, F. Salin, J. S.Coe, P. Bado, and G. Mourou. Characteristics of an actively mode-locked 2-ps Ti-sapphire laser operating in the 1 − μm wavelength regime. *Opt. Lett.*, 16:85–87, 1991.

[340] F. Salin, J. Squier, G. Mourou, and G. Vaillancourt. Millijoule femtosecond pulse amplification in $Ti:Al_2O_3$ at multi-khz repetition rates. In *Ultrafast Phenomena VIII*, pages 203–205, Berlin Heidelberg, 1993. Springer-Verlag.

[341] H. J. Polland, T. Elsaesser, W. Kaiser, M. Kussler, N. J. Marx, B. Sens, and K. Drexhage. Picosecond dye laser emission in the infrared. *Appl. Phys.*, B32:53–57, 1983.

[342] J. Klebniczki, Zs. Bor, and G. Szabo. Theory of traveling wave amplified spontaneous emission. *Appl. Phys. B*, 46:151–155, 1988.

[343] J. Hebling and J. Kuhl. Generation of femtosecond pulses by amplified spontaneous emission. *Opt. Lett.*, 14:278–280, 1989.

[344] P. Chernev and V. Petrov. Numerical study of traveling wave amplified spontaneous emission. *Opt. Quant. Electr.*, 23:45–52, 1991.

[345] G. Werner, W. Rudolph, P. Heist, and P. Dorn. Simulation of femtosecond pulse traveling wave amplification. *J. Opt. Soc. Am. B*, B9:1571–1578, 1992.

[346] J. Hebling and J. Kuhl. Generation of tunable fs pulses by traveling wave amplification. *Opt. Commun.*, 73:375–379, 1989.

[347] M. Haner and W. S. Warren. Generation of arbitrarily shaped picosecond optical pulses using an integrated electro-optic waveguide modulator. *Applied Optics*, 26:3687–3694, 1987.

[348] C. E. Cook. Pulse compression - key to more efficient radar transmission. *Proc. IRE*, 48:310–320, 1960.

[349] T. K. Gustafson, J. P. Taran, H. A. Haus, J. R. Lifsitz, and P. L. Kelley. Self modulation, self steepening, and spectral development of light in small-scale trapped filaments. *Physical Review*, 177:306–313, 1969.

[350] E. B. Traecy. Compression of picosecond light pulses. *Phys. Lett.*, 28A:34–38, 1969.

[351] A. Laubereau. External frequency modulation and compression of ps pulses. *Phys. Lett.*, 29A:539–540, 1969.

[352] A. M. Weiner, J. P. Heritage, and E. M. Kirschner. High resolution femtosecond pulse shaping. *J. Opt. Soc. Am. B*, 5:1563–1572, 1988.

[353] D. Grischkowsky and A. C. Balant. Optical pulse compression based on enhanced frequency chirping. *Appl. Phys. Lett.*, 41:1–3, 1981.

[354] D. Marcuse. *Light Transmission Optics*. Van Nostrand Reinhold, New York, 1982.

[355] L. F. Mollenauer, J. P. Gordon, and N. N. Islam. Soliton propagation in long fibers with periodically compensated loss. *IEEE J. Quantum Electron.*, QE-22:157–173, 1986.

[356] W. J. Tomlinson, R. H. Stolen, and C. V. Shank. Compression of optical pulses chirped by self phase modulation in optical fibers. *J. Opt. Soc. Am.*, B1:139–143, 1984.

[357] F. Calogero and A. Degasperis. *Spectral Transforms and Solitons*. North-Holland, Amsterdam, 1982.

[358] R. Meinel. Generation of chirped pulses in optical fibers suitable for an effective pulse compression. *Opt. Commun.*, 47:343–346, 1983.

[359] B. Wilhelmi. Potentials and limits of femtosecond pulse compression (in german). *Ann. Phys.*, 43:355–368, 1986.

[360] K. Tai and A. Tomita. 1100 x optical fiber pulse compression using grating pair and soliton effect at 1319nm. *Appl. Phys. Lett.*, 48:1033–1035, 1986.

[361] J. G. Fujimoto, A. M. Weiner, and E. P. Ippen. Generation and measurement of optical pulses as short as 16 fs. *Appl. Phys. Lett.*, 44:832–834, 1984.

[362] V. A. Vysloukh and T. A. Matveyeva. Feasibility of cascade compression of pulses in the near infrared. *Sov. J. Quantum Electron.*, 16:665–667, 1986.

[363] E. Bourkhoff, W. Zhao, R. I. Joseph, and D. N. Christodoulides. Evolution of fs pulses in single mode fibers having higher order nonlinearity and dispersion. *Opt. Lett.*, 12:272–274, 1987.

[364] A. Hasegawa and F. Tappert. Transmission of stationary nonlinear optical pulses in dispersive dielectric fibers. *Appl. Phys. Lett.*, 23:142–144, 1973.

[365] L. F. Mollenauer, R. H. Stolen, and J. P. Gordon. Experimental observation of picosecond pulse narrowing and solitons in optical fibers. *Phys. Rev. Lett.*, 45:1095, 1980.

[366] L. F. Mollenauer, R. H. Stolen, J. P. Gordon, and W. J. Tomlinson. Extreme picosecond pulse narrowing by means of soliton effect in single-mode fibers. *Opt. Lett.*, 8:289, 1983.

[367] V. Petrov, W. Rudolph, and B. Wilhelmi. Compression of high energy femtosecond light pulses by self phase modulation in bulk media. *J. Mod. Opt.*, 36:587–595, 1989.

[368] C. Rolland and P. B. Corkum. Compression of high power optical pulses. *J. Opt. Soc. Am. B*, 5:641–647, 1988.

[369] J. Agostinello, G. Harvey, T. Stone, and C. Gabel. Optical pulse shaping with a grating pair. *Appl. Opt.*, 18:2500–2504, 1979.

[370] R. N. Thurston, J. P. Heritage, A. M. Weiner, and W. J. Tomlinson. Analysis of picosecond pulse shape synthesis by spectral masking in

a grating pulse compressor. *IEEE J. Quantum Electron.*, QE-22:682–696, 1986.

[371] A. M. Weiner, J. P. Heritage, and J. A. Saleh. Encoding and decoding of femtosecond pulses. *Opt. Lett.*, 13:300–302, 1988.

[372] K. Naganuma, K. Mogi, and H. Yamada. General method for ultrashort light pulse chirp measurement. *IEEE J. of Quantum Electronics*, QE-25:1225–1233, 1989.

[373] S. Szatmari and F. P. Schaefer. Generation of high power UV femtosecond pulses. In T. Yajima, K. Yoshihara, C. B. Harris, and S. Shionoya, editors, *Ultrafast Phenomena VI*, pages 82–86, Berlin, Heidelberg, 1988. Springer Verlag.

[374] E. S. Kintzel and C. Rempel. Near surface second harmonic generation for autocorrelation measurements in the UV. *Appl. Phys.*, B28:91–95, 1987.

[375] R. Deich, F. Noack, W. Rudolph, and V. Postovalos. Two-photon luminescence in CsI(Na) under UV femtosecond pulse excitation. *Solid state communications*, 74:269–273, 1990.

[376] H. S. Albrecht, P. Heist, J. Kleinschmidt, D. van Lap, and T. Schroeder. Single-shot measurement of femtosecond pulses using the optical kerr effect. *Meas. Sci. Technol.*, 4:1–7, 1993.

[377] R. Trebino and D. J. Kane. Using phase retrieval to measure the intensity and phase of ultrashort pulses: frequency-resolved optical gating. *Journal Opt. Soc. Am. A.*, 10:1101–1111, 1993.

[378] D. J. Kane and R. Trebino. Single-shot measurement of the intensity and phase of an arbitrary ultrashort pulse using frequency-resolved gating. *Optics Lett.*, 18:823–825, 1993.

[379] H. P. Weber and H. G. Danielmeyer. Multimode effects in intensity correlations measurements. *Phys. Rev. A*, 2:2074–2078, 1970.

[380] J. A. Giordmaine, P. M. Rentzepis, S. L. Shapiro, and K. W. Wecht. Two-photon excitation of fluorescence by picosecond light pulses. *Applied Physics Letters*, 11:216–218, 1967.

[381] S. P. LeBlanc, G. Szabo, and R. Sauerbrey. Femtosecond single-shot phase-sensitive autocorrelator for the ultraviolet. *Opt. Lett.*, 16:1508–1510, 1991.

[382] J. Jansky, G. Corradi, and R. N. Gyuzalian. On a possibility of analyzing the temporal characteristics of short light pulses. *Optics Comm.*, 23:293–298, 1977.

[383] C. Rempel and W. Rudolph. Single shot autocorrelator for femtosecond pulses. *Experimentell Technik der Physik*, 37:381–385, 1989.

[384] R. N. Gyuzalian, S. B. Sogomonian, and Z. G. Horvath. Background-free measurement of time behavior of an individual picosecond laser pulse. *Optics Comm.*, 29:239, 1979.

[385] F. Krausz, T. Juhasz, J. S. Bakos, and Cs. Kuti. Microprocessor based system for measurement of the characteristics of ultrashort laser pulses. *J. Phys. E: Sci. Instrum.*, 19:1027–1029, 1986.

[386] F. Salin, P. Georges, G. Roger, and A. Brun. Single-shot measurement of a 52-fs pulse. *Applied Optics*, 26:4528–4531, 1987.

[387] R. Danielius, V. Sirutkaitis, A. Stabinis, G. Valikulis, and A. Yankauskas. Characterization of phase modulated ultrashort pulses using single-shot autocorrelator. *Optics Comm.*, 105:67–71, 1994.

[388] F. Salin, P. Georges, G. Le Saux, G. Roger, and A. Brun. Autocorrelation interferometrique monocoup d'impulsions femtosecondes. *Revue d'Optique*, 1988.

[389] V. Wong and I. A. Walmsley. Analysis of ultrashort pulse-shape measurement using linear interferometers. *Optics Lett.*, 19:287–289, 1994.

[390] Y. Ishida, K. Nagunama, T. Yajima, and L. H. Lin. Ultrafast self-phase modulation in a colliding pulse mode-locked ring dye laser. In D. H. Auston and K. B. Eisenthal, editors, *Ultrafast Phenomena IV*, pages 69–71. Springer-Verlag, 1984.

[391] Y. Ishida, K. Nagunama, and T. Yajima. Self-phase modulation in hybridly mode-locked cw dye lasers. *IEEE J. Quantum Electron.*, QE-21:69–77, 1985.

[392] Chi Yan and Jean-Claude Diels. 3D imaging with ultrashort pulses. *Appl. Opt.*, 34:6869–6873, 1991.

[393] J.-C. Diels, J. J. Fontaine, and W. Rudolph. Ultrafast diagnostics. *Revue Phys. Appl.*, 22:1605–1611, 1987.

[394] B. S. Prade, J. M. Schins, E. T. J. Nimbering, M. A. Franco, and A. Mysyrowicz. A simple method for the determination of the intensity and phase of ultrashort optical pulses. *Optics Comm.*, 113:79–84, 1994.

[395] M. A. Duguay and A. Savage. Picosecond optical sampling oscilloscope. *Optics Comm.*, 9:212–215, 1975.

[396] W. Koenig, H. K. Dunn, and L. Y. Lacy. The sound spectrograph. *J. Acoust. Soc. Amer.*, 18:19–49, 1946.

[397] D. Israelevitz and J.S. Lim. A new direct algorithm for image reconstruction from fourier transform magnitude. *IEEE Trans. Acoust. Speech Signal Process.*, ASSP-35:511–519, 1987.

[398] J. R. Fienup. Reconstruction of a complex-valued object from the modulus of its fourier transform using a support constraint. *J. Opt. Soc. Amer. A.*, 4:118–123, 1987.

[399] Scott Diddams, Xin Miao Zhao, and Jean-Claude Diels. Pulse measurements without optical nonlinearities. In *SPIE Proceedings, Generation, Amplification and measurement of ultrashort laser pulses*, volume 2116, Los Angeles, CA, 1994. SPIE.

[400] K. D. Li, A. S. Hou, E. Ozbay, B. A. Auld, and D. M. Bloom. 2-picosecond, GaAs photodiode optoelectronic circuit for optical correlation applications. *Appl. Phys. Lett.*, 61:3104–3106, 1992.

[401] J. L. A. Chilla and O. E. Martinez. Direct determination of the amplitude and the phase of femtosecond light pulses. *Opt. Lett.*, 16:39, 1991.

[402] M. T. Kauffman, W. C. Banyai, A. A. Godil, and D. M. Bloom. A time-to-frequency converter for measuring picosecond optical pulses. *Applied Physics Letters*, 64:270–272, 1994.

[403] M. L. Riaziat, G. F. Virshup, and J. N. Eckstein. Optical wavelength shifting by traveling-wave electrooptic modulation. *IEEE Photonics Technology Letters*, 5:1002–1005, 1993.

[404] J. A. Valdmanis and G. A. Mourou. Subpicosecond electrooptic sampling: principles and applications. *IEEE Journal of Quantum Electron.*, QE-22:69–78, 1986.

[405] J. Krueger. Problems course, Friedrich-Schiller-University Jena, 1988.

[406] W. Rudolph, J. Puls, F. Henneberger, and D. Lap. Femtosecond studies of transient nonlinearities in wide-gap II-VI semiconductor compounds. *Phys. Stat. sol. (b)*, 159:49–53, 1990.

[407] M. T. Portella, J.-Y. Rigot, R. W. Schoenlein, C. V. Shank, and J. E. Cunningham. k-space carrier dynamics in GaAs. In C. B. Harris, E. P. Ippen, G. A. Mourou, and A. H. Zewail, editors, *Ultrafast Phenomena VII*, pages 285–287, Berlin, 1990. Springer-Verlag.

[408] L. Oudar, A. Migus, D. Hulin, G. Grillon, J. Etchepare, and A. Antonetti. Femtosecond orientational relaxation of photoexcited carriers in GaAs. *Phys. Rev. Lett.*, 53:384–387, 1984.

[409] D. McMorrow, W. T. Lotshaw, and G. A. Kenney-Wallace. Femtosecond optical studies on the origin of the nonlinear responses in simple liquids. *IEEE J. of Quantum Electronics*, QE-24:443–454, 1988.

[410] H. J. Eichler, P. Guenther, and D. W. Pohl. *Laser-Induced Dynamic Gratings*. Springer, Berlin, Heidelberg, 1986.

[411] J.-C. Diels, W.-C. Wang, and H. Winful. Dynamics of the nonlinear four-wave mixing interaction. *Applied Physics*, B26:105–110, 1981.

[412] H. Mahr and M. D. Hirsch. An optical upconversion light gate with picosecond resolution. *Optics Commun.*, 13:96–99, 1975.

[413] W. Rudolph and J.-C. Diels. Femtosecond time resolved fluorescence. In A. E. Siegman, editor, *Picosecond Phenomena V*, pages 71–74, Berlin, 1986. Springer-Verlag.

[414] J. Shah. Ultrafast luminescence specroscopy using sum frequency generation. *IEEE J. of Quantum Electron.*, QE-24:276–288, 1988.

[415] I. D. Abella, N. Q. Kurnit, and S. R. Hartmann. Photon echoes. *Phys. Rev.*, 141:391–406, 1966.

[416] E. L. Hahn. Spin echoes. *Phys. Rev.*, 80:580–594, 1950.

[417] C. V. Heer. Focusing of carr-purcell photon echoes, and collisional effects. *Phys. Rev. A*, 13:1908–1920, 1976.

[418] A. I. Alekseef and I. V. Evseev. Photon echo polarization in a gas medium. *Soviet Physics JETP*, 29:1139–1143, 1969.

[419] P. C. Becker, H. C. Fragnito, C. H. Brito-Cruz, R. L. Fork, J. E. Cunningham, J. E. Henry, and C. V. Shank. Femtosecond photon echoes from band-to-band transitions in GaAs. *Phys. Rev. Lett.*, 61:647–649, 1988.

[420] C. V. Shank, P. C. Becker, H. L. Fragnito, and R. L. Fork. Femtosecond photon echoes. In T. Yajima, K Yoshihara, C. B. Harris, and S. Shionoya, editors, *Ultrafast Phenomena VI*, pages 344–348, Berlin, 1988. Springer-Verlag.

[421] J.-Y. Rigot, M. T. Portella, R. W. Schoenlein, C. V. Shank, and J. E. Cunningham. Two-dimensional carrier-carrier screening studied with femtosecond photon echoes. In C. B. Harris, E. P. Ippen, G. A. Mourou, and A. H. Zewail, editors, *Ultrafast Phenomena VII*, pages 239–243, Berlin, 1990. Springer-Verlag.

[422] J.-C. Diels, W. C. Wang, and R. K. Jain. Experimental demonstration of a new technique to measure ultrashort dephasing times. In K. B. Eisenthal, R. M. Hochstrasser, W. Kaiser, and A. Laubereau, editors, *Ultrafast Phenomena III*, pages 120–122, Berlin, 1982. Springer-Verlag.

[423] J.-C. Diels. Feasibility of measuring phase relaxation time with subpicosecond pulses. *IEEE Journal of Quantum Electron.*, QE-16:1020–1021, 1980.

[424] S. Besnainou, J.-C. Diels, and J. Stone. Molecular multiphoton excitation of phase coherent pulse pairs. *J. of Chem. and Phys.*, 81:143–149, 1984.

[425] S. Ruhman, A. Joly, and K. Nelson. Coherent molecular vibrational motion observed in the time domain through impulsive stimulated raman scattering. *IEEE Journal of Quantum Electron.*, 34:460–468, 1988.

[426] A. Weiner, D. Leaird, G. Wiederrecht, M. Banet, and K. Nelson. Spectroscopy with shaped femtosecond pulses:styles for the 1990s. In *SPIE Proceedings 1209*, pages 185–195, Los Angeles, CA, 1990.

[427] H. T. Grahn, H. J. Maris, and J. Tauc. Picosecond ultrasonics. *IEEE J. of Quantum Electron.*, 25:2562–2568, 1989.

[428] D. Grischkowsky, M. M. T. Loy, and P. F. Liao. Adiabatic following model for two-photon transitions: nonlinear mixing and pulse propagation. *Phys. Rev. A*, 12:2514–2533, 1975.

[429] J. Mlynek, W. Lange, H. Harde, and H. Burggraf. High-resolution coherence spectroscopy using pulse trains. *Phys. Rev. A*, 24:1099–1101, 1981.

[430] C. R. Stroud. The classical limit of an atom. In W. T. Grandy and P. W. Milonni, editors, *Physics and Probability: a symposium in honor of E. T. Jaynes*, pages 1–9. Cambridge University Press, 1993.

[431] A. ten Wolde, L. D. Noordam, H. G. Muller, A. Lagendijk, and H. B. van Linden van den Heuvell. Observation of radially localized atomic wave packets. *Phys. Rev. Lett.*, 61:2099–2102, 1988.

[432] J. A. Yeazell and Jr C. R. Stroud. Observation of fractional revivals in the evolution of a Rydberg atomic wave packet. *Phys. Rev. A*, 4:5153–5156, 1991.

[433] G. Alber and P. Zoller. Laser-induced excitation of electronic Rydberg wave packets. *Contemporary Physics*, 32:185–189, 1991.

[434] J. A. Yeazell, M. Mallalieu, J. Parker, and Jr C. R. Stroud. Observation of the collapse and revival of a Rydberg electronic wave packet. *Phys. Rev. Lett.*, 64:2007–2010, 1990.

[435] J. A. Yeazell and Jr C. R. Stroud. Observation of spatially localized atomic electron wave packets. *Phys. Rev. Lett.*, 60:1494–1497, 1988.

[436] Z. D. Gaeta, M. Noel, and C. R. Stroud. Excitation of the classical-limit state of an atom. *Phys. Rev. Lett.*, to be published, 1994.

[437] N. F. Scherer, R. J. Carlson, A. Matro, M. Du, A. J. Ruggiero, V. Romero-Rochin, J. A. Cina, G. F. Fleming, and S. A. Rice. Fluorescence-detected wave packet interferometry: time resolved molecular spectroscopy with sequences of femtosecond phase-locked pulses. *J. Chem. Phys.*, 95:1487–1511, 1991.

[438] R. M. Bowman, M. Dantus, and A. H. Zewail. Femtosecond transition state spectroscopy of iodine: from strongly bound to repulsive surface dynamics. *Chem. Phys. Lett.*, 161:297–302, 1989.

[439] F. W. Wise, M. J. Rosker, and C. L. Tang. Oscillatory femtosecond relaxation of photoexcited organic molecules. *J. Chem. Phys.*, 86:2827–2832, 1987.

[440] T. Baumert, R. Thalweiser, and G. Gerber. Femtosecond two-photon ionization spectroscopy of the B state of Na_3 clusters. *Chem. Phys. Lett.*, 209:29–32, 1993.

[441] C. H. Brito-Cruz, R. L. Fork, W. H. Knox, and C. V. Shank. Spectral hole burning in large molecules probed with 10 fs optical pulses. *Chem. Phys. Lett.*, 132:341–344, 1986.

[442] M. J. Rosker, M. Dantus, and A. H. Zewail. Femtosecond real-time probing of reactions, i. the technique. *J. Chem. Phys.*, 89:6113–6127, 1988.

[443] A. H. Zewail. Laser femtochemistry. *Science*, 242:1645–1653, 1988.

[444] R. B. Bernstein and A. H. Zewail. Femtosecond real-time probing of reactions, iii. inversion to the potential from femtosecond transition-state spectroscopy experiments. *J. Chem. Phys.*, 90:829–842, 1989.

[445] A. H. Zewail. The birth of molecules. *Scientific American*, 262:76–82, 1990.

[446] P. Cong, A. Mokhtari, and A. H. Zewail. Femtosecond probing of persistent wave wave packet motion in dissociativereactions: up to 40 ps. *Chemical Phys. Lett.*, 172:109–113, 1990.

[447] J. R. Grover and E. A. Walters. Optimization of weak neutral dimers in nozzle beams. *J. Phys. Chem.*, 90:6201–6206, 1986.

[448] J. R. Grover, E. A. Walters, D. L. Arneberg, and C. J. Santandrea. Competitive production of weakly bound heterodimers in free jet expansions. *Chem. Phys. Lett.*, 146:305–307, 1988.

[449] N. F. Scherer, L. R. Khundkar, R. B. Bernstein, and A. H. Zewail. Real-time picosecond clocking of the collision complex in a bimolecular reaction: The birth of OH from $H + CO_2$. *J. Chem. Phys.*, 87:1451–1453, 1987.

[450] E. D. Potter, J. L. Herek, S. Petersen, Q. Liu, and A. H. Zewail. Femtosecond laser control of a chemical reaction. *Nature*, 355:66–68, 1992.

[451] Y. Yan, R. M. Whitnell, K. R. Wilson, and A. H. Zewail. Femtosecond chemical dynamics in solution: wavepacket evolution and caging of i_2. *Chem. Phys. Lett.*, to be published, 1993.

[452] I. Benjamin and K. R. Wilson. Proposed experimental probes of chemical reaction molecular dynamics in solution:icn photodissociation. *J. Chem. Phys.*, 90:4176–4197, 1989.

[453] L. L. Lee, Y. S. Li, and K. R. Wilson. Reaction dynamics from liquid structure. *J. Chem. Phys.*, 95:2458–2464, 1991.

[454] P .C. Becker, D. Lee, M. R. Xavier de Barros, A. M. Johnson, A. G. Prosser, R. D. Feldman, R. F. Austin, and R. E. Behringer. Femtosecond dynamic exciton bleaching in room temperature II-VI quantum wells. *IEEE J. of Quantum Electron.*, 28:2535–2542, 1992.

[455] W. Z. Lin, R. W. Schoenlein, J. G. Fujimoto, and E. P. Ippen. Femtosecond absorption saturation studies of hot carriers in GaAs and AlGaAs. *IEEE J. of Quantum Electron.*, 24:267–275, 1988.

[456] H. Haug and S. W. Koch. *Quantum theory of the optical and electronic properties of semiconductors*. World Scientific, Singapore, 1990.

[457] W. Kuett, W. Albrecht, and H. Kurz. Generation of coherent phonons in condensed media. *IEEE J. Of Quantum Electron.*, 42:2434–2444, 1992.

[458] M. C. Downer, R. L. Fork, and C. V. Shank. Femtosecond imaging of melting and evaporation of a photoexcited silicon surface. *J. of Optical Soc. Am. B*, 2:595–599, 1985.

[459] N. Bloembergen, A. M. Malvezzi, and J. M. Lin. Second harmonic generation in reflection from crystalline GaAs under intense picosecond laser irradiation. *Appl. Phys. Lett.*, 45:1019–1021, 1984.

[460] S. V. Govorkov, I. L. Shumay, W. Rudolph, and T. Schroeder. Time-resolved second-harmonic study of femtosecond laser-induced disordering of GaAs surfaces. *Optics Lett.*, 16:1013–1015, 1991.

[461] P. Saeta, J. K. Wang, Y. Siegal, and N. Bloembergen. *Phys. Rev. Lett.*, 67:1023–1025, 1991.

[462] K. Sokolowski-Tinten, H. Schultz, J. Bialkowski, and D. von der Linde. Two distinct transitions in ultrafast solid-liquid phase transformations of GaAs. *Appl. Phys.*, A53:227–234, 1991.

[463] T. Schroeder, W. Rudolph, S. V. Govorkov, and I. L. Shumay. Femtosecond laser induced melting of GaAs probed by optical second-harmonic generation. *Appl. Phys.*, A51:49–51, 1990.

[464] R. R. Alfano, editor. *Biological events probed by ultrafast laser spectroscopy*. Academic Press, New York, 1982.

[465] R. W. Schoenlein, L. A. Peteanu, R. A. Mathies, and C. V. Shank. The first step in vision: femtosecond isomerization of rhodopsin. *Science*, 254:412–415, 1991.

[466] R. W. Schoenlein, L.-A. Poteanis, R. A. Mathias, and C. V. Shank. Femtosecond isomerization of rhodopsin. Ultrafast Phenomena in Spectroscopy, UPS'91, October 1991. Bayreuth, Germany.

[467] G. R. Fleming and R. van Grondelle. The primary steps of photosynthesis. *Physics Today*, 47:48–55, 1994.

[468] W. Zinth, C. Lauterwasser, U. Finkele, P. Hamm, S. Schmidt, and W. Kaiser. Molecular processes in the primary reaction of photosynthetic reaction centers. In J.-L. Martin, A. Migus, G. A. Mourou, and A. H. Zewail, editors, *Ultrafast Phenomena VIII*, pages 535–538, Antibes, France, 1994. Springer Verlag, Berlin.

[469] J. Deisenhofer and H. Michel. The photosynthetic reaction center from the purple bakterium rhodopseudomonas viridis. *EMBO Journal*, 8:2419–2157, 1989.

[470] W. Zinth, S. Schmidt, T. Arlt, H. Huber, T. Naegele, M. Meyer, and H. Scheer. Direct observation of the accessory bacteriochlorophyll in the primary electron transfer in bacterial reaction centers. In G. A. Mourou and A. H. Zewail, editors, *Ultrafast Phenomena IX*, Dana Point, CA, 1994. Springer Verlag, Berlin.

[471] D. H. Auston, K. P. Cheung, J. A. Valdmanis, and D. A. Kleinman. Cherenkov radiation from femtosecond optical pulses in electrooptic media. *Phys. Rev. Lett.*, 35:1555–1558, 1984.

[472] D. A. Kleinman and D. H. Auston. Theory of electrooptic shock radiation in nonlinear optical media. *IEEE J. Quantum Electron.*, QE-20:964–970, 1984.

[473] B. B. Hu, X. C. Zhang, D. H. Auston, and R. R. Smith. Free-space radiation from electrooptic crystals. *Appl. Phys. Lett.*, 56:506–508, 1990.

[474] Ch. Fattinger and D. Grischkowsky. Point source terahertz optics. *Appl. Phys. Lett.*, 53:1480–1482, 1988.

[475] F. E. Doany, D. Grischkowsky, and C. C. Chi. Carrier life time versus ion-implantation dose in silicon on sapphire. *Appl. Phys. Lett.*, 50:460–462, 1987.

[476] X. C. Zhang and D. H. Auston. Optoelectronic measurements of semiconductor surfaces with femtosecond optics. *J. Appl. Phys.*, 71:326–338, 1992.

[477] G. Mourou, C. V. Stancampiano, A. Antonetti, and A. Orszag. Picosecond microwave pulses generated with subpicosecond laser driven switch. *Appl. Phys. Lett.*, 39:295–296, 1981.

[478] X. C. Zhang, B. B. Hu, J. T. Darrow, and D. H. Auston. Generation of FIR electromagnetic pulse from semiconductor surfaces. *Appl. Phys. Lett.*, 56:1011–1013, 1990.

[479] B. I. Green, P. N. Saeta, D. R. Dykaar, S. Schmitt-Rink, and S. L. Chuang. Far-infrared light generation at semiconductor surfaces and its spectroscopic applications. *IEEE J. of Quantum Electron.*, QE-28:2302–2312, 1992.

[480] P. K. Benicewitz, J. P. Roberts, and A. J. Taylor. Scaling of teraherz radiation from large aperture biased photoconductors. *J. of the Optical Soc. Am.*, 11:2533–2546, 1994.

[481] M. van Exter and D. Grischkowsky. Characterization of an optoelectronic terahertz beam system. *IEEE J. of Quantum Electron.*, QE-28:1684–1691, 1990.

[482] Ch. Fattinger and D. Grischkowsky. A Cherenkov source for freely propagating THz beams. *IEEE J. of Quantum Electron.*, QE-25:2608–2610, 1988.

[483] D. Grischkowsky, S. Keiding, M. van Exter, and C. Fattinger. Far infrared time-domain spectroscopy with teraherz beams of dielectrics and semiconductors. *J. Opt. Soc. B*, 7:2006–2015, 1990.

[484] H. Harde and D. Grischkowski. Coherent transients excited by subpicosecond pulses of teraherz radiation. *J. Opt. Soc. Am. B*, 8:1642–1651, 1991.

[485] R. W. Falcone and M. M. Murnane. Proposal for a fs x-ray light source. In D. T. Atwood and J. Bokor, editors, *Short wavelength coherent radiation: Generation and Applications*, pages 81–85. AIP vol. 147, 1986.

[486] D. Kuehlke, U. Herpers, and D. von der Linde. Soft x-ray emission from subps produced laser-induced plasmas. *Appl. Phys. Lett.*, 50:1785–1787, 1987.

[487] G. Kuehnle, F. P. Schaefer, S. Szatmari, and G. D. Tsakiris. X-ray production by irradiation of solid targets with subps excimer laser pulses. *Appl. Phys.*, B47:361–366, 1988.

[488] J. D. Kmetec. Ultrafast laser generation of hard x-rays. *IEEE J. of Quantum Electron.*, QE-28:2382–2387, 1992.

[489] J. A. Cobble, G. A. Kyrala, A. A. Hauser, A. J. Taylor, C. C. Gomez, N. D. Delamater, and G. T. Schappert. Kilovolt x-ray spectroscopy of subps-laser-excited source. *Phys. Rev. A*, 39:454–457, 1989.

[490] H. M. Milchberg, I. Lyubomirsky, and C. G. Durfee III. Factors controlling the x-ray pulse emission from an intense laser heated solid. *Phys. Rev. Lett.*, 67:2654–2657, 1991.

[491] O. R. Wood, W. T. Silfvast, H. W. K. Tom, W. H. Fork, C. H. Brito-Cruz, M. C. Downer, and P. J. Maloney. Effect of laser pulse duration on short wavelength emission from femtosecond and picosecond laser induced Ta plasmas. *Appl. Phys. Lett.*, 53:654–656, 1988.

[492] M. M. Murnane, H. C. Kapteyn, M. D. Rosen, and R. Falcone. Ultrafst x-ray pulses from laser produced plasmas. *Science*, 25:531, 1991.

[493] H. T. Grahn, H. J. Maris, and J. Tauc. Picosecond ultrasonics. *IEEE J. Quantum Electron.*, QE-25:2562–2569, 1989.

[494] V. F. Sapega, V. I. Belitsky, and A. J. Shields. Resonant one-acoustic-phonon raman scattering in multiple quantum wells. *Solid State Commun.*, 84:1039, 1992.

[495] D. H. Auston. Picosecond optoelectronic switching and gating in silicon. *Appl. Phys. Lett.*, 26:101–103, 1975.

[496] R. A. Lawton and Scavannec. Pulsed laser application to sampling oscilloscope. *Electron. Lett.*, 11:138, 1975.

[497] M. B. Ketchen, D. Grischkowsky, T. C. Chen, C. C. Li, I. N. Duling III, N. J. Halas, J. M. Halbout, J. A. Kash, and C. P. Li. Generation of subpicosecond electrical pulses on coplanar transmission lines. *Appl. Phys. Lett.*, 48:751–753, 1986.

[498] M. Y. Frankel, J. F. Whitaker, and G. A. Mourou. Optoelectronic transient characteristic of ultrafast devices. *IEEE J. Quantum Electron.*, 28:2313–2324, 1992.

[499] D. H. Auston. Impulse response of photoconductors in transmission lines. *IEEE J. Quantum Electron.*, QE-19:639–648, 1983.

[500] D. H. Auston. Ultrafast optoelectronics. In W. Kaiser, editor, *Ultrashort Laser Pulses and Application*, pages 183–233. Springer, 1988.

[501] M. C. Nuss, D. W. Kisker, P. R. Smith, and T. E. Harvey. Efficient generation of 480 fs electrical pulses on transmission lines by photoconductive switching in metalorganic chemical vapor deposited cdte. *Appl. Phys. Lett.*, 54:57–59, 1989.

[502] S. Gupta, M. Y. Frankel, J. A. Valdmanis, J. Whitaker, G. Mourou, F. W. Smith, and A. R. Calawa. Subpicosecond carrier lifetime in GaAs grown by molecular beam epitaxy at very low temperatures. *Appl. Phys. Lett.*, 59:3276–3278, 1991.

[503] G. A. Mourou. High-speed electronics. In B. Kallback and H. Beneking, editors, *Springer series Electronics and Photonics*. Springer, 1986.

[504] D. R. Grisckkowsky, M. B. Ketchen, C. C. Chi, I. N. Duling III, N. J. Halas, J. M. Halbout, and P. G. May. Capacitance-free generation and

detection of subpicosecond electrical pulses on coplanar transmission lines. *IEEE J. Quantum Electron.*, QE-24:221–225, 1988.

[505] M. Y. Frankel, Shantanu Gupta, J. A. Valdmanis, and G. A. Mourou. Terahertz attenuation and dispersion characteristics of coplanar transmission lines. *IEEE Trans. Microwave Theory Tech.*, 39:910–915, 1991.

[506] P. Paulus, B. Wedding, A. Gaseh, and D. Jaeger. Bistability and solitons observed in nonlinear ring resonators. *Phys. Lett.*, 102A:89–92, 1984.

[507] K. Burnett, V. C. Reed, and P. L. Knight. Atoms in ultra-intense laser fields. *J. Phys. B: At. Mol. Opt. Phys.*, 26:561–598, 1993.

[508] Xin Miao Zhao, Scott Diddams, and Jean-Claude Diels. Applications of ultrashort pulses. In F. J. Duarte, editor, *Tunable Lasers Applications*. Marcel Dekker, New York, N. Y., 1995.

[509] A. P. Bruckner. Some applications of picosecond optical range gating. *SPIE*, 94, 1976.

[510] J. L. Martin, A. Antonetti, A. Astier, J. Etchepare, G. Grillon, and A. Migus. Subpicosecond spectroscopic techniques in biological materials. *SPIE*, 211, 1979.

[511] J. L. Martin, Y. Lecarpentier, A. Antonetti, and G. Grillon. Picosecond laser stereometry light scattering measurements on biological material. *Medical & Biological Engineering & Computing*, 18, 1980.

[512] J.-C. Diels and J. J. Fontaine. Imaging with short optical pulses. *Optics and Lasers in Engineering*, 4:145–162, 1983.

[513] J. J. Fontaine, J.-C. Diels, Ching Yue Wang, and H. Sallaba. Subpicosecond time domain reflectrometry. *Optics Lett.*, 6:495–498, 1981.

[514] J. G. Fujimoto, S. De Silvestri, E. P. Ippen, C. A. Puliafito, R. Margolis, and A. Oscroff. Femtosecond optical ranging in biological systems. *Opt. Lett.*, 11:150–152, 1986.

[515] R. C. Younquist, S. Carr, and D. E. Davies. Optical coherence-domain reflectometry: a new optical evaluation technique. *Optics Lett.*, 12:158–160, 1987.

[516] E. A. Swanson, J. A. Izatt, M. R. Hee, D. Huang, C. P. Liu, J. S. Schuman, G. A. Puliafito, and J. G. Fujimoto. In vivo retinal imaging by optical coherence tomography. *Optics Lett.*, 18:1864–1866, 1993.

[517] S. Ertefai and A. E. Profio. Spectral transmittance and contrast in breast diaphanography. *Med. Phys.*, 12:393–400, 1985.

[518] K. M. Yoo, Qirong Xing, and R. R. Alfano. Imaging objects hidden in highly scattering media using femtosecond second-harmonic-generation cross-correlation time gating. *Optics Letters*, 16:1019–1021, 1991.

[519] M. D. Duncan, R. Mahon, L. L. Tankersley, and J. Reintjes. Time-gated imaging through scattering media using stimulated Raman amplification. *Optics Lett.*, 16:1868–1870, 1991.

[520] M. Bashkansky, C. L. Adler, and J. Reintjes. Coherently amplified Raman polarization gate for imaging through scattering media. *Optics Lett.*, 19:350–352, 1994.

[521] J. Izatt, M. R. Hee, and G. M. Owen. Optical coherence microscopy in scattering media. *Optics Lett.*, 19:590–592, 1994.

[522] M. Kempe and W. Rudolph. Microscopy with ultrashort light pulses. *Nonlinear Optics*, 7:129–151, 1994.

[523] A. Ishimaru. *Wave propagation and scattering in Random media*. Academic Press, New York, 1978.

[524] J. A. Moon, R. Mahon, M. D. Duncan, and J. Reintjes. Resolution limits for imaging through turbid media with diffuse light. *Optics Lett.*, 18:1591–1593, 1993.

[525] J. C. Hebden and D. T. Delpy. Enhanced time resolved imaging with a diffusion model of photon transport. *Optics Lett.*, 19:311–313, 1994.

[526] In R. R. Alfano, editor, *OSA Proceedings on advances in optical imaging and photon migration*, volume 21, Washington, 1994. Optical Society of America.

[527] N. Abramson. Light-in-flight recording: high speed holographic motion pictures of ultrafast phenomena. *Appl. Opt.*, 22:215–232, 1983.

[528] N. Abramson. Light-in-flight recording. 2: compensation for the limited speed of the light used for observation. *Appl. Opt.*, 23:1481–1492, 1984.

[529] H. J. Gerritsen. *Proc. Soc. Photo-Opt. Instrum. Eng.*, 519:250, 1984.

[530] H. Chen, Y. Chen, D. Dilworth, E. Leith, J. Lopez, and J. Valdmanis. Two-dimensional imaging through diffusing media using 150-fs gated electronic holography techniques. *Opt. Lett.*, 16:487–489, 1991.

[531] E. Leith, C. Chen, Y. Chen, D. Dilworth, J. Lopez, J. Rudd, P.C. Sun, J. Valdmanis, and G. Vossler. Imaging through scattering media with holography. *J. Opt. Soc. Am. A*, 9:1148–1153, 1992.

[532] T. Wilson and C. J. R. Sheppard. *Theory and practice of scanning optical microscopy.* Academic Press, London, 1984.

[533] W. Denk, J. M. Strickler, and W. W. Webb. Two-photon laser scanning fluorescence microscopy. *Science*, 248:73–76, 1990.

[534] J. R. Lakowicz. Fluorescence lifetime sensing. *Laser Focus World*, 28-5:60–80, 1992.

[535] T. Sawatari. Optical heterodyne scanning microscope. *Applied Optics*, 12:2768–2772, 1973.

[536] M. Kempe and W. Rudolph. Scanning microscopy through thick layers based on linear correlation. *Optics Lett.*, 19:1919–1921, 1994.

[537] J. M. Schmitt, M. Yadloswky, and R. F. Bonner. Subsurface imaging of living skin with optical coherence microscopy. *Dermatology*, to be published, 1995.

[538] M. Kempe and W. Rudolph. Heterodyne microscopy for biomedical applications. 15:365, 1995.

[539] E. J. Post. Sagnac effect. *Rev. Mod. Phys.*, 39:475–493, 1967.

[540] E. O. Schulz-Dubois. Alternative interpretation of rotation rate sensing by ring laser. *IEEE J. Quant. Electr.*, QE-2:299–305, 1966.

[541] J. R. Wilkinson. *Ring Lasers.* Pergamon, New York, 1987.

[542] M. L. Dennis, J.-C. Diels, and M. Mohebi. Study of a carbon dioxide ring laser gyroscope. *Applied Physics B*, B54:278–287, 1992.

[543] D. Gnass, N. P. Ernsting, and F. P. Schaefer. Sagnac effect in the colliding-pulse-mode-locked dye ring laser. *Applied Physics B*, B-53:119–120, 1991.

[544] Xin Miao Zhao, Jean-Claude Diels, A. Braun, X. Liu, D. Du, G. Korn, G. Mourou, and Juan Elizondo. Use of self-trapped filaments in air to trigger lightning. In P. F. Barbara, W. H. Knox, G. A. Mourou, and A. H. Zewail, editors, *Ultrafast Phenomena IX*, pages 233–235, Dana Point, CA, 1994. Springer Verlag, Berlin.

[545] A. Braun, C. Y. Chien, S. Coe, and G. Mourou. Long range, high resolution laser radar. *Opt. Comm.*, 105:63–66, 1993.

[546] A. Braun, G. Korn, X. Liu, D. Du, J. Squier, and G. Mourou. Self-channeling of high-peak-power fs laser pulses in air. *Opt. Lett.*, 20:73–75, 1994.

[547] L. F. Mollenauer. Soliton transmission speeds greatly multiplied by sliding-frequency guiding filters. *Optics and Photonics news*, pages 15–19, 1994.

[548] J. P. Gordon and H. A. Haus. Random walk of coherently amplified solitons in optical fiber. *Opt. Lett.*, 11:665–667, 1986.

[549] A. Braun, G. Korn, X. Liu, D. Du, J. Squier, and G. Mourou. Self-channeling of high-peak-power femtosecond laser pulses in air. *Optics Lett.*, 20:73–75, 1995.

[550] R. Y. Chiao, E. Garmire, and C. H. Townes. Self-trapping of optical beams. *Physics Review Letter*, 13:479–482, 1964.

[551] S. A. Akhmanov, A. P. Sukhorukov, and A. I. Kovrigin. Self focusing and self trapping of intense light beams in a nonlinear medium. *Sov. Phys JETP*, 23:1025–1033, 1966.

[552] M. Giglio and A. Vendramini. Soret-type motion of macromolecules in solution. *Phys. Rev. Lett.*, 38:26–29, 1977.

[553] E. Freysz, M. Afifi, A. Ducasse, B. Pouligny, and J. R. Lalanne. Critical microemulsions as optically nonlinear media. *Journal of the Optical Society of America B*, 1:433, 1984.

[554] Y. R. Shen and M. M. Loy. Theoretical interpretation of small-scale filaments of light originating from moving focal spots. *Physical Review A*, 3:2099–2105, 1971.

[555] G. K. L. Wong and R. Y. Shen. Transient self-focusing in a nematic liquid crystal in the isotropic phase. *Physical Review Letter*, 32:527–530, 1974.

[556] Jr. J. A. Fleck, J. R. Morris, and M. D. Feit. Time-dependent propagation of high energy laser beams through the atmosphere. *Applied Physics*, 10:129–160, 1976.

[557] M. Lax, J. H. Batteh, and G. P. Agrawal. Channeling of intense electromagnetic beams. *J. of Applied Physics*, 52:109–125, 1981.

[558] P. Corkum. Plasma perspective on strong-field multiphoton ionization. *Phys. Rev. Lett.*, 71:1994, 1993.

[559] A. B. Blagoeva, S. B. Dinev, A. A. Dreischuh, and A. Naidenov. Light bullets formation in a bulk media. *IEEE J. Quantum Electronics*, QE-27:2060–2065, 1991.

[560] D. E. Edmundson and R. H. Enns. Fully three-dimensional collisions of bistable light bullets. *Opt. Lett.*, 18:1609–1611, 1993.

[561] N. N. Akhmediev, V. I. Korneev, and R. F. Nabiev. Modulation instability of the ground state of the nonlinear wave equation: Optical machine gun. *Optics Letters*, 15:393–395, 1992.

[562] Xin Miao Zhao, Patrick Rambo, and Jean-Claude Diels. Filamentation of femtosecond uv pulses in air. In *CLEO, 1995*, page QThD2, Baltimore, MA, 1995. Optical Society of America.

[563] R. L. Abrams. Waveguide Gas Lasers. In M. L. Stitch, editor, *Laser Handbook*, pages 41–88. North-Holland, Amsterdam, 1979.

[564] Nandini Mukherjee. *Coherent resonant interaction and harmonic generation in atomic vapors*. PhD thesis, University of North Texas, Denton, Texas, 1987.

Index

aberration
    chromatic, 59
    spherical, 65
absorber
    two-photon, 264
absorption, 111, 112
absorption coefficient, 111, 116, 162
absorption cross section, 237
acoustic pulse, 465, 475
amplification, 111, 112, 310
    chirped pulse (CPA), 324
amplification coefficient, 116
amplified spontaneous emission (ASE), 318, 332
amplifier, 309
    inhomogeneously broadened, 318
    multi-pass, 333
    multi-stage, 331
    regenerative, 335
    traveling wave (TWA), 337
amplitude response, 26
angular dispersion, 75
approximation
    harmonic oscillator, 182
    paraxial, 30
    rate equation, 112, 114, 122, 236
    slowly varying envelope, 16, 110, 140
area theorem, 162, 163
astigmatism, 275
autocorrelation, 66, 256
    intensity, 366
    interferometric, 369
        of linearly chirped pulse, 369
autocorrelator
    intensity, 376
    interferometric, 379
    single shot, 377
average frequency, 4, 171

ballistic component, 490
band gap, 454
beam propagation, 30, 92
beam waist, 31
beat note, 392, 496
Bloch equations, 159, 167
broadening
    homogeneous, 161, 166
    inhomogeneous, 163, 166, 427
    spectral, 352
    temporal, 352

carrier diffusion, 296
carrier frequency, 3
carrier lifetime, 297

INDEX

cavity
    linear, 278
    stability, 274
cavity dumping, 252
chemical reaction, 451
Cherenkov radiation, 466
chirp, 121, 206, 238, 322
chirp amplification, 136
chirp compensation, 344
chirp reversal, 416
chirped, 4, 165
coherence, 153
coherent interaction, 183
compression, 213, 235, 246, 343–360
    in bulk materials, 358
    Gaussian pulse, 351
    parametric oscillator, 285
    second harmonic generation, 288
    soliton, 356
compression factor, 344, 352
compression mechanisms, 215
compressor
    fiber-grating, 347, 356
    ideal, 346
    quadratic, 346
confocal parameter, 31
coplanar strip line, 482
coplanar transmission line, 468
correlation, 103
    with background, 373
    background-free, 373
    cross-, 206
    intensity, 365
        higher-order, 367
    interferometric, 367
    time domain, 396
correlation function, 46
correlation time, 103

correlator, 67
counterpropagating pulses, 211
critical power, 146, 323
cross-correlation, 66
    intensity, 366
cross-section
    emission, 319
    interaction, 115

deconvolution of data, 403
degenerate four-wave mixing, 415
density matrix equation, 102
dielectric constant, 14, 23, 101
dielectric multilayers, 51
diffraction, 34
diffraction integral, 33, 63
diffusion
    measurement of, 415
dipole, 104
dipole moment, 158
dispersion, 18, 34, 214
    angular, 75, 87
    cavity, 235
    grating, 82
    higher-order, 79
    mirror, 57
    prism, 78
    second-order, 80
dispersion length, 350
dispersion relation, 14
dispersive length, 20
dissipative system, 102
duration-bandwidth product, 9, 11, 120

eigenstate, 104, 179
electric field
    pseudo-, 160
electric pulse, 478

electro-optic sampling, 482
electron–hole pair, 455
energy conservation, 162
envelope, 3
    slowly varying, 15
exciton, 456
extraordinary wave, 126

Fabry–Perot interferometer, 53
fiber, 173, 300
    single-mode, 347
filament, 506
fluorescence, 447
    fs time-resolved, 419
focus, 59
Fourier spectrometer, 47
frequency modulation, 237
frequency resolved optical gating (FROG), 388
frequency shift, 24, 171

gain, 213
    linear, 23
    small signal, 318
gain coefficient, 162, 310
gain factor, 312
    saturated, 246
gain media, 313
gain narrowing, 316
gain saturation, 247, 256
gain switching, 291
Gaussian beam, 31, 34, 61, 87, 149, 241, 271
Gaussian pulse, 9, 10, 19, 34, 141
    compression, 346
    interferometric autocorrelation, 370
Gires–Tournois interferometer, 55
grating

population, 229, 231
    transient, 413
grating vector, 414
group velocity, 15, 47, 59, 70
group velocity dispersion (GVD), 16, 21, 62
    in amplifiers, 315
Guoy phase shift, 273
gyroscope, 496

heterodyning, 488
hole burning, 450
holography, 492

idler pulse, 135
imaging, 485
    through scatterers, 490
incoherence, 154
incoherent radiation, 43, 48
index of refraction, 43
interaction matrix, 192
inversion, 111, 114, 158
isomerization, 461

Kerr effect, 214, 238
Kerr gate, 487
Kramers–Kronig relation, 51

laser
    distributed feedback, 292
    dye, 211
    fiber, 300–303
    hybridly mode-locked, 259
    mode-locked, 209
    passively mode-locked, 247
    Raman soliton, 300
    semiconductor, 294
    short cavity, 293
    synchronously pumped, 254
    Ti:sapphire, 211

INDEX

lens, 59–68
light-matter interaction, 101
line broadening
    homogeneous, 106
    inhomogeneous, 107
    natural, 105
line shape factor, 113
linewidth, 108, 157, 246
local oscillator, 206
loss, 213
    linear, 23

Mach–Zehnder interferometer, 425
matrix
    ray, 92
    ray-pulse, 94
    resonator (ABCD), 271
melting
    laser-induced, 460
Michelson interferometer, 45–53, 66, 157, 367
microscopy, 493
mirror, 54, 68
    nonlinear, 226, 302
mode
    longitudinal, 209
mode-locking, 209
    active, 210
    additive, 217, 268
    hybrid, 210, 257
    passive, 210
    self-, 212
    starting of, 244
mode spacing, 240, 246, 254
modes
    cavity, 271
molecular system, 187
molecular vibration, 430, 446
multilevel system, 183

multilevel transitions, 179
multiphoton, 154, 176
multiphoton excitation, 189
multiphoton transition, 191
multiple quantum well (MQW), 226, 296, 456

n_2, 138
negative feedback, 286
nonlinear interaction length, 350
nonlinear refractive index, 138

occupation number, 104
optical matrix, 92
ordinary wave, 126

parametric interaction, 134
parametric oscillator, 137, 281–287
phase filter, 362
phase matching, 127, 134, 135
phase memory, 113
phase modulation, 29, 40, 115, 214
    cross, 142
    nonlinear, 138
phase modulator, 211, 344
phase response, 26
phase velocity, 17, 59, 70, 174
phonon, 431, 458, 476
photoconductive switch, 478
photon echo, 421
photon flux, 6
plasma, 472
polarization, 12, 101, 104, 113, 123
    linear, 12
    nonlinear, 12, 139
    pseudo, 158
    pseudo-, 174
    resonant, 109
polarization rotation

transient, 411
prism, 70–82
probe pulse, 402
pulse
    π, 166, 172, 177, 203, 421
    2π, 165, 172, 177, 421
    steady-state, 172, 173, 218
    zero-area, 186, 410, 425
pulse duration, 7
pulse energy, 6, 221
pulse formation, 212
pulse intensity, 6
pulse power, 6
pulse propagation, 36, 92
pulse shape, 52, 383
pulse shaping, 110, 112, 117
    in amplifiers, 314
    mechanisms of, 225
    through spectral filtering, 360
pulse spectra, 383
pump laser, 329
pump-probe experiment, 401, 404
pump pulse, 318, 401

Rabi frequency, 159
    complex, 181
    four-photon, 200
    two-photon, 194
Raman process, 198
Raman scattering, 430
range gating, 486
rate equations, 161
Rayleigh range, 31
reconstruction
    amplitude and phase, 380
    phase, 397
rectification
    optical, 467
reduced wave equation, 13

refractive index, 111
    effective, 348
regenerative feedback, 256
relaxation, 105
    intraband, 410, 455, 456
    momentum, 411, 425
relaxation time
    cross, 109
    energy, 104, 106
        measurement, 409
    longitudinal, 104
    phase, 104, 106, 112, 122, 153, 159
        measurement, 421, 428
    transverse, 104
resolution
    depth, 495
    temporal, 403, 406
resonator, 270
response time, 45
rotational spectroscopy, 470
rotational state, 187
round-trip model, 217, 248
Rydberg state, 442

satellite pulse, 346
saturable absorber, 117, 211, 231, 244
saturation, 117, 161, 227
    absorption, 214
    amplifier, 313, 314
    gain, 214
    standing wave, 228
saturation energy, 116
saturation intensity, 118, 230
Schrödinger equation, 17, 349
    nonlinear, 505

INDEX

581

second harmomic generation (SHG), 125, 287, 373
    type I, 126
    type II, 132
seeding, 257
self-focusing, 139, 146–149, 173, 215
    in amplifiers, 323
self-focusing length, 147
self-induced transparency, 165
self-lensing, 239–244
self-phase modulation, 139, 213, 302, 322, 344
self-starting, 246
self-steepening, 142
self-trapping, 146
shock wave, 467
signal pulse, 135
soliton, 220, 260, 501
    frequency-shift, 504
    fundamental, 357
    spatial, 505
    spatial and temporal, 506
soliton laser, 269
soliton propagation, 349
space-time analogy, 32
spatial frequency, 88
spectral amplitude, 3
spectral broadening
    in amplifiers, 321
spectral width, 8
spectrogram, 389
spectroscopy
    transient absorption, 407
stability range, 250
Stark shift, 194, 199
steady-state condition, 218

steady-state pulse, 247
stimulated Raman scattering
    impulsive, 430
susceptibility, 124, 438
synchronous pumping, 136, 301

telescope, 85
terahertz radiation, 466
three-level system, 193
tilt of pulse fronts, 71
transfer function, 26, 49, 361
    frequency filter, 220
    GVD element, 219
    output coupler, 219
    prism and grating, 345
    resonator, 248
    saturable gain and loss, 219
transient reflectivity, 459
transient transmission, 462
two-level system, 104, 158, 166, 177
two-photon stimulated emission, 233
two-photon transition, 417

vibration, 185
vibrational state, 187

walk-off, 127
wave equation, 12, 37
wave packet
    angular, 444
    dephasing, 454
    radial, 442
    revival, 444
white light, 45
white light continuum, 144, 333, 410

x-ray, 465
x-ray pulse, 472

Optics and Photonics (Formerly Quantum Electronics)

Edited by
Paul F. Liao, *Bell Communications Research, Inc., Red Bank, New Jersey*
Paul L. Kelly, *Tufts University, Medford, Massachusetts*
Ivan Kaminow, *AT&T Bell Laboratories, Holmdel, New Jersey*

N. S. Kapany and J. J. Burke, *Optical Waveguides*
Dietrich Marcuse, *Theory of Dielectric Optical Waveguides*
Benjamin Chu, *Laser Light Scattering*
Bruno Crosignani, Paolo DiPorto and Mario Bertolotti, *Statistical Properties of Scattered Light*
John D. Anderson, Jr., *Gasdynamic Lasers: An Introduction*
W. W. Duly, CO_2 *Lasers: Effects and Applications*
Henry Kressel and J. K. Butler, *Semiconductor Lasers and Heterojunction LEDs*
H. C. Casey and M. B. Panish, *Heterostructure Lasers: Part A. Fundamental Principles; Part B. Materials and Operating Characteristics*
Robert K. Erf, editor, *Speckle Metrology*
Marc D. Levenson, *Introduction to Nonlinear Laser Spectroscopy*
David S. Kliger, editor, *Ultrasensitive Laser Spectroscopy*
Robert A. Fisher, editor, *Optical Phase Conjugation*
John F. Reintjes, *Nonlinear Optical Parametric Processes in Liquids and Gases*
S. H. Lin, Y. Fujimura, H. J. Neusser and E. W. Schlag, *Multiphoton Spectroscopy of Molecules*
Hyatt M. Gibbs, *Optical Bistability: Controlling Light with Light*
D. S. Chemla and J. Zyss, editors, *Nonlinear Optical Properties of Organic Molecules and Crystals, Volume 1, Volume 2*
Marc D. Levenson and Saturo Kano, *Introduction to Nonlinear Laser Spectroscopy, Revised Edition*
Govind P. Agrawal, *Nonlinear Fiber Optics*
F. J. Duarte and Lloyd W. Hillman, editors, *Dye Laser Principles: With Applications*
Dietrich Marcuse, *Theory of Dielectric Optical Waveguides, 2nd Edition*
Govind P. Agrawal and Robert W. Boyd, editors, *Contemporary Nonlinear Optics*
Peter S. Zory, Jr. editor, *Quantum Well Lasers*
Gary A. Evans and Jacob M. Hammer, editors, *Surface Emitting Semiconductor Lasers and Arrays*
John E. Midwinter, editor, *Photonics in Switching, Volume I, Background and Components*
John E. Midwinter, editor, *Photonics in Switching, Volume II, Systems*
Joseph Zyss, editor, *Molecular Nonlinear Optics: Materials, Physics, and Devices*
F. J. Duarte, editor, *Turnable Lasers Handbook*
Jean-Claude Diels and Wolfgang Rudolph, *Ultrashort Laser Pulse Phenomena: Fundamentals, Techniques, and Applications on a Femtosecond Time Scale*

Yoh-Han Pao, Case Western Reserve University, Cleveland, Ohio, Founding Editor 1972–1979